MW00760082

Entropy in Image Analysis II

Entropy in Image Analysis II

Editor

Amelia Carolina Sparavigna

MDPI • Basel • Beijing • Wuhan • Barcelona • Belgrade • Manchester • Tokyo • Cluj • Tianjin

Editor
Amelia Carolina Sparavigna
Polytechnic University of Turin
Italy

Editorial Office
MDPI
St. Alban-Anlage 66
4052 Basel, Switzerland

This is a reprint of articles from the Special Issue published online in the open access journal *Entropy* (ISSN 1099-4300) (available at: https://www.mdpi.com/journal/entropy/special_issues/entropy_image_analysisII).

For citation purposes, cite each article independently as indicated on the article page online and as indicated below:

LastName, A.A.; LastName, B.B.; LastName, C.C. Article Title. *Journal Name* **Year**, *Article Number, Page Range.*

ISBN 978-3-03943-160-1 (Hbk)
ISBN 978-3-03943-161-8 (PDF)

© 2020 by the authors. Articles in this book are Open Access and distributed under the Creative Commons Attribution (CC BY) license, which allows users to download, copy and build upon published articles, as long as the author and publisher are properly credited, which ensures maximum dissemination and a wider impact of our publications.

The book as a whole is distributed by MDPI under the terms and conditions of the Creative Commons license CC BY-NC-ND.

Contents

About the Editor

Amelia Carolina Sparavigna (Dr.) is a physics researcher working mainly in the field of condensed matter physics and image processing. She graduated from the University of Turin in 1982 and obtained a Ph.D. in Physics at Politecnico of Turin in 1990. Since 1993, she has carried out teaching and research activities at the Politecnico. Her scientific research covers the fields of thermal transport and Boltzmann equation, liquid crystals, and the related image processing of polarized light microscopy. She has proposed new methods of image processing inspired by physical quantities, such as coherence lengths. Her recent works mainly concern the problem of image segmentation. She is also interested in the history of physics and science. The papers that she has published in international journals are mainly on the topics of phonon thermal transport, the elastic theory of nematic liquid crystals, and the texture transitions of liquid crystals, investigated by means of image processing.

Editorial

Entropy in Image Analysis II

Amelia Carolina Sparavigna

Department of Applied Science and Technology, Polytechnic University of Turin, 10129 Turin, Italy;
amelia.sparavigna@polito.it

Received: 10 August 2020; Accepted: 14 August 2020; Published: 15 August 2020

Keywords: image entropy; image processing; image encryption; medical imaging; neural engineering; computer vision; crowd motion detection; security

Image analysis is a fundamental task for any application where extracting information from images is required. The analysis needs numerical and analytical methods that are highly sophisticated, particularly for those applications in medicine, security, and other fields where the results of the processing consist of data of vital importance. This fact is evident from all the articles composing this Special Issue of Entropy in which authors have widely tested methods to verify their results. On the other hand, being specifically involved in numerous applications, image analysis is producing a large number of approaches and related algorithms in which the variety is clearly exemplified by the case studies proposed in this issue. Let us stress that, in its progression, an important stimulus and cross-fertilization among publications was observed with the editor's great pleasure.

Let us describe the articles of the issue shortly.

In Reference [1], we can find a problem of medical imaging based on the ultrasound addressed. It is the analysis of the severity of Duchenne Muscular Dystrophy. Ultrasound imaging enables routine examinations for which entropy represents a great help in visualizing related changes. Using small-window entropy, the imaging technique exhibits higher diagnostic performance than conventional methods. Article [2] is also considering ultrasound imaging. The authors' aim was that of discriminating the normal muscles from neuropathic muscles in children affected by Pompe disease. The method is using a texture-feature parametric imaging that simultaneously considers microstructure and macrostructure. In Reference [3], a compression method is proposed, which can be very useful in telemedicine applications. The addressed and solved problem is that of having a lossless compression of images of malaria-infected red blood cells. In fact, a remote diagnosis of malaria infection could receive a great benefit from efficient compression of high-resolution images.

In Reference [4], we are again in the field of medical imaging. In the article, a blind image quality assessment (BIQA) method for evaluating magnetic resonance images is introduced. Images are first preprocessed to reach acceptable local intensity differences. Quality is expressed by the entropy coming from a thresholding in sequence. Image Quality Assessment appears in Reference [5] as well. The authors are approaching the problem of training IQA, using deep neural networks. Since the image quality is highly sensitive to changes in entropy, a new data expansion method based on this remarkable quantity is proposed.

A security scheme for medical imaging is the subject of article [6], which is presenting a medical image stego-hiding scheme, named BOOST by the authors. It uses a pseudorandom byte output technique based on the nuclear spin generator. The security analyses show that BOOST can be used for secure medical record communication. An image encryption scheme appears in Reference [7] as well. The scheme is based on quasi-resonant Rossby/drift wave triads and Mordell elliptic curves. A dynamic substitution box is employed for the plain image. The security of the proposed scheme was tested and compared with other popular schemes. Article [8] proposes an algorithm for medical color images encryption. It uses chaotic systems to protect medical images against attacks. The algorithm has two main parts: a high-speed permutation process and an adaptive diffusion. Entropy obtained after experiments tells that the algorithm is suitable for this type of image.

Chaos and hyper-chaos, combined with DNA coding, appears in an image encryption algorithm proposed in Reference [9]. The first stage involves three rounds of scrambling. Then a diffusion algorithm is applied to the plaintext image, and the intermediate ciphertext image is partitioned. The final encrypted image is formed by using DNA operation. Additionally, in Reference [10], we find an image encryption based on a hyperchaotic system proposed with a pixel-level filtering obtained by means of variable kernels. A global bit-level scrambling is also conducted to change the values and positions of pixels simultaneously. At the end of the process, a DNA-level diffusion is used as well.

Another security problem and related analysis of an image chaotic encryption algorithm is given in Reference [11]. The proposed algorithm generates Latin-bit cubes and uses them for image chaotic encryption. The algorithm also uses different Latin cube combinations to scramble the diffusion image. In Reference [12], the encryption scheme is based on double chaotic S-boxes. A compound chaotic system, Sine-Tent map, is proposed to widen the chaotic range and improve the chaotic performance. Data hiding is another very crucial research topic in information security [13]. In this article, the authors are proposing a high-capacity data-hiding scheme for absolute moment block truncation coding (AMBTC) decompressed images. For the secret data string, a unique encoding and decoding dictionary is involved, and it is used in embedding and extraction stages.

The issue of image retrieval based on a convolutional neural network (CNN) has been considered by the authors of Reference [14]. The article is proposing a feature distribution entropy to measure the difference of regional distribution information in the feature maps from CNNs. Experiments have been conducted on public datasets. Another application of entropy is available in Reference [15] to improve the methods of image binarization for automatic text recognition in images acquired in uncontrolled lighting conditions. The preprocessing of images is made by means of a local entropy filtering.

Computer vision is the subject of Reference [16]. The article is proposing a novel network structure, which is involving a pipeline guidance strategy for the detection of human key-points. The use of a pipelined guidance allows one to find a balance between the convolution calculations and the communication time in order to improve the training speed of the network. In addition to the computer vision in the issue, we can find neural engineering research. In Reference [17], an application of the continuous wavelet transform and convolutional neural network for brain-computer interface is proposed. The article includes a novel motor imagery classification scheme with the aim of capturing highly informative electroencephalogram images.

Two forms of normalized entropy are used in Reference [18] for evaluating the evolution of neuro-aesthetic variables, displayed by portrait paintings, from Early Renaissance to Mannerism. The variables included symmetry, balance, and contrast (chiaroscuro) as well as intensity and spatial complexities. In Reference [19], the application is an image-based denomination recognition of Pakistani currency notes. The authors propose a procedure in two steps that extracts a currency note from the image background via local entropy and range filters. Then, the aspect ratio of the extracted currency note is calculated to determine its denomination. In Reference [20], a methodology based on weld segmentation and entropy is proposed such as an evaluation by conventional and convolution neural networks to assess the quality of welds. Compared to conventional neural networks, the method does not require image preprocessing. The performed experiments show that the best results are achieved using convolution neural networks. The image evaluation for engine flame is the subject of the investigation proposed in article [21]. In it, we find the detectability of the related infrared radiation. The influence of the earth background interference on plume radiation detection is investigated and discussed in detail.

Let us conclude with an application that can be very useful for controlling the movements of a crowd. In Reference [22], we find an article proposing a method for salient crowd motion detection based on direction entropy and a repulsive force network. This work focuses on the manner by which it is possible to detect the salient regions effectively. Let us observe that the proposed method could be integrated in any crowd check-point, such as during a pandemic, for helping in the control of disease spread.

Acknowledgments: We express our thanks to the authors of the above contributions, and to the journal Entropy and MDPI for their support during this work.

Conflicts of Interest: The author declare no conflict of interest.

References

1. Yan, D.; Li, Q.; Lin, C.-W.; Shieh, J.-Y.; Weng, W.-C.; Tsui, P.-H. Clinical Evaluation of Duchenne Muscular Dystrophy Severity Using Ultrasound Small-Window Entropy Imaging. *Entropy* **2020**, *22*, 715. [CrossRef]
2. Chiou, H.-J.; Yeh, C.-K.; Hwang, H.-E.; Liao, Y.-Y. Efficacy of Quantitative Muscle Ultrasound Using Texture-Feature Parametric Imaging in Detecting Pompe Disease in Children. *Entropy* **2019**, *21*, 714. [CrossRef]
3. Dong, Y.; Pan, W.D.; Wu, D. Impact of Misclassification Rates on Compression Efficiency of Red Blood Cell Images of Malaria Infection Using Deep Learning. *Entropy* **2019**, *21*, 1062. [CrossRef]
4. Obuchowicz, R.; Oszust, M.; Bielecka, M.; Bielecki, A.; Piórkowski, A. Magnetic Resonance Image Quality Assessment by Using Non-Maximum Suppression and Entropy Analysis. *Entropy* **2020**, *22*, 220. [CrossRef]
5. Guan, X.; He, L.; Li, M.; Li, F. Entropy Based Data Expansion Method for Blind Image Quality Assessment. *Entropy* **2020**, *22*, 60. [CrossRef]
6. Stoyanov, B.; Stoyanov, B. BOOST: Medical Image Steganography Using Nuclear Spin Generator. *Entropy* **2020**, *22*, 501. [CrossRef]
7. Ullah, I.; Hayat, U.; Bustamante, M.D. Image Encryption Using Elliptic Curves and Rossby/Drift Wave Triads. *Entropy* **2020**, *22*, 454. [CrossRef]
8. Moafimadani, S.S.; Chen, Y.; Tang, C. A New Algorithm for Medical Color Images Encryption Using Chaotic Systems. *Entropy* **2019**, *21*, 577. [CrossRef]
9. Wan, Y.; Gu, S.; Du, B. A New Image Encryption Algorithm Based on Composite Chaos and Hyperchaos Combined with DNA Coding. *Entropy* **2020**, *22*, 171. [CrossRef]
10. Wu, J.; Shi, J.; Li, T. A Novel Image Encryption Approach Based on a Hyperchaotic System, Pixel-Level Filtering with Variable Kernels, and DNA-Level Diffusion. *Entropy* **2020**, *22*, 5. [CrossRef]
11. Zhang, Z.; Yu, S. On the Security of a Latin-Bit Cube-Based Image Chaotic Encryption Algorithm. *Entropy* **2019**, *21*, 888. [CrossRef]
12. Zhu, S.; Wang, G.; Zhu, C. A Secure and Fast Image Encryption Scheme Based on Double Chaotic S-Boxes. *Entropy* **2019**, *21*, 790. [CrossRef]
13. Yeh, J.-Y.; Chen, C.-C.; Liu, P.-L.; Huang, Y.-H. High-Payload Data-Hiding Method for AMBTC Decompressed Images. *Entropy* **2020**, *22*, 145. [CrossRef]
14. Liu, P.; Gou, G.; Guo, H.; Zhang, D.; Zhao, H.; Zhou, Q. Fusing Feature Distribution Entropy with R-MAC Features in Image Retrieval. *Entropy* **2019**, *21*, 1037. [CrossRef]
15. Michalak, H.; Okarma, K. Improvement of Image Binarization Methods Using Image Preprocessing with Local Entropy Filtering for Alphanumerical Character Recognition Purposes. *Entropy* **2019**, *21*, 562. [CrossRef]
16. Hong, F.; Lu, C.; Liu, C.; Liu, R.; Jiang, W.; Ju, W.; Wang, T. PGNet: Pipeline Guidance for Human Key-Point Detection. *Entropy* **2020**, *22*, 369. [CrossRef]
17. Lee, H.K.; Choi, Y.-S. Application of Continuous Wavelet Transform and Convolutional Neural Network in Decoding Motor Imagery Brain-Computer Interface. *Entropy* **2019**, *21*, 1199. [CrossRef]
18. Correa-Herran, I.; Aleem, H.; Grzywacz, N.M. Evolution of Neuroaesthetic Variables in Portrait Paintings throughout the Renaissance. *Entropy* **2020**, *22*, 146. [CrossRef]
19. Anwar, H.; Ullah, F.; Iqbal, A.; Ul Hasnain, A.; Ur Rehman, A.; Bell, P.; Kwak, D. Invariant Image-Based Currency Denomination Recognition Using Local Entropy and Range Filters. *Entropy* **2019**, *21*, 1085. [CrossRef]
20. Haffner, O.; Kučera, E.; Drahoš, P.; Cigánek, J. Using Entropy for Welds Segmentation and Evaluation. *Entropy* **2019**, *21*, 1168. [CrossRef]

21. Li, X.; Wang, J.; Li, M.; Peng, Z.; Liu, X. Investigating Detectability of Infrared Radiation Based on Image Evaluation for Engine Flame. *Entropy* **2019**, *21*, 946. [CrossRef]
22. Zhang, X.; Lin, D.; Zheng, J.; Tang, X.; Fang, Y.; Yu, H. Detection of Salient Crowd Motion Based on Repulsive Force Network and Direction Entropy. *Entropy* **2019**, *21*, 608. [CrossRef]

 © 2020 by the author. Licensee MDPI, Basel, Switzerland. This article is an open access article distributed under the terms and conditions of the Creative Commons Attribution (CC BY) license (http://creativecommons.org/licenses/by/4.0/).

Article

Clinical Evaluation of Duchenne Muscular Dystrophy Severity Using Ultrasound Small-Window Entropy Imaging

Dong Yan [1], Qiang Li [1], Chia-Wei Lin [2], Jeng-Yi Shieh [3], Wen-Chin Weng [4,5,*] and Po-Hsiang Tsui [6,7,8,*]

[1] School of Microelectronics, Tianjin University, Tianjin 300072, China; yd_beiyang@tju.edu.cn (D.Y.); liqiang@tju.edu.cn (Q.L.)

[2] Department of Physical Medicine and Rehabilitation, National Taiwan University Hospital Hsin-Chu Branch, Hsin-Chu 30059, Taiwan; chiaweionly@gmail.com

[3] Department of Physical Medicine and Rehabilitation, National Taiwan University Hospital, Taipei 100229, Taiwan; jyshieh@ntu.edu.tw

[4] Department of Pediatrics, National Taiwan University Hospital, and College of Medicine, National Taiwan University, Taipei 100233, Taiwan

[5] Department of Pediatric Neurology, National Taiwan University Children's Hospital, Taipei 100226, Taiwan

[6] Department of Medical Imaging and Radiological Sciences, College of Medicine, Chang Gung University, Taoyuan 33302, Taiwan

[7] Medical Imaging Research Center, Institute for Radiological Research, Chang Gung University and Chang Gung Memorial Hospital at Linkou, Taoyuan 33302, Taiwan

[8] Department of Medical Imaging and Intervention, Chang Gung Memorial Hospital at Linkou, Taoyuan 33305, Taiwan

[*] Correspondence: wcweng@ntu.edu.tw (W.-C.W.); tsuiph@mail.cgu.edu.tw (P.-H.T.)

Received: 12 March 2020; Accepted: 25 June 2020; Published: 28 June 2020

Abstract: Information entropy of ultrasound imaging recently receives much attention in the diagnosis of Duchenne muscular dystrophy (DMD). DMD is the most common muscular disorder; patients lose their ambulation in the later stages of the disease. Ultrasound imaging enables routine examinations and the follow-up of patients with DMD. Conventionally, the probability distribution of the received backscattered echo signals can be described using statistical models for ultrasound parametric imaging to characterize muscle tissue. Small-window entropy imaging is an efficient nonmodel-based approach to analyzing the backscattered statistical properties. This study explored the feasibility of using ultrasound small-window entropy imaging in evaluating the severity of DMD. A total of 85 participants were recruited. For each patient, ultrasound scans of the gastrocnemius were performed to acquire raw image data for B-mode and small-window entropy imaging, which were compared with clinical diagnoses of DMD by using the receiver operating characteristic curve. The results indicated that entropy imaging can visualize changes in the information uncertainty of ultrasound backscattered signals. The median with interquartile range (IQR) of the entropy value was 4.99 (IQR: 4.98–5.00) for the control group, 5.04 (IQR: 5.01–5.05) for stage 1 patients, 5.07 (IQR: 5.06–5.07) for stage 2 patients, and 5.07 (IQR: 5.06–5.07) for stage 3 patients. The diagnostic accuracies were 89.41%, 87.06%, and 72.94% for ≥stage 1, ≥stage 2, and ≥stage 3, respectively. Comparisons with previous studies revealed that the small-window entropy imaging technique exhibits higher diagnostic performance than conventional methods. Its further development is recommended for potential use in clinical evaluations and the follow-up of patients with DMD.

Keywords: Duchenne muscular dystrophy; entropy; ultrasound; backscattered signals

1. Introduction

Information entropy of ultrasound imaging recently receives much attention in the diagnosis of Duchenne muscular dystrophy (DMD). DMD is the most common muscular disorder caused by mutations in the dystrophin gene [1] and results in absent or insufficient functional dystrophin, which leads to reduced sarcolemma stability and rendering the muscle fibers vulnerable to mechanical stretching-induced injury [2]. As a consequence, repeated contraction leads to necrosis and the regeneration of muscle fibers, which are gradually replaced by fat and fibrous tissue. This disease is primarily an X-linked condition affecting males; however, some female carriers exhibit symptoms of the disorder, but usually with a milder phenotype [3]. Because of regional and ethnic differences, the estimated incidence is approximately 1 in 5000–10,000 live male births [4–6]. Boys with DMD may exhibit symptoms such as abnormal gait, weakened proximal muscles, and calf muscle pseudohypertrophy at age 3–5 years [7,8]. Patients with DMD inevitably develop a loss of mobility, respiratory and cardiac deterioration as a consequence of the dystrophic changes of muscle, and typically die from respiratory and cardiac complications by the age of 30 [9].

Currently, no curative therapy is available for treating DMD; therefore, early detection and effective health care, rehabilitation, and psychosocial management are essential [10–12]. However, considerable progress has been made recently in terms of genetic approaches [13]; some drugs have also been conditionally approved for the treatment of patients with DMD [14]. This implies that reliable and noninvasive approaches to evaluating the progression of DMD and treatment efficacy are required. The North Star Ambulatory Assessment [15] and timed function tests, including the 6-minute walk test, time to climb 4 stairs, time to stand or 10-meter walk [16] are typical assessments of function during the ambulatory period. Although these outcome measures are clinically meaningful and valuable, their sensitivity is often limited by the effort and mood of children with DMD without objective assessment of muscle pathologic change [17].

Ultrasound and magnetic resonance imaging (MRI) are two of the commonest and widely used noninvasive methods for muscle tissue examination [18,19]. Compared with MRI, ultrasound imaging enables friendlier and safer routine scans and follow-up for pediatric patients [20]. Studies have revealed that quantitative muscle ultrasound can detect DMD progression [21,22]. Fat infiltration and fibrosis formation increase the intensity of the backscattered echo [23], indicating that ultrasound backscattering may provide useful information associated with changes in muscle microstructures for DMD diagnosis. Considering the random nature of ultrasound backscattering, the probability distribution of the backscattered envelope (the echo amplitude) has been explored and demonstrated to be useful in characterizing tissues [24]. Previously, the Nakagami statistical distribution was applied to modeling the backscattered statistics as an evaluation method of ambulatory function in patients with DMD [25]. However, the prerequisite for using the statistical distribution is that the echo data must follow the model used [26]; it is difficult to satisfy the aforementioned condition in practice, because the properties of backscattered signals depend on system characteristics, software settings, and signal/image processing. This limitation has encouraged researchers to pursue a more flexible solution for describing backscattering information, without considering the distribution nature of the echo data.

Among all possibilities, Shannon entropy (a measure of information uncertainty [27]) fulfills the aforementioned requirement. Hughes first introduced the concept of entropy in the field of ultrasound imaging, indicating that entropy can be used to quantitatively characterize changes in the microstructures of scattering media [28–30]. Furthermore, entropy has been reported to be a non-model-based statistical parameter that is proportional to the Nakagami parameter and correlates with backscattered statistics [31]. In particular, information entropy has been applied to ultrasound parametric imaging, allowing the use of the small-window technique to visualize the statistical properties of backscattered signals for tissue characterization with improved image resolution [32,33]. For these reasons, we explored the feasibility of using ultrasound small-window entropy imaging in evaluating the severity of the dystrophic process in patients with DMD.

2. Materials and Methods

2.1. Study Population

The Institutional Review Board of National Taiwan University Hospital (NTUH) approved the study and allowed the reuse of the database collected in a previous study [34]. A total of 85 participants ($n = 85$) aged between 2 and 24 years provided written informed consent, and the experimental methods were conducted according to the approved guidelines. All DMD patients ($n = 73$) were recruited from the joint clinics of neuromuscular disorders in the Department of Pediatrics, NTUH. The clinical manifestations of each patient were consistent with DMD, and diagnoses had been confirmed according to muscle biopsies (revealing absent dystrophin) and/or genetic testing. On the basis of a review report [10], the severity of DMD was classified into three stages: stage 1 (presymptomatic, early ambulatory, and late ambulatory), stage 2 (early non-ambulatory), and stage 3 (late non-ambulatory). Seventy-three patients ($n = 73$) were recruited (stage 1: $n = 41$; stage 2: $n = 20$; stage 3: $n = 12$). Twelve children ($n = 12$) with no history of weakness or neuromuscular disorders were also recruited as controls. The demographic data of participants and stage definitions are summarized in our previous study [34].

2.2. Ultrasound Examination

A commercial clinical ultrasound system (Model 3000; Terason, Burlington, MA, USA) equipped with a linear array transducer (Model 12L5A; Terason) was used for ultrasound scans on the patients with DMD. The central frequency, pulse length, and beam width of the transducer were 7 MHz, 0.7 mm, and 1.2 mm, respectively. All participants underwent a standard-care ultrasound examination of the gastrocnemius using the sagittal scanning approach. For each participant, three valid scans (i.e., no acoustic shadowing artifacts and the exclusion of large vessels in the region of analysis) were performed by a skilled physician. The focus and depth were set at 2 and 4 cm, respectively. The gain index of the Terason system was set at 6, corresponding to a signal-to-noise ratio of approximately 30 dB, which was obtained from the calibrations performed in the previous study [35]. Raw image data obtained from each valid scan, consisting of 128 backscattered radiofrequency (RF) signals at a sampling rate of 30 MHz, were used for offline data processing in MATLAB, including ultrasound B-mode and small-window entropy imaging.

2.3. Entropy Imaging Algorithm

The algorithms for ultrasound B-mode and entropy imaging [32] are illustrated in Figure 1. For the data of each raw image, the absolute values of the Hilbert transform of backscattered RF signals were calculated to obtain the envelope image. Using the logarithm-compressed envelope, which provides different grayscales according to its value at a dynamic range of 40 dB, the B-mode image was formed. The uncompressed envelope data were used for small-window entropy imaging according to the following steps: (a) a small-square window was set up in the upper-left corner of the data with a side length of one time the pulse length of the transducer (0.7 mm) to collect uncompressed envelope data; (b) the envelope data were normalized, and the probability distribution of the envelope data within the window was constructed using a statistical histogram (bins = 50) for estimating the Shannon entropy, using the following equation:

$$H_C = - \sum_{i=1}^{n} w(y_i) \log_2 [w(y_i)] \tag{1}$$

where y_i is the discrete random variable of the envelope data, $w(y_i)$ represents the probability value, and n indicates the number of bins; then, the estimated entropy value was assigned as a new pixel corresponding to the window location; (c) subsequently, the window, with a window overlap ratio of 50%, to provide a tradeoff between the parametric image resolution and computational time [32], was slid throughout the entire envelope image to calculate local entropy values (according to the step (b)) for generating a parametric map; (d) a two-dimensional linear interpolation was performed to obtain an entropy parametric map, with the same size as the uncompressed envelope data [36], which was displayed in a pseudocolor and superimposed onto the B-mode image, to reveal both the anatomical and backscattering information. Finally, the region of interest (ROI) corresponding to the gastrocnemius was manually chosen on the image to calculate the average entropy value.

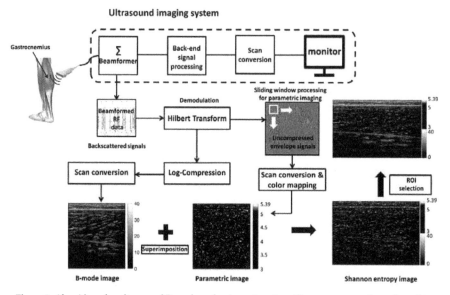

Figure 1. Algorithms for ultrasound B-mode and entropy imaging. The uncompressed envelope data were processed using the sliding window technique. The side length of the window was set as one time the pulse length, to acquire local data points for estimating the entropy values. RF: radiofrequency.

The ROI selection was handled by a pediatric neurologist. To reduce the bias in averaging the entropy values in the ROI, choosing an ROI that satisfies the coverage of the whole gastrocnemius was used as a basic rule in this study. For each participant, the final entropy value was obtained by the average of three valid scans.

2.4. Statistical Analysis

The envelope amplitude and entropy values, as a function of DMD stage, are expressed as vertical box and dotted plots, which exhibit the median, interquartile range (IQR; being equal to the difference between the third quartile and the first quartile), data distribution, and outliers. The Spearman rank correlation coefficient r and the probability value p were calculated for evaluating the correlation between the parameter values (envelope amplitude and entropy) and DMD stage. The receiver operating characteristic (ROC) curve with a 95% confidence interval (CI) was used to evaluate the performances for diagnosing different DMD stages. The ROC curve was created by plotting the true positive rate against the false positive rate at various threshold settings. The optimal cutoff value for diagnosing each DMD stage was determined by the point maximizing the Youden function, which is the difference between true positive rate and false positive rate over all possible cutoff values [37]. The area under the ROC curve (AUROC), sensitivity, specificity, accuracy, and other statistical results were then reported. A p-value of <0.05 was considered statistically significant. All statistical analyses were performed using SigmaPlot Version 12.0 (Systat Software, Inc., CA, USA).

3. Results

Typical images representing different DMD stages are depicted in Figure 2. The dotted lines indicate the ROIs corresponding to the gastrocnemius for ultrasound entropy imaging. Observations on the entropy values obtained from three valid scans for each individual subject showed that the proposed rule for ROI selection ensured the maximum difference of entropy between three valid scans ≤0.02. The image brightness increased from the control group to stage 3, indicating that the entropy value and the probability distribution of ultrasound backscattered signals vary with the severity of DMD (Figure 3).

Figure 4a,b exhibits the box plots with dot density, which reveal the positions of each envelope amplitude and entropy data point. Evidently, the envelope amplitude increased as the DMD stage advanced ($r = 0.49$; $p < 0.05$); the median (IQR) was 102.32 (IQR: 75.90–125.44) for the control, 178.99 (IQR: 158.22–218.73) for stage 1, 271.08 (IQR: 236.65–363.11) for stage 2, and 204.97 (IQR: 135.08–300.83) for stage 3. Ultrasound entropy also increased as DMD stages progressed ($r = 0.76$; $p < 0.05$); the median (IQR) was 4.99 (IQR: 4.98–5.00) for the control, 5.04 (IQR: 5.01–5.05) for stage 1, 5.07 (IQR: 5.06–5.07) for stage 2, and 5.07 (IQR: 5.06–5.07) for stage 3. The AUROCs (95% CI) for diagnosing different DMD stages are shown in Figure 4c,d. The AUROCs obtained from using the B-scan to calculate the envelope amplitude were 0.91 (0.79–1), 0.76 (0.66–0.86), and 0.54 (0.36–0.72) for ≥stage 1, ≥stage 2, and ≥stage 3, respectively (the diagnostic accuracies were 85.88% for ≥stage 1, 75.29% for ≥stage 2, and 52.94% for ≥stage 3), and those of entropy were 0.96 (0.89–1), 0.91 (0.85–0.97), and 0.80 (0.68–0.91) for ≥stage 1, ≥stage 2, and ≥stage 3, respectively (the diagnostic accuracies were 89.41% for ≥stage 1, 87.06% for ≥stage 2, and 72.94% for ≥stage 3). Tables 1 and 2 show the other statistical results obtained from the ROC analysis, including cutoff value, sensitivity, specificity, positive likelihood ratio, negative likelihood ratio, positive predictive value, and negative predictive value, representing that ultrasound entropy imaging outperformed conventional B-scan in evaluating the severity of DMD.

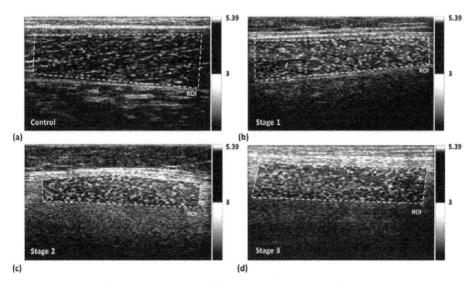

Figure 2. Typical images measured at different Duchenne muscular dystrophy (DMD) stages. (a) normal control; (b) stage 1; (c) stage 2; (d) stage 3. The dotted lines indicate the regions of interest (ROIs) corresponding to the gastrocnemius. Image brightness increased between the control group and the stage 3 group, representing a corresponding entropy value increase.

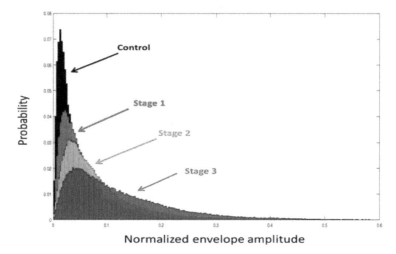

Figure 3. Probability distributions of ultrasound backscattered signals measured in the ROIs for different DMD stages. The probability distribution was described using a statistical histogram (bins = 50); it varied with the severity of DMD.

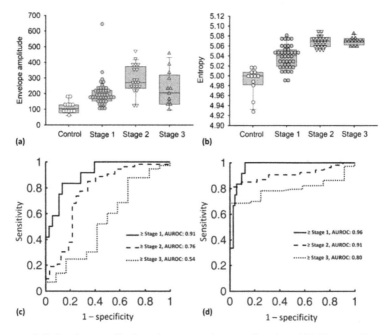

Figure 4. (**a**,**b**) Envelope amplitude and entropy values as a function of DMD stage. Data were expressed by vertical box and dotted plots, which revealed the median, interquartile range (IQR), data distribution, and outliers. The entropy value increased as the DMD stage advanced ($r = 0.76$; $p < 0.05$), and the envelope amplitude also showed a similar trend ($r = 0.49$; $p < 0.05$). (**c**) and (**d**) AUROCs for diagnosing different DMD stages using the B-mode and entropy images. Compared with the B-scan, ultrasound entropy imaging could detect early stage DMD with improved diagnostic performance; it also performed well in detecting the difference between ambulatory and non-ambulatory stages.

Table 1. Clinical performance of ultrasound B-mode imaging (envelope amplitude) in evaluating the severity of DMD.

Clinical Severity	≥Stage 1	≥Stage 2	≥Stage 3
Cutoff value	128.22	212.86	183.04
Sensitivity	83.33	77.36	52.05
Specificity	86.3	71.88	58.33
Accuracy	85.88	75.29	52.94
LR+	6.08	2.75	1.25
LR-	0	0.32	0.82
PPV, %	50	82	88.37
NPV, %	96.92	65.71	16.67
AUROC (95% CI)	0.91 (0.79–1)	0.76 (0.66–0.86)	0.54 (0.36–0.72)

LR+: positive likelihood ratio, LR−: negative likelihood ratio, PPV: positive predictive value, NPV: negative predictive value, AUROC: area under the receiver operating characteristics curve.

Table 2. Clinical performance of ultrasound entropy imaging in evaluating the severity of DMD.

Clinical Severity	≥Stage 1	≥Stage 2	≥Stage 3
Cutoff value	5.01	5.05	5.05
Sensitivity	100	84.91	68.49
Specificity	87.67	90.63	100
Accuracy	89.41	87.06	72.94
LR+	8.11	9.06	6.55
LR-	0	0.17	0.32
PPV, %	57.14	93.75	100
NPV, %	100	78.38	34.29
AUROC (95% CI)	0.96 (0.89–1)	0.91 (0.85–0.97)	0.80 (0.68–0.91)

4. Discussion

In DMD, the progression of two critical periods should be noted: the first refers to the period before amyotrophia occurs and the second is when patients lose their ambulation. However, free-acting capability is apparently a critical index representing the effect of DMD on patients and their families. For this reason, the major aim of DMD treatment is the prolongation of walking function [38]. In most cases, neuromuscular specialists assess and characterize each patient's unique disease trajectory using validated assessment tools and their clinical experience, aiming to establish a patient's expected clinical course [10]. Because of advances in medical technologies, ultrasound imaging has become the preferred method for clinicians to follow up DMD patients, because it serves as a real-time point-of-care tool. Ultrasound imaging can also provide further quantitative information associated with the echo intensity to aid DMD evaluations [39,40]. To satisfy the requirement for evaluating the walking function of patients with DMD, an emerging research trend involves using statistical distributions to model the backscattered statistics for characterizing tissue microstructures, which are highly correlated with the behavior of ultrasound backscattering [24].

As reviewed in the Introduction, ultrasound Nakagami parametric imaging has been applied to imaging the backscattered statistics measured from the gastrocnemius [25]. Variation in the Nakagami parameter from 0 to 1 indicates a change in the envelope statistics from a pre-Rayleigh to a Rayleigh distribution; a Nakagami parameter higher than 1 indicates that the backscattered statistics represent post-Rayleigh distributions [25]. The Nakagami parameter increased and remained close to 1 when the DMD progressed to stage 4; the performance of diagnosing the walking function of the patients with DMD was also acceptable (AUROC: 0.89; accuracy: 85.52%; sensitivity: 76.31%; specificity: 94.73%) [25]. In this study, we used information entropy, a simpler and more effective parameter, as the non-model-based solution to analyze the uncertainty and complexity of ultrasound backscattered signals. We also followed the algorithm of ultrasound parametric imaging to construct the entropy map to image the backscattering information of DMD. According to our findings, the entropy value was a monotonically increasing function of the DMD stage, and gradually entered a plateau phase when DMD was at stages 2 and 3, representing that distinguishing stages 2 and 3 is difficult. However, clinical treatments and managements of DMD actually need early detection and evaluations of the walking function (at stages 1 and 2), and ultrasound entropy may be a qualified imaging biomarker to satisfy the above purposes. Evidently, ultrasound entropy imaging was able to detect early stage DMD with improved diagnostic performance (AUROC: 0.96; accuracy: 89.41%; sensitivity: 100%; specificity: 87.67%). Moreover, ultrasound entropy values accurately detected the difference between ambulatory and non-ambulatory stages (AUROC: 0.91; accuracy: 87.06%; sensitivity: 84.91%; specificity: 90.63%). Several studies have revealed that fatty and connective tissues in the muscles of patients with DMD cause strong echoes [21,22,40]. This can be regarded as the behavior of constructive wave interference to induce changes in the backscattered statistics from the pre-Rayleigh to the Rayleigh distribution, corresponding to the increase in the signal uncertainty [31,33]. This can explain why the Nakagami and entropy values increase with the severity of DMD. More importantly, ultrasound entropy imaging exhibited improved diagnostic performance for DMD evaluations compared with the Nakagami parameter proposed previously.

The improved diagnostic performance of ultrasound entropy imaging may be attributed to the suppression of boundary artifacts during sliding window processing. Entropy and Nakagami images differ in that entropy allows the use of a small sliding window for parametric imaging. This advantage enables an effective reduction in the appearance of boundary artifacts in ultrasound parametric imaging constructed using the sliding window processing technique, as explained in detail previously [32]. Briefly, as the sliding window moves across the interface, the window acquires not only the backscattered signals returned from the interface, but also those from the tissue parenchyma. The difference in the echo amplitude of the interface and tissue parenchyma tends to lead to underestimation of the parameter, generating a boundary artifact. The simplest approach to suppressing boundary artifacts is to use a small window for parametric imaging. However, the distribution parameters typically require a relatively large window for parametric imaging, because a sufficient sample size is necessary for stable parameter estimation. Unlike the distribution parameters, information entropy is a relative measure of signal uncertainty, not a model-based parameter or an absolute physical estimate. Therefore, entropy calculated using fewer data points acquired from a small window for scatterer characterization is allowed and feasible [32]. On the other hand, the offline processing time for ultrasound entropy imaging of one image raw data was < 1 sec (operating environment: Windows 10; RAM: 8 GB; CPU: Intel® Core™ i3-8100 at 3.6 GHz), implying the possibility of real-time capability if the algorithm is combined with an ultrasound system.

The limited dynamic range of information entropy is the major limitation in practice, although this does not affect the statistical significance of the results. The dynamic range of information entropy must be enlarged to facilitate the color mapping and visualization of parametric imaging as well as to improve its sensitivity. Probably, using phantoms with different scatterer properties to find a theoretically dynamic range of entropy as the calibration reference for parameter normalization may be a feasible method. Moreover, using ultrasound entropy imaging to follow up on patients with DMD is an area for future research, which could provide possibilities to predict the progression of walking function in these patients. Prior to using entropy imaging as a reliable follow-up tool of DMD, a large-scale clinical validation is still necessary. Using different datasets for training and tests are also needed to investigate the power of prediction through ultrasound entropy imaging.

5. Conclusions

In this study, we investigated the performance of ultrasound small-window entropy imaging in evaluating the clinical severity of symptoms in patients with DMD. The results revealed that entropy imaging can visualize changes in the information uncertainty of ultrasound backscattered signals during the progression of DMD. In particular, entropy value performed well in detecting the early stages of DMD; entropy values were also highly correlated with walking function in patients with DMD.

Compared with conventional B-scan and model-based methods, entropy imaging, constructed using the small-window technique, exhibits great potential, and we recommended its further development for the clinical evaluations and follow-up of patients with DMD.

Author Contributions: Conceptualization, W.-C.W. and P.-H.T.; methodology, P.-H.T.; software, D.Y.; validation, D.Y.; formal analysis, D.Y.; investigation, C.-W.L. and J.-Y.S.; resources, Q.L. and W.-C.W.; data curation, W.-C.W.; writing—original draft preparation, D.Y. and Q.L.; writing—review and editing, P.-H.T. and W.-C.W.; visualization, P.-H.T.; supervision, P.-H.T.; project administration, W.-C.W.; funding acquisition, P.-H.T. All authors have read and agreed to the published version of the manuscript.

Funding: This work was supported by the Ministry of Science and Technology in Taiwan (Grant Nos. MOST 106-2221-E-182-023-MY3 and 108-2218-E-182-001) and the Chang Gung Memorial Hospital at Linkou in Taiwan (Grant Nos. CMRPD1J0171 and CMRPD1H0381).

Conflicts of Interest: The authors declare no conflict of interest.

References

1. Emery, A.E. The muscular dystrophies. *Lancet* **2002**, *359*, 687–695. [CrossRef]
2. Hoffman, E.P.; Brown, R.H., Jr.; Kunkel, L.M. Dystrophin: The protein product of the Duchenne muscular dystrophy locus. *Cell* **1987**, *51*, 919–928. [CrossRef]
3. Ryder, S.; Leadley, R.M.; Armstrong, N.; Westwood, M.; de Kock, S.; Butt, T.; Jain, M.; Kleijnen, J. The burden, epidemiology, costs and treatment for Duchenne muscular dystrophy: An evidence review. *Orphanet. J. Rare. Dis.* **2017**, *12*, 79. [CrossRef]
4. Romitti, P.A.; Zhu, Y.; Puzhankara, S.; James, K.A.; Nabukera, S.K.; Zamba, G.K.; Ciafaloni, E.; Cunniff, C.; Druschel, C.M.; Mathews, K.D.; et al. STARnet. Prevalence of Duchenne and Becker muscular dystrophies in the United States. *Pediatrics* **2015**, *135*, 513–521. [CrossRef]
5. Moat, S.J.; Bradley, D.M.; Salmon, R.; Clarke, A.; Hartley, L. Newborn bloodspot screening for Duchenne muscular dystrophy: 21 years experience in Wales (UK). *Eur. J. Hum. Genet.* **2013**, *21*, 1049–1053. [CrossRef]
6. Mendell, J.R.; Shilling, C.; Leslie, N.D.; Flanigan, K.M.; al-Dahhak, R.; Gastier-Foster, J.; Kneile, K.; Dunn, D.M.; Duval, B.; Aoyagi, A.; et al. Evidence-based path to newborn screening for Duchenne muscular dystrophy. *Ann. Neurol.* **2012**, *71*, 304–313. [CrossRef]
7. Guiraud, S.; Chen, H.; Burns, D.T.; Davies, K.E. Advances in genetic therapeutic strategies for Duchenne muscular dystrophy. *Exp. Physiol.* **2015**, *100*, 1458–1467. [CrossRef]
8. Emery, A.E. Historical Duchenne muscular dystrophy—Meryon's disease. *Neuromuscul Disord* **1993**, *3*, 263–266. [CrossRef]
9. Bach, J.R.; O'Brien, J.; Krotenberg, R.; Alba, A.S. Management of end stage respiratory failure in Duchenne muscular dystrophy. *Muscle Nerve* **1987**, *10*, 177–182. [CrossRef]
10. Birnkrant, D.J.; Bushby, K.; Bann, C.M.; Apkon, S.D.; Blackwell, A.; Brumbaugh, D.; Case, L.E.; Clemens, P.R.; Hadjiyannakis, S.; Pandya, S.; et al. DMD Care Considerations Working Group. Diagnosis and management of Duchenne muscular dystrophy, part 1: Diagnosis, and neuromuscular, rehabilitation, endocrine, and gastrointestinal and nutritional management. *Lancet Neurol.* **2018**, *17*, 251–267. [CrossRef]
11. Birnkrant, D.J.; Bushby, K.; Bann, C.M.; Alman, B.A.; Apkon, S.D.; Blackwell, A.; Case, L.E.; Cripe, L.; Hadjiyannakis, S.; Olson, A.K.; et al. DMD Care Considerations Working Group. Diagnosis and management of Duchenne muscular dystrophy, part 2: Respiratory, cardiac, bone health, and orthopaedic management. *Lancet Neurol.* **2018**, *17*, 347–361. [CrossRef]
12. Birnkrant, D.J.; Bushby, K.; Bann, C.M.; Apkon, S.D.; Blackwell, A.; Colvin, M.K.; Cripe, L.; Herron, A.R.; Kennedy, A.; Kinnett, K.; et al. DMD Care Considerations Working Group. Diagnosis and management of Duchenne muscular dystrophy, part 3: Primary care, emergency management, psychosocial care, and transitions of care across the lifespan. *Lancet Neurol.* **2018**, *17*, 445–455. [CrossRef]
13. Guglieri, M.; Bushby, K. Molecular treatments in Duchenne muscular dystrophy. *Curr. Opin. Pharmacol.* **2010**, *10*, 331–337. [CrossRef]
14. Guiraud, S.; Davies, K.E. Pharmacological advances for treatment in Duchenne muscular dystrophy. *Curr. Opin. Pharmacol.* **2017**, *34*, 36–48. [CrossRef] [PubMed]

15. Mazzone, E.S.; Messina, S.; Vasco, G.; Main, M.; Eagle, M.; D'Amico, A.; Doglio, L.; Politano, L.; Cavallaro, F.; Frosini, S.; et al. Reliability of the North Star Ambulatory Assessment in a multicentric setting. *Neuromuscul. Disord.* **2009**, *19*, 458–461. [CrossRef] [PubMed]

16. McDonald, C.M.; Henricson, E.K.; Han, J.J.; Abresch, R.T.; Nicorici, A.; Elfring, G.L.; Atkinson, L.; Reha, A.; Hirawat, S.; Miller, L.L. The 6-minute walk test as a new outcome measure in Duchenne muscular dystrophy. *Muscle Nerve* **2010**, *41*, 500–510. [CrossRef]

17. Shklyar, I.; Pasternak, A.; Kapur, K.; Darras, B.T.; Rutkove, S.B. Composite biomarkers for assessing Duchenne muscular dystrophy: An initial assessment. *Pediatr. Neurol.* **2015**, *52*, 202–205. [CrossRef]

18. Pillen, S.; Arts, I.M.; Zwarts, M.J. Muscle ultrasound in neuromuscular disorders. *Muscle Nerve* **2008**, *37*, 679–693. [CrossRef]

19. Kinali, M.; Arechavala-Gomeza, V.; Cirak, S.; Glover, A.; Guglieri, M.; Feng, L.; Hollingsworth, K.G.; Hunt, D.; Jungbluth, H.; Roper, H.P.; et al. Muscle histology vs MRI in Duchenne muscular dystrophy. *Neurology* **2011**, *76*, 346–353. [CrossRef]

20. Pillen, S.; Verrips, A.; van Alfen, N.; Arts, I.M.; Sie, L.T.; Zwarts, M.J. Quantitative skeletal muscle ultrasound: Diagnostic value in childhood neuromuscular disease. *Neuromuscul. Disord.* **2007**, *17*, 509–516. [CrossRef]

21. Zaidman, C.M.; Wu, J.S.; Kapur, K.; Pasternak, A.; Madabusi, L.; Yim, S.; Pacheck, A.; Szelag, H.; Harrington, T.; Darras, B.T.; et al. Quantitative muscle ultrasound detects disease progression in Duchenne muscular dystrophy. *Ann. Neurol.* **2017**, *81*, 633–640. [CrossRef] [PubMed]

22. Shklyar, I.; Geisbush, T.R.; Mijialovic, A.S.; Pasternak, A.; Darras, B.T.; Wu, J.S.; Rutkove, S.B.; Zaidman, C.M. Quantitative muscle ultrasound in Duchenne muscular dystrophy: A comparison of techniques. *Muscle Nerve* **2015**, *51*, 207–213. [CrossRef] [PubMed]

23. Pillen, S.; Tak, R.O.; Zwarts, M.J.; Lammens, M.M.; Verrijp, K.N.; Arts, I.M.; van der Laak, J.A.; Hoogerbrugge, P.M.; van Engelen, B.G.; Verrips, A. Skeletal muscle ultrasound: Correlation between fibrous tissue and echo intensity. *Ultrasound Med. Biol.* **2009**, *35*, 443–446. [CrossRef] [PubMed]

24. Destrempes, F.; Cloutier, G. A critical review and uniformized representation of statistical distributions modeling the ultrasound echo envelope. *Ultrasound Med. Biol.* **2010**, *36*, 1037–1051. [CrossRef]

25. Weng, W.C.; Tsui, P.H.; Lin, C.W.; Lu, C.H.; Lin, C.Y.; Shieh, J.Y.; Lu, F.L.; Ee, T.W.; Wu, K.W.; Lee, W.T. Evaluation of muscular changes by ultrasound Nakagami imaging in Duchenne muscular dystrophy. *Sci. Rep.* **2017**, *7*, 4429. [CrossRef]

26. Shankar, P.M. A compound scattering pdf for the ultrasonic echo envelope and its relationship to K and Nakagami distributions. *IEEE Trans. Ultrason. Ferroelectr. Freq. Control* **2003**, *50*, 339–343. [CrossRef]

27. Shannon, C.E. A mathematical theory of communication. *Bell Syst. Technol. J.* **1948**, *27*, 379–423. [CrossRef]

28. Hughes, M.S. Analysis of ultrasonic waveforms using Shannon entropy. *IEEE Ultrason. Symp. Proc.* **1992**, *1*, 1205–1209.

29. Hughes, M.S. Analysis of digitized waveforms using Shannon entropy. *J. Acoust. Soc. Am.* **1993**, *93*, 892. [CrossRef]

30. Hughes, M.S. Analysis of digitized waveforms using Shannon entropy. II. High speed algorithms based on Green's functions. *J. Acoust. Soc. Am.* **1994**, *95*, 2582–2588. [CrossRef]

31. Tsui, P.H. Ultrasound detection of scatterer concentration by weighted entropy. *Entropy* **2015**, *17*, 6598–6616. [CrossRef]

32. Tsui, P.H.; Chen, C.K.; Kuo, W.H.; Chang, K.J.; Fang, J.; Ma, H.Y.; Chou, D. Small-window parametric imaging based on information entropy for ultrasound tissue characterization. *Sci. Rep.* **2017**, *7*, 41004. [CrossRef] [PubMed]

33. Zhou, Z.; Tai, D.I.; Wan, Y.L.; Tseng, J.H.; Lin, Y.R.; Wu, S.; Yang, K.C.; Liao, Y.Y.; Yeh, C.K.; Tsui, P.H. Hepatic steatosis assessment with ultrasound small-window entropy imaging. *Ultrasound Med. Biol.* **2018**, *44*, 1327–1340. [CrossRef] [PubMed]

34. Weng, W.C.; Lin, C.W.; Shen, H.C.; Chang, C.C.; Tsui, P.H. Instantaneous frequency as a new approach for evaluating the clinical severity of Duchenne muscular dystrophy through ultrasound imaging. *Ultrasonics* **2019**, *94*, 235–241. [CrossRef] [PubMed]

35. Zhou, Z.; Wu, W.; Wu, S.; Jia, K.; Tsui, P.H. Empirical mode decomposition of ultrasound imaging for gain independent measurement on tissue echogenicity: A feasibility study. *Appl. Sci.* **2017**, *7*, 324. [CrossRef]

36. Ma, H.Y.; Lin, Y.H.; Wang, C.Y.; Chen, C.N.; Ho, M.C.; Tsui, P.H. Ultrasound window-modulated compounding Nakagami imaging: Resolution improvement and computational acceleration for liver characterization. *Ultrasonics* **2016**, *70*, 18–28. [CrossRef]
37. Unal, I. Defining an optimal cut-point value in ROC Analysis: An alternative approach. *Comput. Math. Methods Med.* **2017**, *2017*, 3762651. [CrossRef]
38. Matthews, E.; Brassington, R.; Kuntzer, T.; Jichi, F.; Manzur, A.Y. Corticosteroids for the treatment of Duchenne muscular dystrophy. *Cochrane Database Syst. Rev.* **2016**, *5*, CD003725. [CrossRef] [PubMed]
39. Jansen, M.; van Alfen, N.; Nijhuis van der Sanden, M.W.; van Dijk, J.P.; Pillen, S.; de Groot, I.J. Quantitative muscle ultrasound is a promising longitudinal follow-up tool in Duchenne muscular dystrophy. *Neuromuscul. Disord.* **2012**, *22*, 306–317. [CrossRef]
40. Zaidman, C.M.; Connolly, A.M.; Malkus, E.C.; Florence, J.M.; Pestronk, A. Quantitative ultrasound using backscatter analysis in Duchenne and Becker muscular dystrophy. *Neuromuscul. Disord.* **2010**, *20*, 805–809. [CrossRef]

© 2020 by the authors. Licensee MDPI, Basel, Switzerland. This article is an open access article distributed under the terms and conditions of the Creative Commons Attribution (CC BY) license (http://creativecommons.org/licenses/by/4.0/).

Article

Efficacy of Quantitative Muscle Ultrasound Using Texture-Feature Parametric Imaging in Detecting Pompe Disease in Children

Hong-Jen Chiou [1,2,3], Chih-Kuang Yeh [4], Hsuen-En Hwang [5] and Yin-Yin Liao [6,*]

1   Division of Ultrasound and Breast Imaging, Department of Radiology, Taipei Veterans General Hospital, Taipei 11217, Taiwan
2   School of Medicine, National Yang Ming University, Taipei 11221, Taiwan
3   National Defense Medical Center, Taipei 11490, Taiwan
4   Department of Biomedical Engineering and Environmental Sciences, National Tsing Hua University, Hsinchu 30013, Taiwan
5   Department of Radiology, Taipei Veterans General Hospital, Taipei 11217, Taiwan
6   Department of Biomedical Engineering, Hungkuang University, Taichung 43302, Taiwan
*   Correspondence: g9612536@gmail.com; Tel.: +886-4-26318652

Received: 18 June 2019; Accepted: 21 July 2019; Published: 22 July 2019

Abstract: Pompe disease is a hereditary neuromuscular disorder attributed to acid α-glucosidase deficiency, and accurately identifying this disease is essential. Our aim was to discriminate normal muscles from neuropathic muscles in children affected by Pompe disease using a texture-feature parametric imaging method that simultaneously considers microstructure and macrostructure. The study included 22 children aged 0.02–54 months with Pompe disease and six healthy children aged 2–12 months with normal muscles. For each subject, transverse ultrasound images of the bilateral rectus femoris and sartorius muscles were obtained. Gray-level co-occurrence matrix-based Haralick's features were used for constructing parametric images and identifying neuropathic muscles: autocorrelation (AUT), contrast, energy (ENE), entropy (ENT), maximum probability (MAXP), variance (VAR), and cluster prominence (CPR). Stepwise regression was used in feature selection. The Fisher linear discriminant analysis was used for combination of the selected features to distinguish between normal and pathological muscles. The VAR and CPR were the optimal feature set for classifying normal and pathological rectus femoris muscles, whereas the ENE, VAR, and CPR were the optimal feature set for distinguishing between normal and pathological sartorius muscles. The two feature sets were combined to discriminate between children with and without neuropathic muscles affected by Pompe disease, achieving an accuracy of 94.6%, a specificity of 100%, a sensitivity of 93.2%, and an area under the receiver operating characteristic curve of 0.98 ± 0.02. The CPR for the rectus femoris muscles and the AUT, ENT, MAXP, and VAR for the sartorius muscles exhibited statistically significant differences in distinguishing between the infantile-onset Pompe disease and late-onset Pompe disease groups ($p < 0.05$). Texture-feature parametric imaging can be used to quantify and map tissue structures in skeletal muscles and distinguish between pathological and normal muscles in children or newborns.

Keywords: Pompe disease; children; quantitative muscle ultrasound; texture-feature parametric imaging

---

1. Introduction

Pompe disease is a hereditary disorder that affects the neuromuscular system and is attributed to acid α-glucosidase (GAA) deficiency. The typical manifestations of the disorder involve generalized weakness of muscles in addition to cardiomyopathy, which finally end in death [1]. In general, the first muscles to be affected in this disorder are the proximal lower limb muscles as well as the paraspinal trunk muscles [2,3]. Because skeletal muscle cells exhibit glycogen granule accumulations and various degenerative changes, muscle cells are replaced by fibrous tissues and fat cells, thus disrupting the corresponding muscular architecture [4]. These processes consequently include conditions under which infants present with floppy baby syndrome. In neuromuscular disorders, some of the regularly executed diagnostic procedures are as follows: genetic analysis, muscle enzyme activity measurement, muscle biopsy procedures, and electromyography (EMG) [5,6].

Although EMG is often used for identifying neuromuscular diseases, its accuracy in children varies between 10% and 98% [7]. Magnetic resonance imaging (MRI) and ultrasound are among the imaging modalities that facilitate noninvasive illustration of the muscular anatomy; these modalities are consequently being increasingly integrated into neuromuscular disease diagnosis procedures [8–10]. Compared with MRI, ultrasound is a more child-friendly modality as it is rapid and obviates the requirement for sedation. By measuring muscle echo intensity as well as muscle thickness, ultrasound has the ability to detect muscular-disorder-induced structural changes [11–14]. Ultrasound typically represents normal muscle tissue as a low-echo-intensity structure. In contrast, muscles subjected to fat infiltration exhibit increased ultrasound beam reflections, and such reflections have a relatively bright appearance [15]. Typically, Heckmatt's qualitative criteria based on muscle and bone echogenicities have been used for evaluating the degree of muscle abnormality in ultrasound [8,9,16]. However, these criteria have a drawback: as age increases, the echo intensity of muscles increases as well; this trend is attributable to age-related muscle replacement by fibrous tissues and fat cells [9]. Changes in system settings can result in muscles appearing as brighter structures, and such structures are likely to be misconstrued as pathological changes [9].

For detecting muscle pathology severity and identifying structural changes of muscles, quantitative muscle ultrasound can be used, which is reliable for obtaining such identification [9,12,17–19]. Texture analysis primarily reflects changes in a muscle's structural echogenicity. Histograms can be used to visualize the frequency of occurrence of gray levels; accordingly, in computer programs for analyzing ultrasound images, the following statistics that constitute typical image texture parameters are extensively used for identifying abnormalities: first, second, and run-length statistics [20–22]. Shannon entropy has also been used as a measure of the texture information by analyzing the probability distribution for ultrasound backscattered signals [23–25]. Previous studies have used some linear and first-order descriptors to characterize myopathic muscles for identifying Duchenne muscular dystrophy, a disorder that is typified by homogeneously increased echogenicity levels [26,27]. The feasibility of using Shannon entropy to characterize tissues has been explored in monitoring the progress of Duchenne muscular dystrophy [28]. First-order statistics and Shannon entropy only capture the image's non-spatial information, so they cannot fully characterize neuropathic muscles in ultrasound B-mode images [22,29]. Neuropathic processes are often associated with heterogeneous echogenicity levels in muscles that can be attributed to muscle architecture disruptions induced by the underlying pathological condition [9,13,30].

Molinari et al. reported higher-order statistics to be superior to first-order features in terms of classifying muscle images [22]. Gray-level co-occurrence matrix (GLCM) is a second-order statistical method of texture analysis [31]. GLCM-derived Haralick's features have been applied to detect changes in the structures of pathological muscle tissues in ultrasound [19,22,32,33]. To enable the Shannon entropy to quantify the configurational information of an image, a GLCM has also been used to characterize the configuration of image pixels and then reflect the characterization in the computation of Shannon entropy [29,31]. We previously presented a texture-based imaging approach that involves the application of Haralick's texture features to simultaneously preserve local and global texture information [34]. In this study, we probed the diagnostic accuracy of texture-feature parametric imaging in discriminating normal muscles from neuropathic muscles affected by Pompe disease in children. Because muscle weakness in Pompe disease is typically noticed first in the lower limbs [2,3], each child's rectus femoris muscle and sartorius muscle were examined in this study. Seven Haralick's texture feature parameters with various image spatial information were evaluated and used to establish corresponding parametric images.

This paper's remaining sections are structured as follows. In Section 2, the acquired materials and the executed methods in this study are introduced. A description of the executed clinical tests is provided in Section 3. Section 4 presents the study's major findings and the conclusions drawn regarding the potential applications of our proposed texture-feature parametric imaging approach in muscle ultrasound.

2. Materials and Methods

2.1. Participants

The Institutional Review Board associated with Taipei Veterans General Hospital granted approval of the research protocol (approval number 2015-08-008B). We acquired informed consent from the legal representatives of the children examined in this study. The study included 22 patients aged 0.02–54 months with Pompe disease and 6 healthy children aged 2–12 months with normal muscles. We separated Pompe disease into two categories: infantile-onset Pompe disease (IOPD; occurring at the age of <1 year with progressive cardiac hypertrophy, hypotonia, and respiratory distress) and late-onset Pompe disease (LOPD; occurring between 1 year of age and adulthood or at the age of <1 year without cardio hypertrophy) [35]. We collected GAA mutation, activity/performance, and pathological data for the patients. The serum expression levels of the following enzymes were examined for the patients: creatine kinase (CK), alanine transaminase (ALT), lactate dehydrogenase (LDH), and aspartate transaminase (AST).

2.2. Ultrasound Examinations

Several ultrasound machines in our radiology department were used to perform muscle ultrasound examinations including: an Aixplorer system (Supersonic Imagine SA, Aix-en Provence, France), S2000 system (Siemens-Acuson, Mountain View, CA, USA), S3000 system (Siemens-Acuson, Mountain View, CA, USA), and LOGIQ E9 system (GE, Wauwatosa, WI, USA). These machines were equipped with linear broadband transducers operating at 5–14, 5–14, 4–9, and 4–15 MHz, respectively. The spatial resolution of these ultrasound systems ranged from 0.5 to 1 mm. Each subject was examined by the same examiner using one of these ultrasound machines. The system settings were not fixed but adjusted individually. For each subject, transverse ultrasound B-mode images of bilateral rectus femoris and sartorius muscles were obtained. For each muscle, one B-mode image was selected that included as much of the muscle as possible. Therefore, four muscle ultrasound images were measured for each subject. A doctor experienced in the analysis of muscle ultrasound images used Adobe Photoshop software (Adobe Systems, Mountain View, CA, USA) to manually outline the muscle contour, avoiding the surrounding fascia. The maximum transverse diameter of the rectus femoris and sartorius muscles in the participants ranged from 2 to 4 cm.

2.3. Texture-Feature Parametric Imaging

The GLCM is a second-order statistics method used for extracting texture features from gray-level images, which is based on information about gray levels in pairs of pixels [31]. The GLCM between gray levels i and j is defined as:

$$C_{ij}|(\delta, \theta) = \frac{P_{ij}|(\delta, \theta)}{\sum\limits_{i=0}^{N_g-1} \sum\limits_{j=0}^{N_g-1} P_{ij}|(\delta, \theta)} \tag{1}$$

where the matrix element $P_{ij}|(\delta, \theta)$ represents the number of occurrences between gray levels i and j, to describe the frequency of occurrence of two pixels at a particular distance (δ) and angle (θ). The sum in the denominator represents the total number of occurrences of gray levels i and j within the window, and N_g is the quantized number of gray level. The number of rows and columns in the GLCM is equal to N_g. The ultrasound B-mode images were 8-bit gray-level images (256 gray levels), so we used 8 for the gray level quantization (N_g) to increase the speed of computation and reduce noise [36]. The means for the columns and rows of the GLCM are, respectively, defined as:

$$\mu_x = \sum_{i=0}^{N_g-1} \sum_{j=0}^{N_g-1} i \cdot C_{ij} \tag{2}$$

and

$$\mu_y = \sum_{i=0}^{N_g-1} \sum_{j=0}^{N_g-1} j \cdot C_{ij} \tag{3}$$

We investigated seven texture features to quantitatively evaluate the textural characteristics of the muscles on ultrasound B-mode images. The seven texture features are defined as follows.

Autocorrelation (AUT) is used for measuring repeating patterns of gray levels in an image. A higher AUT signifies a greater amount of regularity as well as the fineness/coarseness of texture.

$$AUT = \sum_{i=0}^{N_g-1} \sum_{j=0}^{N_g-1} (i \cdot j) \cdot C_{ij} \tag{4}$$

Contrast (CON) is used for measuring the disparity that exists between the highest and lowest values of a pixel set, with a lower CON value being typical for a block that is locally homogeneous [36].

$$CON = \sum_{n=0}^{N_g-1} n^2 \cdot \left\{ \sum_{i=0}^{N_g-1} \sum_{j=0}^{N_g-1} C_{ij} \big| |i-j| = n \right\} \tag{5}$$

Energy (ENE) measures repetitions of pairs of pixels and is dominated by the frequency of gray-level transitions to the power of two. ENE, also known as angular second moment, is a measure of the homogeneity of an image [36]. A homogeneous image results in a higher ENE value, whereas a heterogeneous region results in a lower ENE value.

$$ENE = \sum_{i=0}^{N_g-1} \sum_{j=0}^{N_g-1} C_{ij}^2 \tag{6}$$

Entropy (ENT), also defined as GLCM-based improved Shannon entropy, is developed to enable the Shannon entropy to quantify the spatial information of an image [29]. ENT measures the randomness of a gray-level distribution. The ENT value is expected to be high if the gray levels are distributed randomly throughout the image.

$$ENT = -\sum_{i=0}^{N_g-1} \sum_{j=0}^{N_g-1} C_{ij} \cdot \log\left(C_{ij}\right) \tag{7}$$

Maximum probability (MAXP) measures the maximum value in a pixel pair. When the occurrence of the most predominant pixel pair is high, the MAXP is high.

$$MAXP = \max\{C_{ij}\} \forall (i, j) \tag{8}$$

Variance (VAR) measures the heterogeneity degree and is associated with the standard deviation within an image. The VAR value increases as the difference between gray-level values and the corresponding global means increases [34].

$$VAR = \sum_{i=0}^{N_g-1} \sum_{j=0}^{N_g-1} (i - \mu_x)^2 \cdot C_{ij} + \sum_{i=0}^{N_g-1} \sum_{j=0}^{N_g-1} (j - \mu_y)^2 \cdot C_{ij} \tag{9}$$

Cluster prominence (CPR) characterizes the tendency of pixels to cluster and is a measure of asymmetry. When the CPR value is high, the image is asymmetric [36].

$$CPR = \sum_{i=0}^{N_g-1} \sum_{j=0}^{N_g-1} \{i + j - \mu_x - \mu_y\}^4 \cdot C_{ij} \tag{10}$$

We chose a displacement vector of $\delta = 1$ pixel in our analyses. We provided four displacement operators, which can be used to generate GLCMs along four different directions (i.e., $\theta = 0°, 45°, 90°,$ and $135°$). A total of four GLCMs can be obtained because a GLCM can be generated along four directions. For constructing a texture-feature parametric image, we applied a 13×13 pixel sliding window to the ultrasound B-mode image to evaluate each local texture feature. The local texture feature was computed by averaging the four texture feature values obtained from the four GLCMs within the sliding window. Note that we selected the 13×13 pixel sliding window size because it is larger than the system resolution and could characterize variations in the local muscle structure [37]. When moving the sliding window throughout the ultrasound B-mode image, we used 1 pixel steps; in each movement step, we considered the new center pixel of the window as the local texture feature. This approach produced a texture-feature parametric image in the form of a map of texture feature values. The texture-feature parametric image was smaller than the ultrasound B-mode image because the pixel values at the borders in the ultrasound B-mode image were ignored. For each texture-feature parametric image, the muscle region was manually determined on the basis of the corresponding B-mode image; the relevant texture feature parameter was averaged for the entirety of the internal region of the contour.

2.4. Statistical Analysis

If a relatively large set of features is used for classification processes, high coefficients of correlation between two or more features necessitate the selection and integration of multiple feature attributes to improve classification performance [38]. In general, data optimality, independence, reliability, and discrimination must be included in the criteria established for the selection of significant features in classification processes [38]. Accordingly, in this study, we used the Student's *t*-test to evaluate the level of significance of the differences between normal muscles and pathological muscles affected by Pompe disease. We assumed a derived *p*-value of <0.05 as signifying a statistically significant difference. During our comparison of *p*-values, we adjusted the level of significance by adopting the Holm–Bonferroni method.

For feature selection, stepwise regression analysis was used to obtain the best candidate final regression model. Stepwise regression is a systematic approach to build a multilinear model by including and eliminating individual features, alternating between backward and forward [39]. The backward–forward selection begins with an initial model, and then the explanatory power of incrementally larger and smaller models is compared through F-statistics of significance. A feature to be added or removed from the set of features is chosen based on the estimated p-values of the F-statistics. The algorithm consists of the following steps [39]:

(1) At the beginning, the initial model is an empty model, and the entrance and exit tolerances for the p-values of F-statistics are 0.05 and 0.10, respectively.
(2) If any feature is not in the model and the feature has a p-value less than the entrance tolerance, add the feature with the smallest p-value to the model and repeat this step; otherwise, proceed to the next step.
(3) If any feature in the model has a p-value greater than the exit tolerance, remove the feature with the largest p-value and return to step 2; otherwise, end.

The procedure automatically stops when no feature in the model can be removed and all the next best candidates cannot be retained in the model. Then, a stable set of features is attained. Although the stepwise model has the possibility of reaching a local optimal solution, it is still widely used because of its simplicity and efficacy.

We subsequently used Fisher's linear discriminant analysis (FLDA) to integrate selected texture feature parameters for classifying normal and pathological muscles. FLDA is a supervised classification method as it requires a class label, and is used when groups are known a priori [40]. The FLDA process involves five steps [40]:

(1) The d-dimensional mean vectors for the different classes from the dataset are computed.
(2) The within-class and between-class scatter matrices are calculated.
(3) The eigenvectors and corresponding eigenvalues for the scatter matrices are estimated. An eigenvalue indicates the length or magnitude of the eigenvector.
(4) The eigenvectors of the corresponding k largest eigenvalues are selected to form a $d \times k$ dimensional matrix W, where the eigenvectors are the columns of this matrix.
(5) The W eigenvector matrix is used to transform the original dimensional dataset into the lower dimensional dataset. This can be summarized by the matrix multiplication: $Y = X \times W$, where X is the original $n \times d$-dimensional dataset, and Y is the transformed $n \times k$-dimensional dataset in the new subspace.

Because the selected features contained more information about our data distribution, we were interested in retaining only those eigenvectors with the highest eigenvalues to obtain the optimal feature set. The first feature set (F1) was defined as the combination of selected features for classifying the rectus femoris muscles. The second feature set (F2) was defined as the combination of selected features for classifying the sartorius muscles. The third feature set (F3) was defined as the combination of the parameters in F1 and F2.

We used receiver operating characteristic (ROC) curve analysis to evaluate the performance of the feature sets in discriminating normal muscles from pathological muscles. Sensitivity and $1 -$ specificity pairs typically constitute an ROC curve, with every point along the curve representing a sensitivity/specificity pair that is related to an established decision threshold [41]. Sensitivity measures the percentage of pathological muscles that have been correctly classified. Specificity is a measure of the proportion of normal muscles that have been correctly classified. The area under the ROC curve (Az) could additionally be considered a potential feature.

3. Results

The characteristics and descriptive statistics of the Pompe disease and normal groups are listed in Table 1. In the Pompe disease group, five patients were newborns confirmed to have IOPD, and 17 patients were diagnosed as having LOPD.

Table 1. Characteristics of participants.

Variable	Normal (*n* = 6)	Pompe Disease (*n* = 22)	
		LOPD	**IOPD**
Male:Female	4:2	12:5	2:3
Age (Mean)	12.14 months	21.82 months	0.04 months
GAA activity by DBS (Mean ± SD)	-	0.40 ± 0.22 μm/L/h	0.08 ± 0.03 μm/L/h
LDH level (Mean ± SD)	-	459.1 ± 271.3 U/L	511.2 ± 90.6 U/L
CK level (Mean ± SD)	-	314.7 ± 329.7 U/L	661.0 ± 384.9 U/L
ALT level (Mean ± SD)	-	51.7 ± 50.4 U/L	41.4 ± 18.1 U/L
AST level (Mean ± SD)	-	90.8 ± 88.2 U/L	94.4 ± 17.8 U/L

Note: ALT: alanine transferase; AST: aspartate transferase; CK: creatine kinase; DBS: dried blood spot; GAA: glucosidase alpha acid; IOPD: infantile-onset Pompe disease; LDH: lactate dehydrogenase; LOPD: late-onset Pompe disease; U/L: units per liter.

Figure 1a depicts a B-mode image of a normal rectus femoris muscle: clear borders with low echo intensity. The muscle region, delineated by the dashed white line in Figure 1, was extracted to form a texture-feature parametric image (Figure 1b); seven parametric images based on the seven texture features were created (Figure 1c–i). Figure 2 depicts the B-mode image of a pathological rectus femoris muscle (i.e., the muscle of a patient with Pompe disease); the image has blurry borders and increased internal echoes (Figure 2a,b). On the basis of this image, we derived seven texture-feature parametric images (Figure 2c–i). We compared a normal sartorius muscle with the sartorius muscle of a patient with Pompe disease and found that the pathological sartorius muscle exhibited a higher echo intensity level (Figures 3a and 4a). Figures 3b and 4b present images depicting the muscle boundaries of the normal sartorius muscle and the sartorius muscle of the patient with Pompe disease, respectively, and Figures 3c–i and 4c–i depict the corresponding texture-feature parametric images. The images displayed in all figures were formed with a dynamic range of 60 dB and composed of shades of gray, varying from black at the weakest intensity to white at the strongest. The results show that the shading in the AUT, CON, ENE, ENT, MAXP, and VAR images differed between the normal and pathological muscles, with a greater amount of white shading for the pathological muscle than for the normal muscle. The intensity in the CPR image was lower for pathological muscle than for normal muscle.

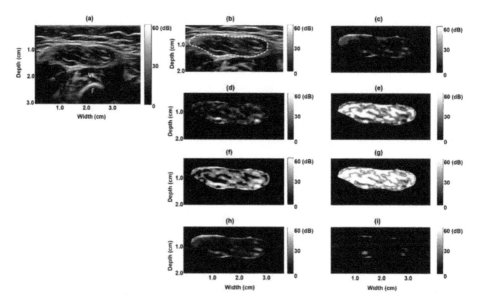

Figure 1. Texture-feature parametric imaging of a normal rectus femoris muscle in a 12 month old boy. (**a**) Original B-mode image, (**b**) extracted rectus femoris muscle region (indicated by the white dashed line) in the B-mode image, (**c**) autocorrelation image, (**d**) contrast image, (**e**) energy image, (**f**) entropy image, (**g**) maximum probability image, (**h**) variance image, and (**i**) cluster prominence image. F: femur bone reflection, VI: vastus intermedius muscle.

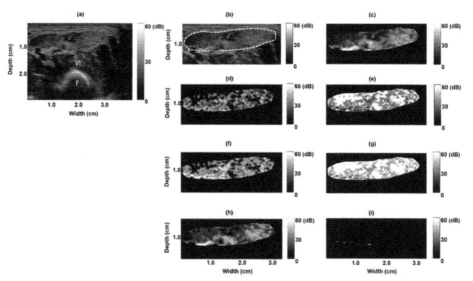

Figure 2. Texture-feature parametric imaging of a pathological rectus femoris muscle in a 10 day old boy with infantile-onset Pompe disease. (**a**) Original B-mode image, (**b**) extracted rectus femoris muscle region (indicated by the white dashed line) in the B-mode image, (**c**) autocorrelation image, (**d**) contrast image, (**e**) energy image, (**f**) entropy image, (**g**) maximum probability image, (**h**) variance image, and (**i**) cluster prominence image.

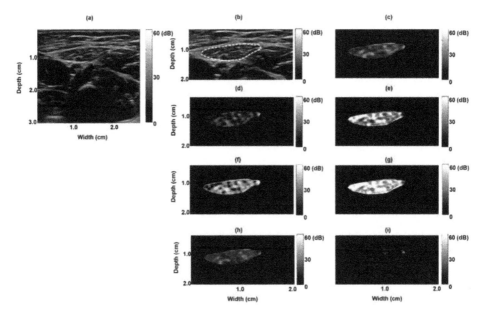

Figure 3. Texture-feature parametric imaging of a normal sartorius muscle in a 12 month old boy. (**a**) Original B-mode image, (**b**) extracted sartorius muscle region (indicated by the white dashed line) in the B-mode image, (**c**) autocorrelation image, (**d**) contrast image, (**e**) energy image, (**f**) entropy image, (**g**) maximum probability image, (**h**) variance image, and (**i**) cluster prominence image.

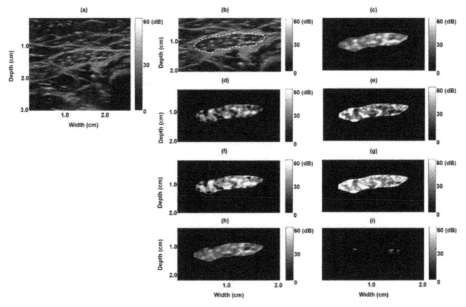

Figure 4. Texture-feature parametric imaging of a pathological sartorius muscle in a five month old boy with late-onset Pompe disease. (**a**) Original B-mode image, (**b**) extracted sartorius muscle region (indicated by the white dashed line) in the B-mode image, (**c**) autocorrelation image, (**d**) contrast image, (**e**) energy image, (**f**) entropy image, (**g**) maximum probability image, (**h**) variance image, and (**i**) cluster prominence image.

Box plots were used to represent the CON, AUT, ENE, ENT, MAXP, CPR, and VAR distributions for normal and pathological rectus femoris muscles (Figure 5), providing a quantitative description of all texture feature parameters. We found that the AUT, VAR, and CPR estimates were appropriate for distinguishing normal rectus femoris muscles from pathological rectus femoris muscles. For normal and pathological rectus femoris muscles, the average AUT, VAR, and CPR estimates were 3.91 ± 1.13 and 5.62 ± 1.78 ($p = 0.0004$), 9.17 ± 2.30 and 15.60 ± 5.51 ($p < 0.0001$), and 8.12 ± 2.44 and 4.06 ± 2.53 ($p < 0.0001$), respectively. However, the average CON, ENE, ENT, and MAXP estimates for normal and pathological rectus femoris muscles were associated with p-values >0.05.

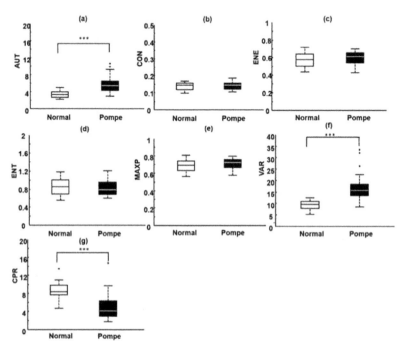

Figure 5. Box plots of the distributions of the seven parameters for normal rectus femoris muscles and pathological rectus femoris muscles affected by Pompe disease. (**a**) AUT: autocorrelation; (**b**) CON: contrast; (**c**) ENE: energy; (**d**) ENT: entropy; (**e**) MAXP: maximum probability; (**f**) VAR: variance; (**g**) CPR: cluster prominence; *** $p < 0.001$.

Box plots were also used to represent the CON, AUT, ENE, ENT, MAXP, CPR, and VAR distributions for normal and pathological sartorius muscles (Figure 6). The AUT, ENE, VAR, and CPR estimates exhibited statistically significant differences and thus could be used for distinguishing normal sartorius muscles from pathological sartorius muscles. In contrast, the CON, ENT, and MAXP estimates did not differ significantly. For normal and pathological sartorius muscles, the average AUT, ENE, VAR, and CPR estimates were 6.00 ± 2.18 and 8.01 ± 2.64 ($p = 0.0133$), 0.40 ± 0.07 and 0.48 ± 0.08 ($p = 0.0011$), 15.04 ± 3.84 and 21.62 ± 7.64 ($p = 0.0002$), and 6.55 ± 2.29 and 4.25 ± 2.46 ($p = 0.0071$), respectively.

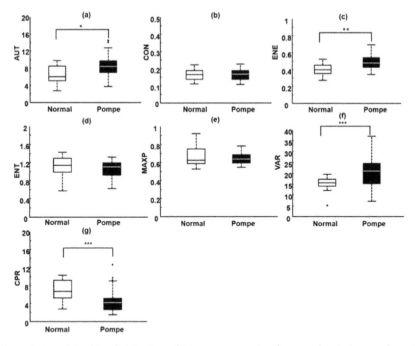

Figure 6. Box plots of the distributions of the seven parameters for normal sartorius muscles and pathological sartorius muscles affected by Pompe disease. (**a**) AUT: autocorrelation; (**b**) CON: contrast; (**c**) ENE: energy; (**d**) ENT: entropy; (**e**) MAXP: maximum probability; (**f**) VAR: variance; (**g**) CPR: cluster prominence; * $p < 0.05$; ** $p < 0.01$; and *** $p < 0.001$.

In stepwise regression, we selected VAR and CPR as the optimal feature set for classifying normal and pathological rectus femoris muscles, whereas ENE, VAR, and CPR were selected for the optimal feature set to distinguish between normal and pathological sartorius muscles. The FLDA was used for searching for a linear combination of the selected features that best distinguished between normal and pathological muscles. VAR and CPR for the rectus femoris muscles constituted F1; ENE, VAR, and CPR for the sartorius muscles constituted F2; and a combination of the parameters in F1 and F2 constituted F3. The classification performances of these feature sets were evaluated using ROC analysis. We found F3 produced the best performance (Figure 7 and Table 2), with the highest *Az* (0.98 ± 0.02) and 100% specificity, whereas F1 and F2 produced 83.3% and 91.7% specificity, respectively.

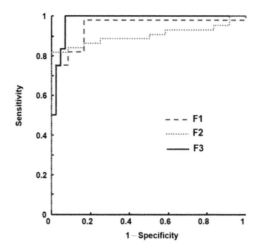

Figure 7. Receiver operating characteristic (ROC) curves of each feature set. F1: comprising the variance and cluster prominence for rectus femoris muscles. F2: comprising the energy, variance, and cluster prominence for sartorius muscles. F3: constituting a combination of F1 and F2.

Table 2. Individual performance assessed by the area under receiver operating characteristic curve (*Az*) values (mean ± standard error and 95% confidence intervals), accuracy, specificity, sensitivity, positive predictive value (PPV), and negative predictive value (NPV) of each feature set in discriminating between children with and without neuropathic muscles for Pompe disease.

Feature Sets Performance	F1 *	F2	F3
Accuracy (%)	94.6	85.7	94.6
Specificity (%)	83.3	91.7	100
Sensitivity (%)	97.7	84.1	93.2
PPV (%)	95.6	97.6	85.7
NPV (%)	90.9	78.6	100
Az (mean ± standard error)	0.95 ± 0.03	0.90 ± 0.04	0.98 ± 0.02
Az (95% CI)	0.88–1.00	0.82–0.98	0.95–1.00

* F1: comprising the variance and cluster prominence for rectus femoris muscles. F2: comprising the energy, variance, and cluster prominence for sartorius muscles. F3: constituting a combination of F1 and F2.

We observed that the CPR estimates for rectus femoris muscles and the AUT, ENT, MAXP, and VAR estimates for sartorius muscles were different between the IOPD and LOPD groups (Table 3). These parameters were associated with *p*-values <0.05 for the IOPD and LOPD groups.

Table 3. Mean, SD, and *p*-values derived from Student's *t*-test of significant texture feature parameters for the infantile-onset Pompe disease (IOPD) group and the late-onset Pompe disease (LOPD) group.

Texture Feature Parameters	IOPD	LOPD	
	Mean ± SD		*p*-Value
CPR for rectus femoris muscles *	6.00 ± 2.18	8.01 ± 2.64	<0.0001
AUT for sartorius muscles	5.90 ± 1.49	8.63 ± 2.60	0.0002
ENT for sartorius muscles	0.89 ± 0.16	1.05 ± 0.17	0.0151
MAXP for sartorius muscles	0.69 ± 0.07	0.62 ± 0.06	0.0176
VAR for sartorius muscles	16.52 ± 4.07	23.12 ± 7.84	0.0071

*AUT: autocorrelation; CPR: cluster prominence; ENT: entropy; MAXP: maximum probability; VAR: variance.

4. Discussion

Muscle ultrasound is a beneficial method for diagnosing patients with suspected muscle diseases or neuromuscular disorders. Ultrasound changes observed in diseased muscles include increased echogenicity within muscle substance, atrophic change in muscles, and loss of bone echo. Several studies have demonstrated that qualitative and quantitative ultrasound methods can be used to assess the presence and degree of muscle pathology [12–19]. Muscles that are determined to be normal exhibit a relatively hypoechoic appearance; however, on ultrasound images, different muscles exhibit distinct appearances (distinct normal ranges of echo intensity), and this is attributed to different fibrous tissue proportions and muscle fiber orientations [9]. Many conditions affect muscle ultrasound signal intensity, such as differences in patient age, system settings, and imaging modality. Although qualitative rating scales can be applied in ultrasound systems, they are subjective as they depend on the examiner's expertise [8,9,16]. For these reasons, an adequate quantitative ultrasound method for evaluating neuromuscular disorders must be able to describe changes in muscle microstructures during fatty infiltration and be independent of system settings. Studies have confirmed that texture-feature parametric imaging can be a useful approach for characterizing breast masses or fatty livers [21,34]. Microstructure and macrostructure echo information is considered simultaneously in this approach to minimize texture analysis errors due to artifact interference. Texture-feature parametric imaging achieves image dynamic range consistency by applying normalization processes, thus overcoming dependence on system settings.

In this study that included children with normal muscles and those with muscles affected by Pompe disease, we used B-mode ultrasound. The resulting B-mode images depicted clearly visible boundaries of normal muscles. This visibility is attributable to the highly reflective nature of the epimysia. Normal sartorius muscles were found to be generally more homogeneously hyperechoic than normal rectus femoris muscles. Consequently, we characterized the sartorius muscles and rectus femoris muscles in the Pompe disease and normal groups of children separately. The rectus femoris muscles and sartorius muscles in the children with Pompe disease exhibited increased echogenicity. Scholten et al. reported that muscle echo intensity levels increase with age in adults; in contrast, age was found to have no effect on muscle echo intensity in children [4]. Accordingly, connective tissue and fat infiltration could be the most likely explanation for the observed augmentation of muscle echo intensity in children. We noted that for both the rectus femoris muscles and sartorius muscles, the AUT, VAR, and CPR estimates exhibited statistically significant differences in distinguishing normal muscles from pathological muscles. Compared with normal muscles, pathological muscles had a higher AUT, reflecting a higher degree of fineness/coarseness; a higher VAR, representing a higher degree of heterogeneity; and a lower CPR, demonstrating a higher degree of symmetry.

The optimal feature sets were obtained using stepwise regression and FLDA. F1 (i.e., comprising VAR and CPR for rectus femoris muscles) yielded high sensitivity, which can improve the diagnosis of rectus femoris muscles affected by Pompe disease. This feature set exhibited weak specificity (less than 85.0%), which can influence the identification of normal rectus femoris muscles. We additionally noted a similar phenomenon when F2 (i.e., comprising ENE, VAR, and CPR for sartorius muscles) was used to classify normal and pathological sartorius muscles. F2 had low sensitivity (84.1%) because some pathological sartorius muscles resembled normal muscles in terms of echogenicity. A possible reason for this finding is that normal sartorius muscles exhibit a similar structure: they are divided by hyperechoic transverse tendinous inscriptions into segments. We subsequently combined F1 and F2 into F3 to improve the detection of Pompe disease, achieving a specificity of 100% and a sensitivity of 93.2%. This implies that the optimal texture feature parameter sets for rectus femoris and sartorius muscles are independent and complementary; therefore, ensuring their appropriate combination can enhance Pompe disease classification.

We found that some texture feature parameters for rectus femoris muscles and sartorius muscles were significantly different between the IOPD and LOPD groups. This result is consistent with the findings of Hwang et al., who used a muscle ultrasound scoring system based on modified

Heckmatt's qualitative criteria to distinguish IOPD from LOPD, achieving 100.0% sensitivity and 84.0% specificity [42]. They proposed that the echogenicity of muscle tissues in newborns and infants can increase because newborns and infants have small muscle fibers and a relatively high proportion of endomysial and perimysial connective tissues. Their findings revealed that the muscle ultrasound score is correlated with the serum levels of laboratory parameters in the diagnosis of IOPD. However, the qualitative scores obtained from subjective assessments can vary dramatically and affect the reliability of the results. Therefore, we suggest that disease severity can be estimated using changes in the texture feature parameters of muscles in patients with IOPD. Although a fluorometric GAA activity assay based on dried blood spots is the predominant method for diagnosing Pompe disease, it does not effectively distinguish between IOPD and LOPD or false-positive cases with pseudodeficiency mutation [42,43]. In future research, texture-feature parametric imaging will be a useful method for differentiating IOPD from LOPD and as a correlate of changes in clinical parameters.

Although this study offers valuable insight into Pompe disease identification using quantitative muscle ultrasound, it has some limitations. The first limitation is the small sample size; the sample must be increased to improve the effectiveness of identifying Pompe disease severity. Second, although all texture-feature parametric images have the same dynamic range to ensure consistency among ultrasound machines, further research on standardization approaches among scanning protocols and ensuring the reproducibility of measured values is warranted. The sliding window size used for constructing texture-feature parametric images should be dependent on different ultrasound equipment and the different ages of subjects. Third, inter- and intra-reader agreement regarding the texture feature parameters of rectus femoris and sartorius muscles should be considered during data collection.

In conclusion, our study demonstrated that texture-feature parametric imaging can be used to quantify and map tissue structures in skeletal muscles and to differentiate pathological from normal muscles in children. Such imaging is therefore a potentially useful diagnostic tool for IOPD.

Author Contributions: Experimental design: H.-J.C., C.-K.Y and Y.-Y.L.; methodology: C.-K.Y. and Y.-Y.L.; data collection: H.-J.C. and H.-E.H.; data analysis: H.-E.H. and Y.-Y.L.; manuscript preparation: Y.-Y.L. and H.-J.C. All the authors have read and approved the final manuscript.

Funding: This study was supported by Taiwan's Ministry of Science and Technology (Grant Nos. MOST 107-2635-E-241-002) and Hungkuang University and Kuang Tien General Hospital, Taiwan (Grant Nos. HK-KTOH-106-02).

Conflicts of Interest: The authors declare no conflict of interest.

References

1. Geel, T.M.; McLaughlin, P.M.; De Leij, L.F.; Ruiters, M.H.; Niezen-Koning, K.E. Pompe disease: Current state of treatment modalities and animal models. *Mol. Genet. Metab.* **2007**, *92*, 299–307. [CrossRef] [PubMed]
2. Di Rocco, M.; Buzzi, D.; Tarò, M. Glycogen storage disease type II: Clinical overview. *Acta. Myol.* **2007**, *26*, 42–44. [PubMed]
3. Cupler, E.J.; Berger, K.I.; Leshner, R.T.; Wolfe, G.I.; Han, J.J.; Barohn, R.J.; Kissel, J.T. Consensus treatment recommendations for late-onset Pompe disease. *Muscle Nerve* **2012**, *45*, 319–333. [CrossRef] [PubMed]
4. Scholten, R.R.; Pillen, S.; Verrips, A.; Zwarts, M.J. Quantitative ultrasonography of skeletal muscles in children: Normal values. *Muscle Nerve* **2003**, *27*, 693–698. [CrossRef] [PubMed]
5. Aydinli, N.; Baslo, B.; Caliskan, M.; Ertaş, M.; Ozmen, M. Muscle ultrasonography and electromyography correlation for evaluation of floppy infants. *Brain Dev.* **2003**, *25*, 22–24. [CrossRef]
6. Mellies, U.; Lofaso, F. Pompe disease: A neuromuscular disease with respiratory muscle involvement. *Respir. Med.* **2009**, *103*, 477–484. [CrossRef]
7. Rabie, M.; Jossiphov, J.; Nevo, Y. Electromyography (EMG) accuracy compared to muscle biopsy in childhood. *J. Child Neurol.* **2007**, *22*, 803–808. [CrossRef]
8. Zuberi, S.M.; Matta, N.; Nawaz, S.; Stephenson, J.B.; McWilliam, R.C.; Hollman, A. Muscle ultrasound in the assessment of suspected neuromuscular disease in childhood. *Neuromuscul. Disord.* **1999**, *9*, 203–207. [CrossRef]

9.   Pillen, S.; Arts, I.M.; Zwarts, M.J. Muscle ultrasound in neuromuscular disorders. *Muscle Nerve* **2008**, *37*, 679–693. [CrossRef]

10.   Carlier, R.Y.; Laforet, P.; Wary, C.; Mompoint, D.; Laloui, K.; Pellegrini, N.; Annane, D.; Carlier, P.G.; Orlikowski, D. Whole-body muscle MRI in 20 patients suffering from late onset Pompe disease: Involvement patterns. *Neuromuscul. Disord.* **2011**, *21*, 791–799. [CrossRef]

11.   Brockmann, K.; Becker, P.; Schreiber, G.; Neubert, K.; Brunner, E.; Bönnemann, C. Sensitivity and specificity of qualitative muscle ultrasound in assessment of suspected neuromuscular disease in childhood. *Neuromuscul. Disord.* **2007**, *17*, 517–523. [CrossRef] [PubMed]

12.   Pillen, S.; Verrips, A.; Van Alfen, N.; Arts, I.M.; Sie, L.T.; Zwarts, M.J. Quantitative skeletal muscle ultrasound: Diagnostic value in childhood neuromuscular disease. *Neuromuscul. Disord.* **2007**, *17*, 509–516. [CrossRef] [PubMed]

13.   Pillen, S. Skeletal muscle ultrasound. *Eur. J. Transl. Myol.* **2010**, *1*, 145–155. [CrossRef]

14.   Zaidman, C.M.; Van Alfen, N. Ultrasound in the assessment of myopathic disorders. *J. Clin. Neurophysiol.* **2016**, *33*, 103–111. [CrossRef] [PubMed]

15.   Reimers, K.; Reimers, C.D.; Wagner, S.; Paetzke, I.; Pongratz, D.E. Skeletal muscle sonography: A correlative study of echogenicity and morphology. *J. Ultrasound Med.* **1993**, *12*, 73–77. [CrossRef] [PubMed]

16.   Heckmatt, J.Z.; Leeman, S.; Dubowitz, V. Ultrasound imaging in the diagnosis of muscle disease. *J. Pediatr.* **1982**, *101*, 656–660. [CrossRef]

17.   Pohle, R.; Fischer, D.; Von Rohden, L. Computer-supported tissue characterization in musculoskeletal ultrasonography. *Ultraschall. Med.* **2000**, *21*, 245–252. [PubMed]

18.   Pillen, S.; Van Dijk, J.P.; Weijers, G.; Raijmann, W.; De Korte, C.L.; Zwarts, M.J. Quantitative gray-scale analysis in skeletal muscle ultrasound: A comparison study of two ultrasound devices. *Muscle Nerve* **2009**, *39*, 781–786. [CrossRef]

19.   König, T.; Steffen, J.; Rak, M.; Neumann, G.; von Rohden, L.; Tönnies, K.D. Ultrasound texture-based CAD system for detecting neuromuscular diseases. *Int. J. Comput. Assist Radiol. Surg.* **2015**, *10*, 1493–1503. [CrossRef]

20.   Gaitini, D.; Baruch, Y.; Ghersin, E.; Veitsman, E.; Kerner, H.; Shalem, B.; Yaniv, G.; Sarfaty, C.; Azhari, H. Feasibility study of ultrasonic fatty liver biopsy: Texture vs. attenuation and backscatter. *Ultrasound Med. Biol.* **2004**, *30*, 1321–1327. [CrossRef]

21.   Liao, Y.Y.; Yang, K.C.; Lee, M.J.; Huang, K.C.; Chen, J.D.; Yeh, C.K. Multifeature analysis of an ultrasound quantitative diagnostic index for classifying nonalcoholic fatty liver disease. *Sci. Rep.* **2016**, *6*, 35083. [CrossRef] [PubMed]

22.   Molinari, F.; Caresio, C.; Acharya, U.R.; Mookiah, M.R.; Minetto, M.A. Advances in quantitative muscle ultrasonography using texture analysis of ultrasound images. *Ultrasound Med. Biol.* **2015**, *41*, 2520–2532. [CrossRef] [PubMed]

23.   Shannon, C.E. A mathematical theory of communication. *Bell Syst. Tech. J.* **1948**, *27*, 379–423. [CrossRef]

24.   Hughes, M.S. Analysis of ultrasonic waveforms using Shannon entropy. *IEEE Ultrason. Symp. Proc.* **1992**, *1*, 1205–1209.

25.   Lin, Y.H.; Liao, Y.Y.; Yeh, C.K.; Yang, K.C.; Tsui, P.H. Ultrasound entropy imaging of nonalcoholic fatty liver disease: Association with metabolic syndrome. *Entropy* **2018**, *20*, 893. [CrossRef]

26.   Jansen, M.; Van Alfen, N.; Nijhuis-Van Der Sanden, M.W.; Van Dijk, J.P.; Pillen, S.; De Groot, I.J. Quantitative muscle ultrasound is a promising longitudinal follow-up tool in Duchenne muscular dystrophy. *Neuromuscul. Disord.* **2012**, *22*, 306–317. [CrossRef] [PubMed]

27.   Koppaka, S.; Shklyar, I.; Rutkove, S.B.; Darras, B.T.; Anthony, B.W.; Zaidman, C.M.; Wu, J.S. Quantitative ultrasound assessment of Duchenne muscular dystrophy using edge detection analysis. *J. Ultrasound Med.* **2016**, *35*, 1889–1897. [CrossRef] [PubMed]

28.   Hughes, M.S.; Marsh, J.N.; Wallace, K.D.; Donahue, T.A.; Connolly, A.M.; Lanza, G.M.; Wickline, A.S. Sensitive ultrasonic detection of dystrophic skeletal muscle in patients with Duchenne muscular dystrophy using an entropy-based signal receiver. *Ultrasound Med. Biol.* **2007**, *33*, 1236–1243. [CrossRef]

29.   Gao, P.; Li, Z.; Zhang, H. Thermodynamics-based evaluation of various improved Shannon entropies for configurational information of gray-level images. *Entropy* **2018**, *20*, 19. [CrossRef]

30.   Maurits, N.M.; Bollen, A.E.; Windhausen, A.; De Jager, A.E.; Van Der Hoeven, J.H. Muscle ultrasound analysis: Normal values and differentiation between myopathies and neuropathies. *Ultrasound Med. Biol.* **2003**, *29*, 215–225. [CrossRef]

31. Haralick, R.M.; Shanmugam, K.; Dinstein, I. Textural features for image classification. *IEEE Trans. Syst. Man. Cybern.* **1973**, *SMC-3*, 610–621. [CrossRef]
32. Martínez-Payá, J.J.; Ríos-Díaz, J.; Del Baño-Aledo, M.E.; Tembl-Ferrairó, J.I.; Vazquez-Costa, J.F.; Medina-Mirapeix, F. Quantitative muscle ultrasonography using textural analysis in Amyotrophic lateral sclerosis. *Ultrason. Imaging* **2017**, *39*, 357–368. [CrossRef] [PubMed]
33. Matta, T.T.D.; Pereira, W.C.A.; Radaelli, R.; Pinto, R.S.; Oliveira, L.F. Texture analysis of ultrasound images is a sensitive method to follow-up muscle damage induced by eccentric exercise. *Clin. Physiol. Funct. Imaging* **2018**, *38*, 477–482. [CrossRef] [PubMed]
34. Liao, Y.Y.; Wu, J.C.; Li, C.H.; Yeh, C.K. Texture feature analysis for breast ultrasound image enhancement. *Ultrason. Imaging* **2011**, *33*, 264–278. [CrossRef] [PubMed]
35. Chien, Y.H.; Hwu, W.L.; Lee, N.C. Pompe disease: Early diagnosis and early treatment make a difference. *Pediatr. Neonatol.* **2013**, *54*, 219–227. [CrossRef] [PubMed]
36. Yang, X.; Tridandapani, S.; Beitler, J.J.; Yu, D.S.; Yoshida, E.J.; Curran, W.J.; Liu, T. Ultrasound GLCM texture analysis of radiation-induced parotid-gland injury in head-and-neck cancer radiotherapy: An in vivo study of late toxicity. *Med. Phys.* **2012**, *39*, 5732–5739. [CrossRef] [PubMed]
37. Valckx, F.M.J.; Thijssen, J.M. Characterization of echographic image texture by cooccurrence matrix parameters. *Ultrasound Med. Biol.* **1997**, *23*, 559–571. [CrossRef]
38. Cheng, H.D.; Shan, J.; Ju, W.; Guo, Y.; Zhang, L. Automated breast cancer detection and classification using ultrasound images: A survey. *Pattern Recognit.* **2010**, *43*, 299–317. [CrossRef]
39. Pope, P.T.; Webster, J.T. The use of an F-statistic in stepwise regression procedures. *Technometrics* **1972**, *14*, 327–340. [CrossRef]
40. Hastie, T.; Buja, A.; Tibshirani, R. Penalized discriminant–analysis. *Ann. Stat.* **1995**, *23*, 73–102. [CrossRef]
41. Hanley, J.A.; McNeil, B.J. The meaning and use of the area under a receiver operating characteristic (ROC) curve. *Radiology* **1982**, *143*, 29–36. [CrossRef] [PubMed]
42. Hwang, H.E.; Hsu, T.R.; Lee, Y.H.; Wang, H.K.; Chiou, H.J.; Niu, D.M. Muscle ultrasound: A useful tool in newborn screening for infantile onset pompe disease. *Medicine* **2017**, *96*, e8415. [CrossRef] [PubMed]
43. Chien, Y.H.; Lee, N.C.; Thurberg, B.L.; Chiang, S.C.; Zhang, X.K.; Keutzer, J.; Huang, A.C.; Wu, M.H.; Huang, P.H.; Tsai, F.J.; et al. Pompe disease in infants: Improving the prognosis by newborn screening and early treatment. *Pediatrics* **2009**, *124*, e1116–e1125. [CrossRef] [PubMed]

 © 2019 by the authors. Licensee MDPI, Basel, Switzerland. This article is an open access article distributed under the terms and conditions of the Creative Commons Attribution (CC BY) license (http://creativecommons.org/licenses/by/4.0/).

Article

Impact of Misclassification Rates on Compression Efficiency of Red Blood Cell Images of Malaria Infection Using Deep Learning

Yuhang Dong [1], W. David Pan [1,*] and Dongsheng Wu [2]

[1] Department of Electrical and Computer Engineering, University of Alabama in Huntsville, Huntsville, AL 35899, USA; yd0009@uah.edu

[2] Department of Mathematical Sciences, University of Alabama in Huntsville, Huntsville, AL 35899, USA; dw0001@uah.edu

* Correspondence: pand@uah.edu

Received: 12 October 2019; Accepted: 27 October 2019; Published: 30 October 2019

Abstract: Malaria is a severe public health problem worldwide, with some developing countries being most affected. Reliable remote diagnosis of malaria infection will benefit from efficient compression of high-resolution microscopic images. This paper addresses a lossless compression of malaria-infected red blood cell images using deep learning. Specifically, we investigate a practical approach where images are first classified before being compressed using stacked autoencoders. We provide probabilistic analysis on the impact of misclassification rates on compression performance in terms of the information-theoretic measure of entropy. We then use malaria infection image datasets to evaluate the relations between misclassification rates and actually obtainable compressed bit rates using Golomb–Rice codes. Simulation results show that the joint pattern classification/compression method provides more efficient compression than several mainstream lossless compression techniques, such as JPEG2000, JPEG-LS, CALIC, and WebP, by exploiting common features extracted by deep learning on large datasets. This study provides new insight into the interplay between classification accuracy and compression bitrates. The proposed compression method can find useful telemedicine applications where efficient storage and rapid transfer of large image datasets is desirable.

Keywords: lossless compression; pattern classification; machine learning; malaria infection; entropy; Golomb–Rice codes

1. Introduction

Malaria occurs in nearly 100 countries worldwide, imposing a huge toll on human health and heavy socioeconomic burdens on developing countries [1]. The agents of malaria are mosquito-transmitted *Plasmodium* parasites. Microscopy is the gold standard for diagnosis; however, manual blood smear evaluation depends on time-consuming, error-prone, and repetitive processes requiring skilled personnel [2]. Ongoing research has therefore focused on computer-assisted Plasmodium characterization and classification from digitized blood smear images [3–7]. Traditional algorithms labeled images using manually designed feature extraction, with drawbacks in both time-to-solution and accuracy [4]. Newly proposed methods aim to apply automated learning to large-size wholeslide images. Leveraging high-performance computing, deep machine learning algorithms could potentially drive true artificial intelligence in malaria research. Concurrently, the convergence of mobile computing, the Internet, and biomedical instrumentation now allows the worldwide transfer of biomedical images for telemedicine applications. Consultation or screening by specialists located in geographically different locations is now possible.

Among recent works on computer-aided diagnosis of malaria infection, two types of images have found prevalent use: light microscopic images and wholeslide images. Recent advances in computing power, improved cloud based services and robust algorithms have enabled the widespread use of wholeslide images [8–13]. Higher resolutions can help identify the specific species and the degree of infection. Most of the prior studies utilize light microscopic images [14–24]. While machine learning algorithms have been applied to light microscopic images with relatively low-resolution image processing, higher resolutions would be necessary to identify the specific species and the degree of infection [25].

A notable challenge in such applications is the storage and rapid transfer of massive wholeslide image datasets. Efficient lossless compression methods will be much sought after for malaria infection images. Lossless compression for images has the obvious advantage of suffering no quality loss over lossy methods. Traditional image compression methods seek to minimize the correlation inside the image. For large image datasets, especially medical images that share lots of commonality, the inter-image correlation should also be taken into consideration. Deep learning based neural networks can be trained on samples within the same class to learn the common features shared by these samples. In our prior work [26], we proposed a coding scheme for red blood cell images by using stacked autoencoders, where the reconstruction residues were entropy-coded to achieve lossless compression. Specifically, we trained two separate stacked autoencoders to automatically learn the discriminating features from input images of infected and non-infected cells. Subsequently, the residues of these two classes of images were coded by two independent Golomb–Rice encoders. Simulation results showed that this deep learning approach can provide more efficient compression than several state-of-the-art methods. However, this work assumes that the class labels for the input images are known in advance with perfect classification, which is typically not the case in practice. Hence in this paper, we introduce a more realistic framework where the input images are first classified before being compressed using autoencoders. We study how the accuracy of the classifiers would affect the overall compression ratios for two-class image dataset compression. Note that for traditional lossless compression methods, misclassified samples were not a problem since images were compressed individually. But for compressors based on deep learning methods such as stacked autoencoders, misclassified images fed into autoencoders trained for the other class can lead to very large residues, which could degrade the compression performance. For a more in-depth study, we conduct theoretical analysis based on probabilistic distributions of the prediction residues, and derive formulas for compressed bit rates as a function of classification accuracies. We then use synthesized data based on the models to verify the theoretical results. Next, we use real malaria infection image datasets to evaluate the relations between classification accuracies and compressed bit rates.

In the following, we provide a literature survey on the existing work on joint data compression and classification. While most work in the literature studies data compression and pattern classification separately, some papers [27–29] address joint compression and classification, albeit without an in-depth treatment of the interplay between classification and compression. An algorithm on discrete cosine transform (DCT)-based classification scheme was presented in [27] for fractal based image compression, where three classes of image blocks were defined: smooth class, diagonal/sub-diagonal edge class and horizontal/vertical edge class. Two lowest horizontal and vertical DCT coefficients of the given block were used for classification. This reduces the searching space, therefore accelerating the fast fractal encoding process. The author assumed that the classifier was perfect, so no discussion about how the classification accuracy would affect the algorithm was given. A lifting based system was proposed in [28] for Joint Photographic Experts Group (JPEG) 2000 compression to control the trade-off between compression and classification performance. While the paper claims that good classification performance was typically obtained at the expense of some compression performance degradation, no detailed analysis of the interplay between classification and compression was provided. Both [29] and [30] worked on electrocardiogram (ECG) system. A quad-level vector (QLV) was proposed in [29] to support both classification flow and compression flow, in order to achieve better performance with low computational complexity. Wavelet-based features were used in [30] for classification with Support Vector Machine (SVM), where wavelet transform and run length coding were used for compression.

Neither of these two papers mentioned the interaction between classification flow and compression flow. Furthermore, several papers [31–33] address classification of hyperspectral images (HSI) or multispectral image (MSI) in order to improve the compression performance. Several classification trees were constructed in [31] to study the relationship between compression rate and classification accuracy for lossy compression on HSI. The results showed that high compression rates could be achieved without degrading classification accuracy too much. HSI were also used in [32], where several lossy compression methods were compared on how they would impact classification using pixel-based support vector machine (SVM). Compression of MSI was achieved in [33] by segmentation of image into regions of homogeneous land covers. The classification was conducted via tree-structured vector quantization, and residues were coded using transform coding techniques. The method proposed in [34] is similar to that in [32]. Pixel classification and sorting scheme in wavelet domain was used for image compression. Pixels were classified into several quantized contexts, so as to exploit the intra-band correlation in wavelet domain. Compression and classification of images were combined in [35]. The compressed image incorporated implicit classification information, which can be used directly for low-level classification. Some other researchers [36–38] worked with vector quantizer based classifiers to improve compression performance. On the other hand, researchers use neural network [39–42] for joint classification/compression. A classifier based on wavelet and Fourier descriptor features was employed in [39] to promote lossless image compression. The neural network in [40] was accelerated by compressing image data with an algorithm based on the discrete cosine transform. Singular Value Decomposition (SVD) was used in [41] as compression method that can reduce the size of fingerprint images, while improving the classification accuracy. Two unsupervised data reduction techniques, Autoencoder and self-organizing maps, were compared in [42] to identify malaria from blood smear images.

To the best of our knowledge, there is no in-depth study on the interplay between misclassification rate and compression ratio for lossless image compression methods, in particular, for compression methods based on deep-learning based pattern classification. In this work, to achieve efficient compression of red blood cell images, we use autoencoders to learn the correlations of the image pixels, as well as the correlations among similar images. We train separate autoencoders for images belonging to different classes. Autoencoders can automatically generate hierarchical feature vectors, which reflect common features shared by the images from the same class. We can then recover the original images from the feature vectors. By coding the residues, we can achieve lossless compression on the images. We study how misclassification rate affects the overall compression efficiency.

2. Materials and Methods

2.1. Construction of the Dataset of Malaria-Infected Red Blood Cell Images

As the result of collaborative research with a group of pathologists from the Medical School of the University of Alabama at Birmingham, we built a dataset of red blood cell (RBC) images extracted from a wholeslide image (WSI) with 100× magnification [43]. The images belong to either one of the following two classes: malaria infected cells and normal cells. Figure 1 shows the glass slide of thin blood smear and the scanned WSI under its highest resolution. The WSI was divided into more than 80,000 image tiles, each with 284 × 284 pixels. Image morphological transforms were applied onto each tile to separate cell samples from the background, as shown in Figure 1 [44]. Some overlapped cells can be separated using Hough circle transform [45]. Finally, all samples were resized into 50 × 50 images, with some examples shown in Figure 2. The entire dataset can be found on our website [46]. For simplicity, we only used red channel for training neural network.

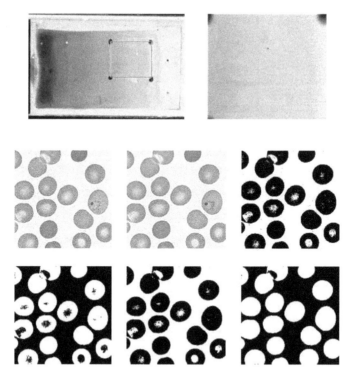

Figure 1. Wholeslide image of malaria-infected RBCs and normal cells. The top left image is the original glass slide after staining. The rectangle delineated in green was cropped out to be the image on the right. After zooming in the area with 100× magnification, we can see the normal cells and infected cells (with the parasites in the ring form) in the leftmost image in the second row. The remaining five grayscale images are the result of step-by-step processing of the leftmost image in the second row. First, the color image is converted into a grayscale image. Then a thresholding operation removes irrelevant info and converts the image into a binary image. The next two steps fills the isolated pixels in both foreground and background. After filling all the holes, we finally got the binary mask. Applying the mask onto the color image, we can extract each single cell image as shown in Figure 2.

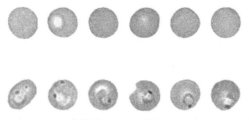

Figure 2. Some example segmented RBC images. (Upper row) normal cells and (lower row) infected cells.

2.2. Lossless Compression Using Autoencoders

An autoencoder is an artificial neural network that performs unsupervised learning [47], which consists of an encoder and a decoder. The encoder converts the high dimensional input data into a low dimensional feature vector. By reversing this process, the decoder attemps to recover the original data, typically with loss. Back propagation is used when traing the autoencoder to minimize the loss. A more complicated network can be built by stacking several autoencoders together, which will generate a more hierarchical representations of the input data. A fine-tuned autoencoder is able to perform data dimensionality reduction, while extracting features shared by the input data. Thus autoencoders can be used for lossless compression, if the differences between the input data and the reconstructed version are retained and coded efficiently. The flow chart of using stacked autoencoders (SAE) on malaria-infected RBC images is shown in Figure 3.

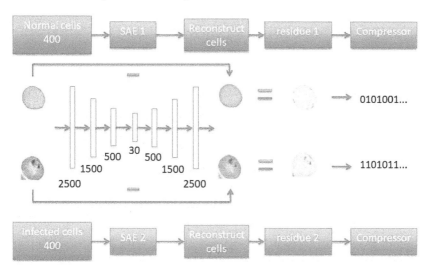

Figure 3. Using autoencoder to compress a 50 × 50 image to a 30-point vector, together with the residue. The residue will be coded using Golomb–Rice Code.

Two separate stacked autoencoders (SAE) were assigned to images belonging to normal and infected cell classes, respectively, each with 400 samples. Since cell images in the same class share more common features, higher compression efficiency can be acquired than using one SAE for all samples. Each SAE consists of an encoder and a decoder. A cell image of 50 × 50 was reshaped into a vector of 2500 points, and then fed into encoder. The encoder consists of four layers: The input layer takes in 2500-point vectors, which are reduced by the remaining encoder layers to 1500, 500 and 30 points respectively. Therefore, the stacked autoencoder reduces the input vector into a very low-dimension vector of only 30 entries. Then the decoder attempts to reconstruct the original image from the 30-point vector. The training of the entire autoencoder takes many iterations in order to reduce the difference between the reconstructed image and the original image to a very small value. The resulting residues, along with the 30-point vector are coded to ensure the compression is lossless. Specifically, the residues are compressed efficiently using the Golomb–Rice Code [48].

Unlike most conventional lossless image compression methods such as JPEG2000 [49], which exploits correlations within a single images to be compressed, the autoencoder based method is able to extract common features among a group of similar images. This will allow for potentially more efficient compression on these similarly looking images in a dataset.

2.3. Golomb–Rice Coding

If the autoencoder is well trained on the input dataset, the differences (residues) between the reconstructed images and original images tend to center around zero. If the residues are converted to non-negative integers using the following equation:

$$Output = \begin{cases} -2 \cdot Input - 1, & \text{if } Input < 0; \\ 2 \cdot Input, & \text{otherwise,} \end{cases}$$

then the resulting non-negative values n can be approximated by the geometrical distribution with the following probability mass function parameterized by p:

$$\text{Prob}(n) = p^n(1-p), \tag{1}$$

where p is a real number within the range of $(0,1)$. Golomb–Rice codes are optimal to compress the geometrically distributed source with $p^m = \frac{1}{2}$, where m is a coding parameter.

The entropy $H(p)$, and expected value $E[n]$ of n's are given below.

$$H(p) = \frac{-(1-p) \cdot \log_2(1-p) - p \cdot \log_2 p}{p}, \tag{2}$$

$$E[n] = \sum_{n=0}^{\infty} np^n(1-p) = \frac{p}{1-p}. \tag{3}$$

Using Equation (3), the parameter p can be estimated from the sample mean as follows:

$$p \approx \frac{E(n)}{1 + E(n)}. \tag{4}$$

The Golomb–Rice coding procedure can be summarized by the following steps:

1. Each non-negative integer n to be coded is decomposed into two numbers, q and r, where $n = mq + r$, q is the quotient of (n/m), and r is the remainder.
2. Unary-coding q by generating q "1"s, followed by a "0".
3. Coding of r depends on if m is a power of two:

   - If $m = 2^s$, r can be simply represented using an s-bit binary code.
   - If m is not power of two, the following thresholds should be calculated first:

$$A = \lceil \log_2 m \rceil, \text{and } B = \lfloor \log_2 m \rfloor. \tag{5}$$

   If $0 \le r \le (2^A - m - 1)$, then r is represented by a B-bit binary code; Otherwise, if $(2^A - m) \le r \le (m-1)$, then $[r + (2^A - m)]$ is represented by a A-bit binary code.

If $m = 2^s$, then s can be estimated from the sample mean of the input data as

$$s \approx max \left\{ 0, \left\lceil \log_2 \frac{E(n)}{2} \right\rceil \right\}, \tag{6}$$

and the average codeword length (ACWL) of the Golomb–Rice codes is:

$$ACWL = E[q] + 1 + s, \tag{7}$$

where $E[q]$ is the expected value of the quotients q.

2.4. Joint Classification and Compression Framework

Previously, we used autoencoders to exploit the correlations of similar images to achieve high compression on red blood cell images [26]. For this sake, two separate autoencoders were trained using images known in advance to belong to one of the two classes (either normal cells, or malaria infected cells). However, the compression performance suffers if the images fed to the autoencoders actually come from different classes, which is typically the case, where classifiers are not perfect. Therefore, in this work, we study a more realistic framework, as shown in Figure 4, where the input images are first classified before being compressed using autoencoders. So after classification, each class may have some samples that are incorrectly classified. In the following, we conduct an analysis on how the accuracy of the classifiers would affect the overall compression ratios.

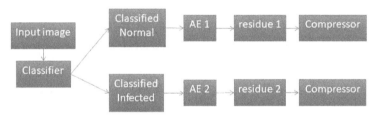

Figure 4. A more realistic framework taking into account misclassification of input images.

2.5. Theoretical Analysis

We employ a binary channel model as illustrated in Figure 5 to characterize the four possible cases of cell image classification, with the meanings of the symbols explained in Table 1. Since there are only two possible classes of input images, we have the source probabilities summing up to unity:

$$P(S0) + P(S1) = 1. \tag{8}$$

Similarly, the misclassification rates ($P(C1|S0)$ and $P(C0|S1)$) are related to correct classification rates as:

$$P(C1|S0) + P(C0|S0) = 1, \tag{9}$$
$$P(C1|S1) + P(C0|S1) = 1. \tag{10}$$

The source probabilities and the conditional probabilities can be estimated from the image datasets and the pattern classifiers used. We can then derive the joint probabilities of the four possible cases of image classification as listed in Table 1. For example, the joint probability of a cell being normal and correctly classified can be calculated as

$$P(S0, C0) = P(C0|S0) \cdot P(S0). \tag{11}$$

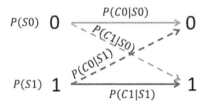

Figure 5. A binary state transition model for cell image classifications. The symbols "1" and "0" to the left represent input source images belonging to either one of two possible classes (infected and normal cells, respectively). The symbols "1" and "0" to the right represent the type of the images an input image is classified into. Arrows represent transitions, e.g., the transition from "1" to "1" means an infected cell is correctly classified. In contrast, the transition from "1" to "0" means an infected cell is incorrectly classified as a normal cell, where the misclassification rate can be described by the conditional probability $P(C0|S1)$ for each class. See Table 1 for the meanings of other probabilities involved.

Table 1. Meanings of the probabilities involved in the binary channel model.

Symbols	Meaning
$P(S0)$	Source probability of a normal cell image
$P(S1)$	Source probability of an infected cell image
$P(C0\|S0)$	Conditional probability of a normal cell being correctly classified
$P(C1\|S0)$	Cond. prob. of a normal cell being incorrectly classified as an infected cell
$P(C0\|S1)$	Cond. prob. of an infected cell being incorrectly classified as a normal cell
$P(C1\|S1)$	Cond. prob. of an infected cell being correctly classified
$P(S0, C0)$	Joint probability of a cell being normal and correctly classified
$P(S0, C1)$	Joint prob. of a cell being normal but incorrectly classified as an infected cell
$P(S1, C0)$	Joint prob. of a cell being infected but incorrectly classified as a normal cell
$P(S1, C1)$	Joint prob. of a cell being infected and correctly classified

Following the joint image classification/compression framework in Figure 4, subsequent to image classification, we use stacked autoencoders to generate residues. As shown in Figure 6, corresponding to different cases of image classifications (S_i, C_j), we can distinguish four distinct probabilistic distributions of residues R_{ij}. where $i, j = 0, 1$.

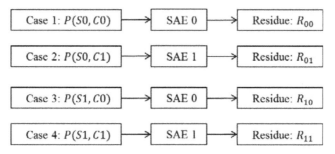

Figure 6. Image compression using stacked autoencoders (SAEs) after pattern classification. "SAE0" and "SAE1" stand for stacked autoencoders trained for normal and infected cells, respectively. R_{ij}, where $i, j = 0, 1$, denotes the probability distributions of the residues to be entropy coded using Golomb–Rice codes.

Given that the input images are either for normal cells or infected cells, the following two conditional entropies, $H0$ and $H1$, can provide estimates of the compressed bitrates. Specifically,

$$H0 = P(C0|S0)H(R_{00}) + P(C1|S0)H(R_{01}) \tag{12}$$
$$= [1 - P(C1|S0)]H(R_{00}) + P(C1|S0)H(R_{01}), \tag{13}$$

which is a function of the misclassification rate $P(C1|S0)$. Similarly,

$$H1 = P(C0|S1)H(R_{10}) + P(C1|S1)H(R_{11}) \tag{14}$$
$$= P(C0|S1)H(R_{10}) + [1 - P(C0|S1)]H(R_{11}), \tag{15}$$

which is also a function of the misclassification rate $P(C0|S1)$.

The overall bitrate (BR) in theory can be obtained as follows by probabilistically combining the individual bitrates for the four cases. The individual bitrates can be represented by the entropies of the residues $H(R_{ij})$ since lossless compression is used.

$$BR = \sum_{i=0}^{1} \sum_{j=0}^{1} P(Si, Cj) H(R_{ij}). \tag{16}$$

We can see that the overall bitrate can also be obtained by probabilistically combining the conditional entropies $H0$ and $H1$ in Equations (13) and (15) as follows:

$$BR = H0 \cdot P(S0) + H1 \cdot P(S1), \tag{17}$$

which shows that the overall bitrate is a function of the misclassification rates.

In practice, the residue sources can be modeled by the geometric distributions with varying parameters p_{ij} (corresponding to one of the four possible cases of image classifications (Si, C_j)). That is, the probability mass functions of the residue sources are

$$\text{Prob}(n) = p_{ij}^n (1 - p_{ij}), \tag{18}$$

where n denotes the values of residues, and $i, j = 0, 1$. Therefore, we can use Equation (2) to replace $H(R_{ij})$ with the entropy of the geometric source:

$$H(R_{ij}) = \frac{-(1 - p_{ij}) \cdot \log_2(1 - p_{ij}) - p_{ij} \cdot \log_2 p_{ij}}{1 - p_{ij}}. \tag{19}$$

Furthermore, we can derive the following formula for estimating the average codeword lengths (ACWL in bits, which is the practically achievable bitrates) over all four cases when we employ Golomb–Rice codes to compress the residues.

$$ACWL_{Overall} = \sum_{i=0}^{1} \sum_{j=0}^{1} P(Si, Cj) \cdot ACWL(R_{ij}) \tag{20}$$

$$= \sum_{i=0}^{1} \sum_{j=0}^{1} P(Cj|Si) \cdot P(Si) \cdot ACWL(R_{ij}), \tag{21}$$

where $ACWL(R_{i,j})$ denotes the average codeword length of Golomb–Rice coding the residue source R_{ij}, which can be estimated by using Equation (7). We can see that the overall average codeword length is a function of the misclassification rates $P(C1|S0)$ and $P(C0|S1)$.

3. Results and Discussion

For the purpose of visualizing this relation revealed by the foregoing theoretic analysis, we simply assume that the cells are equally likely to be either normal or infected, i.e., $P(S0) = P(S1) = \frac{1}{2}$. Note here the theoretical results obtained in the previous section can handle other more general situations, e.g., the there will be more normal cells than infected cells, or the two misclassification rates are different. However, making the above simplifying assumptions can allow for 2D plotting of the relations between compression performance and a single misclassification rate.

We use two image datasets (with 400 images for each class) to estimate the compression performance. We first train two stacked autoencoders, one for normal cells and the other for infected cells. Then we vary the misclassification rates from 0.01 to 0.2 with a step size of 0.01. We then formulate the mixed images datasets according to the misclassification rates. For example, if the misclassification rate $P(C1|S0) = P(C0|S1) = 0.1$, then we will feed an image dataset consisting of 360 normal cells and 40 infected cells to the stacked autoencoders trained to compress normal cell images. Similarly, another image dataset consisting of 360 infected cells and 40 normal cells will be fed to the other stacked autoencoders trained to compress infected cell images.

3.1. Conditional Entropies Versus Misclassification Rates

We first use Equations (13) and (15) to obtain the empirical entropies of the residues (conditional upon whether the inputs are normal or infected cells) as an estimate of the compressed bitrates.

The results are plotted in Figure 7. We can see that the infected cells tend to be "easier" to compress than the normal cells. This can be attributed to the fact that infected cells share some common features, e.g., the existence of the ring form characteristic of parasite infection. While the autoencoders have been trained effectively capture the common features of the input images belonging to the same class, more and more "wrong" inputs from the other class due to misclassification lead to larger prediction residues, which translate to larger entropies, or lower compression. Thus for both classes of input images, we can see the apparent trend of lower and lower compression performance with an increasing misclassification rate, as expected.

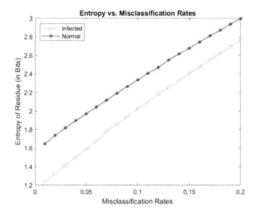

Figure 7. Estimated conditional entropies of the residues as a function of misclassification rates.

3.2. Joint Entropy Versus Misclassification Rates

Here we still assume that the cells are equally likely to be either normal or infected, i.e., $P(S0) = P(S1) = \frac{1}{2}$, but allow the misclassification rates $P(C1|S0)$, $P(C0|S1)$ to change freely within the range. Based on Equation (16), we can plot a 3D surface as shown in Figure 8. We can see the general trend remains the same as the conditional entropies: when misclassification rates increase, the joint entropy (overall bitrates in theory) also increase.

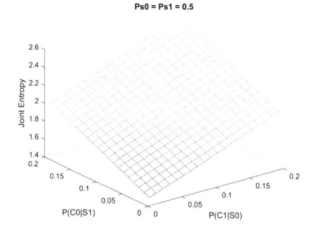

Figure 8. The joint entropy as a function of misclassification rates.

3.3. Average Codeword Lengths Versus Misclassification Rates

We use Golomb–Rice codes to compress the residues and use Equation (21) to calculate the average codeword lengths (ACWL in bits, which is the practically achievable bitrates) over all four cases (as shown in Figure 6). Figure 9 shows the relation between the overall ACWL (bitrates) and the misclassification rates. Again, the curve clearly shows the general trend of increased bitrates (less compression) when the misclassification rate increases, which is what we expected. In the following, we compare the compression performance of deep learning based method with some popular lossless image compression methods.

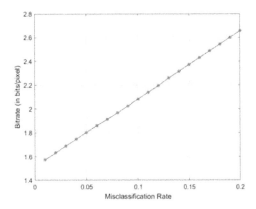

Figure 9. The overall average Golomb–Rice codeword lengths as a function of misclassification rates.

3.4. Comparisons with Mainstream Lossless Compression Methods

We compare with four well known lossless image compression methods. A brief introduction to these methods is given below.

- JPEG2000 [49] is an image compression standard designed to improve the performance of JPEG compression standard, albeit at the cost of increased computational complexity. Instead of using DCT in JPEG, JPEG2000 uses discrete wavelet transform (DWT).
- JPEG-LS is a lossless image compression standard. JPEG-LS improves the compression by using more context pixels (pixels already encoded) to predict the current pixel [50]. We use the codec based on the LOCO-I algorithm [51].
- CALIC (Context-based, adaptive, lossless image codec) uses a large number of contexts to condition a non-linear predictor, which makes it adaptive to varying source statistics [52].
- WebP [53] is an image format currently developed by Google. WebP is based on block prediction, and a variant of LZ77-Huffman coding is used for entropy coding.

The comparison results are shown in Figure 10. We can see that our method significantly outperforms other four conventional compression methods, which are not sensitive to the change of the misclassification rates. This is because these standard methods are designed to be as generic as possible, without taking advantage of the correlations among images belonging to the same classes, which can be captured by sufficiently trained autoencoders. Here we take into account practical scenarios where there will be mismatch between the input images and the autoencoders of the corresponding class. For example, the autoencoders pre-trained to compress infected cell images would suffer from degrading performance as more and more normal cell images (due to increasing misclassification rates) are mixed with the infected cells as the input. However, even at a very low misclassification rate of 20% (which a reasonably good pattern classifier can easily do better in terms of accuracy), the curve Figure 10 shows the deep learning based method still has better performance than the four other methods.

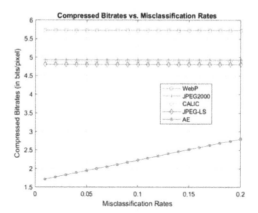

Figure 10. Comparison of bitrates for varying misclassification rates.

The result highlights the advantage of our data-specific approach of "train once and then compress many times", where deep learning seems to be very effective in extracting common features within the dataset, thereby providing more efficient data compression. Nonetheless, in practical implementations of an end-to-end compression/decompression system, the parameters of the stack autoencoders already trained have to be provided as side information to the decoder to ensure lossless decompression. Fortunately, this one-time cost of bitrates for the side information can be amortized over a large number of images to be compressed in the dataset. The other side information is the 30-point vector for each image at the output of the autoencoder at the last stage. Again, the bits needed for coding the vector is a one-time cost for the entire image, representing an negligible increase in the average bitrates (in bits/pixel).

It should also be noted that this deep learning based approach has some limitations. First, the approach is more suitable for achieving good compression on average over an entire dataset, where images can be grouped into different classes by a reasonably well trained classifier. The images within the same class share some common features, which can be exploited to achieve higher compression than would be possible by considering only individual image statistics. Therefore, this joint classification/compression approach is not intended for compression of individual images, for which mainstream lossless compression methods are more suitable, since they optimize their performance based on individual image statistics. Second, training stacked autoencoders on large dataset tend to be expensive computationally. Therefore, the high computational cost will only justify the "train once and then compress many times" approach applied on the entire dataset. Finally, the autoencoder parameters (e.g., the weights and biases of each layer) have to be made available to the decoder as a side information. Therefore, the advantage of the deep learning based method would be more pronounced for large datasets, where the impact of the side information overhead on the overall bitrates will become less noticeable for the entire dataset.

In the literature, existing work on deep learning for image compression is fairly sparse, mostly with the goal of achieving low bit rates and higher visual quality for lossy compression. For example, Toderici et al. proposed a general framework for variable-rate image compression based on convolutional and deconvolutional long short-term memory (LSTM) recurrent networks [54]. They reported better visual quality than JPEG2000 and WebP on 32 × 32 thumbnail images. Their follow-up work in [55] proposed a hybrid of Gated Recurrent Unit (GRU) and ResNet as a full-resolution lossy image compression methods. Jiang et al. [56] proposed an end-to-end lossy compression framework consisting of two convolutional neural networks (CNNs) for image compaction, albeit still requiring the main compression engine to be a standard compression method such as JPEG. Li et al. proposed a CNN-based content-weighted lossy compression method, which outperforms traditional methods on low bit rate images [57]. Generative Adversarial Networks (GANs) were used in [58] for lossy image compression, achieving good reconstructed image quality at very low bit rates (e.g., below 0.1 bit per pixel). In contrast, this work focuses on lossless compression. Our results shows that autoencoders are capable of capturing inter-image correlations in a large datasets, which are beneficial to efficient lossless compression of the entire dataset. It would be a good research direction to study how to integrate autoencoders with other deep learning architectures such as CNNs and GANs to exploit also local image statistics, as well as recurrent neural networks (RNNs) and LSTM networks to take advantage of pixel dependence within an image.

4. Conclusions

In this paper, we study how the performance of lossless compression on red blood cell images is affected by an imperfect classifier in a realistic setting where images are first classified prior to being compressed using deep learning methods based on stacked autoencoders. We provide an in-depth analysis on the impact of misclassification rates on the overall image compression performance and derive formulas for both empirical entropy and average codeword lengths based on Golomb–Rice codes for residues. These formulas provide new insight into how the overall compression efficiency are affected by different source probability and misclassification rates. We also use malaria infection image datasets to evaluate the relations between misclassification rates and actually obtainable compressed bit rates. The results show the advantage of our data driven approach of "train the neural network once and then compress the data many times", where deep learning seems to be very effective in extracting common features within the dataset, thereby providing more efficient data compression than conventional methods, even at elevated misclassification rates. This special feature will be useful when only some important parts (regions of interest) of a large high-resolution (e.g., a wholeslide image) are required for lossless compression, while the rest (e.g., the background) only need lossy compression, or can simply be discarded. In the case of computer assisted malaria diagnosis, pathologists are mainly interested in red blood cell images. So we can classify the infected and normal cells, which can lead to more efficient compression of an entire image datasets. Thus, the proposed compression method can find useful applications in telemedicine where efficient storage and rapid transfer of large image datasets is sought after. As future work, we aim to study the compression performance and computational efficiencies of an end-to-end classification/compression system, taking into account the overhead associated with the descriptions of the neural network structure and feature vectors.

Author Contributions: Conceptualization, Y.D. and W.D.P.; Methodology, Y.D., W.D.P., and D.W.; Software, Y.D.; Validation, Y.D., W.D.P., and D.W.; Formal Analysis, Y.D., W.D.P., and D.W.; Investigation, Y.D., W.D.P., and D.W.; Resources, Y.D. and W.D.P.; Data Curation, Y.D. and W.D.P.; Writing—Original Draft Preparation, Y.D. and W.D.P.; Writing—Review & Editing, Y.D., W.D.P., and D.W.; Visualization, Y.D. and W.D.P.; Supervision, W.D.P.; Project Administration, W.D.P.; Funding Acquisition, W.D.P. and D.W.

Funding: The first and the second author received no external funding for this research. The support for the third author might be better described by the Acknowledgments section, with the statement in its entirety provided by the funding agency.

Acknowledgments: Dongsheng Wu's research has been supported in part by Mission Support and Test Services, LLC, with the U.S. Department of Energy, National Nuclear Security Administration, NA-10 Office of Defense Programs, and the Site-Directed Research and Development Program. The United States Government retains and the publisher, by accepting the article for publication, acknowledges that the United States Government retains a non-exclusive, paid-up, irrevocable, world-wide license to publish or reproduce the published content of this manuscript, or allow others to do so, for United States Government purposes. The U.S. Department of Energy will provide public access to these results of federally sponsored research in accordance with the DOE Public Access Plan (http://energy.gov/downloads/doe-publicaccess-plan). The views expressed in the article do not necessarily represent the views of the U.S. Department of Energy or the United States Government.

Conflicts of Interest: The authors declare no conflict of interest.

References

1. Chan, C. *World Malaria Report*; Technical Report; World Health Organization: Geneva, Switzerland, 2015.
2. Kettelhut, M.M.; Chiodini, P.L.; Edwards, H.; Moody, A. External quality assessment schemes raise standards: Evidence from the UKNEQAS parasitology subschemes. *J. Clin. Pathol.* **2003**, *56*, 927–932. [CrossRef] [PubMed]
3. Delahunt, C.B.; Mehanian, C.; Hu, L.; McGuire, S.K.; Champlin, C.R.; Horning, M.P.; Wilson, B.K.; Thompon, C.M. Automated microscopy and machine learning for expert-level malaria field diagnosis. In Proceedings of the 2015 IEEE Global Humanitarian Technology Conference (GHTC), Seattle, WA, USA, 8–11 October 2015; pp. 393–399.
4. Muralidharan, V.; Dong, Y.; Pan, W.D. A comparison of feature selection methods for machine learning based automatic malarial cell recognition in wholeslide images. In Proceedings of the 2016 IEEE-EMBS International Conference on Biomedical and Health Informatics (BHI), Las Vegas, NV, USA, 24–27 February 2016; pp. 216–219.
5. Park, H.S.; Rinehart, M.T.; Walzer, K.A.; Chi, J.T.A.; Wax, A. Automated Detection of P. falciparum Using Machine Learning Algorithms with Quantitative Phase Images of Unstained Cells. *PLoS ONE* **2016**, *11*, e0163045. [CrossRef] [PubMed]
6. Sanchez, C.S. Deep Learning for Identifying Malaria Parasites in Images. Master's Thesis, University of Edinburgh, Edinburgh, UK, 2015.
7. Quinn, J.A.; Nakasi, R.; Mugagga, P.K.B.; Byanyima, P.; Lubega, W.; Andama, A. Deep Convolutional Neural Networks for Microscopy-Based Point of Care Diagnostics. In Proceedings of the International Conference on Machine Learning for Health Care, Los Angeles, CA, USA, 19–20 August 2016.
8. Center for Devices and Radiological Health. *Technical Performance Assessment of Digital Pathology Whole Slide Imaging Devices*; Technical Report; Center for Devices and Radiological Health: Silver Spring, MD, USA, 2015.
9. Farahani, N.; Parwani, A.V.; Pantanowitz, L. Whole slide imaging in pathology: Advantages, limitations, and emerging perspectives. *Pathol. Lab. Med. Int.* **2015**, 23–33.
10. University of Alabama at Birmingham. PEIR-VM. Available online: http://peir-vm.path.uab.edu/about.php (accessed on 6 May 2019).
11. Cornish, T.C. An Introduction to Digital Wholeslide Imaging and Wholeslide Image Analysis. Available online: https://docplayer.net/22756037-An-introduction-to-digital-whole-slide-imaging-and-whole-slide-image-analysis.html (accessed on 6 May 2019).
12. Al-Janabii, S.; Huisman, A.; Nap, M.; Clarijs, R.; van Diest, P.J. Whole Slide Images as a Platform for Initial Diagnostics in Histopathology in a Medium-sized Routine Laboratory. *J. Clin. Pathol.* **2012**, *65*, 1107–1111. [CrossRef] [PubMed]
13. Pantanowitz, L.; Valenstein, P.; Evans, A.; Kaplan, K.; Pfeifer, J.; Wilbur, D.; Collins, L.; Colgan, T. Review of the current state of whole slide imaging in pathology. *J. Pathol. Inform.* **2011**, *2*, 36. [CrossRef] [PubMed]
14. Tek, F.B.; Dempster, A.G.; Kale, I. Computer vision for microscopy diagnosis of malaria. *Malar. J.* **2009**, *8*, 1–14. [CrossRef]
15. World Health Organization. Microscopy. Available online: http://www.who.int/malaria/areas/diagnosis/microscopy/en/ (accessed on 6 May 2019).
16. Halim, S.; Bretschneider, T.R.; Li, Y.; Preiser, P.R.; Kuss, C. Estimating malaria parasitaemia from blood smear images. In Proceedings of the IEEE International Conference on Control, Automation, Robotics and Vision, Singapore, 5–8 December 2006; pp. 1–6.
17. Das, D.; Ghosh, M.; Chakraborty, C.; Pal, M.; Maity, A.K. Invariant Moment based feature analysis for abnormal erythrocyte segmentation. In Proceedings of the International Conference on Systems in Medicine and Biology (ICSMB), Kharagpur, India, 16–18 December 2010; pp. 242–247.
18. Das, D.K.; Ghosh, M.; Pal, M.; Maiti, A.K.; Chakraborty, C. Machine learning approach for automated screening of malaria parasite using light microscopic images. *J. Micron* **2013**, *45*, 97–106. [CrossRef]
19. Tek, F.B.; Dempster, A.G.; Kale, I. Parasite detection and identification for automated thin blood film malaria diagnosis. *J. Comput. Vis. Image Underst.* **2010**, *114*, 21–32. [CrossRef]
20. Di Ruberto, C.; Dempster, A.; Khan, S.; Jarra, B. Analysis of infected blood cell images using morphological operators. *J. Comput. Vis. Image Underst.* **2002**, *20*, 133–146. [CrossRef]

21. Ross, N.E.; Pritchard, C.J.; Rubin, D.M.; Duse, A.G. Automated image processing method for the diagnosis and classification of malaria on thin blood smears. *Med. Biol. Eng. Comput.* **2005**, *44*, 427–436. [CrossRef]

22. Makkapati, V.V.; Rao, R.M. Segmentation of malaria parasites in peripheral blood smear images. In Proceedings of the IEEE International Conference on Acoustics, Speech, and Signal Processing, Taipei, Taiwan, 19–24 April 2009; pp. 1361–1364.

23. Tek, F.B.; Dempster, A.G.; Kale, I. Malaria parasite detection in peripheral blood images. In Proceedings of the British Machine Vision Conference 2006, Edinburgh, UK, 4–7 September 2006.

24. Linder, N.; Turkki, R.; Walliander, M.; Mårtensson, A.; Diwan, V.; Rahtu, E.; Pietikäinen, M.; Lundin, M.; Lundin, J. A Malaria Diagnostic Tool Based on Computer Vision Screening and Visualization of Plasmodium falciparum Candidate Areas in Digitized Blood Smears. *PLoS ONE* **2014**, *9*, e104855. [CrossRef] [PubMed]

25. Pan, W.D.; Dong, Y.; Wu, D. Classification of Malaria-Infected Cells Using Deep Convolutional Neural Networks. In *Machine Learning—Advanced Techniques and Emerging Applications*; Farhadi, H., Ed.; IntechOpen: London, UK, 2018.

26. Shen, H.; Pan, W.D.; Dong, Y.; Alim, M. Lossless compression of curated erythrocyte images using deep autoencoders for malaria infection diagnosis. In Proceedings of the IEEE Picture Coding Symposium (PCS), Nuremberg, Germany, 4–7 December 2016; pp. 1–5. [CrossRef]

27. Duh, D.J.; Jeng, J.H.; Chen, S.Y. DCT based simple classification scheme for fractal image compression. *Image Vis. Comput.* **2005**, *23*, 1115–1121. [CrossRef]

28. Fahmy, G.; Panchanathan, S. A lifting based system for optimal compression and classification in the JPEG2000 framework. In Proceedings of the IEEE International Symposium on Circuits and Systems (ISCAS 2002), Phoenix-Scottsdale, AZ, USA, 26–29 May 2002; Volume 4.

29. Kim, H.; Yazicioglu, R.F.; Merken, P.; Van Hoof, C.; Yoo, H.J. ECG signal compression and classification algorithm with quad level vector for ECG holter system. *IEEE Trans. Inf. Technol. Biomed.* **2010**, *14*, 93–100. [PubMed]

30. Jha, C.K.; Kolekar, M.H. Classification and Compression of ECG Signal for Holter Device. In *Biomedical Signal and Image Processing in Patient Care*; IGI Global: Hershey, PA, USA, 2018; pp. 46–63.

31. Minguillón, J.; Pujol, J.; Serra, J.; Ortimo, I. Influence of lossy compression on hyperspectral image classification accuracy. *WIT Trans. Inf. Commun. Technol.* **2000**, *25*. [CrossRef]

32. Garcia-Vilchez, F.; Muñoz-Marí, J.; Zortea, M.; Blanes, I.; González-Ruiz, V.; Camps-Valls, G.; Plaza, A.; Serra-Sagristà, J. On the impact of lossy compression on hyperspectral image classification and unmixing. *IEEE Geosci. Remote Sens. Lett.* **2011**, *8*, 253–257. [CrossRef]

33. Gelli, G.; Poggi, G. Compression of multispectral images by spectral classification and transform coding. *IEEE Trans. Image Process.* **1999**, *8*, 476–489. [CrossRef]

34. Peng, K.; Kieffer, J.C. Embedded image compression based on wavelet pixel classification and sorting. *IEEE Trans. Image Process.* **2004**, *13*, 1011–1017. [CrossRef]

35. Oehler, K.L.; Gray, R.M. Combining image classification and image compression using vector quantization. In Proceedings of the IEEE Data Compression Conference (DCC'93), Snowbird, UT, USA, 30 March–2 April 1993; pp. 2–11.

36. Oehler, K.L.; Gray, R.M. Combining image compression and classification using vector quantization. *IEEE Trans. Pattern Anal. Mach. Intell.* **1995**, *17*, 461–473. [CrossRef]

37. Li, J.; Gray, R.M.; Olshen, R. Joint image compression and classification with vector quantization and a two dimensional hidden Markov model. In Proceedings of the Data Compression Conference, DCC'99, Snowbird, UT, USA, 29–31 March 1999; pp. 23–32.

38. Baras, J.S.; Dey, S. Combined compression and classification with learning vector quantization. *IEEE Trans. Inf. Theory* **1999**, *45*, 1911–1920. [CrossRef]

39. Ayoobkhan, M.U.A.; Chikkannan, E.; Ramakrishnan, K.; Balasubramanian, S.B. Prediction-Based Lossless Image Compression. In Proceedings of the International Conference on ISMAC in Computational Vision and Bio-Engineering 2018 (ISMAC-CVB), Palladam, India, 16–17 May 2018; pp. 1749–1761.

40. Fu, D.; Guimaraes, G. Using Compression to Speed Up Image Classification in Artificial Neural Networks. Available online: http://www.danfu.org/files/CompressionImageClassification.pdf (accessed on 6 October 2019).

41. Andono, P.N.; Supriyanto, C.; Nugroho, S. Image compression based on SVD for BoVW model in fingerprint classification. *J. Intell. Fuzzy Syst.* **2018**, *34*, 2513–2519. [CrossRef]

42. Mohanty, I.; Pattanaik, P.A.; Swarnkar, T. Automatic Detection of Malaria Parasites Using Unsupervised Techniques. In Proceedings of the International Conference on ISMAC in Computational Vision and Bio-Engineering 2018 (ISMAC-CVB), Palladam, India, 16–17 May 2018; pp. 41–49.

43. Whole Slide Image Data. Available online: http://peir-vm.path.uab.edu/debug.php?slide=IPLab11Malaria (accessed on 6 May 2019).

44. Dong, Y.; Jiang, Z.; Shen, H.; Pan, W.D.; Williams, L.A.; Reddy, V.V.; Benjamin, W.H.; Bryan, A.W. Evaluations of deep convolutional neural networks for automatic identification of malaria infected cells. In Proceedings of the 2017 IEEE EMBS International Conference on Biomedical & Health Informatics (BHI), Orlando, FL, USA, 16–19 February 2017; pp. 101–104.

45. Duda, R.O.; Hart, P.E. Use of the Hough transformation to detect lines and curves in pictures. *Commun. ACM* **1972**, *15*, 11–15. [CrossRef]

46. Link to the Dataset Used. Available online: http://www.ece.uah.edu/~dwpan/malaria_dataset/ (accessed on 6 May 2019).

47. Hinton, G.E.; Salakhutdinov, R.R. Reducing the dimensionality of data with neural networks. *Science* **2006**, *313*, 504–507. [CrossRef] [PubMed]

48. Golomb, S. Run-length encodings (Corresp.). *IEEE Trans. Inf. Theory* **1966**, *12*, 399–401. [CrossRef]

49. JPEG2000 Home Page. Available online: https://jpeg.org/jpeg2000/ (accessed on 6 May 2019).

50. JPEG-LS Home Page. Available online: https://jpeg.org/jpegls/ (accessed on 6 May 2019).

51. Weinberger, M.J.; Seroussi, G.; Sapiro, G. The LOCO-I lossless image compression algorithm: Principles and standardization into JPEG-LS. *IEEE Trans. Image Process.* **2000**, *9*, 1309–1324. [CrossRef] [PubMed]

52. Wu, X.; Memon, N. CALIC—A context based adaptive lossless image codec. In Proceedings of the 1996 IEEE International Conference on Acoustics, Speech, and Signal Processing (ICASSP-96), Atlanta, GA, USA, 9 May 1996; Volume 4, pp. 1890–1893.

53. WebP Home Page. Available online: https://developers.google.com/speed/webp/ (accessed on 6 May 2019).

54. Toderici, G.; O'Malley, S.M.; Hwang, S.J.; Vincent, D.; Minnen, D.; Baluja, S.; Covell, M.; Sukthankar, R. Variable Rate Image Compression with Recurrent Neural Networks. *arXiv* **2015**, arXiv:1511.06085.

55. Toderici, G.; Vincent, D.; Johnston, N.; Hwang, S.; Minnen, D.; Shor, J.; Covell, M. Full Resolution Image Compression with Recurrent Neural Networks. In Proceedings of the IEEE Conference on Computer Vision and Pattern Recognition (CVPR), Honolulu, HI, USA, 21–26 July 2017; pp. 5435–5443.

56. Jiang, F.; Tao, W.; Liu, S.; Ren, J.; Guo, X.; Zhao, D. An End-to-End Compression Framework Based on Convolutional Neural Networks. *IEEE Trans. Circuits Syst. Video Technol.* **2018**, *28*, 3007–3018. [CrossRef]

57. Li, M.; Zuo, W.; Gu, S.; Zhao, D.; Zhang, D. Learning Convolutional Networks for Content-Weighted Image Compression. In Proceedings of the 2018 IEEE/CVF Conference on Computer Vision and Pattern Recognition, Salt Lake City, UT, USA, 18–23 June 2018; pp. 3214–3223.

58. Agustsson, E.; Tschannen, M.; Mentzer, F.; Timofte, R.; Van Gool, L. Generative Adversarial Networks for Extreme Learned Image Compression. In Proceedings of the IEEE Conference on Computer Vision and Pattern Recognition (CVPR) Workshops, Salt Lake City, UT, USA, 18–23 June 2018; pp. 2587–2590.

 © 2019 by the authors. Licensee MDPI, Basel, Switzerland. This article is an open access article distributed under the terms and conditions of the Creative Commons Attribution (CC BY) license (http://creativecommons.org/licenses/by/4.0/).

Article

Magnetic Resonance Image Quality Assessment by Using Non-Maximum Suppression and Entropy Analysis

Rafał Obuchowicz [1], Mariusz Oszust [2], Marzena Bielecka [3,*], Andrzej Bielecki [4] and Adam Piórkowski [5]

[1] Department of Diagnostic Imaging, Jagiellonian University Medical College, 19 Kopernika Street, 31-501 Cracow, Poland; rafalobuchowicz@su.krakow.pl

[2] Department of Computer and Control Engineering, Rzeszow University of Technology, W. Pola 2, 35-959 Rzeszow, Poland; marosz@kia.prz.edu.pl

[3] Faculty of Geology, Geophysics and Environmental Protection, AGH University of Science and Technology, al. Mickiewicza 30, 30-059 Cracow, Poland

[4] Faculty of Electrical Engineering, Automation, Computer Science and Biomedical Engineering, AGH University of Science and Technology, al. Mickiewicza 30, 30-059 Cracow, Poland; azbielecki@gmail.com

[5] Department of Biocybernetics and Biomedical Engineering, AGH University of Science and Technology, al. Mickiewicza 30, 30-059 Cracow, Poland; pioro@agh.edu.pl

* Correspondence: bielecka@agh.edu.pl

Received: 5 January 2020; Accepted: 13 February 2020; Published: 16 February 2020

Abstract: An investigation of diseases using magnetic resonance (MR) imaging requires automatic image quality assessment methods able to exclude low-quality scans. Such methods can be also employed for an optimization of parameters of imaging systems or evaluation of image processing algorithms. Therefore, in this paper, a novel blind image quality assessment (BIQA) method for the evaluation of MR images is introduced. It is observed that the result of filtering using non-maximum suppression (NMS) strongly depends on the perceptual quality of an input image. Hence, in the method, the image is first processed by the NMS with various levels of acceptable local intensity difference. Then, the quality is efficiently expressed by the entropy of a sequence of extrema numbers obtained with the thresholded NMS. The proposed BIQA approach is compared with ten state-of-the-art techniques on a dataset containing MR images and subjective scores provided by 31 experienced radiologists. The Pearson, Spearman, Kendall correlation coefficients and root mean square error for the method assessing images in the dataset were 0.6741, 0.3540, 0.2428, and 0.5375, respectively. The extensive experimental evaluation of the BIQA methods reveals that the introduced measure outperforms related techniques by a large margin as it correlates better with human scores.

Keywords: blind image quality assessment; magnetic resonance images; entropy; non-maximum suppression

1. Introduction

The ubiquity of advancements in imaging has brought significant attention of medical specialists due to the role of the quality of displayed content in diagnosis [1–3]. The quality of Magnetic Resonance (MR) images depends on used hardware parts, software techniques, as well as human errors involving patient noncompliance or operator mistakes [4–8]. Therefore, the development of automatic image

Entropy **2020**, *22*, 220; doi:10.3390/e22020220

quality assessment (IQA) methods for MR scans is particularly important since the contamination of acquired images may compromise subsequent diagnosis and treatment. Moreover, such methods may support a selection of algorithms for image processing or parameters of imaging systems. Hopefully, a time-consuming examination of images by trained medical specialists can be avoided. Furthermore, the lack of reproducibility of subjective tests and personal quality preferences impeding scores of small groups encourages the use of automatic and repeatable IQA methods. The IQA measures are divided into three categories: Full-reference (FR), reduced-reference (RR), and no-reference or blind (BIQA) methods [9]. The full-reference methods compare input images with their non-distorted versions. However, most medical imaging systems do not produce pristine images, limiting the application range of FR methods [10]. The reduced reference techniques, in turn, require only a part of the information on the pristine image, and blind IQA methods assess images without any external information. Therefore, the development of BIQA approaches is desired.

Among the applications of FR-IQA methods to MR images, Baselice et al. [10] compared results of denoising approaches using Mean Square Error (MSE) with the Structural Similarity Index (SSIM) [11]. Jang et al. [12] employed the SSIM and the Root-Mean-Square Error (RMSE) for an evaluation of BIQA methods on synthetically distorted MR scans. In the work of Chow and Rajagopal [13], Noise Quality Measurement (NQM) with Feature SIMilarity (FSIM) were applied to evaluate a BIQA method. Recognition of a supportive role of FR measures in the assessment of medical images and the need for creating new datasets are among findings of that work [13]. The reduced-reference techniques are not used for the assessment of MR images. In the literature, several BIQA methods have been introduced. Interestingly, as the Signal-to-Noise Ratio (SNR) and Contrast-to-Noise Ratio (CNR) are frequently used for the assessment of medical images [14], they are often criticized due to the need of indication of clearly defined regions with tissue and background [14–16]. Considering MR images, Chow and Rajagopal [13] adapted Blind/Referenceless Image Spatial Quality Evaluator (BRISQUE) [17] by training it on MR images instead of natural images. The method fits the Mean Substracted Contrast Normalization (MSCN) of an image to the Generalized Gaussian Distribution (GGD). Similarly, the GGD was used by Jang et al. [12]. In that work, the characteristics of MR scans were taken into account by employing a multidirectional-filtering of images. In the work of Yu et al. [16], four BIQA methods, i.e., BRISQUE, Natural Image Evaluator (NIQE), Blind Image Integrity Notator using DCT statistics (BLIINDS-II), and Blind Image Quality Index (BIQI), were trained on the SNR scores. Their correlation with the SNR was investigated by Zhang et al. [18]. The BIQA methods for the assessment of brain MR scans were employed by Sandilya and Nirmala [19] and Osadebey et al. [20]. In the first work, the reconstructed scans were assessed with BRISQUE, while in the second approach, binary images of brain scans were evaluated considering noise, lightness, contrast, sharpness, and texture details.

The literature review reveals that the lack of BIQA approaches designed for MR scans is caused mainly by the lack of IQA databases of such images with subjective scores. Moreover, natural images differ from MR images concerning characteristics of used imaging systems for their registration, the complexity of captured structures, or noise. Considering the popularity of IQA methods designed for natural images, some of BIQA approaches for MR scans adapted or modified them. However, there exist many concepts in the IQA of natural images that are not yet utilized for MR images and they should be examined. Therefore, in this paper, apart from the novel BIQA approach designed for MR scans, a set of representative IQA methods is evaluated. Furthermore, a dataset with MR scans and subjective scores used in the evaluation is released.

The introduced method, ENtropy-based Magnetic resonance Image Quality Assessment measure (ENMIQA), takes into account thresholded local intensity differences obtained by using the non-maximum suppression (NMS) [21,22] operation and calculates the entropy of a sequence of extrema numbers.

The extrema represent a set of filtered versions of an input image. Then, entropy is used for quality prediction.

The major contributions of this work are a novel method for the quality assessment of MR images and a comprehensive evaluation of the measure against the state-of-the-art IQA techniques on a dataset of MR images assessed by a large group of experienced radiologists.

The remainder of this paper is organized as follows. In Section 2, the approach is introduced. Then, in Section 3, it is evaluated against the related BIQA methods. Finally, in Section 4, the paper is concluded.

2. Proposed Image Quality Measure

In the introduced method, ENMIQA, an input image I is filtered to determine pixels that represent local intensity extrema. To determine which pixels should be selected, the NMS operation [21,22] is performed. However, to provide a more thorough examination instead of selecting pixels that are of greater or lesser intensity value than its surrounding neighbors, in this work, a sequence of intensity thresholds $\mathfrak{T} = [1, 2, \ldots, S]$, $S \in \mathbb{Z}^+$, is introduced. The NMS uses the threshold $t \in \mathfrak{T}$ to indicate the local extrema. Consequently, image I for each threshold t is represented by the number of found local extrema $I(t)$. This can be written as:

$$I(t) = \sum_{a=1}^{M} \sum_{b=1}^{N} T(a, b, t),$$ (1)

where a pair (a, b) denotes the pixel location within an image of the size $M \times N$ and $T(a, b, t)$ is a test in which the NMS is calculated using the proposed threshold t. The test is obtained as follows:

$$T(a, b, t) = \begin{cases} 1, & \text{if } \forall_{(i,j)} I(a, b) > I(a+i, b+j) + t, \\ 1, & \text{else if } \forall_{(i,j)} I(a, b) < I(a+i, b+j) - t, \\ 0, & \text{otherwise,} \end{cases}$$ (2)

where $(i, j) \in \{(0, 1), (0, -1), (1, 0), (-1, 0)\}$. The pair of indices (i, j) forms the neighborhood of 3×3 pixels around the location (a, b). Finally, a sequence of sums $I(\mathfrak{T}) = [I(t = 1), I(t = 2), \ldots, I(t = S)]$ is obtained. Then, it is divided by the image size to normalize the values. To determine the quality of the input image I, entropy of $I(\mathfrak{T})$ is calculated.

Entropy is the fundamental concept of Shannon information theory [23,24]. It is usually considered in the framework of measure theory. Assuming that space X with a probabilistic measure μ and a countable partition $\mathfrak{P}$ of X are given [25], the entropy h is:

$$h(\mu, \mathfrak{P}) = \sum_{P \in \mathfrak{P}} s(\mu(P)),$$ (3)

where $s: [0, 1] \rightarrow [0, \infty)$ can be expressed as $s(x) = -x \log x$ for $0 < x \leq 1$ and $s(0) = 0$. Note that entropy equals zero if and only if there exists such $P \in \mathfrak{P}$ that $\mu(P) = 1$. If X contains R elements, then $\mathfrak{P} = \{P_1, \ldots, P_R\}$. Furthermore, if μ is based on counting measure, then Equation (3) has the following form:

$$h(\mu, \mathfrak{P}) = -\sum_{i=1}^{R} k_i \log k_i,$$ (4)

where $k_i = \frac{m_i}{m}$, m_i and m are the numbers of elements in P_i and X, respectively. Entropy defined by Equation (4) reaches its maximum for the uniform distribution of the measure μ on the family $\mathfrak{P}$. Such defined entropy refers to the amount of information on (X, μ) introduced by $\mathfrak{P}$. Consequently, the inversely proportional relationship between entropy and information is often applied in practice.

In this paper, entropy analysis is used for the IQA of two-dimensional MR images. In such a context, it can be employed for measuring disorders. In MR scans of internal organs, single isolated impulses with higher or lower intensity concerning a local neighborhood are common in distorted images. Thus, the greater the value of the threshold t in the NMS, the greater the probability that the detected intensity irregularities are disorders that decrease the quality of an image. The observed discriminative capabilities of entropy regarding images of different qualities justify its use for the IQA of MR images. In this work, Equation (4) is directly used as a quality measure, assuming that a set X is expressed as $\{(I, t), t \in \mathfrak{T}\}$ and $\mathfrak{T}$ determines the partition of X. The main computational steps of the method are shown in Figure 1.

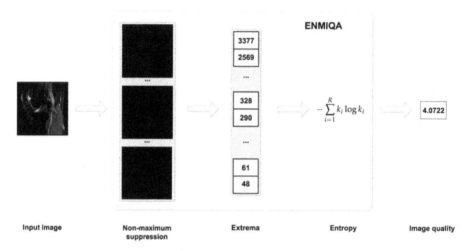

| Input image | Non-maximum suppression | Extrema | Entropy | Image quality |

Figure 1. Image processing steps towards the calculation of image quality in ENtropy-based Magnetic resonance Image Quality Assessment measure (ENMIQA).

Figure 2 presents two MR images of different quality and the influence of t on the local extrema. As shown, the proposed method determines more extrema in images with more distortions.

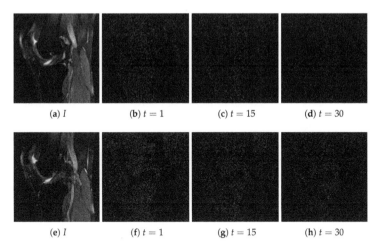

(a) I (b) $t = 1$ (c) $t = 15$ (d) $t = 30$

(e) I (f) $t = 1$ (g) $t = 15$ (h) $t = 30$

Figure 2. Two magnetic resonance (MR) images of different quality and the determined local extrema for $t = 1, 15, 30$.

3. Results and Discussion

In this section, a dataset that contains MR images with associated subjective scores is introduced. Then, the performance of ENMIQA against ten state-of-the-art related methods is evaluated using a typical methodology and discussed. Finally, the influence of parameters of ENMIQA on its performance is provided.

3.1. Experimental Data

The introduced ENMIQA and related techniques are evaluated on a dataset that contains MR images and subjective scores collected in tests with human subjects. The dataset consists of 70 T2-weighted MR images (T2w) extracted from the lumbar and cervical spine, brain, hip, knee, and wrist sequences in axial, sagittal, and coronal planes. The sequences were obtained for a group of 51 patients of 27–41 years old (26 men and 25 women). The study protocol was designed according to the guidelines of the Declaration of Helsinki and the Good Clinical Practice Declaration Statement. The data safety was ensured by removing the personal details from images. Written acceptance for conducting the study was obtained from the Ethics Committee of Jagiellonian University (no. 1072.6120.15.2017). To produce images with different quality for the IQA purposes, shortened sequences were acquired using Process Analytical Technology (PAT) I software (Siemens) and employing the GeneRalized Autocalibrating Partially Parallel Acquisitions (GRAPPA) 3 in which 25% of the echoes were acquired with 60% signal reduction regarding the original acquisition mode [26,27]. Then, images with distortion types that were not present in all examined body parts were rejected. The obtained dataset is characterized in Table 1. There are 15, 9, and 11 image pairs captured in sagittal, axial, and coronal planes, respectively. The size of the images ranges from 192×320 to 512×512. The subjective scores for images were obtained in a group of 31 experienced radiologists with more than six years of diagnostic reading residency. Each radiologist assessed two images of the same part of the body at once, spending a minute on the assessment of the pair. The images were scored from 1 to 5, with a higher score associated with better quality. The examination was repeated until all images in the dataset were assessed. Then, scores for images were averaged and the mean opinion score

(MOS) was obtained. The number of radiologists that took part in the subjective tests was large enough to ensure that personal quality preferences do not impair the MOS. However, the number of images in the database depended on the number of medical professionals and the time spent on the examination. Exemplary images from the dataset can be seen in Figure 3.

Table 1. Summary of images used in experiments.

Body Part	No. of Image Pairs	Axial Plane	Sagittal Plane	Coronal Plane
Lumbar and cervical spine	7	2	5	0
Knee	7	2	4	1
Shoulder	8	2	2	4
Wrist	3	0	0	3
Hip	2	1	1	0
Pelvis	2	0	0	2
Elbow	1	1	0	0
Ankle	1	0	1	0
Brain	4	1	2	1
Total pairs	35	9	15	11

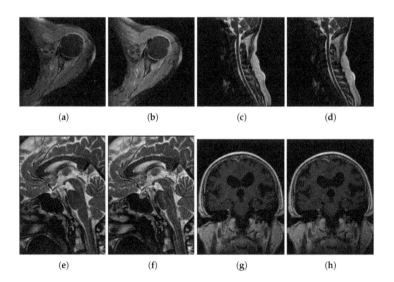

(a)	(b)	(c)	(d)
(e)	(f)	(g)	(h)

Figure 3. Exemplary MR images used in experiments.

3.2. Evaluation Methodology

According to the popular protocol for the performance evaluation of IQA measures, objective scores Q for images in a database are compared with subjective scores (i.e., MOS) $\mathfrak{S}$ collected for them in tests with human subjects. Typically, the four criteria are used to characterize IQA measure [28]: Pearson correlation coefficient (PLCC), Spearman Rank order Correlation Coefficient (SRCC), Kendall Rank order Correlation Coefficient (KRCC), and Root Mean Square Error (RMSE). The PLCC and RMSE are calculated for the vector Q_p obtained via a nonlinear mapping between objective scores Q and subjective scores $\mathfrak{S}$ using fitted parameters of the regression model $\beta = [\beta_1, \beta_2, \ldots, \beta_5]$, i.e., $Q_p = \beta_1 \left(\frac{1}{2} - \frac{1}{1+\exp(\beta_2(Q-\beta_3))} \right) + \beta_4 Q + \beta_5$. The PLCC is obtained as:

$$\text{PLCC}(Q_p, \mathfrak{S}) = \frac{\bar{Q}_p^T \tilde{\mathfrak{S}}}{\sqrt{\bar{Q}_p^T \bar{Q}_p \tilde{\mathfrak{S}}^T \tilde{\mathfrak{S}}}}, \tag{5}$$

where $\bar{Q}_p$ and $\tilde{\mathfrak{S}}$ are mean-removed vectors. The SRCC is calculated as:

$$\text{SRCC}(Q, \mathfrak{S}) = 1 - \frac{6 \sum_{i=1}^{m} d_i^2}{m(m^2 - 1)}, \tag{6}$$

where d_i is the difference between i-th image in vectors of scores and m denotes the number of images in the dataset. The KRCC is obtained as:

$$\text{KRCC}(Q, \mathfrak{S}) = \frac{m_c - m_d}{0.5m(m-1)}, \tag{7}$$

where m_c, m_d are the number of concordant and discordant pairs, respectively. The RMSE, in turn, is obtained as:

$$\text{RMSE}(Q_p, \mathfrak{S}) = \sqrt{\frac{(Q_p - \mathfrak{S})^T (Q_p - \mathfrak{S})}{m}}. \tag{8}$$

3.3. Comparative Evaluation

The ENMIQA is compared against the following ten related BIQA measures: SNRTOI [18], BPRI [29], ILNIQE [30], QENI [31], SISBLIM [32], metricQ [33], SSEQ [34], SINDEX [35], MEON [36], and DEEPIQ [37]. The SNRTOI [18] was implemented by authors of this paper, while other methods were run using their publicly available Matlab implementations. All compared methods, similarly to ENMIQA, do not require training. However, MEON and DEEPIQ represent recently introduced deep learning approaches and are already trained by their authors. The ENMIQA run with $S = 30$ in experiments and other measures used their default parameters. In cases in which a method was designed to process color images, three identical channels were used as an input. The performance of the methods and their approaches to image quality modeling and prediction are shown in Table 2.

Table 2. Evaluation and characteristics of compared blind image quality assessment (BIQA) measures. The best value for each performance criterion is written in bold.

Method	PLCC	SRCC	KRCC	RMSE	Approach to Image Quality Modeling and Prediction
ENMIQA	**0.6741**	**0.3540**	**0.2428**	**0.5375**	Thresholded NMS and entropy
BPRI	0.3440	0.1515	0.1120	0.6832	Distortion-specific metrics and pseudo-reference image
DEEPIQ	0.4039	0.3030	0.2037	0.6657	RankNet trained on quality-discriminable image pairs
ILNIQE	0.3465	0.1796	0.1162	0.6826	Multivariate Gaussian model of pristine images
MEON	0.0439	0.1247	0.0771	0.7272	End-to-end deep neural network with subtasks
MetricQ	0.3075	0.2300	0.1520	0.6924	Singular value decomposition of local image gradient matrix
QENI	0.2886	0.2385	0.1587	0.6967	Self-similarity of local features and saliency models
SINDEX	0.3307	0.2802	0.1962	0.6869	Global and local phase information
SNRTOI	0.2262	0.1828	0.1245	0.7088	Signal-to-nose ratio
SSEQ	0.2903	0.0855	0.0487	0.6963	Distortion classification using local entropy
SISBLIM	0.5733	0.2885	0.1820	0.5962	Free energy theory based fusion of distortion-specific metrics

As reported, the measure introduced in this paper, ENMIQA, outperforms related techniques by a large margin in terms of all four performance indices. Depending on the considered index, it is followed by SISBLIM (PLCC and RMSE) and DEEPIQ (SRCC and KRCC). To show the performance of the measures for images of body parts largely represented in the database, the PLCC calculated for their subsets is reported in Figure 4. Here, ENMIQA obtains greater PLCC than it can be seen for the remaining methods for images of the lumbar and cervical spine, knee, shoulder, and wrist. It is slightly worse than BPRI for brain images. Interestingly, it seems that the recently introduced BPRI is suitable for such images, despite being the second worse technique regarding the entire database and the fourth-best technique in ranking based on the individual body parts. The worse results of methods designed for the assessment of natural images, as well as by complex deep learning approaches, can be justified by the specifics of MR images in which a large portion of the area is covered by organs or tissue while the background is usually dark and may contain noise. In natural images, such empty or nearly empty spaces are seldom found. Furthermore, popular BIQA methods are often trained to recognize typical distortion types (e.g., BPRI, ILNIQE, MEON, SSEQ, or DEEPIQ). Interestingly, methods trained on images contaminated with Gaussian noise can, to some extent, correctly predict the quality of MR images since Gaussian noise manifests itself in magnitude images as a Rician distribution of pixel intensities [38]. This is confirmed by weaker performance of the SNRTOI, which, being an SNR derivative, is often used by radiologists as supporting information on the captured images. The reported results for other methods seem to justify the need for the development of measures designed for the IQA of MR images.

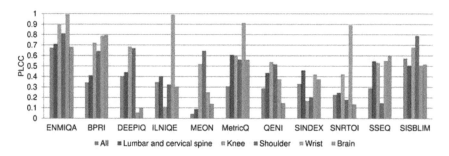

Figure 4. Pearson correlation coefficient (PLCC) performance of the BIQA methods for subsets of images of common body parts.

To evaluate the statistical significance of the obtained errors in the prediction of IQA methods, hypothesis tests based on the prediction residuals of each IQA measure after non-linear mapping were conducted using F-statistic [28]. The F-test is based on an assumption of the Gaussianity of residuals and determines whether the two compared sample sets come from the same distribution, based on the ratio of their variances. The test is often used for the comparison of IQA measures [28]. Therefore, at first, the Jarque–Bera (JB) statistic to determine whether residuals come from a normal distribution was used [39]. In the JB test, the null hypothesis is that the vector of residuals of NR measure follows a normal distribution while the alternative hypothesis is that it does not follow it. Since for all compared measures the null hypothesis was not rejected at the 5% significance level, the F-statistic could be reliably employed. In the F-test, the null hypothesis is that the vectors of residuals of two IQA measures come from the same distribution with the same variance and are statistically indistinguishable (95% confidence). The alternative hypothesis is that the vectors are statistically distinguishable and have different variances. Before the calculation of the F-statistic, a vector of residuals of a measure was used to fit a normal distribution and 1000 samples were drawn from it. The tests revealed that the residual variance of ENMIQA is statistically smaller than those of all compared IQA methods with confidence greater than 95%. This is also indicated by the ratio in all cases. The obtained JB statistics for measures and ratios of the residual variances of algorithms to the ENMIQA are presented in Table 3.

Table 3. Ratios of residual variances of methods to ENMIQA and the Jarque–Bera (JB) statistics. Smaller values of JB statistics denote smaller deviations from the Gaussianity. All measures follow a normal distribution.

Method	Ratio	JB Statistic
ENMIQA	1.0000	0.8523
BPRI	0.6189	2.8999
DEEPIQ	0.6510	1.3870
ILNIQE	0.6201	3.9911
MEON	0.5462	3.8930
MetricQ	0.6032	2.8356
QENI	0.5952	2.7040
SINDEX	0.6124	3.2580
SNRTOI	0.5751	1.7389
SSEQ	0.5958	3.5343
SISBLIM	0.8128	0.1254

3.4. Computational Complexity

The computational complexity of ENMIQA depends on the size of processed image ($N \times M$), the length of the sequence of thresholds S, and the size of the neighborhood used for the NMS ($k = 3 \times 3$). Therefore, its computational complexity is of the order of $O(NMSk^2)$.

The introduced dataset was used to analyze the computational complexity of methods in terms of the average time taken to assess an image. The methods were run on a 2.2 GHz Intel Core CPU with 8 GB RAM using Matlab 2019b environment. Table 4 reports obtained timings. As shown, ENMIQA is slower than MEON, SINDEX, and SNRTOI, but it is faster than the remaining seven measures. The fastest methods (i.e., SINDEX and SNRTOI) are characterized by inferior IQA performance, and taking into account the results for more promising techniques, the introduced ENMIQA is relatively fast and provides the superior quality prediction of MR images.

Table 4. Time–cost comparison of BIQA measures (in seconds).

Method	ENMIQA	BPRI	DEEPIQ	ILNIQE	MEON	MetricQ	QENI	SINDEX	SNRTOI	SSEQ	SISBLIM
Runtime	0.2151	0.2524	2.439	9.299	0.1853	0.4813	1.212	0.0479	0.0069	0.9140	1.629

3.5. Influence of Parameters

The ENMIQA is governed by the sequence of thresholds $\mathfrak{T} = [1, 2, \ldots, S]$, $S \in \mathbb{Z}^+$ used by the non-maximum suppression. Therefore, it is worth to determine how stable is its performance for various S. The S is the greatest threshold in the sequence and indicates its length. The PLCC performance of the method on the entire database, ranging S from 5 to 100 with the step of 5 is shown in Figure 5a. The previously introduced evaluation methodology was applied on the entire dataset to allow a coherent comparison with already reported results of other IQA methods (see Section 3.3). Considering the value of the threshold S, it can be set in between 20 and 60 without a visible drop in the prediction performance. Since ENMIQA exhibits a stable performance across the values of S, $S = 30$ used in experiments is justified.

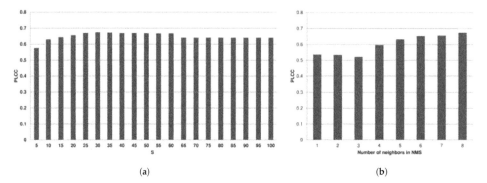

(a)　　　　　　　　　　　　　　　　　　(b)

Figure 5. Influence of the threshold S (**a**) and the number of neighboring pixels in the non-maximum suppression (NMS) (**b**) on the PLCC performance of ENMIQA.

The non-maximum suppression selects a pixel with the extreme value, taking into account its eight neighbors and the threshold t. Since a pixel has 8 neighbors, it is reasonable to use its full neighborhood (the size of 8). However, the suppression can be modified to accept a lesser number of neighboring pixels that are used to indicate the local extrema (see Equation (2)). Therefore, in Figure 5b, the impact of the number of neighbors on the PLCC results of ENMIQA is shown. Here, if the number of used neighbors while determining the local extrema is lower than 8, the performance of the method visibly deteriorates. Hence, the entire pixel neighborhood should be considered by ENMIQA with the NMS. Interestingly, even with a smaller neighborhood the approach still offers a promising performance.

4. Conclusions

In this work, a new BIQA measure for the evaluation of MR images is proposed. The method uses the non-maximum suppression with a sequence of thresholds to detect local intensity extrema in MR images. A relationship between the number of extrema and entropy is investigated. Consequently, a new measure is introduced and experimentally validated against ten representative BIQA techniques on a database that contains MR images assessed by a large group of experienced medical professionals. The experimental comparison reveals that ENMIQA outperforms the-state-of-the-art measures by a large margin in terms of four performance criteria, confirming its suitability for the quality prediction of MR images.

To facilitate the replicability of the reported findings, as well as the applicability of the measure, the Matlab code of ENMIQA and the dataset are available at http://marosz.kia.prz.edu.pl/ENMIQA.html.

Author Contributions: Conceptualization, R.O., M.O., M.B., A.B., and A.P.; methodology, M.O.; software, M.O., M.B., A.B., and A.P.; validation, R.O., M.O., M.B., A.B., and A.P.; investigation, R.O., M.O., M.B., A.B., and A.P.; writing and editing, R.O., M.O., M.B., A.B., and A.P. All authors have read and agreed to the published version of the manuscript.

Funding: This research received no external funding.

Conflicts of Interest: The authors declare no conflict of interest.

References

1. Hameed, M.H.; Umer, F.; Khan, F.R.; Pirani, S.; Yusuf, M. Assessment of the diagnostic quality of the digital display monitors at the dental clinics of a university hospital. *Inform. Med. Unlocked* **2018**, *11*, 83–86. doi:10.1016/j.imu.2018.02.002. [CrossRef]

2. Bielecka, M. Syntactic-geometric-fuzzy hierarchical classifier of contours with application to analysis of bone contours in X-ray images. *Appl. Soft Comput.* **2018**, *69*, 368–380. doi:10.1016/j.asoc.2018.04.038. [CrossRef]

3. Bielecka, M.; Bielecki, A.; Korkosz, M.; Skomorowski, M.; Wojciechowski, W.; Zieliński, B. Application of Shape Description Methodology to Hand Radiographs Interpretation. In *Computer Vision and Graphics*; Bolc, L., Tadeusiewicz, R., Chmielewski, L.J., Wojciechowski, K., Eds.; Springer: Berlin/Heidelberg, Germany, 2010; pp. 11–18.

4. Kustner, T.; Liebgott, A.; Mauch, L.; Martirosian, P.; Bamberg, F.; Nikolaou, K.; Yang, B.; Schick, F.; Gatidis, S. Automated reference-free detection of motion artifacts in magnetic resonance images. *Magn. Reson. Mater. Phys. Biol. Med.* **2018**, *31*, 243–256. [CrossRef]

5. Senel, L.K.; Kilic, T.; Gungor, A.; Kopanoglu, E.; Guven, H.E.; Saritas, E.U.; Koc, A.; Cukur, T. Statistically Segregated k-Space Sampling for Accelerating Multiple-Acquisition MRI. *IEEE Trans. Med Imaging* **2019**, *38*, 1701–1714. doi:10.1109/TMI.2019.2892378. [CrossRef] [PubMed]

6. Knoll, F.; Hammernik, K.; Kobler, E.; Pock, T.; Recht, M.P.; Sodickson, D.K. Assessment of the generalization of learned image reconstruction and the potential for transfer learning. *Magn. Reson. Med.* **2019**, *81*, 116–128. doi:10.1002/mrm.27355. [CrossRef] [PubMed]

7. Chow, L.S.; Rajagopal, H.; Paramesran, R. Correlation between subjective and objective assessment of magnetic resonance (MR) images. *Magn. Reson. Imaging* **2016**, *34*, 820–831. [CrossRef] [PubMed]

8. Chen, S.; Hu, P.; Gu, Y.; Pang, L.; Zhang, Z.; Zhang, Y.; Meng, X.; Cao, T.; Liu, X.; Fan, Z.; et al. Impact of patient comfort on diagnostic image quality during PET/MR exam: A quantitative survey study for clinical workflow management. *J. Appl. Clin. Med. Phys.* **2019**, *20*, 184–192. doi:10.1002/acm2.12664. [CrossRef]

9. Chandler, D.M. Seven Challenges in Image Quality Assessment: Past, Present, and Future Research. *ISRN Signal Process.* **2013**, *2013*, 905685. doi:10.1155/2013/905685. [CrossRef]

10. Baselice, F.; Ferraioli, G.; Pascazio, V. A 3D MRI denoising algorithm based on Bayesian theory. *Biomed. Eng. OnLine* **2017**, *16*, 25. doi:10.1186/s12938-017-0319-x. [CrossRef]

11. Wang, Z.; Bovik, A.C.; Sheikh, H.R.; Simoncelli, E.P. Image Quality Assessment: From Error Visibility to Structural Similarity. *IEEE Trans. Image Process.* **2004**, *13*, 600–612. doi:10.1109/tip.2003.819861. [CrossRef]

12. Jang, J.; Bang, K.; Jang, H.; Hwang, D.; Initiative, A.D.N. Quality evaluation of no-reference MR images using multidirectional filters and image statistics. *Magn. Reson. Med.* **2018**, *80*, 914–924. [CrossRef] [PubMed]

13. Chow, L.S.; Rajagopal, H. Modified-BRISQUE as no reference image quality assessment for structural MR images. *Magn. Reson. Imaging* **2017**, *43*, 74–87. doi:doi.org/10.1016/j.mri.2017.07.016. [CrossRef] [PubMed]

14. Welvaert, M.; Rosseel, Y. On the Definition of Signal-To-Noise Ratio and Contrast-To-Noise Ratio for fMRI Data. *PLOS ONE* **2013**, *8*, 1–10. doi:10.1371/journal.pone.0077089. [CrossRef] [PubMed]

15. Dietrich, O.; Raya, J.G.; Reeder, S.B.; Reiser, M.F.; Schoenberg, S.O. Measurement of signal-to-noise ratios in MR images: influence of multichannel coils, parallel imaging, and reconstruction filters. *J. Magn. Reson. Imaging* **2007**, *26*, 375–385. [CrossRef] [PubMed]

16. Yu, S.; Dai, G.; Wang, Z.; Li, L.; Wei, X.; Xie, Y. A consistency evaluation of signal-to-noise ratio in the quality assessment of human brain magnetic resonance images. *BMC Med. Imaging* **2018**, *18*, 17. doi:10.1186/s12880-018-0256-6. [CrossRef] [PubMed]

17. Mittal, A.; Moorthy, A.K.; Bovik, A.C. No-Reference Image Quality Assessment in the Spatial Domain. *IEEE Trans. Image Process.* **2012**, *21*, 4695–4708. doi:10.1109/TIP.2012.2214050. [CrossRef]

18. Zhang, Z.; Dai, G.; Liang, X.; Yu, S.; Li, L.; Xie, Y. Can Signal-to-Noise Ratio Perform as a Baseline Indicator for Medical Image Quality Assessment. *IEEE Access* **2018**, *6*, 11534–11543. doi:10.1109/ACCESS.2018.2796632. [CrossRef]

19. Sandilya, M.; Nirmala, S.R. Determination of reconstruction parameters in Compressed Sensing MRI using BRISQUE score. In Proceedings of the 2018 International Conference on Information, Communication, Engineering and Technology (ICICET), Pune, India, 29–31 August 2018; pp. 1–5. doi:10.1109/ICICET.2018.8533865. [CrossRef]

Entropy **2020**, *22*, 220

20. Osadebey, M.; Pedersen, M.; Arnold, D.; Wendel-Mitoraj, K. No-reference quality measure in brain MRI images using binary operations, texture and set analysis. *IET Image Process.* **2017**, *11*, 672–684. doi:10.1049/iet-ipr.2016.0560. [CrossRef]

21. Neubeck, A.; Van Gool, L. Efficient Non-Maximum Suppression. In Proceedings of the 18th International Conference on Pattern Recognition (ICPR'06), Hong Kong, China, 20–24 August 2006; pp. 850–855. doi:10.1109/ICPR.2006.479. [CrossRef]

22. Hosang, J.; Benenson, R.; Schiele, B. Learning non-maximum suppression. In Proceedings of the IEEE Conference on Computer Vision and Pattern Recognition, Honolulu, HI, USA, 21–26 July 2017; pp. 4507–4515.

23. Shannon, C.E. A Mathematical Theory of Communication. *Bell Syst. Tech. J.* **1948**, *27*, 379–423. doi:10.1002/j.1538-7305.1948.tb01338.x. [CrossRef]

24. Jiménez-García, J.; Romero-Oraá, R.; García, M.; López-Gálvez, M.I.; Hornero, R. Combination of Global Features for the Automatic Quality Assessment of Retinal Images. *Entropy* **2019**, *21*, 311. [CrossRef]

25. Śmieja, M.; Tabor, J. Entropy of the Mixture of Sources and Entropy Dimension. *IEEE Trans. Inf. Theory* **2012**, *58*, 2719–2728. doi:10.1109/TIT.2011.2181820. [CrossRef]

26. Deshmane, A.; Gulani, V.; Griswold, M.A.; Seiberlich, N. Parallel MR imaging. *J. Magn. Reson. Imaging* **2012**, *36*, 55–72. doi:10.1002/jmri.23639. [CrossRef] [PubMed]

27. Breuer, F.A.; Kellman, P.; Griswold, M.A.; Jakob, P.M. Dynamic autocalibrated parallel imaging using temporal GRAPPA (TGRAPPA). *Magn. Reson. Med.* **2005**, *53*, 981–985. doi:10.1002/mrm.20430. [CrossRef] [PubMed]

28. Sheikh, H.R.; Sabir, M.F.; Bovik, A.C. A Statistical Evaluation of Recent Full Reference Image Quality Assessment Algorithms. *IEEE Trans. Image Process.* **2006**, *15*, 3440–3451. doi:10.1109/tip.2006.881959. [CrossRef] [PubMed]

29. Min, X.; Gu, K.; Zhai, G.; Liu, J.; Yang, X.; Chen, C.W. Blind Quality Assessment Based on Pseudo-Reference Image. *IEEE Trans. Mult.* **2018**, *20*, 2049–2062. doi:10.1109/TMM.2017.2788206. [CrossRef]

30. Zhang, L.; Zhang, L.; Bovik, A.C. A Feature-Enriched Completely Blind Image Quality Evaluator. *IEEE Trans. Image Process.* **2015**, *24*, 2579–2591. doi:10.1109/TIP.2015.2426416. [CrossRef]

31. Oszust, M. No-Reference quality assessment of noisy images with local features and visual saliency models. *Inf. Sci.* **2019**, *482*, 334–349. doi:doi.org/10.1016/j.ins.2019.01.034. [CrossRef]

32. Gu, K.; Zhai, G.; Yang, X.; Zhang, W. Hybrid No-Reference Quality Metric for Singly and Multiply Distorted Images. *IEEE Trans. Broadcast.* **2014**, *60*, 555–567. doi:10.1109/TBC.2014.2344471. [CrossRef]

33. Zhu, X.; Milanfar, P. Automatic Parameter Selection for Denoising Algorithms Using a No-Reference Measure of Image Content. *IEEE Trans. Image Process.* **2010**, *19*, 3116–3132. doi:10.1109/TIP.2010.2052820. [CrossRef]

34. Liu, L.; Liu, B.; Huang, H.; Bovik, A.C. No-reference image quality assessment based on spatial and spectral entropies. *Signal Process. Image Commun.* **2014**, *29*, 856–863. doi:10.1016/j.image.2014.06.006. [CrossRef]

35. Leclaire, A.; Moisan, L. No-Reference Image Quality Assessment and Blind Deblurring with Sharpness Metrics Exploiting Fourier Phase Information. *J. Math. Imaging Vis.* **2015**, *52*, 145–172. doi:10.1007/s10851-015-0560-5. [CrossRef]

36. Ma, K.; Liu, W.; Zhang, K.; Duanmu, Z.; Wang, Z.; Zuo, W. End-to-End Blind Image Quality Assessment Using Deep Neural Networks. *IEEE Trans. Image Process.* **2018**, *27*, 1202–1213. doi:10.1109/TIP.2017.2774045. [CrossRef] [PubMed]

37. Ma, K.; Liu, W.; Liu, T.; Wang, Z.; Tao, D. dipIQ: Blind Image Quality Assessment by Learning-to-Rank Discriminable Image Pairs. *IEEE Trans. Image Process.* **2017**, *26*, 3951–3964. [CrossRef] [PubMed]

38. Cardenas-Blanco, A.; Tejos, C.; Irarrazaval, P.; Cameron, I. Noise in magnitude magnetic resonance images. *Concepts Magn. Reson. Part A* **2008**, *32A*, 409–416. doi:10.1002/cmr.a.20124. [CrossRef]

39. Jarque, C.M.; Bera, A.K. Efficient tests for normality, homoscedasticity and serial independence of regression residuals. *Econ. Lett.* **1980**, *6*, 255–259. doi:10.1016/0165-1765(80)90024-5. [CrossRef]

© 2020 by the authors. Licensee MDPI, Basel, Switzerland. This article is an open access article distributed under the terms and conditions of the Creative Commons Attribution (CC BY) license (http://creativecommons.org/licenses/by/4.0/).

Article

Entropy Based Data Expansion Method for Blind Image Quality Assessment

Xiaodi Guan [1,2], Lijun He [1], Mengyue Li [1] and Fan Li [1,*]

[1] School of Information and Communications Engineering, Xi'an Jiaotong University, Xi'an 710049, China; gxd1997@stu.xjtu.edu.cn (X.G.); lijunhe@mail.xjtu.edu.cn (L.H.); lmy1882923@stu.xjtu.edu.cn (M.L.)

[2] Guangdong Xi'an Jiaotong University Academy, Foshan 528300, China

* Correspondence: lifan@mail.xjtu.edu.cn

Received: 20 November 2019; Accepted: 30 December 2019; Published: 31 December 2019

Abstract: Image quality assessment (IQA) is a fundamental technology for image applications that can help correct low-quality images during the capture process. The ability to expand distorted images and create human visual system (HVS)-aware labels for training is the key to performing IQA tasks using deep neural networks (DNNs), and image quality is highly sensitive to changes in entropy. Therefore, a new data expansion method based on entropy and guided by saliency and distortion is proposed in this paper. We introduce saliency into a large-scale expansion strategy for the first time. We regionally add distortion to a set of original images to obtain a distorted image database and label the distorted images using entropy. The careful design of the distorted images and the entropy-based labels fully reflects the influences of both saliency and distortion on quality. The expanded database plays an important role in the application of a DNN for IQA. Experimental results on IQA databases demonstrate the effectiveness of the expansion method, and the network's prediction effect on the IQA databases is found to be improved compared with its predecessor algorithm. Therefore, we conclude that a data expansion approach that fully reflects HVS-aware quality factors is beneficial for IQA. This study presents a novel method for incorporating saliency into IQA, namely, representing it as regional distortion.

Keywords: deep neural network; entropy; data expansion; blind image quality assessment; saliency and distortion; human visual system; declining quality

1. Introduction

With the current state of development of multimedia technology, a large number of videos and images are being generated and processed every day, which are often subject to quality degradation. As a fundamental technology for various image applications, image quality assessment (IQA) has always been an important issue. The aim of IQA is to automatically estimate image quality to assist in the handling of low-quality images during the capture process. IQA methods can be divided into three major classes, namely, full-reference IQA (FR-IQA) [1,2], reduced-reference IQA (RR-IQA) [3], and no-reference IQA (NR-IQA), based on whether reference images are available. In most cases, no reference version of a distorted image is available; consequently, it is both more realistic and increasingly important to develop an NR-IQA model that can be widely applied [4]. NR-IQA models are also called blind IQA (BIQA) models. Notably, deep neural networks (DNNs) have performed well in many computer vision tasks [5–8], which encouraged researchers to use the formidable feature representation power of DNNs to perform end-to-end optimized BIQA, an approach called DNN-based BIQA. These methods use some prior knowledge from the IQA domain, such as the relationship among entropy, distortion and image quality, to attempt to solve IQA tasks using the powerful learning ability of neural networks. Accordingly, there is a strong need for DNN-based BIQA models in various cases where image quality is crucial.

However, attempts to use DNNs for the BIQA task were limited due to the conflicting characteristics of DNNs and IQA [9]. DNNs require massive amounts of training data to comprehensively learn the relationships between image data and score labels; however, classical IQA databases are much smaller than the computer vision datasets available for deep learning. An IQA database is composed of a series of distorted images and corresponding subjective score labels. Because obtaining a large number of reliable human-subjective labels is a time-consuming process, the construction of IQA databases requires many volunteers and complex, long-term experiments. Therefore, expanding the available number of distorted image samples and labels that fully reflect human visual system (HVS)-aware quality factors for training is a key problem for DNN-based BIQA.

Based on the baseline datasets considered for expansion, the current DNN-based BIQA methods can be divided into two general approaches. The first approach is to use the images in an existing IQA dataset as the parent samples; we call this approach small-scale expansion. In this case, the goal of expansion is achieved by dividing the distorted images from the IQA dataset into small patches and assigning to each patch a separate quality label that conforms to human visual perception. The second strategy is to expand the number of distorted images by using another, non-IQA dataset as the parent dataset; we call this approach large-scale expansion. In this approach, nondistorted images from outside the IQA dataset are first selected; then, distortion is added to these images based on the types of distortion present in the IQA dataset to construct new distorted images on a large scale. Then, the newly generated distorted images are simply labeled with different values that reflect their ranking in terms of human visual perception quality to achieve the goal of expansion.

The small-scale expansion strategy relies on division. The initial algorithm [10] assigns the score labels of the parent images to the corresponding small patches and then uses a shallow CNN to perform end-to-end optimization. The small patches and their labels are input directly to the network during training, and the predicted scores for all the small patches are averaged to obtain the overall image score during prediction. However, this type of expansion is not strictly consistent with the principles of the HVS. Previous studies have shown that saliency exerts a crucial influence on human-perceived quality; thus, saliency should be considered in IQA together with distortion and content [11–13]. These studies have shown that the human eye tends to focus on certain regions when assessing an image's visual quality and that different regions have different influences on the perceived quality of a distorted image. Therefore, it is not appropriate for all patches from a single image to be assigned identical quality labels because local perceptual quality is not always consistent with global perceptual quality [14,15]: the uneven spatial distortion distribution will result in varying local scores for different image patches. Thus, many works have attempted to consider this aspect of the problem. The saliency factor was first considered in DNN-based BIQA algorithms. The authors of [16,17] still assigned identical initial quality labels to the small patches, but the predicted scores for all small patches were eventually multiplied by different weights based on their saliency to obtain the overall image scores, thereby weakening the influence of patches with inaccurate labels in nonsalient regions on the overall image quality. In [18,19], strategies based on proxy quality scores [18] and an objective error map [19] were used to further improve the accuracy of the labels for different patches. All these strategies further increased the accuracy of this type of expansion and led to better predictions, confirming that the joint consideration of the influence of saliency and distortion on image quality more comprehensively reflects HVS-related perceptual factors. However, division strategies have obvious inherent drawbacks. First, because expansion is applied only to the existing distorted images in the IQA database (the expansion parent), the diversity of the training sample contents is not increased. The different levels of quality influenced by saliency and distortion must already be present in the training dataset, but it is difficult to claim that a typical small IQA database can comprehensively represent the influence of HVS factors on quality; hence, such methods are easily susceptible to overfitting. Second, there is a tradeoff between the extent of expansion achieved and the patch size. When the patch size is too small, each individual patch will no longer contain sufficient distorted semantic information for IQA, thus inevitably destroying the correlations between image patches. In contrast, a large patch size results in smaller-scale expansion, meaning that only a shallow network can be used for training. Moreover, the generated saliency-based patch weights will show large deviations from the real salient regions.

To avoid dividing the images in the IQA database while still not requiring human labeling, the large-scale expansion strategy instead involves creating new distorted images by adding distortion to a large number of high-definition images obtained from outside the IQA database. Separate values that reflect the overall quality level are assigned to each distorted image obtained from each original parent image. Because the labels of the newly generated images are not direct quality scores, the expanded database is used only to pretrain the DNN, which is then fine-tuned on the IQA database. This approach alleviates the training pressure placed on the small IQA dataset and successfully avoids the drawbacks of division encountered in the small-scale strategy because the number of labeled training images is expanded by a large amount, increasing the diversity of the training sample content. Such unrestricted, large-scale expansion also makes it possible to use deeper networks; in fact, a deep model pretrained on an image recognition task could also be used to further enhance the effect. This large-scale expansion approach was developed over the past two years, and it showed a much better effect than small-scale expansion algorithms. However, large-scale expansion also has some significant shortcomings. Although the newly added images with quality-level labels are consistent with human perception, they reflect only HVS-aware quality factors; distortion and the joint effects of saliency and distortion are not considered. Moreover, large-scale expanded datasets are typically prepared to assist in specific IQA tasks. The more similar the extended pretraining dataset is to the original IQA dataset for the target task, the more effectively it can support the IQA task. In this case, a "similar" dataset is an expanded dataset that fully reflects the influences of the HVS-related perceptual factors (saliency and distortion) as embodied in the IQA task of interest. The current algorithms [15,20] that use this approach mainly follow the lead of RankIQA [20]: they generate a series of distorted image versions by adding different levels of distortion to each original parent image (with uniform distortion for each image region) and assign different numerical-valued labels to them to reflect the overall quality level. Consequently, the quality degradation of each distorted image depends only on the level of the distortion added to the whole image. As a result, HVS-aware quality factors are not well embedded into the expanded database. Using this type of extended dataset to pretrain the network will simply cause it to learn that a greater level of distortion leads to greater quality degradation; the network will be unable to discern that salient regions are more important than nonsalient regions and that different regions contribute differently to the overall image quality. Obviously, this type of expansion does not result in an ideal pretraining dataset for IQA.

In this paper, we introduce saliency into the large-scale expansion method, with the aim of constructing DNN-based BIQA models that will be effective in various cases where image quality is crucial. The objective is to be able to automatically estimate image quality to assist in handling low-quality images during the capture process. Moreover, by virtue of the introduction of saliency, our proposed model can achieve better prediction accuracy for large-aperture images (with clear foregrounds and blurred backgrounds), which are currently popular. We propose a new approach for incorporating saliency into BIQA that is perfectly compatible with the large-scale data expansion approach to ensure the full consideration of HVS-related factors in the mapping process. Specifically, we introduce saliency factors through regional distortion, thereby conveniently combining saliency and distortion factors during the expansion of each image to generate a series of distorted image versions. Then, we use the information entropy to rank these images based on their quality to complete the labeling process. By constructing a more efficient pretraining tool for DNN-based BIQA, we improve the prediction performance of the final model. We use our generated large-scale dataset to pretrain a DNN (VGG-16) and then use the original small IQA dataset to fine-tune the pretrained model. Extensive experimental results obtained by applying the final model to four IQA databases demonstrate that compared with existing BIQA models, our proposed BIQA method achieves state-of-the-art performance, and it is effective on both synthetic and authentic distorted images. Therefore, we conclude that a data expansion approach that fully reflects HVS-aware quality factors is beneficial for IQA. This study presents a novel method for incorporating saliency into IQA tasks, namely, representing it as regional distortion.

Our contributions can be summarized as follows: (1) We introduce saliency into the large-scale expansion method in a manner that fully reflects the influence of HVS-aware factors on image quality, representing a new means of considering saliency in IQA. With the incorporation of the saliency factor, the proposed data expansion method overcomes the main drawback of its predecessor algorithm, RankIQA [20], which enables the learning of only the quality decline caused by the overall distortion level. Our approach enables the construction of an efficient pretraining dataset for DNN-based BIQA tasks and results in improved prediction accuracy compared to previous BIQA methods. (2) We propose a new data expansion method that fully reflects HVS-aware factors by generating distorted images based on both distortion and saliency and assigning labels based on entropy. This method successfully embeds the joint influence of saliency and distortion into a large-scale expanded distorted image dataset.

The remainder of this paper is organized as follows. Section 2 describes the important factors that affect image quality and explores how those factors affect human judgments of image quality. Section 3 introduces the proposed expansion method and describes its use in IQA in detail. Section 4 reports the experimental results and presents corresponding discussions. Finally, Section 5 offers conclusions.

2. Exploration of Functional HVS Aspects for Image Quality

As stated above, the main requirement for the expanded dataset is that it should be as similar as possible to the original IQA dataset. Therefore, to identify some features of good BIQA model design, we analyzed the influence of the three functional aspects of the HVS on human visual IQA. To improve the reliability of the results, all images considered below were taken from the IQA dataset, which consists of distorted images and subjective quality score labels that are often used as criteria based on the human visual perception mechanism.

2.1. The Influence of Saliency on Image Quality

As previously discussed, saliency is an important factor that influences image quality because when people observe an image, they tend to focus on the regions that contain the most relevant information in the visual scene. Previous HVS evaluation experiments with eye trackers [11–13,21] showed that the visual importance of different local regions varies when humans are estimating the visual quality of a whole image.

To conduct a detailed analysis of the substantial impact of saliency on quality, we analyzed several images with different visual quality scores. The images shown in panels (a) and (b) of Figure 1 are derived from the LIVE Challenge dataset [22], an authentic distortion database in which the labels represent the mean opinion score (MOS) and take values in the range of [0, 100], with higher values indicating better quality; in this dataset, multiple nonuniform distortions typically appear in each image. Images (a) and (b) contain identical levels of blurring in the salient and nonsalient regions, respectively. However, image (b) has a much better visual quality label in the database than image (a) does. These examples show that the level of distortion in the salient regions of an image is more likely to determine the final quality rank than is the level of distortion in nonsalient regions. Humans can more easily perceive distortions in the salient regions and thus assign lower quality scores to images with such distortions. When the foreground area of an image is distorted, the visual quality score of the whole image immediately decreases, regardless of whether the background region is distorted. Thus, the quality in the salient regions is closely related to the final quality score of the whole image.

(a)   (b)

Figure 1. Distorted images from the LIVEC dataset: (**a**) MOS = 50.1882; (**b**) MOS = 72.4574.

By contrast, the effect in nonsalient regions is the opposite. As shown in Figure 1, the low level of distortion in the nonsalient regions of image (a) does not prevent the quality degradation caused by distortions in the salient region. This phenomenon is widespread, especially in synthetic distortion databases. The images shown in panels (a)–(d) of Figure 2 are from the LIVE [23] dataset, which is a synthetic distortion database that contains 29 reference images and 779 distorted images derived from them. The corresponding difference mean opinion score (DMOS) labels for these images, representing subjective quality scores, lie in the range of [0, 100], with a lower value indicating better visual quality. Images (a)–(d) contain no distortion in the salient regions and exhibit varying distortion intensities in the nonsalient regions. However, all of these images have the highest possible DMOS value of 0. This indicates that distortion in nonsalient regions attracts little attention and has little effect on the quality of the entire image.

(a)   (b)   (c)   (d)

Figure 2. Reference images (DMOS = 0) from the LIVE database: (**a**) painthouse; (**b**) caps; (**c**) monarch; (**d**) stream.

2.2. The Influence of Content on Image Quality

We will now discuss the crucial impact of content on IQA. We present detailed figures for observation. Among the existing IQA databases, LIVE is the most commonly used. Its 29 reference images were distorted using five types of distortion: JPEG2000 (JP2K), JPEG, white noise in the RGB components (WN), Gaussian blur (GB), and transmission errors in the JPEG2000 bit stream using a fast-fading Rayleigh channel model (FF). Moreover, different levels of distortion were added to each reference image using the same distortion type to ensure that the quality of the distorted images of the same distortion type covers the entire quality range. To draw our conclusions, we selected 4 of the 29 reference images ("painthouse", "caps", "monarch" and "stream", as shown in Figure 2) as well as the distorted images derived from these 4 reference images, as shown in Figure 3. Only 4 of the distortion types (JP2K, JPEG, WN, and GB), all of which are commonly used in IQA databases, are considered here. For each distortion type, we observed the distortion parameter and the DMOS label for each distorted image derived from the 4 reference images. We first generated a scatter plot showing the distortion parameters and the corresponding DMOS quality labels. Then, for each reference image, we artificially fit a smooth curve to these scatter points to observe the trend of variation relating the image quality and the perceived level of distortion.

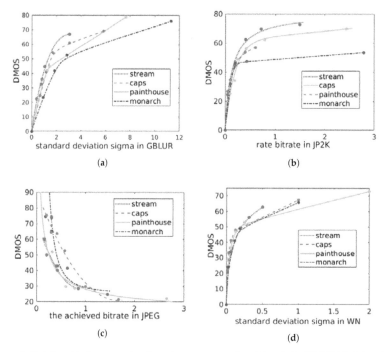

Figure 3. Relationship between the distortion parameter (x-axis) and the DMOS label (y-axis) for different distortion types. Each *x*-axis represents the distortion parameter for the corresponding distortion type. Each scatter point represents one sample in the LIVE dataset [23]. The scatter points representing the distorted images derived from the same reference image were separately fitted to a smooth curve. Different colors indicate different images: (**a**) GB; (**b**) JP2K; (**c**) JPEG; (**d**) WN.

First, we observe that there are no rating biases associated with the reference image contents; all of the reference images, each with different contents, are assigned the same quality score (DMOS = 0) in the public IQA database. This phenomenon is clearly reflected in Figure 3; the starting points of all curves in the same figure are consistent (because the x-axis of the figure for JPEG distortion represents the achieved bitrate, this characteristic is not reflected in this figure). Second, we find that as the level of distortion added to the image increases, images with different contents have different quality degradation curves. In other words, different image contents have different capacities for hiding distortion. For example, image (d) in Figure 2 has a dark content, and the slight distortion in it is unacceptable to the human eye. However, a further increase in the distortion level does not strongly affect an observer's understanding of the content of "monarch"; therefore, the rate of quality degradation is slow.

2.3. The Influence of Distortion on Image Quality

The degree of distortion seriously affects the image quality. This conclusion is obvious and beyond doubt for all four distortion types displayed in Figure 3. For the same type of distortion, all images with different contents exhibit the same behavior: when the distortion added to the whole image is uniform, the higher the level of distortion (distinguishable by the human eye) added to the whole image is, the lower the image quality is. This means that a negative correlation exists between the level of distortion and the image quality. The same conclusion can be drawn from Figure 3.

2.4. Conclusion: The Joint Influence of Saliency and Distortion on Image Quality Given the Same Type of Distortion for Each Image

The above analysis provides the following inspiration. When a DNN learns a mapping between distorted images and quality scores, it is actually learning different curves for different image contents. This suggests that it should be possible to construct an expanded training database to improve the DNN-based prediction performance on BIQA tasks by simply adding synthetic distortions to a baseline database containing a large number of original images, exactly as was done to construct the LIVE database. As discussed above, the original external content is not subject to any rating biases; therefore, we should find a parent database that consists of multiple types of images with no distortion and then distort them using several distortion types. For each type of distortion, we could generate a series of distorted images of different qualities for each original image such that the generated distorted images would reflect the joint effects of saliency and distortion on image quality. Then, we could apply a reasonable two-stage training method. First, pairs of images of different qualities from the expanded dataset would be sent to the DNN to pretrain the network to learn the quality ranking of distorted images with the same content. Then, the smaller original IQA database could be used to fine-tune the DNN, which would already be trained to perform quality ranking, to refine the mapping of distorted images to quality scores for each type of content. Then, the DNN should be able to output high-precision score prediction results. The authors of the RankIQA algorithm [20] accomplished this task using four distortion types (JP2K, JPEG, WN, and GB) because these four types can be implemented by means of MATLAB functions and frequently appear in IQA datasets; they generated a series of distorted images from the contents of each original image separately. However, in their expanded dataset, the degradation of image quality with a given distortion type for each parent image depends only on the overall distortion level; the joint influence of distortion and saliency on image quality is not reflected. Thus, if we could fully capture the influence of both saliency and distortion in the expanded dataset, the performance should improve.

3. Proposed Method

As mentioned above, our main goal is to construct a newly expanded dataset to support DNN-based BIQA tasks. We introduce saliency into the large-scale expansion strategy for the first time by creating distorted images based on the joint consideration of both saliency and distortion. Finally, we label the images based on the information entropy. The degradation of image quality in our new expanded dataset not only is related to the distortion level (as in RankIQA [20]) but also fully reflects the joint influence of distortion and saliency on image quality. We use this large-scale expanded dataset to pretrain a DNN and then use the original small IQA dataset to fine-tune the pretrained DNN. After fine-tuning, we obtain the final BIQA model. The flow chart of our proposed method is shown in Figure 4.

In this section, we present a detailed description of our method, which is divided into two main stages: dataset expansion and the use of the expanded dataset. First, we introduce our novel method of incorporating saliency into the large-scale dataset expansion process for IQA. Then, we describe the dataset generation process: image expansion based on saliency and distortion and image labeling guided by the information entropy. Finally, we describe how the expanded dataset is used in the IQA task, which involves a two-step training process to ensure that the DNN fully learns how HVS-aware factors influence image quality.

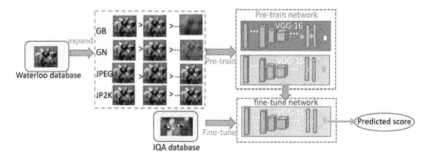

Figure 4. Pipeline of the proposed data expansion method for IQA. Based on the Waterloo database, we generate a large-scale expanded dataset, and this expanded dataset is then used to pretrain a double-branch network. Then, the original IQA dataset is used to fine-tune a single branch of the network to output quality scores.

3.1. The Usage of Saliency in IQA

The incorporation of saliency into the expansion procedure is a key step because we want to consciously capture the influences of both saliency and distortion when generating distorted images. Previous algorithms [16–19] introduced saliency into the IQA task by assigning different weights to different regions of a distorted image when predicting the final score. Such saliency usage is suitable for small-scale expansion but cannot be applied in the case of large-scale expansion. Moreover, there is no opportunity to add saliency factors to the existing distorted image versions generated for RankIQA (large-scale expansion), for which several images with different distortion intensities were created and labeled by quality rank. Because each label is a simple number that represents the overall quality level, using regional saliency weights is insufficient. Moreover, the salient regions in any given image may shift under different distortion levels; examples of this attentional shift based on distortion are shown in Figure 5. As the level of distortion increases, the salient areas also shift. Thus, we can see that differently distorted images with the same content should have different local saliency weight values. This saliency shift further increases the difficulty of adding saliency into the existing distorted images generated for RankIQA. Therefore, finding a new way to introduce saliency into the large-scale expansion process for IQA is crucial.

On the one hand, the characteristics of the large-scale expansion strategy are as follows: the time-consuming psychometric approach is not employed to obtain subjective score labels, and each distorted image derived from a given image by applying a given type of distortion has only a simple numerical label that represents its level of quality. On the other hand, Section 2.1 shows that the influences of salient and nonsalient regions on quality are quite different. Based on the two considerations above, we are inspired to introduce saliency into an expanded dataset in the form of regional distortion. We can generate multiple distorted images by adding distortion to high-resolution reference images. Among these distorted images, some will be subjected to global distortion of the original images, some will be distorted only in the salient regions of the reference images, and others will be distorted only in the nonsalient regions. Because the locality of the distortion (both regional and global) in the extended set of distorted images will be different, these images will have different perceptual qualities. Next, instead of asking volunteers to provide subjective scores, we can sort the distorted images based on their information entropy and assign simple numerical labels that represent their quality ranking. In this way, the combined effects of both saliency and distortion on quality will be reflected in the expanded dataset.

Figure 5. Saliency shift caused by different levels of blur. Two images with the same content but different distortion levels are shown on the left. The corresponding saliency maps are shown on the right. (**a**) "painthouse" with low level's distortion; (**b**) the saliency map of (**a**); (**c**) "painthouse" with high level's distortion; (**d**) the saliency map of (**c**).

To implement the approach proposed above, we performed two preparatory steps. First, we needed to choose a saliency model. From among the many possible saliency models, we selected [24] because it emphasizes the identification of the entire salient area. Second, we needed to establish a measure of how the impact factor affects the quality (as discussed in Section 2). In addition to the information entropy, this will be another important measure for guiding the image generation and labeling processes during our expansion procedure. Based on these two preparatory steps, we introduce the details of our expansion method below.

3.2. Generating Images for the Expansion Dataset

We selected the Waterloo database [25], which includes a large number of high-resolution images (4744), as the parent database to be used in the expansion process. Using MATLAB, we added distortion to these images to construct a large-scale expanded dataset containing a total of $4744 \times 4 \times 9$ distorted images. Here, the factor of 4 arises from the 4 types of distortion (JP2K, JPEG, WN, and GB) applied to each parent image; we adopted these four distortion types because they are found in most available IQA databases. The factor of 9 arises from the fact that for each distortion type, a total of nine distorted images of different qualities were generated, using a total of five distortion levels for each distortion type. We summarize this information in Table 1. Please note that because we used MATLAB to simulate the types of distortion present in the LIVE dataset, the distortion functions and distortion factors used may be different from those used in LIVE; therefore, the parameters in Table 1 are slightly different from those in Figure 3. Next, with the help of Figure 6, we will explain how we used the five distortion levels and different saliency models to generate nine distorted versions of each parent image.

Table 1. Important indicators and parameters involved in the expansion process. Distortion levels 1–5 are defined based on the distortion parameters used in the expansion procedure. The five distortion parameters presented for each distortion type are listed in order from level 1 to level 5.

Parent Database	Distortion Type	Distortion Level of 1–5 Involved	Expanded Numbers
	GB	Gaussian filter factor: 7, 15, 39, 91, 199.	9
Waterloo [25]	WN	Gaussian white noise: the mean is 0, and the variance is $2^{-10}, 2^{-7.5}, 2^{-5.5}, 2^{-3.5}, 2^{0}$.	9
	JP2K	Quality factor: normalized variance at 43, 12, 7, 4, 0.	9
	JPEG	compression ratio factors: 0.46, 0.16, 0.07, 0.04, 0.02	9

As an example, we chose one original parent image ("shrimp" from the Waterloo database), and the image shown in panel (b) is its saliency map, generated as described in [24]. Due to space constraints, only the nine distorted images generated using GB distortion are shown in Figure 6. Please note that nine corresponding distorted image versions were also generated for each of the other three distortion types from each original parent image. As Figure 6 shows, during the expansion procedure, we used the method introduced in [24] to extract the saliency map of each original parent image. Then, according to the saliency map, we defined the region with pixel values greater than 30 as the salient region and defined the remaining area as the nonsalient region. Each image was thus divided into two parts, the salient region and the nonsalient region. Then, we independently added different levels of distortion to these two regions of the original image and spliced the results to obtain a distorted image. The distortion levels applied to the salient and nonsalient regions to generate the nine distorted images are shown in the GB distortion level column (e.g., "level 0 + level 1" for image (c) means that this image was generated by adding GB distortion of level 0 to the salient region and GB distortion of level 1 to the nonsalient region of image (a)). The definitions of distortion levels 1–5 for each distortion type can be found in Table 1, and a level of 0 means no distortion.

Our expanded set of distorted images fully reflects the influence of HVS-aware quality factors. The nine distorted image versions generated from each parent image contain different levels of distortion across the entire image region, thus representing the influence of the overall distortion level on quality. In addition, some distorted images have different levels of distortion in the salient and nonsalient regions, thus representing the joint influence of saliency and distortion on quality. We ranked the nine distorted images of the same distortion type generated from each original image separately. The corresponding distorted image versions of decreasing quality can fully reflect the quality degradation caused by various HVS-aware factors.

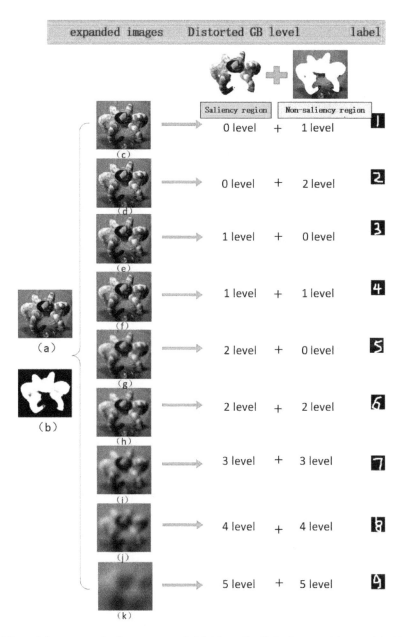

Figure 6. Image examples from our generated database. The first column contains an original image and its saliency map. The nine distorted images generated using GB distortion are displayed; each distortion was generated by adding particular levels of distortion to the salient and nonsalient regions of the original image. These distortion levels are displayed alongside the corresponding distorted images, and the corresponding image quality labels are given in the last column. Please note that the definitions of the salient regions and the distortion levels can be found in Section 3.2. Nine distorted images were generated in this way for all four distortion types, although only the results of GB expansion are shown in this figure. (**a**) the original version of "shrimp"; (**b**) the saliency map of (**a**); (**c–k**) the nine distorted versions of (**a**) under GB distortion type.

3.3. Entropy-Based Image Quality Ranking of the Expanded Dataset

After generating the distorted images, we next assigned quality labels to them. We know that each image in an IQA database will have an assigned quality score generated through a time-consuming psychometric experiment, an option that is unavailable to us, and is, in fact, unnecessary. Labels that simply reflect the quality ranking are sufficient to create the needed effect (as discussed in detail in Section 3.4). We refer to the nine distorted images of the same distortion type generated from the same parent image as a group; thus, there are a total of 4744 × 4 groups in our expanded dataset. We sorted the nine distorted images in each group separately by quality using the information entropy defined on the basis of Shannon's theorem because the information entropy is a measure that reflects the richness of the information contained in an image. The larger the information entropy of an image is, the richer its information and the better its quality. Moreover, the information entropy value is sensitive to image distortion and quality. Distortion in the salient region will lead to a significant reduction in the entropy value. Therefore, the information entropy is a suitable basis for our labeling procedure. The formula is as follows:

$$H = -\sum_{i=0}^{255} p_i * log p_i \tag{1}$$

where H represents the information entropy of the image and p_i represents the proportion of pixels with a grayscale value of i in the grayscale version of the image. The ordering of the information entropy values reflects the quality ranking of a group of images. We used this formula to calculate the information entropy of each of the nine distorted images in one group and ranked these nine images in order of their information entropy values. Accordingly, labels 1–9 were assigned to represent the image quality ranking. As mentioned above, there are a total of 4744 × 4 groups in our expanded dataset. We use letters c to k to denote the distorted image versions generated to compose each group (where c represents the distorted image generated by adding no distortion to the salient region and level 1 distortion to the nonsalient region of the original image). For each of these nine distorted image versions, we calculated the average entropy for the corresponding 4744 × 4 images, as shown in Table 2. The information entropy ranking results for most groups are consistent with the average order listed in Table 2. For each group, the labels for distorted images c to k range from 1 to 9, representing their sequentially decreasing quality. For example, for the nine images in Figure 6, their entropy sequentially decreases in the order in which they are displayed; the labels range from 1 to 9. Some groups also exist in which the information entropy order is different from the average order displayed in Table 2; in most such cases, the entropy values of images d (in which only the nonsalient region is distorted at level 2) and e (in which only the salient region is distorted at level 1) are reversed. However, we still sort image e below image d in quality to emphasize the importance of the salient region.

Table 2. The average information entropy values for all images corresponding to the same distorted image version across all groups. Each value is calculated as the average information entropy of the corresponding image version in each of the 4744 × 4 groups.

Distorted Version	c	d	e	f	g	h	i	j	k
H(entropy)	7.5129	7.5087	7.5079	7.5074	7.4920	7.4850	7.3964	7.1882	6.8059

These information entropy results are consistent with the previous conclusions regarding how HVS factors affect image quality. Images with only background distortion have higher quality indices than those with foreground distortion and whole-region distortion because distortion in only nonsignificant regions leads to only weak quality degradation due to the smaller entropy of nonsalient regions. Consequently, images with only foreground distortion and with overall distortion at the same level are of similar quality. In addition, as we discussed in Section 2, the quality of the salient regions is highly consistent with that of the whole image. Please note that for a few landscape images in the Waterloo database, which have no obvious salient regions, we treated the entire image as the salient region to avoid negative effects. Although no convincing quality score labels could be extracted for these images, we were still able to use the expanded database for our BIQA task by adopting a Siamese network and a corresponding training method, as discussed in the next section.

3.4. Using the Expansion Dataset for the IQA Task

Now, we will introduce the use of our new expanded dataset. Our training process consists of two steps: pretraining on the expanded dataset and fine-tuning on the IQA database. We trained a model based on VGG-16 [26], with the number of neurons in the output layer modified to 1. In our expanded database, for each original image, there are nine distorted images with corresponding labels from 1–9 that represent their quality ranking for each distortion type. We followed the training setup used by the authors of RankIQA [20]. During pretraining, to train the network on the quality ranking task, we used a double-branch version of VGG-16 (called a Siamese network) with shared parameters and a hinge loss. We show a schematic diagram of the pretraining process in Figure 7 and explain the training process in conjunction with the figure. Each input to the network consists of two images and two labels: a pair of images of different quality that are randomly selected from among the nine distorted images in one group. The image with the lower label (indicating higher quality) is always sent to the x_1 branch, and the other image is sent to the x_2 branch. When the outputs of the two branches are consistent with the order of the two labels, meaning that the network correctly ranks the two images by quality, the loss is 0. Otherwise, the loss is not 0, and the parameters will be adjusted (by decreasing the gradient of the higher branch and increasing the gradient of the lower branch) as follows:

$$\frac{\partial L}{\partial \theta} = \begin{cases} 0 & \text{if } (f(x_2;\theta) - f(x_1;\theta)) \leq 0, \\ \frac{\partial f(x_2;\theta)}{\partial \theta} - \frac{\partial f(x_1;\theta)}{\partial w} & \text{otherwise.} \end{cases} \tag{2}$$

where θ represents the network parameters. Thus, the loss function is continuously optimized by comparing the outputs of the two branches, and eventually, the training of the quality ranking model is complete. Because any two of the nine distorted images in a group may be paired to form the input, the network is efficiently forced to learn the joint influence of saliency and distortion on image quality. After pretraining, either network branch can produce a value for an input image (because the two branches share parameters), and the quality ranking of different input images will be reflected by the order of their corresponding output values.

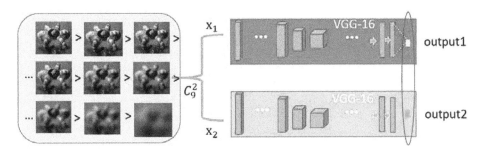

Figure 7. Pretraining process. The left side presents a series of distorted images of decreasing quality in the same group, and the right side presents a two-branch VGG-16 network where the two branches share parameters and a loss function.

We have found that this pretrained model is nearly identical to the IQA model and can effectively judge the effects of saliency and distortion on quality. However, the output of this network is not a direct image quality score. Only when multiple different images are input to obtain different output values does the order of these values reflect the order of the images in terms of quality. Therefore, to facilitate the comparison of our model with other BIQA models and transform the network output into a direct quality score, our method includes an IQA-database-based fine-tuning step. From the pretrained model, we extract one branch to obtain a single VGG-16 network and perform training on the original IQA dataset to complete the fine-tuning process. In each round of training, the input to the network is one image, and the corresponding quality score is the label in the IQA database; thus, the network learns an accurate mapping from distorted images to scores. Again following the approach of RankIQA, we use the sum of the squared errors as the loss function during fine-tuning.

4. Experiments and Results

4.1. Datasets and Evaluation Protocols

We used two types of datasets in our experiments: a non-IQA dataset used for the generation of the large-scale expanded pretraining dataset and several IQA datasets for performing fine-tuning. As the non-IQA dataset that was used to generate new distorted images, we adopted the Waterloo Exploration Database [25], which includes 4744 high-resolution images. The diversity of the image scenes and the clarity of the images make this database suitable for our purposes. As the IQA datasets, we used three synthetic IQA databases (i.e., databases containing synthetic distortions), namely, LIVE [23], CSIQ [27], and LIVE MD [28], and one authentic IQA database, namely, LIVE Challenge (LIVEC) [22], in which the distortion present in each image may be a complex combination of multiple types (such as camera shaking and overexposure) to test our model's generalization capability and its scope of application.

As the evaluation measures, we selected two metrics that are commonly used in the BIQA domain, namely, the Spearman rank order correlation coefficient (SROCC) and the Pearson linear correlation coefficient (PLCC). Given N input images, the SROCC is calculated as follows:

$$SROCC = 1 - \frac{6\sum_{i=0}^{N}(p_i - q_i)^2}{N(N^2 - 1)} \tag{3}$$

first, the N ground-truth scores and N predicted scores are ranked separately. Accordingly, p_i denotes the i-th value in the ordered list of predicted scores, and q_i denotes the i-th value in the ordered list of ground-truth scores. Therefore, the SROCC measures the monotonicity of the predictions. The PLCC is calculated as follows:

$$PLCC = \frac{\sum_{i=0}^{N}(u_i - \overline{u})(v_i - \overline{v})}{\sqrt{\sum_{i=0}^{N}(u_i - \overline{u})^2}\sqrt{\sum_{i=0}^{N}(v_i - \overline{v})^2}} \tag{4}$$

where u_i and v_i are the predicted score and ground-truth score, respectively, for the i-th image and u and v are the averages of the N predicted scores and the N ground-truth scores, respectively. Therefore, the PLCC measures the accuracy of the predictions. It can be seen from the formulas that the SROCC and PLCC both lie in the range of [0, 1] and that a larger value indicates a stronger correlation between the two columns of variables.

4.2. Experimental Setup

In Section 3.4, we introduced some information on the training process. Here, we provide more details and explain the reason for the selected experimental settings. To evaluate the performance improvement achieved by our algorithm in comparison with its predecessor algorithm RankIQA [20], we adopted the same network used in RankIQA—the VGG-16 architecture [26]—and changed the number of neurons in the output layer to 1 because our objective is not regression but rather a classification task. During both pretraining and fine-tuning, we randomly cropped a single subimage with dimensions of 224 × 224 from each training image to be used as the input in each epoch. During testing, we randomly sampled 30 224 × 224 subimages from one image and adopted the average of the corresponding 30 predicted outputs as the final score for this image. The quality ranking of the nine distorted images in a group was determined on the basis of an overall comparison of the full image region. Although a size of 224 × 224 is not sufficient to cover the entire image, it does cover more than 1/3 of the full image size; thus, this requirement does not destroy the quality ranking of the input images. Also for consistency with RankIQA [20], we adopted the Caffe [29] framework for training. The entire pretraining process consisted of 50,000 iterations, while the fine-tuning process consisted of 20,000 iterations. Additionally, L2 weight decay was used throughout the entire training process.

4.3. Performance Comparison

We compared the performance of our method on several IQA databases with the performance of various state-of-the-art FR-IQA and NR-IQA methods, including the FR-IQA methods PSNR, SSIM [1] and FSIMc [2]; the traditional NR-IQA methods BRISQUE [30], CORNIA [31], IL-NIQE [32] and FRISQUEE [33]; and the DNN-based NR-IQA methods CNN [10], RankIQA [20], BIECON [18], and DIQA [19]); as well as a DNN-based NR-IQA method that incorporates saliency, DIQaM [16]. We also compared our method with other well-known DNN models. Three networks (AlexNet [34], ResNet50 [35] and VGG-16, initialized from ImageNet) were also directly fine-tuned on each IQA database and treated as baselines. We used the final version of our DNN model, which was pretrained on the expanded dataset and then fine-tuned on the IQA dataset, to obtain image quality scores. The SROCC and PLCC were then calculated between the predicted quality scores (the output of our fine-tuned model) and the quality labels of the distorted images in the IQA database. The results are shown in Table 3, where the best three performance results are highlighted in bold. We divided the distorted images and their corresponding score labels into two groups, using 80% for training and 20% for testing. For all databases, the contents of the training and test sets did not overlap. This division process was repeated ten times. To avoid the influence of randomness on the evaluation of the prediction effect, the results of averaging the SROCC and PLCC scores over all ten runs are reported in Table 3.

Table 3. Comparison of the SROCC and PLCC scores on the four IQA datasets.

Types	Algorithms	LIVE		CSIQ		LIVE MD		LIVEC	
		SROCC	PLCC	SROCC	PLCC	SROCC	PLCC	SROCC	PLCC
FR	PSNR	0.876	0.872	0.806	0.800	0.725	0.815	N/A	N/A
	SSIM	0.913	0.945	0.834	0.861	0.845	0.882	N/A	N/A
	FSIMc	0.963	0.960	0.913	0.919	0.863	0.818	N/A	N/A
NR	CORNIA	0.942	0.943	0.714	0.781	0.900	0.915	0.618	0.662
	BRISQUE	0.939	0.942	0.775	0.817	0.897	0.921	0.607	0.645
	IL-NIQE	0.902	0.908	0.821	0.865	0.902	0.914	0.594	0.589
	FRIQUEE	0.948	0.962	0.669	0.704	0.925	0.940	0.720	0.720
	AlexNet	0.942	0.933	0.647	0.681	0.881	0.899	0.765	0.788
	VGG-16	0.952	0.949	0.762	0.814	0.884	0.900	0.753	0.794
	ResNet50	0.950	0.954	0.876	0.905	0.909	0.920	0.809	0.826
	CNN	0.956	0.953	0.683	0.754	0.933	0.927	0.516	0.536
	RANK	0.981	0.982	0.861	0.893	0.908	0.929	0.641	0.675
	BIECON	0.961	0.960	0.815	0.823	0.909	0.933	0.663	0.705
	DIQaM	0.960	0.972	0.869	0.894	0.906	0.931	0.606	0.601
	DIQA	0.970	0.972	0.844	0.880	0.920	0.933	0.703	0.704
	ours	0.978	0.983	0.893	0.916	0.935	0.947	0.818	0.837

Red: the highest. Blue: the second. Green: the third.

First, it can be clearly seen that our proposed model achieves the highest PLCC and SROCC scores on almost all tested databases, indicating that the proposed data expansion method for DNN-based BIQA has the best overall effect for both synthetic and authentic distortion databases and is largely consistent with the subjective judgments made by humans. Moreover, we can see that compared with its predecessor algorithm, RankIQA, our method achieves better results on all of the datasets listed, especially on LIVEC, because the introduction of the saliency factor causes the model to somewhat depend on the distortion type consistency between the expanded dataset and the original IQA database. The performance improvement on CSIQ is also considerable, possibly because the reference images in this dataset include many examples with clear foregrounds but blurred backgrounds.

Table 3 also concisely presents a comparison of the different types of methods that can be applied to IQA tasks. We can see that our method is superior to any of the classical NR methods due to the strong autonomous learning capability of CNNs. Among the deep learning methods, many models performed poorly on the LIVEC dataset because their training process requires reference images, which do not exist in the LIVEC dataset. By contrast, our fine-tuning process does not require reference images. Moreover, from the results of the directly fine-tuned baselines listed above, we can see not only that a good algorithm can perform well but also that the convolutional computing ability of a relatively deep and large network such as ResNet50, which has a total of 50 layers, is advantageous. However, our approach of introducing an expanded dataset makes it easy to use a smaller network, which incurs lower computational costs, to achieve results similar to those of a larger network.

Because the SROCC and PLCC results were averaged over ten runs in our experiment, we also present the standard deviations of these results to illustrate the stability of our model's predictive performance. Table 4 shows the standard deviations of the PLCC and SROCC scores over ten runs for RankIQA and our method. Because our method uses the same training procedure as RankIQA but differs in the use of the expanded dataset, RankIQA is a suitable choice for comparison. The other BIQA methods (whose specific experimental data are unavailable and which are also less suitable as methods for comparison) are not shown in Table 4. As Table 4 shows, the standard deviations of the results of our algorithm are smaller than those of RankIQA. This finding indicates that our algorithm achieves not only better prediction performance but also higher stability. Moreover, it is interesting to find that the performance on the LIVE MD dataset sometimes fluctuates across different divisions of the training and test datasets. These fluctuations may occur because some of the images contained in this dataset have unclear foregrounds, and when these images appear in the training set, they may induce a reduction in performance. Nevertheless, the average result is high. Therefore, the standard deviations of the prediction results further reflect the effectiveness of our proposed data expansion method.

Table 4. Standard deviations of the SROCC and PLCC scores for RankIQA and our method.

Standard Deviation	RankIQA		Our Method	
	SROCC	PLCC	SROCC	PLCC
LIVE	0.0106	0.0152	0.0057	0.0095
CSIQ	0.0131	0.0119	0.0125	0.0093
LIVE MD	0.0584	0.0454	0.0482	0.0303
LIVEC	0.0340	0.0309	0.0073	0.0106

4.4. Scatter Plots

To further visualize the consistency between our method's final predicted scores and the subjective human perception scores in the IQA databases, we show scatter plots of the scores predicted by our model (pretrained on the expanded dataset and fine-tuned on the corresponding IQA database) versus the ground-truth labels (DMOSs/MOSs) in Figure 8. This is another way of expressing the information in Table 3 and clearly shows the agreement between the predicted scores and the ground-truth values. Scatter plots of the results obtained on each of the four IQA databases (LIVE, CSIQ, LIVE MD and LIVEC) are shown in Figure 8. In these scatter plots, each point represents an image sample, the x-axis represents the DMOS/MOS scores associated with the samples in the dataset, and the y-axis represents the predicted quality scores obtained with our method. Because the four databases use different subjective score labels (i.e., LIVE and LIVE MD use DMOS scores in the range of [0, 100],

LIVEC uses MOS scores in the range of [0, 100], and CSIQ uses DMOS scores in the range of [0, 1]), there are two different x-axis ranges in Figure 8. For the CSIQ database, the x- and y-axis scales range from 0 to 1. For the other three databases, unified scales from 0 to 100 are used.

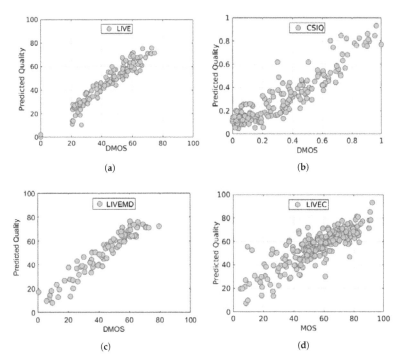

Figure 8. Scatter plots of the quality scores predicted by our method versus the ground-truth subjective DMOSs/MOSs for four datasets: (**a**) LIVE; (**b**) CSIQ; (**c**) LIVE MD; (**d**) LIVEC.

Figure 8 shows that the predicted quality scores output by our method have a monotonic relationship with the ground-truth labels, especially on the LIVE MD and CSIQ datasets. This plot also explains the high correlation coefficients achieved on these two datasets. For the LIVEC dataset, the sample points are not tightly clustered around the isopleth, and the correlation is more obvious when the MOS value is small, which is unexpected due to the multiple distortion types and diversity of scenes. Nevertheless, the sample points for LIVEC are roughly evenly distributed on both sides of the isopleth, which represents great progress compared with the other algorithms. Thus, we can conclude that our expanded dataset provides effective support for IQA and gives the final model the capability to precisely predict human-perceived image quality over a wide range of datasets.

4.5. Ablation Studies

The output of our pretrained model is not a direct image quality score. Only when multiple different images are input and their output values are obtained can the order of these values reflect the quality ranking of the images. To further evaluate the contribution of our expanded dataset and more accurately evaluate the contribution of the incorporation of saliency during the pretraining stage, we applied our pretrained model to various images and compared its predicted outputs to evaluate whether it could precisely rank images by quality. We compared our model with the pretrained RankIQA model, for which only images with whole-region distortion are considered during pretraining. Five image examples from the CSIQ database are shown in Figure 9, and in Table 5, we show the ground-truth label ranking, the order of the output of the pretrained RankIQA model, and the order of the output of our pretrained model for these images. We can see that RankIQA can accurately sort images (d) and (e), which contain whole-region distortion, but fails on images (a) and (c), which have clear foregrounds but blurred backgrounds accounting for nearly half of the entire image. By contrast, after only pretraining on the expanded dataset, our model fully reflects the joint influence of saliency and distortion and thus can perform well on images with both whole-region distortion and only local-region distortion, as is particularly evident from its performance on (a) and (c). We can see that the distortion level in the foreground in (c) is larger than that in (a), but the overall distortion of (c) is less than that of (a) when the entire image is considered. RankIQA [20] will tend to output a better quality score for (c) because the RankIQA model has only "distortion-level" awareness during training; it considers all regions equally in the final prediction. However, because our pretrained model has saliency awareness, it can sort the images correctly, as expected; therefore, our expanded database, which is based on both saliency and distortion and guided by information entropy, is more "similar" to the IQA database and can thus provide more effective assistance for the IQA task.

(a)	(b)	(c)	(d)	(e)

Figure 9. Distorted images from the CSIQ dataset. Their DMOS values increase from (a–e), representing a decrease in image quality.

Table 5. The ranking orders for several images as obtained with the pretrained models of RankIQA and our method. A larger value represents a worse image quality.

Algorithms	(a)	(b)	(c)	(d)	(e)
label ranking	1	2	3	4	5
ranking of RankIQA[20]	3	1	2	4	5
ranking of ours	1	2	3	4	5

4.6. Discussion

4.6.1. Studies on the Generation of Expanded Datasets from Different Parent Databases

In this section, we study the effects of using different parent databases for expansion. To confirm the effectiveness of our selected parent database, we performed tests using different databases as parents for data expansion, including the Waterloo database, which consists of images with rich scene contents, and MSRA-B [36], another classical database that contains 5000 original high-quality images and their saliency maps. The results can be seen in Table 6, where better performance results are highlighted in bold type. When we used MSRA-B as the baseline database to generate a series of distorted images for pretraining, the performance was reduced to a certain degree. This result was unexpected; however, it can be attributed to the insufficient richness of the MSRA-B dataset, which contains only images that are somewhat monotonous and have clear salient regions. By contrast, the complexity of the image content in the IQA datasets varies widely. Therefore, the more content-abundant Waterloo dataset was better suited to our requirements and resulted in higher performance.

Table 6. Comparison of the SROCC and PLCC scores of fine-tuned models pretrained on different expanded datasets.

Database	LIVE		CSIQ		LIVE MD		LIVEC	
	SROCC	PLCC	SROCC	PLCC	SROCC	PLCC	SROCC	PLCC
Waterloo	**0.978**	**0.983**	**0.893**	**0.916**	**0.935**	**0.947**	**0.818**	0.837
MSRA-B	0.975	0.972	0.871	0.896	0.907	0.923	**0.818**	**0.841**

4.6.2. Studies on Generating Different Numbers of Distorted Images for Each Distortion Type

As mentioned in Section 3, we refer to the nine distorted images of the same distortion type that are generated from the same original image as a group. Here, we study the influence of generating different numbers of distorted images per group. We tested three different designs for the distorted images generated in the expansion process. The number "7" refers to a design in which we removed the second (no distortion added to the salient region and level 2 distortion added to the nonsalient region of the original image) and fifth (level 2 distortion added to the salient region and no distortion added to the nonsalient region of the original image) distorted images in each group, resulting in only seven distorted images per group. This approach results in the generation of a total of 4744 × 4 × 7 distorted images. The number "9" refers to the group design represented in Figure 6. This approach results in the generation of a total of 4744 × 4 × 9 distorted images. The number "11" refers to a design in which we added two further distorted images to each group—a distorted image obtained by adding no distortion to the salient region and level 3 distortion to the nonsalient region of the original image—and another distorted image obtained by adding level 3 distortion to the salient region and no distortion to the nonsalient region. These additional images were inserted after the second image and after the sixth image, respectively, of the previously described 9-image group. The results are shown in Table 7, where we highlight the best performance results in bold type. We can see that when these different expanded databases are used for pretraining, as the number of distorted images per group increases, the performance initially increases and then decreases. The highest value is reached in case "9", possibly because of overfitting induced by the larger database, leading to reduced performance. When the number of distorted images per group increases past a certain threshold, the saliency effect becomes invalid and may lead to incorrect sorting. These findings indicate that the training process reaches saturation with the addition of two pairs of local-region distortions. Therefore, we elected to use nine distorted images per group, as shown in Figure 6.

Table 7. Performance differences caused by generating different numbers of distorted images per parent image for each distortion type.

Number	LIVE		CSIQ		LIVE MD		LIVEC	
	SROCC	PLCC	SROCC	PLCC	SROCC	PLCC	SROCC	PLCC
7	0.974	0.971	0.874	0.908	0.931	0.944	**0.825**	0.795
9	**0.978**	**0.983**	**0.893**	**0.916**	**0.935**	**0.947**	0.818	**0.837**
11	0.975	0.976	0.868	0.874	0.935	0.923	0.808	0.807

5. Conclusions

In this paper, we have proposed a new approach for considering saliency in IQA. In this approach, we expand a large-scale distorted image dataset with HVS-aware labels to assist in training a DNN model to more effectively address IQA tasks. The novel feature of the proposed method is that this is the first time that a saliency factor was incorporated into the large-scale expansion strategy by representing saliency the form of a regional distortion. Then, by using the information entropy to rank the generated images by quality, we ensure that the labels in the newly expanded dataset are highly consistent with human perception. The ability to fully consider the various factors affecting image

Entropy **2020**, *22*, 60

quality also solves the overfitting problem. Specifically, the introduction of saliency not only improves the applicability and versatility of the overall model but also overcomes the heavy reliance of the predecessor to our algorithm on the degree of similarity between the distortion types in the expanded dataset and the original IQA database. The final experimental results demonstrate the effectiveness of the proposed method, which outperforms other advanced BIQA methods on several IQA databases.

Author Contributions: All authors have contributed to this work significantly. X.G., and M.L. provided ideas, performed experiments and wrote the manuscript. L.H., and F.L. revised the manuscript. All authors have read and agreed to the published version of the manuscript.

Funding: This research was funded by National Science Foundation of China Project (No. 61701389 and No. 61671365), Joint Foundation of Ministry of Education of China (No. 6141A02022344), and Foshan Science and Technology Bureau Project (No. 2017AG100443).

Conflicts of Interest: The authours declare no conflicts of interest.

References

1. Lin, Z.; Li, H. SR-SIM: A fast and high performance IQA index based on spectral residual. In Proceedings of the 19th IEEE International Conference on Image Processing, Orlando, FL, USA, 30 September–3 October 2013.
2. Kim, J.; Lee, S. Deep learning of human visual sensitivity in image quality assessment framework. In Proceedings of the CVPR, Honolulu, HI, USA, 21–26 July 2017.
3. Golestaneh, S.-A.; Karam, L.-J. Reduced-reference quality assessment based on the entropy of DWT coefficients of locally weighted gradient magnitudes. *IEEE Trans. Image Process.* **2016**, *25*, 5293–5303. [CrossRef] [PubMed]
4. Li, F.; Fu, S.; Liu, Z.-Y.; Qian, X.-M. A cost-constrained video quality satisfaction study on mobile devices. *IEEE Trans. Multimed.* **2018**, *20*, 1154–1168. [CrossRef]
5. Fu, J.; Zheng, H.; Mei, T. Look closer to see better: Recurrent attention convolutional neural network for fine-grained image recognition. In Proceedings of the 2017 IEEE Conference on Computer Vision and Pattern Recognition (CVPR), Honolulu, HI, USA, 21–26 July 2017.
6. Dai, J.; Li, Y.; He, K.; Sun, J. R-FCN: Object detection via region-based fully convolutional networks. *IEEE Trans. Multimed.* **2015**, *17*, 2338–2344.
7. Xie, S.; Tu, Z. Holistically-nested edge detection. In Proceedings of the CVPR, Honolulu, HI, USA, 21–26 July 2017.
8. Rastgoo, R.; Kiani, K.; Escalera, S. Multi-Modal Deep Hand Sign Language Recognition in Still Images Using Restricted Boltzmann Machine. *Entropy* **2018**, *20*, 809. [CrossRef]
9. Yang, X.; Li, F.; Liu, H. A Survey of DNN Methods for Blind Image Quality Assessment. *IEEE Access* **2019**, *7*, 123788–123806. [CrossRef]
10. Kang, L.; Ye, P.; Li, Y.; Doermann, D. Convolutional neural networks for no-reference image quality assessment. In Proceedings of the CVPR, Columbus, OH, USA, 24–27 June 2014; pp. 1733–1740.
11. Vu, E.; Chandler, D.-M. Visual fixation patterns when judging image quality: Effects of distortion type, amount, and subject experience. In Proceedings of the 2008 IEEE Southwest Symposium on Image Analysis and Interpretation, Santa Fe, NM, USA, 24–26 March 2008; pp. 73–76.
12. Yang, X.; Li, F.; Zhang, W.; He, L. Blind Image Quality Assessment of Natural Scenes Based on Entropy Differences in the DCT Domain. *Entropy* **2018**, *20*, 885. [CrossRef]
13. Ren, Y.; Sun, L.; Wu, G.; Huang, W. DIBR-synthesized image quality assessment based on local entropy analysis. In Proceedings of the 2017 International Conference on the Frontiers and Advances in Data Science, Xi'an, China, 23–25 October 2017; pp. 86–90.
14. Deng, J.; Dong, W.; Socher, R.; Li, L.-J.; Li, K.; Li, F.-F. ImageNet: A large-scale hierarchical image database. In Proceedings of the CVPR, Miami, FL, USA, 20–26 June 2009; pp. 248–255.
15. Ma, K.; Liu, W.; Zhang, K. End-to-End blind image quality assessment using deep neural networks. *IEEE Trans. Image Process.* **2018**, *27*, 1202213. [CrossRef]
16. Bosse, S.; Maniry, D.; Müller, K.-R.; Wiegand, T.; Samek, W. Deep neural networks for no-reference and full-reference image quality assessment. *IEEE Trans. Image Process.* **2018**, *27*, 206–219. [CrossRef]

17. Cheng, Z.; Takeuchi, M.; Katto, J. A Pre-Saliency Map Based Blind Image Quality Assessment via Convolutional Neural Networks. In Proceedings of the 2017 IEEE International Symposium on Multimedia (ISM), Taichung, Taiwan, 11–13 December 2017; pp. 77–82.

18. Kim, J.; Lee, S. Fully deep blind image quality predictor. *IEEE J. Sel. Topics Signal Process.* **2017**, *11*, 206–220. [CrossRef]

19. Kim, J.; Nguyen, A.; Lee, S. Deep CNN-based blind image quality predictor. *IEEE Trans. Neural Netw. Learn. Syst.* **2019**, *30*, 11–24. [CrossRef] [PubMed]

20. Liu, X.; Weijer, J.; Bagdanov, A. RankIQA: Learning from ranking for no-reference image quality assessment. In Proceedings of the ICCV, Venice, Italy, 22–29 October 2017; pp. 1040–1049.

21. Liu, H.; Heynderickx, I. Visual attention in objective image quality assessment: Based on eye-tracking data. *IEEE Trans. Circuits Syst. Video Technol.* **2011**, *21*, 971–982.

22. Ghadiyaram, D.; Bovik, A.-C. Massive online crowdsourced study of subjective and objective picture quality. *IEEE Trans. Image Process.* **2016**, *25*, 372–387. [CrossRef] [PubMed]

23. Sheikh, H.; Sabir, M.; Bovik, A.-C. A statistical evaluation of recent full reference image quality assessment algorithms. *IEEE Trans. Image Process.* **2006**, *15*, 3440–3451. [CrossRef]

24. Hou, Q.; Cheng, M.-M.; Hu, X.; Borji, A.; Tu, Z.; Torr, P. Deeply supervised salient object detection with short connections. In Proceedings of the CVPR, Honolulu, HI, USA, 21–26 July 2017; pp. 5300–5309.

25. Ma, K.; Duanmu, Z.; Wu, Q.; Wang, Z.; Yong, J.; Li, H.; Zhang, L. Waterloo exploration database: New challenges for image quality assessment models. *IEEE Trans. Image Process.* **2017**, *26*, 1004–1016. [CrossRef]

26. Simonyan, K.; Zisserman, A. Very deep convolutional networks for large-scale image recognition. *arXiv* **2015**, arXiv:1409.1556.

27. Larson, E.-C.; Chandler, D.-M. Most apparent distortion: Full reference image quality assessment and the role of strategy. *J. Electron. Imag.* **2010**, *19*, 19–21.

28. Jayaraman, D.; Mittal, A.; Moorthy, A.-K.; Bovik, A.-C. Objective quality assessment of multiply distorted images. In Proceedings of the 2012 Conference Record of the Forty Sixth Asilomar Conference on Signals, Systems and Computers (ASILOMAR), Pacific Grove, CA, USA, 4–7 November 2012; pp. 1693–1697.

29. Jia, Y.; Shelhamer, E.; Donahue, J.; Karayev, S.; Long, J.; Girshick, R.; Guadarrama, S.; Darrell, T. Caffe: Convolutional architecture for fast feature embedding. In Proceedings of the ACM International Conference Multimedia, Orlando, FlL, USA, 3–7 November 2014; pp. 675–678.

30. Mittal, A.; Moorthy, A.; Bovik, A.-C. No-reference image quality assessment in the spatial domain. *IEEE Trans. Image Process.* **2012**, *21*, 4695–4708. [CrossRef]

31. Ye, P.; Kumar, J.; Kang, L.; Doermann, D. Unsupervised feature learning framework for no-reference image quality assessment. In Proceedings of the CVPR, Providence, RI, USA, 16–21 June 2012; pp. 1098–1105.

32. Mittal, A.; Soundararajan, R.; Bovik, A.-C. Making a 'completely blind' image quality analyzer. *IEEE Signal Process. Lett.* **2013**, *20*, 209–212. [CrossRef]

33. Ghadiyaram, D.; Bovik, A.-C. Perceptual quality prediction on authentically distorted images using a bag of features approach. *J. Vis.* **2017**, *17*, 32–58. [CrossRef]

34. Krizhevsky, A.; Sutskever, I.; Hinton, G.-E. ImageNet classification with deep convolutional neural networks. In Proceedings of the NIPS, Lake Tahoe, NV, USA, 3–6 December 2012; pp. 1097–1105.

35. He, K.; Zhang, X.; Ren, S.; Sun, J. Deep residual learning for image recognition. In Proceedings of the CVPR, Las Vegas, NV, USA, 27–30 June 2016; pp. 770–778.

36. Liu, T.; Yuan, Z.; Sun, J.; Wang, J.; Zheng, N.; Tang, X.; Shum, H.-Y. Learning to detect a salient object. *IEEE Trans. Pattern Anal. Mach. Intell.* **2011**, *33*, 353–367. [PubMed]

 © 2019 by the authors. Licensee MDPI, Basel, Switzerland. This article is an open access article distributed under the terms and conditions of the Creative Commons Attribution (CC BY) license (http://creativecommons.org/licenses/by/4.0/).

Article

BOOST: Medical Image Steganography Using Nuclear Spin Generator

Bozhidar Stoyanov and **Borislav Stoyanov** *

Konstantin Preslavsky University of Shumen, 9712 Shumen, Bulgaria; b.stoyanov@shu.bg
* Correspondence: borislav.stoyanov@shu.bg

Received: 12 March 2020; Accepted: 21 April 2020; Published: 26 April 2020

Abstract: In this study, we present a medical image stego hiding scheme using a nuclear spin generator system. Detailed theoretical and experimental analysis is provided on the proposed algorithm using histogram analysis, peak signal-to-noise ratio, key space calculation, and statistical package analysis. The provided results show good performance of the brand new medical image steganographic scheme.

Keywords: steganography; nuclear spin generator; medical image; peak signal-to-noise ratio; key space calculation

---

1. Introduction

In this century, with the rapid evolution of data processing and information technologies, web security instruments are becoming more and more relevant. Various health systems are constantly relocating into the cloud and mobile device space. A body of US national rules for the defence of certain medical information must be taken into account for secure communication [1,2]. Many technologies have been introduced in recent years for secure storage and transmission of medical records and information regarding patient identity, such as digital watermarking [3,4], image encryption [5–9], and steganography [10,11].

Nevertheless, most of those schemes depend on some form of cryptography. The aim of cryptography is to create and analyze protocols that prevent individuals or the public from reading private data. In cryptography, an encryption is the method of encoding data. This method converts the original representation of the data, known as input text, into an alternative form known as encrypted text. Only authorized parties can decrypt encrypted data back to input text and access the original data [12]. Unlike cryptography, steganography is the art and science of hiding in plain sight secret data without being detected inside an innocent objects, called containers, so that it can be safely transmitted on a public channel of communication [13,14]. Containers may have the form of video streams, audio records, and digital images.

Image steganography refers to the hiding of user data in an image file [15]. Medical image steganographic schemes play a significant function in contemporary therapeutic procedures. The digital security of medical records and patient data both during communication and at the storage location must be ensured [16]. For medical images, sensitive patient information is embedded as header details defined in the Digital Imaging and Communications in Medicine (DICOM) standard in the image files [17] and should be removed before network transmission.

The efficiency of the steganography methods can be calculated by the three valuable specifications: security, capacity, and visual undetectability [18,19].

Entropy **2020**, *22*, 501; doi:10.3390/e22050501   www.mdpi.com/journal/entropy

Numerous strategies are employed to conceal a variety of input data with respect to medical images. Because of the resistance of increasing statistical attacks, use of chaotic functions in steganography algorithms becomes more popular. Satish et al. [20] introduced Logistic map based spread spectrum image steganography. Jain and Lenka [19] used an asymmetric cryptographic system for secret information hiding in brain images. Jain and Kumar [21] presented a medical record steganography method based on Rivest–Shamir–Adleman cryptosystem and decision tree for data inclusion. Jain et al. [22] described an improved medical image steganographic methodology using a public key cryptosystem and linear feedback shift register (LFSR), and dynamically picked diagonal blocks. Ambika and Biradar [23] proposed a novel technique to hide data in medical images. The scheme uses two level discrete wavelet transformation with a pixel selection by Elephant Herding–Monarch Butterfly algorithm. By using 1D chaotic function, medical image stego algorithm is presented in [24].

The steganography techniques provide the necessary security and privacy in data transmission. In our humble opinion, the main contributions of our work can be summarized as follows:

- We present novel algorithm for pseudorandom byte output using nuclear spin generator (NSG), which has acceptable statistical properties.
- We apply the pseudorandom algorithm to a novel medical image steganography scheme.
- We examine the proposed method, and the data show that it has excellent peak signal-to-noise ratio, strong collision resistance, and desirable security properties that can withstand most common theoretical and statistical attacks.

In Section 2, we present a novel pseudorandom byte output method based on two nuclear spin generators. In Section 3, we introduce the novel medical image steganography algorithm BOOST and complete steganalysis is given. Finally, the article is concluded in Section 4.

2. Pseudorandom Byte Output Algorithm Using Nuclear Spin Generator

Pseudorandom generators are basic primitives used in cryptography algorithms but in our case we apply the random properties of pseudorandom byte generator to steganography algorithm. Pseudorandom generators are software realized methods for extracting sequences of random values.

2.1. Proposed Pseudorandom Byte Output Algorithm

The nuclear spin generator is a high-frequency oscillator which generates and controls the oscillations of the motion of a nuclear magnetization vector in a magnetic field. This system exhibits a large variety of regular and dynamic motions [25–29]. The nuclear spin generator was first described by Sherman [30]. The typical NSG is nonlinear three-dimensional dynamical system given by

$$
\begin{aligned}
\dot{x}(t) &= -\beta x + y \\
\dot{y}(t) &= -x - \beta y(1 - kz) \\
\dot{z}(t) &= \beta(\alpha(1 - z) - ky^2),
\end{aligned}
\tag{1}
$$

where x, y, and z are the components of the nuclear magnetization vector in the X, Y, and Z directions, respectively, and α, β, and k are positive parameters. The nuclear spin generator with initial values $(x, y, z) = (0.12, 0.25, 0.0032)$ and parameters equal to $(\alpha, \beta, k) = (0.15, 0.75, 21.5)$ is plotted in Figures 1 and 2.

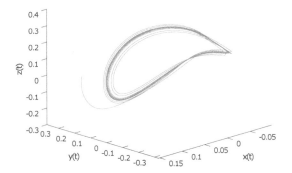

Figure 1. Nuclear spin generator in 3D phase space.

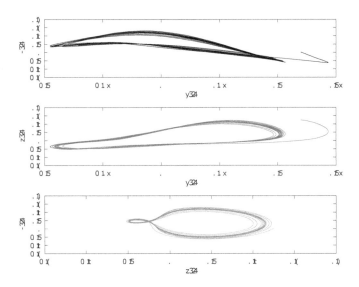

Figure 2. Nuclear spin generator time series.

The novel pseudorandom byte output algorithm is based on the next few steps:

1. The seed values $x(0)$, $y(0)$, and $z(0)$ from Equation (1) are determined. The output byte length L is determined.
2. Equation (1) is iterated for N times.
3. The iteration of the nuclear spin generator continues. As a result, the three floating-point values $x(i)$, $y(i)$, and $z(i)$ are calculated. They are manipulated as follows: $xm(i) = mod(abs(int(x(i) \times$

10^{13}))), 256), $ym(i) = mod(abs(int(y(i) \times 10^{13}))), 256)$, and $zm(i) = mod(abs(int(z(i) \times 10^{13}))), 256)$, where $abs(a)$ returns the modulus of a, $int(a)$ returns the the integer part of a, truncating the value behind the decimal sign, and $mod(a, b)$ returns the reminder after division.

4. Perform XOR operation between xmi, ymi, and zmi to get an output byte.
5. Return to Step 3 until the output byte length L is reached.

2.2. Key Size Analysis

The set of all initial values compose the key size. The key size of the proposed pseudorandom generator has three secret values $x(0)$, $y(0)$, and $z(0)$. As reported by IEEE floating-point standard [31], the computational precision of the 64-bit double-precision number is about 10^{-14}. The key size of the proposed scheme is $(10^{14})^3 = 10^{42} \approx 2^{139}$ bits. This is high enough against mechanisms of exhaustive attack [32].

2.3. Statistical Tests

To estimate unpredictability of the novel nuclear spin equation based pseudo-random byte generator, we used National Institute of Standards and Technology (NIST) statistical software [33] and ENT [34] statistical application. Using the novel pseudorandom byte generator, 3000 sequences of 125,000 bytes were produced.

The NIST package contains 15 statistical tests: frequency, block frequency, cumulative sums forward and reverse, runs, longest run of ones, rank, spectral, non overlapping templates, overlapping templates, universal, approximate entropy, serial first and second, linear complexity, random excursion, and random excursion variant. The application calculates the proportion of streams that pass the particular tests. The range of acceptable proportion is determined using the confidence interval, defined as

$$\hat{p} \pm 3 \sqrt{\frac{\hat{p}(1 - \hat{p})}{m}},$$

where $\hat{p} = 1 - \alpha$ and m is the number of binary tested sequences. NIST recommends that, for these tests, the user should have at least 1000 sequences of 1,000,000 bits each. In our setup, $m = 3000$. Thus, the confidence interval is

$$0.99 \pm 3 \sqrt{\frac{0.99(0.01)}{3000}} = 0.99 \pm 0.0054498.$$

The proportion should lie above 0.9845502 with exception of random excursion and random excursion variant tests. These two tests only apply whenever the number of cycles in a sequence exceeds 500. Thus, the sample size and minimum pass rate are dynamically reduced taking into account the tested sequences.

The distribution of p-values is examined to ensure uniformity. The interval between 0 and 1 is divided into 10 subintervals. The p-values that lie within each subinterval are counted. Uniformity may also be specified through an application of a χ^2 test and the determination of a p-value corresponding to the goodness-of-fit distributional test on the p-values obtained for an arbitrary statistical test, p-value of the p-values. This is implemented by calculating

$$\chi^2 = \sum_{i=1}^{10} \frac{(F_i - s/10)^2}{s/10},$$

where F_i is the number of p-values in subinterval i and s is the sample size. A p-value is computed such that $p\text{-}value_T = IGAMC(9/2, \chi^2/2)$, where $IGAMC$ is the complemented incomplete gamma statistical function. If $p\text{-}value_T \geq 0.0001$, then the sequences can be considered to be uniformly distributed.

Entropy **2020**, *22*, 501

The output values of the first 13 test are in Table 1. The minimum pass rate for each statistical test with the exception of the random excursion variant test is approximately 2953 for a sample size of 3000 binary sequences. The random excursion test outputs eight p-values, which are tabulated in Table 2. The random excursion variant test outputs 18 randomness probability values: p-values, as shown in Table 3. The minimum pass rate for the random excursion variant test is approximately 1788 for a sample size of 1819 binary sequences.

The output results in Tables 1–3 indicate that all p-values are uniformly distributed over the $(0, 1)$ interval. The total numbers of acceptable streams are within the expected confidence levels for all performed tests. Based on the results, the novel pseudo-random byte generator passed without error NIST suite.

The ENT consists of six statistical tests (entropy, optimum compression, χ^2 square, arithmetic mean value, Monte Carlo for π, and serial correlation), which focus on the pseudorandomness of byte sequences. We tested a stream of 375,000,000 bytes of the proposed generator. The value of entropy is 8.0 byte per byte; the optimum compression would reduce the byte file by 0%; χ^2 square is 238.18 (randomly would exceed this value 76.79% of the times; the sequence is random); arithmetic mean value is 127.5040 (very close to 127.5, less then 10%); Monte Carlo for π is 3.141616448 (error 0.00%); and serial correlation coefficient is 0.000003 (less then 0.005 for true random generators). The novel pseudorandom byte generator passed successfully ENT tests.

Based on the excellent test outputs, we can infer that the proposed pseudorandom byte generator has satisfying statistical properties and provides reasonable level of security.

Table 1. National Institute of Standards and Technology (NIST) test suite results.

NIST Test	p-Value	Pass Rate	Results
Frequency	0.633649	2972/3000	Success
Block frequency	0.014996	2964/3000	Success
Cumulative sums forward	0.928857	2976/3000	Success
Cumulative sums reverse	0.053059	2977/3000	Success
Runs	0.215195	2970/3000	Success
Longest run of ones	0.158133	2974/3000	Success
Rank	0.851939	2971/3000	Success
Spectral	0.552383	2955/3000	Success
Non overlapping templates	0.489210	2970/3000	Success
Overlapping templates	0.117661	2967/3000	Success
Universal	0.800626	2971/3000	Success
Approximate entropy	0.092411	2971/3000	Success
Serial first	0.646836	2963/3000	Success
Serial second	0.410055	2970/3000	Success
Linear complexity	0.370821	2974/3000	Success

Table 2. NIST Random excursion test results.

State	p-Value	Pass Rate	Result
-4	0.042839	1793/1819	Success
-3	0.176043	1792/1819	Success
-2	0.958805	1800/1819	Success
-1	0.821611	1791/1819	Success
$+1$	0.905874	1801/1819	Success
$+2$	0.932163	1804/1819	Success
$+3$	0.395583	1798/1819	Success
$+4$	0.695564	1793/1819	Success

Table 3. NIST Random excursion variant test results.

State	p-Value	Pass Rate	Result
−9	0.136979	1804/1819	Success
−8	0.218022	1805/1819	Success
−7	0.458964	1806/1819	Success
−6	0.250128	1805/1819	Success
−5	0.368209	1805/1819	Success
−4	0.210521	1806/1819	Success
−3	0.821611	1805/1819	Success
−2	0.365446	1800/1819	Success
−1	0.475836	1796/1819	Success
+1	0.927657	1804/1819	Success
+2	0.183647	1805/1819	Success
+3	0.457919	1799/1819	Success
+4	0.188110	1795/1819	Success
+5	0.286462	1798/1819	Success
+6	0.750377	1794/1819	Success
+7	0.957844	1793/1819	Success
+8	0.916782	1794/1819	Success
+9	0.542519	1798/1819	Success

3. Medical Image Steganography Using Nuclear Spin Generator

3.1. Embedding Scheme

In this subsection, by using the pseudorandom byte generation algorithm based on the nuclear spin function in Section 2, we present a medical image steganography algorithm named BOOST.

We consider 16 bits DICOM grayscale input images of $n \times n$ size. As input message, we specify the patient information (text based patient medical records with patient identification data). The information includes patient name, patient ID/UID, and doctors remarks. Stego image is the input image with embedded encrypted patient information. The DICOM header data are directly transferred into stego image, based on [35].

The proposed medical image steganography algorithm BOOST consists of the following steps:

1. Iterate for L times the pseudorandom generator based on the nuclear spin generator in Section 2.
2. Apply XOR operation between the pseudorandom byte sequence and all of the input message to produce an encrypted bytes C.
3. Specify the input intervals of gray levels $[a, b]$ of non-black pixels, where a and b determine the boundaries of the container.
4. Index the image pixels by consecutive passing through columns and separate those that fall within the interval $[a, b]$.
5. Convert encrypted data to binary sequence using ASCII table.
6. Consecutively embed the encrypted data into the last bits of the pixels from the interval $[a, b]$
7. The list output pixels is checked to see if their new values are in the input interval. For those pixels that fall outside this range, their value increases by $+2$ if their new values are below the minimum value of the interval or decreases by -2 if the maximum value of the range is exceeded.

3.2. Extraction Scheme

1. Retrieve the number L of embedded bytes, input levels interval $[a, b]$, and the secret key space of the pseudorandom generator based on the nuclear spin generator in Section 2.

2. Index the image pixels by consecutive passing through columns and separate those that fall within the interval $[a, b]$.
3. Consecutively extract the embedded data from the last bits of the pixels from the interval $[a, b]$.
4. Iterate for L times the pseudorandom generator based on the nuclear spin generator in Section 2.
5. Apply XOR operation between the output pseudorandom byte sequence and all of the extracted bytes to produce the input bytes C.

The proposed medical image steganography algorithm was implemented in C++ programming language. Fifteen 16-bit monochrome DICOM images were used for the experimental analysis. The test images were selected from the National Electrical Manufacturers Association (NEMA) medical image database: ftp://medical.nema.org/medical/dicom/DataSets/WG16/Philips/ClassicSingleFrame/. The folder consists of classical 16 bits DICOM grayscale single frame medical images of brains, knees, and livers. An example to illustrate the BOOST is presented in Figure 3.

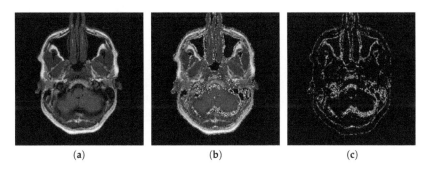

(a)   (b)   (c)

Figure 3. Illustration of embedding a message using the BOOST method and input levels interval $[20, 48]$: (a) the original input image Brain IM_0001; and (b,c) the location of embedded message.

3.3. Steganographic Analysis

An image histogram is an accurate illustration of the tonal value distribution in digital images. This check compares both input and stego image histograms. Histograms, performed using ImageJ2x 2.1.5.0 (http://www.rawak.de/rs2012/), for three input images and their stego images are also shown in Figure 4.

It is considered that the histograms of the stego images are much the same as those of the input images with no evidence of hidden messages in stego images.

Peak Signal-to-Noise Ratio (PSNR) is the proportion between the highest possible value of a signal and the value of distorting noise that affects the accuracy of its representation. It is defined as:

$$PSNR = 10 \log_{10} \frac{(2^d - 1)^2}{MSE} (dB),$$ (2)

where d is the bit depth of the pixel and MSE is the Mean-Square Error between the input and stego images. MSE is defined as:

$$MSE = \frac{1}{mn} \sum_{i=1}^{m} \sum_{j=1}^{n} (P[i, j] - S[i, j])^2,$$ (3)

where $P[i, j]$ and $S[i, j]$ are the ith row and jth column pixel in the input and stego images, respectively.

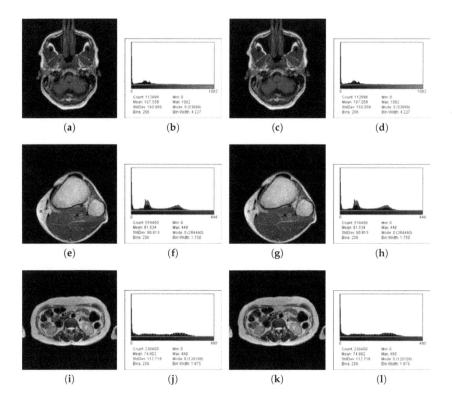

Figure 4. (**a,e,i**) Input images Brain IM_0001, Knee IM_0001, and Liver IM_0001; (**b,f,j**) their histograms; (**c,g,k**) stego images; and (**d,h,l**) their histograms.

In Table 4, we provide the computed values for MSE and PSNR for BOOST algorithm. MSE and PSNR are calculated for images with 1050 bytes (8400 bits), 1042 bytes (8336 bits), and 1119 bytes (8952 bits) embedded. Maximum payload is calculated as a number of non-black pixels.

From results obtained, as shown in Table 4, the PSNR values are extremely high, above 113 dB, which suggests an excellent level of security for the proposed BOOST algorithm.

The Bit Error Rate (BER) is computed as the actual number of bit positions which are changed in the stego image compared with the input image. A value of BER close to 0.0 stands for high efficiency of the steganography algorithm. The Normalized Cross-Correlation (NCC) calculates the cross-correlation in the the frequency domain, depending on the size of the images. Then, it computes the local sums by pre-computing running sums. Use local sums to normalize the cross-correlation to get correlation coefficients. The output matrix holds the correlation coefficients, which can range between −1.0 and 1.0. NCC is defined as:

$$NCC = \frac{\sum_{i=1}^{m} \sum_{j=1}^{n} (P[i,j] \times S[i,j])}{\sum_{i=1}^{m} \sum_{j=1}^{n} (P[i,j])^2}. \tag{4}$$

A value of NCC close to 1.0 represents perfect quality of the stego image.

The Structural SIMilarity (SSIM) index is an algorithm for measuring the similarity between input and stego images [36]. The output SSIM index is a decimal number between −1 and 1. Value 1 indicates excellent structural similarity.

Table 4. Mean-Square Error(MSE) and Peak Signal-to-Noise Ratio (PSNR) results.

Input Image	Image Size	Maximum Payload	Percent Volume	Available Levels	Input Levels	Message (Bytes)	MSE	PSNR (dB)
Brain IM_0001	336 × 336	83,179	73.68	1083	[50, 146]	1050	0.0191	113.5238
Brain IM_0002	336 × 336	83,362	73.84	851	[50, 146]	1050	0.0192	113.4977
Brain IM_0003	336 × 336	83,557	74.01	823	[50, 146]	1050	0.0191	113.5218
Brain IM_0004	336 × 336	83,341	73.82	875	[50, 146]	1050	0.0190	113.5319
Brain IM_0005	336 × 336	83,883	74.30	834	[50, 146]	1050	0.0191	113.5198
Knee IM_0001	720 × 720	249,148	48.06	449	[30, 56]	1042	0.0041	120.1618
Knee IM_0002	720 × 720	250,531	48.33	426	[30, 56]	1042	0.0043	120.0302
Knee IM_0003	720 × 720	251,867	48.59	461	[30, 56]	1042	0.0043	120.0263
Knee IM_0004	720 × 720	256,834	48.54	453	[30, 56]	1042	0.0042	120.0637
Knee IM_0005	720 × 720	260,969	50.34	444	[30, 56]	1042	0.0042	120.0558
Liver IM_0001	480 × 480	109,631	47.58	481	[20, 68]	1119	0.0098	116.4055
Liver IM_0002	480 × 480	112,992	49.04	581	[20, 68]	1119	0.0100	116.3465
Liver IM_0003	480 × 480	114,107	49.53	626	[20, 68]	1119	0.0103	116.2160
Liver IM_0004	480 × 480	115,670	50.20	643	[20, 68]	1119	0.0098	116.4325
Liver IM_0005	480 × 480	116,373	50.51	624	[20, 68]	1119	0.0098	116.4383

In Table 5, we provide the calculated values for BER, NCC, and SSIM for the presented BOOST scheme. From the obtained results shown in Table 5, it is clear that the BER are very close to 0.0 and NCC and SSIM values are almost equal to 1.0. The data indicate that the BOOST scheme provides good quality and excellent structural similarity.

Table 5. Bit Error Rate (BER), Normalized Cross-Correlation (NCC), and SSIM (Structural SIMilarity) results.

Image	BER	NCC	SSIM
Brain IM_0001	0.0012	0.9999971	0.9999787
Brain IM_0002	0.0012	0.9999950	0.9999757
Brain IM_0003	0.0012	0.9999934	0.9999838
Brain IM_0004	0.0012	0,9999968	0.9999769
Brain IM_0005	0.0012	0.9999955	0.9999809
Knee IM_0001	0.00026	0.9999979	0.9999806
Knee IM_0002	0.00027	0,9999982	0.9999794
Knee IM_0003	0.00027	0.9999979	0.9999720
Knee IM_0004	0.00027	0.9999980	0,9999682
Knee IM_0005	0.00026	0.9999976	0.9999581
Liver IM_0001	0.00061	0.9999982	0.9998838
Liver IM_0002	0.00062	0.9999973	0.9998954
Liver IM_0003	0.00064	0.9999970	0.9999311
Liver IM_0004	0.00061	0.9999983	0.9999308
Liver IM_0005	0.00061	0.9999984	0.9999253

The resistance of the BOOST algorithm against cropping attack [37,38] was tested. Cropping is the mechanism by which outer parts of the image are cut. Three stego images (Brain IM_0001, Knee IM_0001, and Liver IM_0001) generated from the BOOST algorithm were subjected to cropping attacks.

The normalized correlation (NC) values were calculated for the stego image and the corresponding cropped image [38]. The output NC results varied between 0.8944 and 1, as shown in Table 6. We see from these results that the proposed BOOST algorithm reasonably resists cropping attack.

Table 6. Normalized correlation (NC) results against cropping attack.

Cropping Attack		Brain IM_0001	Knee IM_0001	Liver IM_0001
Percent	10%	0.999	0.9872	0.9858
	20%	0.981	0.9729	0.9724
	30%	0.8944	0.9455	0.9093

The steganographic analysis undoubtedly shows the good rate of the proposed algorithm. Table 7 summarizes some of the computed values of our proposed scheme with other algorithms.

Table 7. Comparison of our medical image steganography with other techniques.

Algorithm	Minimum Calculated PSNR(dB)	Capacity Bits per Pixel	Maximum Calculated BER
Proposed	113.50	0.74	0.0012
[16] Mantos 2016	103.68	0.5	-
[37] Thiyagarajan 2013	74.36	-	0.004
[22] Jain 2017 Improved	72.17	0.37	-
[39] Elhoseny 2018	57.02	-	0.0

Using the given test results, we can conclude that the presented algorithm BOOST, based on the nuclear spin generator, has satisfying statistical properties and provides a proper safety expectation.

4. Conclusions

We introduce a novel medical image steganographic scheme named BOOST. The presented algorithm uses a novel pseudorandom byte output technique based on the nuclear spin generator. Our security investigation (mean square error, peak signal-to-noise ratio, normalized cross-correlation, and structural similarity) shows that the proposed hiding can be used with success for secure medical record communication.

Author Contributions: B.S. (Bozhidar Stoyanov) and B.S. (Borislav Stoyanov) wrote and edited the manuscript. Both authors have read and agreed to the published version of the manuscript

Funding: The paper was partially supported by the National Scientific Program "Information and Communication Technologies for a Single Digital Market in Science, Education and Security (ICTinSES)", financed by the Ministry of Education and Science, Bulgaria for Bozhidar Stoyanov and Borislav Stoyanov.

Conflicts of Interest: The authors declare no conflict of interest.

References

1. Office for Civil Rights. HIPAA Compliance Assistance. Summary of the HIPAA Privacy Rule. Available online: https://www.hhs.gov/sites/default/files/privacysummary.pdf (accessed on 12 March 2020).
2. Barrows, R.; Clayton, P. Privacy, Confidentiality, and Electronic Medical Records. *J. Am. Med Inform. Assoc.* **1996**, *3*, 139–148. [CrossRef] [PubMed]
3. Niu, X.M.; Lu, Z.M.; Sun, S.H. Digital watermarking of still images with gray-level digital watermarks. *IEEE Trans. Consum. Electron.* **2000**, *46*, 137–145. [CrossRef]

4.   Kutter, M.; Jordan, F.D.; Bossen, F. Digital watermarking of color images using amplitude modulation. *J. Electron. Imaging* **1998**, *7*, 326–332. [CrossRef]
5.   Cao, W.; Zhou, Y.; Chen, C.P.; Xia, L. Medical image encryption using edge maps. *Signal Process.* **2017**, *132*, 96–109. [CrossRef]
6.   Kanso, A.; Ghebleh, M. An efficient and robust image encryption scheme for medical applications. *Commun. Nonlinear Sci. Numer. Simul.* **2015**, *24*, 98–116. [CrossRef]
7.   Abdelfattah, M.; Hegazy, S.F.; Areed, N.F.; Obayya, S.S. Compact optical asymmetric cryptosystem based on unequal modulus decomposition of multiple color images. *Opt. Lasers Eng.* **2020**, *129*, 106063. [CrossRef]
8.   Wang, X.; Zhao, H.; Feng, L.; Ye, X.; Zhang, H. High-sensitivity image encryption algorithm with random diffusion based on dynamic-coupled map lattices. *Opt. Lasers Eng.* **2019**, *122*, 225–238. [CrossRef]
9.   Chen, H.; Liu, Z.; Zhu, L.; Tanougast, C.; Blondel, W. Asymmetric color cryptosystem using chaotic Ushiki map and equal modulus decomposition in fractional Fourier transform domains. *Opt. Lasers Eng.* **2019**, *112*, 7–15. [CrossRef]
10. Huang, L.C.; Tseng, L.Y.; Hwang, M.S. A reversible data hiding method by histogram shifting in high quality medical images. *J. Syst. Softw.* **2013**, *86*, 716–727. [CrossRef]
11. Jiang, N.; Zhao, N.; Wang, L. LSB based quantum image steganography algorithm. *Int. J. Theor. Phys.* **2016**, *55*, 107–123. [CrossRef]
12. Agrawal, M.; Mishra, P. A comparative survey on symmetric key encryption techniques. *Int. J. Comput. Sci. Eng.* **2012**, *4*, 877.
13. Zielińska, E.; Mazurczyk, W.; Szczypiorski, K. Trends in steganography. *Commun. ACM* **2014**, *57*, 86–95. [CrossRef]
14. Chen, H.; Du, X.; Liu, Z.; Yang, C. Optical color image hiding scheme by using Gerchberg–Saxton algorithm in fractional Fourier domain. *Opt. Lasers Eng.* **2015**, *66*, 144–151. [CrossRef]
15. Ibrahim, R.; Kuan, T.S. Steganography Algorithm to Hide Secret Message inside an Image. *Comput. Technol. Appl.* **2011**, *2*, 102–108.
16. Mantos, P.L.K.; Maglogiannis, I. Sensitive Patient Data Hiding using a ROI Reversible Steganography Scheme for DICOM Images. *J. Med Syst.* **2016**, *40*, 156. [CrossRef]
17. National Electrical Manufacturers Association. Digital Imaging and Communications in Medicine (DICOM). Available online: https://www.dicomstandard.org/current/ (accessed on 12 March 2020).
18. Wu, S.; Zhong, S.; Liu, Y. Deep residual learning for image steganalysis. *Multimed. Tools Appl.* **2018**, *77*, 10437–10453. [CrossRef]
19. Jain, M.; Lenka, S.K. Diagonal queue medical image steganography with Rabin cryptosystem. *Brain Inform.* **2016**, *3*, 39–51. [CrossRef]
20. Satish, K.; Jayakar, T.; Tobin, C.; Madhavi, K.; Murali, K. Chaos based spread spectrum image steganography. *IEEE Trans. Consum. Electron.* **2004**, *50*, 587–590. [CrossRef]
21. Jain, M.; Kumar, A. RGB channel based decision tree grey-alpha medical image steganography with RSA cryptosystem. *Int. J. Mach. Learn. Cybern.* **2017**, *8*, 1695–1705. [CrossRef]
22. Jain, M.; Kumar, A.; Choudhary, R.C. Improved diagonal queue medical image steganography using Chaos theory, LFSR, and Rabin cryptosystem. *Brain Inform.* **2017**, *4*, 95–106. [CrossRef]
23. Ambika.; Biradar, R.L. Secure medical image steganography through optimal pixel selection by EH-MB pipelined optimization technique. *Health Technol.* **2020**, *10*, 231–247. [CrossRef]
24. Rajendran, S.; Doraipandian, M. Chaotic Map Based Random Image Steganography Using LSB Technique. *Int. J. Netw. Secur.* **2017**, *19*, 593–598. [CrossRef]
25. Huang, Z. Stationary distribution of stochastic nuclear spin generator systems. *J. Nonlinear Sci. Appl.* **2016**, *9*, 5410–5427. [CrossRef]
26. Sachdev, P.; Sarathy, R. Periodic and chaotic solutions for a nonlinear system arising from a nuclear spin generator. *Chaos Solitons Fractals* **1994**, *4*, 2015–2041. [CrossRef]
27. Molaei, M.; Umut, O. Generalized synchronization of nuclear spin generator system. *Chaos, Solitons Fractals* **2008**, *37*, 227–232. [CrossRef]

28. Nikolov, S.; Nedev, V.; Zlatanov, V. A Numerical Investigation of the Modified Sherman Systems. *Eng. Mech.* **2011**, *18*, 127–142.

29. Nikolov, S.; Bozhov, B.; Nedev, V.; Zlatanov, V. The Sherman system: Bifurcations, regular and chaotic behaviour. *Comptes Rendus De L'Academie Bulg. Des Sci.* **2003**, *56*, 5–19.

30. Sherman, S. A third-order nonlinear system arising from a nuclear spin generator. *Contrib. Differ. Equations* **1963**, *2*, 197–227.

31. IEEE Standard for Floating-Point Arithmetic. IEEE Std 754-2008. 2008; pp. 1–70. Available online: https://ieeexplore.ieee.org/document/4610935 (accessed on 12 March 2020).

32. Alvarez, G.; Li, S. Some basic cryptographic requirements for chaos-based cryptosystems. *Int. J. Bifurc. Chaos* **2006**, *16*, 2129–2151. [CrossRef]

33. Rukhin, A.; Soto, J.; Nechvatal, J.; Smid, M.; Barker, E.; Leigh, S.; Levenson, M.; Vangel, M.; Banks, D.; Heckert, A.; et al. *A Statistical Test Suite for Random and Pseudorandom Number Generators for Cryptographic Application*; NIST Special Publication 800-22: Revision 1a, Lawrence E. Bassham III, Eds.; NIST: Gaithersburg, MD, USA, 2010.

34. Walker, J. A Pseudorandom Number Sequence Test Program. Available online: https://www.fourmilab.ch/random/ (accessed on 12 March 2020).

35. Digital Imaging and Communications in Medicine (DICOM). *Supplement 55: Attribute Level Confidentiality (Including De-Identification)*; Technical Report; National Electrical Manufacturers Association (NEMA): Rosslyn, VA, USA, 2002.

36. Wang, Z.; Bovik, A.C.; Sheikh, H.R.; Simoncelli, E.P. Image Quality Assessment: From Error Visibility to Structural Similarity. *IEEE Trans. Image Process.* **2004**, *13*, 600–612. [CrossRef]

37. Thiyagarajan, P.; Aghila, G. Reversible dynamic secure steganography for medical image using graph coloring. *Health Policy Technol.* **2013**, *2*, 151–161. [CrossRef]

38. Dong, P.; Brankov, J.G.; Galatsanos, N.P.; Yang, Y.; Davoine, F. Digital watermarking robust to geometric distortions. *IEEE Trans. Image Process.* **2005**, *14*, 2140–2150. [CrossRef] [PubMed]

39. Elhoseny, M.; Ramírez-González, G.; Abu-Elnasr, O.M.; Shawkat, S.A.; Arunkumar, N.; Farouk, A. Secure medical data transmission model for IoT-based healthcare systems. *IEEE Access* **2018**, *6*, 20596–20608. [CrossRef]

© 2020 by the authors. Licensee MDPI, Basel, Switzerland. This article is an open access article distributed under the terms and conditions of the Creative Commons Attribution (CC BY) license (http://creativecommons.org/licenses/by/4.0/).

Article

Image Encryption Using Elliptic Curves and Rossby/Drift Wave Triads

Ikram Ullah [1], Umar Hayat [1,*] and Miguel D. Bustamante [2,*]

[1] Department of Mathematics, Quaid-i-Azam University, Islamabad 45320, Pakistan;
ikram.ullah@math.qau.edu.pk

[2] School of Mathematics and Statistics, University College Dublin, Belfield, Dublin 4, Ireland

* Correspondence: umar.hayat@qau.edu.pk (U.H.); miguel.bustamante@ucd.ie (M.D.B)

Received: 4 March 2020; Accepted: 14 April 2020; Published: 16 April 2020

Abstract: We propose an image encryption scheme based on quasi-resonant Rossby/drift wave triads (related to elliptic surfaces) and Mordell elliptic curves (MECs). By defining a total order on quasi-resonant triads, at a first stage we construct quasi-resonant triads using auxiliary parameters of elliptic surfaces in order to generate pseudo-random numbers. At a second stage, we employ an MEC to construct a dynamic substitution box (S-box) for the plain image. The generated pseudo-random numbers and S-box are used to provide diffusion and confusion, respectively, in the tested image. We test the proposed scheme against well-known attacks by encrypting all gray images taken from the USC-SIPI image database. Our experimental results indicate the high security of the newly developed scheme. Finally, via extensive comparisons we show that the new scheme outperforms other popular schemes.

Keywords: quasi-resonant Rossby/drift wave triads; Mordell elliptic curve; pseudo-random numbers; substitution box

1. Introduction

The exchange of confidential images via the internet is usual in today's life, even though the internet is an open source that is unsafe and unauthorized persons can steal useful or sensitive information. Therefore it is essential to be able to share images in a secure way. This goal is achieved by using cryptography. Traditional cryptographic techniques such as data encryption standard (DES) and advanced encryption standard (AES) are not suitable for image transmission because image pixels are usually highly correlated [1,2]. By contrast, DES and AES are ideal techniques for text encryption [3], so researchers are trying to develop such techniques to meet the demand for reliable image delivery.

A number of image encryption schemes have been developed using different approaches [4–14]. Hua et al. [12] developed a highly secure image encryption algorithm, where pixels are shuffled via the principle of the Josephus problem and diffusion is obtained by a filtering technology. Wu et al. [13] proposed a novel image encryption scheme by combining a random fractional discrete cosine transform (RFrDCT) and the chaos-based Game of Life (GoL). In their scheme, the desired level of confusion and diffusion is achieved by GoL and an XOR operation, respectively. "Confusion" entails hiding the relation between input image, secret keys and the corresponding cipher image, and "diffusion" is an alteration of the value of each pixel in an input image [1].

One of the dominant trends in encryption techniques is chaos-based encryption [15–20]. The reason for this dominance is that the chaos-based encryption schemes are highly sensitive to the initial parameters. However, there are certain chaotic cryptosystems that exhibit a lower security level due to the usage of chaotic maps with less complex behavior (see [21]). This problem is addressed in [22] by introducing a cosine-transform-based chaotic system (CTBCS) for encrypting images with higher security. Xu et al. [23] suggested an image encryption technique based on fractional chaotic systems and verified experimentally the higher security of the underlying cryptosystem. Ahmad et al. [24] highlighted certain defects of the above-mentioned cryptosytem by recovering the plain image without the secret key. Moreover, they proposed an enhanced scheme to thwart all kinds of attacks.

The chaos-based algorithms also use pseudo-random numbers and substitution boxes (S-boxes) to create confusion and diffusion [25,26]. Cheng et al. [25] proposed an image encryption algorithm based on pseudo-random numbers and AES S-box. The pseudo-random numbers are generated using AES S-box and chaotic tent maps. The scheme is optimized by combining the permutation and diffusion phases, but the image is encrypted in rounds, which is time consuming. Belazi et al. [26] suggested an image encryption algorithm using a new chaotic map and logistic map. The new chaotic map is used to generate a sequence of pseudo-random numbers for masking phase. Then eight dynamic S-boxes are generated. The masked image is substituted in blocks via aforementioned S-boxes. The substituted image is again masked by another pseudo-random sequence generated by the logistic map. Finally, the encrypted image is obtained by permuting the masked image. The permutation is done by a sequence generated by the map function. This algorithm fulfills the security analysis but performs slowly due to the four cryptographic phases. In [27], an image encryption method based on chaotic maps and dynamic S-boxes is proposed. The chaotic maps are used to generate the pseudo-random sequences and S-boxes. To break the correlation, pixels of an input image are permuted by the pseudo-random sequences. In a second phase the permuted image is decomposed into blocks. Then blocks are encrypted by the generated S-boxes to get the cipher image. From histogram analysis it follows that the suggested technique generates cipher images with a nonuniform distribution.

Similar to the chaotic maps, elliptic curves (ECs) are sensitive to input parameters, but EC-based cryptosystems are more secure than those of chaos [28]. Toughi et al. [29] developed a hybrid encryption algorithm using elliptic curve cryptography (ECC) and AES. The points of an EC are used to generate pseudo-random numbers and keys for encryption are acquired by applying AES to the pseudo-random numbers. The proposed algorithm gets the promising security but pseudo-random numbers are generated via the group law, which is time consuming. In [3], a cyclic EC and a chaotic map are combined to design an encryption algorithm. The developed scheme overcomes the drawbacks of small key space but is unsafe to the known-plaintext/chosen-plaintext attack [30]. Similarly, Hayat et al. [31] proposed an EC-based encryption technique. The stated scheme generates pseudo-random numbers and dynamic S-boxes in two phases, where the construction of S-box is not guaranteed for each input EC. Therefore, changing of ECs to generate an S-box is a time-consuming work. Furthermore, the generation of ECs for each input image makes it insufficient.

Based on the above discussion, we propose an improved image encryption algorithm, based on quasi-resonant Rossby/drift wave triads [32,33] (triads, for short) and Mordell elliptic curves (MECs). The triads are utilized in the generation of pseudo-random numbers and MECs are employed to create dynamic S-boxes. The proposed scheme is novel in that it introduces the technique of pseudo-random numbers generation using triads, which is faster than generating pseudo-random numbers by ECs. Moreover, the scheme does not require to separately generate triads for each input image of the same size. In the present scheme, MECs are used opposite to [31], in the sense that now, for each input image, the generation of a dynamic S-box is guaranteed [34]. Finally, extensive performance analyses and comparisons reveal the efficiency of the proposed scheme.

This paper is organized as follows. Preliminaries are described in Section 2. In Section 3, the proposed encryption algorithm is explained in detail. Section 4 provides the experimental results as well as a comparison between the proposed method and other existing popular schemes. Lastly, conclusions are presented in Section 5.

2. Preliminaries

Barotropic vorticity equation: The barotropic vorticity equation (in the so-called β-plane approximation) is one of the simplest two-dimensional models of the large-scale dynamics of a shallow layer of fluid on the surface of a rotating sphere. It is described in mathematical terms by the partial differential equation

$$\frac{\partial}{\partial t}(\nabla^2\psi - F\psi) + \left(\frac{\partial\psi}{\partial x}\frac{\partial\nabla^2\psi}{\partial y} - \frac{\partial\psi}{\partial y}\frac{\partial\nabla^2\psi}{\partial x}\right) + \gamma\frac{\partial\psi}{\partial x} = 0, \tag{1}$$

where $\psi(x,y,t) \in \mathbb{R}$ represents the geopotential height, γ is the Coriolis parameter, a real constant measuring the variation of the Coriolis force with latitude (x represents longitude and y represents latitude) and F is a non-negative real constant representing the inverse of the square of the deformation radius. We assume periodic boundary conditions: $\psi(x+2\pi,y,t) = \psi(x,y+2\pi,t) = \psi(x,y,t)$ for all $x,y,t \in \mathbb{R}$. In the literature Equation (1) is also known as the Charney–Hasegawa–Mima equation (CHM) [35–39]. This equation accepts harmonic solutions, known as Rossby waves, which are solutions of both the linearized form and the whole (nonlinear) form of Equation (1). A Rossby wave solution is given explicitly by the parameterized function $\psi_{(k,l)}(x,y,t) = \Re\{A\,e^{i(kx+ly-\omega(k,l)t)}\}$, where $A \in \mathbb{C}$ is an arbitrary constant, $\omega(k,l) = -\frac{\gamma k}{k^2+l^2+F}$ is the so-called dispersion relation, and $(k,l) \in \mathbb{Z}^2$ is called the wave vector. For simplicity, we take $\gamma = -1$ and $F = 0$ in what follows [32,33].

Resonant triads: As Equation (1) is nonlinear, modes with different wave vectors tend to couple and exchange energy. If the nonlinearity is weak, this exchange happens to be quite slow and is more efficient amongst groups of modes that are in *resonance*. To the lowest order of nonlinearity in Equation (1), approximate solutions known as resonant triad solutions can be constructed via linear combinations of the form

$$\psi(x,y,t) = \Re\{A_1\,e^{i(k_1x+l_1y-\omega(k_1,l_1)t)} + A_2\,e^{i(k_2x+l_2y-\omega(k_2,l_2)t)} + A_3\,e^{i(k_3x+l_3y-\omega(k_3,l_3)t)}\},$$

where A_1, A_2, A_3 are slow functions of time (they satisfy a closed system of ODEs, not shown here), and the wave vectors $(k_1,l_1), (k_2,l_2)$ and (k_3,l_3) satisfy the Diophantine system of equations:

$$k_1 + k_2 = k_3, \quad l_1 + l_2 = l_3 \quad \text{and} \quad \omega_1 + \omega_2 = \omega_3, \tag{2}$$

for $\omega_i = \omega(k_i,l_i), i = 1,2,3$. A set of three wavevectors satisfying Equations (2) is called a resonant triad. Solutions can be found analytically via a rational transformation to elliptic surfaces (see below).

Quasi-resonant triads and detuning level: If, in (2), the equation $\omega_1 + \omega_2 = \omega_3$ is replaced by the inequality $|\omega_1 + \omega_2 - \omega_3| \leq \delta^{-1}$, for a large positive number δ, then the triad becomes a quasi-resonant triad and δ^{-1} is known as the detuning level of the quasi-resonant triad. It is possible to construct quasi-resonant triads via downscaling of resonant triads that have very large wave vectors [32]. For simplicity, in what follows we simply call a quasi-resonant triad a triad and denote it by Δ. Finally, to avoid over-counting of triads we will impose the condition $k_3 > 0$.

Rational transformation: In [32], wave vectors are explicitly expressed in terms of rational variables X, Y and D as follows:

$$\frac{k_1}{k_3} = \frac{X}{Y^2+D^2}, \quad \frac{l_1}{k_3} = \left(\frac{X}{Y}\right)\left(1 - \frac{D}{Y^2+D^2}\right), \quad \frac{l_3}{k_3} = \frac{D-1}{Y}. \tag{3}$$

In the case $F = 0$, the rational variables X, Y, D lie on an elliptic surface. The transformation is bijective and its inverse mapping is given by:

$$X = \frac{k_3(k_1^2+l_1^2)}{k_1(k_3^2+l_3^2)}, \quad Y = \frac{k_3(k_3l_1 - k_1l_3)}{k_1(k_3^2+l_3^2)}, \quad D = \frac{k_3(k_3k_1 - l_1l_3)}{k_1(k_3^2+l_3^2)}. \tag{4}$$

New parameterization: In [40], Kopp parameterized the resonant triads and in terms of parameters u and t it follows by [40] (Equation (1.22)) that:

$$\frac{k_1}{k_3} = (t^2 + u^2)(t^2 - 2u + u^2)/(1 - 2u), \tag{5}$$

$$\frac{l_3}{k_3} = \left(u(2u - 1) + (t^2 + u^2)(t^2 - 2u + u^2)\right)/(t(1 - 2u)), \tag{6}$$

$$\frac{l_1}{k_3} = (t^2 + u^2)\left((2u - 1) + u(t^2 - 2u + u^2)\right)/(t(1 - 2u)). \tag{7}$$

In 2019, Hayat et al. [33] found a new parameterisation of X, Y and D in terms of auxiliary parameters a, b and hence $\frac{k_1}{k_3}, \frac{l_3}{k_3}$ and $\frac{l_1}{k_3}$ are given by:

$$\frac{k_1}{k_3} = \frac{\left(a^2 + b(2 - 3b) + 1\right)^3}{(a^2 - 3b^2 - 2b + 1)\left(2(11 - 3a^2)b^2 + (a^2 + 1)^2 - 16ab + 9b^4\right)}, \tag{8}$$

$$\frac{l_3}{k_3} = \frac{6(a^2 + a - 1)b^2 - (a + 1)^2(a^2 + 1) + 4ab - 9b^4}{(a^2 - 3b^2 - 1)(a^2 - 3b^2 - 2b + 1)}, \tag{9}$$

$$\frac{l_1}{k_3} = \frac{\left(a^2 + b(2 - 3b) + 1\right)}{(a^2 - 3b^2 - 1)(a^2 - 3b^2 - 2b + 1)\left(2(11 - 2a^2)b^2 + (a^2 + 1)^2 - 16ab + 9b^4\right)} \tag{10}$$
$$\times [a^6 + 2a^5 + a^4(-9b^2 - 6b + 3) - 4a^3(3b^2 + 2b - 1) + 3a^2(3b^2 + 2b - 1)^2$$
$$+ 2a(9b^4 + 12b^3 + 14b^2 - 4b + 1) - (3b^2 + 1)^2(3b^2 + 6b - 1)]$$

Elliptic curve (EC): Let $\mathbb{F}_p$ be a finite field for any prime p, then an EC E_p over $\mathbb{F}_p$ is defined by

$$y^2 \equiv x^3 + bx + c \pmod{p}, \tag{11}$$

where $b, c \in \mathbb{F}_p$. The integers b, c and p are called parameters of an EC. The number of all $(x, y) \in \mathbb{F}_p^2$ satisfying the congruence (11) is denoted by $\#E_p$.

Mordell elliptic curve (MEC): In the special but important case $b = 0$, the above EC is known as an MEC and is represented by

$$y^2 \equiv x^3 + c \pmod{p}. \tag{12}$$

For $p \equiv 2 \pmod 3$, there are exactly $p + 1$ points $(x, y) \in \mathbb{F}_p^2$ satisfying the congruence (12), see [41] for further details.

If points on E_p are ordered according to some total order $\prec$ then E_p is said to be an ordered EC. Recall that total order is a binary relation which possesses the reflexive, antisymmetric and transitive properties. Azam et al. [42] introduced a total order known as a natural ordering on MECs given by

$$(x_1, y_1) \prec (x_2, y_2) \Leftrightarrow \begin{cases} \text{either } x_1 < x_2, \text{ or} \\ x_1 = x_2 \text{ and } y_1 < y_2, \end{cases}$$

and generated efficient S-boxes using the aforesaid ordering. We will use natural ordering to generate S-boxes. Thus from here on E_p stands for a naturally ordered MEC unless it is specified otherwise.

3. The Proposed Encryption Scheme

The proposed encryption scheme is based on pseudo-random numbers and S-boxes. The pseudo-random numbers are generated using quasi-resonant triads. To get an appropriate level of diffusion we need to properly order the Δs. For this purpose we define a binary relation $\lesssim$ as follows.

3.1. Ordering on Quasi-Resonant Triads

Let Δ, Δ' represent the triads $(k_i, l_i), (k'_i, l'_i), i = 1, 2, 3$, respectively, then

$$\Delta \lesssim \Delta' \Leftrightarrow \begin{cases} \text{either } a < a', \text{ or} \\ a = a' \text{ and } b < b', \text{ or} \\ a = a', b = b' \text{ and } k_3 \le k_3', \end{cases}$$

where a, b and a', b' are the corresponding auxiliary parameters of Δ and Δ', respectively.

Lemma 1. *If T denotes the set of Δs in a box of size L, then $\lesssim$ is a total order on T.*

Proof. The reflexivity of $\lesssim$ follows from $a = a, b = b$ and $k_3 = k_3$ and hence $\Delta \lesssim \Delta$. As for antisymmetry we suppose $\Delta \lesssim \Delta'$ and $\Delta' \lesssim \Delta$. Then, by definition $a \le a'$ and $a' \le a$, which imply $a = a'$. Thus we are left with two results: $b \le b'$ and $b' \le b$, which imply $b = b'$. Thus, we obtain the results $k_3 \le k_3'$ and $k_3' \le k_3$, which ultimately give $k_3 = k_3'$. Solving Equations (8)–(10) for the obtained values, we get $k_1 = k_1', l_3 = l_3'$ and from Equation (2) it follows that $l_2 = l_2'$. Consequently $\Delta = \Delta'$ and $\lesssim$ is antisymmetric. As for transitivity, let us assume $\Delta \lesssim \Delta'$ and $\Delta' \lesssim \Delta''$. Then $a \le a'$ and $a' \le a''$, implying $a \le a''$. If $a < a''$, then transitivity follows. If $a = a''$, then $a' = a''$ too. Thus, $b \le b'$ and $b' \le b''$, so $b \le b''$. If $b < b''$, then transitivity follows. If $b = b''$, then $b' = b''$ too. Thus, $k_3 \le k_3'$ and $k_3' \le k_3''$, implying $k_3 \le k_3''$ and hence transitivity follows: $\Delta \lesssim \Delta''$. $\square$

Let $\overset{*}{T}$ stand for the set of Δs ordered with respect to the order $\lesssim$. The main steps of the proposed scheme are explained as follows.

3.2. Encryption

A. Public parameters: In order to exchange the useful information the sender and receiver should agree on the public parameters described as below:

(1)   Three sets: choose three sets $\mathcal{A}_i = [A_i, B_i], i = 1, 2, 3$ of consecutive numbers with unknown step sizes, where the end points $A_i, B_i, i = 1, 2, 3$ are rational numbers.
(2)   A total order: select a total order $\prec$ so that the triads generated by the above-mentioned sets may be arranged with respect to that order.

Suppose that P represents an image of size $m \times n$ to be encrypted, and the pixels of P are arranged in column-wise linear ordering. Thus, for positive integer $i \le mn$, $P(i)$ represents the i-th pixel value in linear ordering. Define S_P as the sum of all pixel values of the image P. Then the proposed scheme chooses the secret keys in the following ways.

B. Secret keys: To generate confusion and diffusion in an image, the sender chooses the secret keys as follows.

(1)   Step size: select positive integers a_i, b_i to construct the step sizes $\alpha_i = \frac{a_i}{b_i}$ of $\mathcal{A}_i, i = 1, 2$. Additionally, choose a non-negative integer a_3 as a step size of $\mathcal{A}_3$ in such a way that $\prod_{i=1}^{3} n_i \ge mn$, where $\#\mathcal{A}_i = n_i$ represents the number of elements in $\mathcal{A}_i$.
(2)   Detuning level: fix some posive integer δ to find the detuning level δ^{-1} allowed for the triads.
(3)   Bound: select a positive integer L such that $|k_i|, |l_i| \le L$ for $i = 1, 2, 3$. This condition is imposed in order to bound the components of the triad wave vectors. Furthermore, choose an integer t to find $r = \lfloor S_P/t \rceil$, where $\lfloor \cdot \rceil$ gives the nearest integer when S_P is divided by t. The reason for choosing such a t is to generate key-dependent S-boxes and the integer r is used to diffuse the components of triads.
(4)   A prime: select a prime $p \ge 257$ such that $p \equiv 2 \pmod 3$ as a secret key for computing nonzero $c \equiv S_P + t \pmod p$ to generate an S-box $\zeta_{E_p}(p, t, S_P)$ on the E_p. The S-box construction technique is made clear in Algorithm 1, and the S-box generated for $p = 1607, t = 182$ and $S = 0$

by Algorithm 1 is shown in Table 1. Furthermore, the cryptographic properties of the said S-box are evaluated in Sections 4.1 and 4.2.

Algorithm 1: Construction of 8×8 S-box.

/* B is a set of points (x, y) satisfying E_p, $B(i)$ is i-th point of B and y_i stands for y-component of point $B(i)$. */

Input : A prime $p \equiv 2 \pmod 3$ and two integers t and S such that $c = S + t$ and $S + t \not\equiv 0 \pmod p$.

Output: An S-box $\zeta_{E_p}(p, t, S)$.

1 $B := \varnothing$;
2 $Y := [0, (p-1)/2]$;
3 $i \leftarrow 0$;
4 **for** $x \in [0, p-1]$ **do**
5   **for** $y \in Y$ **do**
6   **if** $y^2 \equiv x^3 + c \pmod p$ **then**
7              $i \leftarrow i+1$; $B(i) := (x, y)$;
8   **if** $y \neq 0$ **then**
9                  $i \leftarrow i+1$; $B(i) := (x, p - y)$;
10   break;
11      $Y = Y - \{y\}$;
12 $\zeta_{E_p}(p, t, S) = \{y_i \in B(i) : 0 \leq y_i < 256\}$.

Table 1. The obtained S-box $\zeta_{E_{1607}}(1607, 182, 0)$.

220	118	17	158	25	138	33	196	247	252	15	226	135	177	232	83
161	70	107	186	137	236	21	142	131	103	54	58	217	181	201	172
91	84	223	89	29	156	136	14	69	99	164	171	35	188	76	139
153	16	198	227	32	10	115	122	184	61	208	225	213	106	94	56
165	40	245	189	163	239	193	194	129	175	241	141	130	231	215	127
151	199	105	22	148	39	179	173	78	248	81	23	75	55	146	109
195	251	178	170	162	206	228	169	147	28	210	221	80	121	202	77
9	74	197	31	26	154	145	44	47	82	43	60	117	250	88	191
67	8	174	93	1	20	128	53	218	237	96	72	3	65	6	253
150	101	119	87	160	133	108	57	41	64	51	49	185	243	2	249
167	50	205	183	97	114	48	27	246	254	124	92	19	134	159	95
24	224	111	62	116	168	200	86	79	143	126	112	45	71	125	13
5	216	187	222	7	113	238	36	204	52	140	46	240	85	207	4
152	104	235	190	242	68	63	203	230	176	180	59	157	244	66	212
34	90	120	0	30	166	37	255	38	110	211	233	11	155	209	219
192	12	144	73	182	132	98	214	42	102	18	149	123	229	100	234

The positive integers $a_1, b_1, a_2, b_2, a_3, \delta, L, S_P, t$ and p are secret keys. Here it is mentioned that the parameters $a_1, b_1, a_2, b_2, a_3, \delta$ and L are used to generate mn triads in a box of size L. The generation of triads is explained step by step in Algorithm 2. These triads along with keys S_P and t are used to generate the sequence $\beta_{\frac{*}{T}}(t, S_P)$ of pseudo-random numbers.

Algorithm 2: Generating quasi-resonant triads.

```
/* T is a set containing the Quasi-resonant triads, while m and n are the
   dimensions of an input image.                                          */
```

Input : Three sets $\mathcal{A}_i, i = 1, 2, 3$, inverse detuning level δ, bound L, two positive integers m and n.

Output: Quasi-resonant triads

1   $T := \varnothing$;

2   $c_1 \leftarrow 0, c_2 \leftarrow 1$;

3   **for** $a \in \mathcal{A}_1$ **do**

4   **for** $b \in \mathcal{A}_2$ **do**

5          $c_1 \leftarrow c_1 + 1$;

6   Calculate and store the values of $k_1'(c_1), l_3'(c_1)$, and $l_1'(c_1)$ for each pair (a, b) using Equations (8)–(10).

7   **for** $c_2 \in [1, c_1]$ **do**

8   **for** $k_3 \in \mathcal{A}_3$ **do**

9          $k_1 = \lfloor (k_1'(c_2) * k_3) \rceil, l_3 = \lfloor (l_3'(c_2) * k_3) \rceil$ and $l_1 = \lfloor (l_1'(c_2) * k_3) \rceil$;

10         $k_2 = k_3 - k_1, l_2 = l_3 - l_1$ and $\omega_i = k_i / (k_i^2 + l_i^2), i = 1, 2, 3$;

11         $\omega_4 = \omega_3 - \omega_2 - \omega_1$;

12   **if** $|\omega_4| < \delta^{-1}$ and $0 < |k_i|, |l_i| < L, i = 1, 2, 3$ **then**

13            $T := T \cup \{\Delta\}$;

14   **if** #$T=mn$ **then**

15   break;

16   break;

17   Sort T with respect to the ordering $\lesssim$ to get $\overset{*}{T}$.

Thus Δ_j represents the j-th triad in ordered set $\overset{*}{T}$. Moreover, $(k_{ji}, l_{ji}), i = 1, 2, 3$ are the components of Δ_j . In Algorithm 3, the generation of $\beta_{\overset{*}{T}}(t, S_P)$ is interpreted.

Algorithm 3: Generating the proposed pseudo-random sequence.

Input : An ordered set $\overset{*}{T}$, an integer t and a plain image P.

Output: Random numbers sequence $\beta_{\overset{*}{T}}(t, S_P)$.

1   $Tr(j) := |rk_{j1}| + |l_{j1}| + |k_{j2}|$;

2   $\beta_{\overset{*}{T}}(t, S_P)(j) = (Tr(j) + S_P) \pmod{256}$;

The proposed sequence $\beta_{\overset{*}{T}}(t, S_P)$ is cryptographically a good source of pseudo-randomness because triads are highly sensitive to the auxiliary parameters (a, b) [33] and inverse detuning level δ. It is shown in [32] that the intricate structure of clusters formed by triads depends on the chosen δ, and the size of the clusters increases as the inverse detuning level increases. Moreover, the generation of triads is rapid due to the absence of modular operation.

C. Performing diffusion. To change the statistical properties of an input image, a diffusion process is performed. While performing the diffusion, the pixel values are changed using the sequence $\beta_{\overset{*}{T}}(t, S_P)$. Let M_P denote the diffused image for a plain image P. The proposed scheme alters the pixels of P according to:

$$M_P(i) = \beta_{\overset{*}{T}}(t, S_P)(i) + P(i) \pmod{256}. \tag{13}$$

D. Performing confusion. A nonlinear function causes confusion in a cryptosystem, and nonlinear components are necessary for a secure data encryption scheme. The current scheme uses the dynamic S-boxes to produce the confusion in an encrypted image. If C_P stands for the encrypted image of P, then confusion is performed as follows:

$$C_P(i) = \zeta_{E_p}(p, t, S_P)(M_P(i)). \tag{14}$$

Lemma 2. *If $\#\mathcal{A}_i = n_i, i = 1, 2, 3$ and p is a prime chosen for the generation of an S-box, then the time complexity of the proposed encryption scheme is $\max\{\mathcal{O}(n_1 n_2 n_3), p^2\}$.*

Proof. The computation of all possible values of k_1', l_3' and l_1' in Algorithm 2 takes $\mathcal{O}(n_1 n_2)$ time. Similarly the time complexity for generating $\overset{*}{T}$ is $\mathcal{O}(c_1 n_3)$ but c_1 executes $n_1 n_2$ times. Thus the time required by $\overset{*}{T}$ and hence by $\beta_{\overset{*}{T}}(t, S_P)$ is $\mathcal{O}(n_1 n_2 n_3)$. Additionally, Algorithm 1 shows that the proposed S-box can be constructed in $\mathcal{O}(p^2)$ time. Thus the time complexity of the proposed scheme is $\max\{\mathcal{O}(n_1 n_2 n_3), p^2\}$. $\square$

Example 1. *In order to have a clear picture of the proposed cryptosystem, we explain the whole procedure using the following hypothetical 4×4 image. For example, let I represent the plain image of $Lena_{256 \times 256}$, and let P be the subimage of I consisting of the intersection of the first four rows and the first four columns of I as shown in Table 2, whereas the column-wise linearly ordered image P is shown in Table 3.*

Table 2. Plain image P.

162	162	162	163
162	162	162	163
162	162	162	163
160	163	160	159

Table 3. Linear ordering of image P.

$P(1)$	$P(5)$	$P(9)$	$P(13)$
$P(2)$	$P(6)$	$P(10)$	$P(14)$
$P(3)$	$P(7)$	$P(11)$	$P(15)$
$P(4)$	$P(8)$	$P(12)$	$P(16)$

We have $S_P = 2589$ and $c = 247$ and the values of other parameters are described in Section 4.3. The corresponding 16 triads are obtained by Algorithm 2 as shown in Table 4.

Table 4. The corresponding set $\overset{*}{T}$ for image P.

Δ_j	k_1	l_1	k_2	l_2	k_3	l_3	Δ_j	k_1	l_1	k_2	l_2	k_3	l_3
Δ_1	−1128	1152	1529	668	401	1820	Δ_9	−1240	1267	1681	735	441	2002
Δ_2	−1142	1167	1548	676	406	1843	Δ_{10}	−1254	1282	1700	743	446	2025
Δ_3	−1156	1181	1567	685	411	1866	Δ_{11}	−1268	1296	1719	751	451	2047
Δ_4	−1170	1195	1586	694	416	1889	Δ_{12}	−1282	1310	1738	760	456	2070
Δ_5	−1184	1210	1605	701	421	1911	Δ_{13}	−1296	1325	1757	768	461	2093
Δ_6	−1198	1224	1624	710	426	1934	Δ_{14}	−1310	1339	1776	776	466	2115
Δ_7	−1212	1238	1643	719	431	1957	Δ_{15}	−1325	1353	1796	785	471	2138
Δ_8	−1226	1253	1662	726	436	1979	Δ_{16}	−1339	1368	1815	793	476	2161

From $S_P = 2589$ and $t = 2$, it follows that $r = 1295$ and hence by application of Algorithm 3 the terms of $\beta_{\overset{*}{T}}(2, 2589)$ are listed in Table 5. Moreover, the S-box $\zeta_{E_{293}}(293, 2, 2589)$ is constructed by Algorithm 1, giving the mapping $\zeta_{E_{293}}(293, 2, 2589) : \{0, 1, \ldots, 255\} \rightarrow \{0, 1, \ldots, 255\}$, which maps the list $(0, \ldots, 255)$ to the list

(80, 213, 29, 113, 180, 2, 119, 174, 10, 103, 190, 120, 173, 99, 194, 126, 167, 42, 251, 78, 215, 84, 209, 93, 200, 130, 163, 32, 17, 117, 176, 62, 231, 110, 183, 56, 237, 75, 218, 127, 166, 73, 220, 13, 91, 202, 28, 129, 164, 118, 175, 69, 224, 50, 243, 100, 193, 137, 156, 89, 204, 12, 63, 230, 74, 219, 4, 131, 162, 134, 159, 123, 170, 90, 203, 70, 223, 87, 206, 59, 234, 145, 148, 58, 235, 57, 236, 65, 228, 15, 112, 181, 52, 241, 76, 217, 60, 233, 121, 172, 68, 225, 51, 242, 135, 158, 41, 252, 21, 142, 151, 26, 25, 40, 253, 96, 197, 136, 157, 9, 116, 177, 122, 171, 45, 248, 115, 178, 102, 191, 67, 226, 95, 198, 143, 150, 133, 160, 98, 195, 3, 94, 199, 30, 104, 189, 132, 161, 8, 64, 229, 144, 149, 140, 153, 14, 85, 208, 20, 6, 109, 184, 125, 168, 92, 201, 19, 53, 240, 31, 66, 227, 35, 82, 211, 108, 185, 139, 154, 33, 16, 86, 207, 128, 165, 5, 71, 222, 38, 255, 23, 0, 81, 212, 1, 141, 152, 111, 182, 138, 155, 49, 244, 22, 106, 187, 105, 188, 36, 54, 239, 46, 247, 43, 250, 97, 196, 27, 11, 24, 44, 249, 83, 210, 61, 232, 39, 254, 7, 72, 221, 77, 216, 47, 246, 107, 186, 48, 245, 55, 238, 124169, 34, 79, 214, 88, 205, 114, 179, 37, 18, 146, 147, 101, 192).

Table 5. Pseudo-random sequence for plain image P.

$\beta_{\stackrel{*}{T}}(2, 2589)(1) = 188$	$\beta_{\stackrel{*}{T}}(2, 2589)(5) = 126$	$\beta_{\stackrel{*}{T}}(2, 2589)(9) = 65$	$\beta_{\stackrel{*}{T}}(2, 2589)(13) = 3$
$\beta_{\stackrel{*}{T}}(2, 2589)(2) = 108$	$\beta_{\stackrel{*}{T}}(2, 2589)(6) = 47$	$\beta_{\stackrel{*}{T}}(2, 2589)(10) = 241$	$\beta_{\stackrel{*}{T}}(2, 2589)(14) = 180$
$\beta_{\stackrel{*}{T}}(2, 2589)(3) = 29$	$\beta_{\stackrel{*}{T}}(2, 2589)(7) = 224$	$\beta_{\stackrel{*}{T}}(2, 2589)(11) = 162$	$\beta_{\stackrel{*}{T}}(2, 2589)(15) = 115$
$\beta_{\stackrel{*}{T}}(2, 2589)(4) = 206$	$\beta_{\stackrel{*}{T}}(2, 2589)(8) = 144$	$\beta_{\stackrel{*}{T}}(2, 2589)(12) = 83$	$\beta_{\stackrel{*}{T}}(2, 2589)(16) = 35$

Hence by the respective application of Equation (13) and the S-box $\zeta_{E_{293}}(293, 2, 2589)$, the pixel values of diffused image M_P and encrypted image C_P are shown in Tables 6 and 7, respectively.

Table 6. Diffused image M_P.

94	32	227	166
14	209	147	87
191	130	68	22
110	51	243	194

Table 7. Encrypted image C_P.

76	231	254	19
194	54	161	65
0	67	162	209
151	69	34	1

3.3. Decryption

In our scheme the decryption process can take place by reversing the operations of the encryption process. One should know the inverse S-box $\zeta_{E_p}^{-1}(n, t, S_P)$ and the pseudo-random numbers $\beta_{\stackrel{*}{T}}(t, S_P)$. Assume the situation when the secret keys $a_1, b_1, a_2, b_2, a_3, \delta,\ L, S_P, t$ and p are transmitted by a secure channel, so that the set $\stackrel{*}{T}$ is obtained using keys $a_1, b_1, a_2, b_2, a_3, \delta$ and L, and hence the S-box $\zeta_{E_p}^{-1}(p, t, S_P)$ and the pseudo-random numbers $\beta_{\stackrel{*}{T}}(t, S_P)$ can be computed by S_P, t and p. Finally, the receiver gets the original image P by applying the following equations:

$$M_P(i) = \zeta_{E_p}^{-1}(p, t, S_P)(C_P(i)), \tag{15}$$

$$P(i) = M_P(i) - \beta_{\stackrel{*}{T}}(t, S_P)(i) \quad (\text{mod } 256). \tag{16}$$

4. Security Analysis

In this section the cryptographic strength of both the S-box construction technique and encryption scheme are analyzed in detail.

4.1. Evaluation of the Designed S-Box

An S-box with good cryptographic properties ensures the quality of an encryption technique. Generally, some standard tests such as nonlinearity (NL), linear approximation probability (LAP), strict avalanche criterion (SAC), bit independence criterion (BIC) and differential approximation probability (DAP) are used to evaluate the cryptographic strength of an S-box.

The NL [43] and the LAP [44] are outstanding features of an S-box, used to measure the resistance against linear attacks. The NL measures the level of nonlinearity and the LAP finds the maximum imbalance value of an S-box. The optimal value of the nonlinearity is 112. A low value of LAP corresponds to a high resistance. The minimum NL and the LAP values for the displayed S-box are 106 and 0.1484, respectively. This ensures that the proposed S-box is immune to linear attacks. Webster and Tavares [45] developed the concepts of the SAC and the BIC, which are used to find the confusion and diffusion creation potential of an S-box. In other words, the SAC criterion measures the change in output bits when an input bit is altered. Similarly, the BIC criterion explores the correlation in output bits when change in a single input bit occurs. The average values of the SAC and the BIC for the constructed S-box are 0.4951 and 0.4988, respectively, which are close to the optimal value 0.5. Thus, both tests are satisfied by the suggested S-box. The DAP [46] is another important feature used to analyze the capability of an S-box against differential attacks. The lowest value of DAP for an S-box implies the highest security to the differential attacks. Our DAP result is 0.0234, which is good enough to resist differential cryptanalysts.

4.2. Performance Comparison of the S-Box Generation Algorithm

After performing the rigorous analyses, the S-box constructed by the current algorithm is compared with some cryptographically strong S-boxes developed by recent schemes, as shown in Table 8.

Table 8. Comparison table of the proposed S-box $\zeta_{E_{1607}}(1607, 182, 0)$.

S-Boxes	NL	LAP	SAC			BIC			DAP
			(min)	(avg)	(max)	(min)	(avg)	(max)	
Ours	106	0.1484375	0.390625	0.49511719	0.609375	0.47265625	0.49888393	0.52539063	0.0234375
Ref. [31]	104	0.1484375	0.421900	-	0.6094	0.4629	-	0.5430	0.0469
Ref. [47]	104	0.1328125	0.40625	0.49755859	0.625	0.46679688	0.50223214	0.5234375	0.0234375
Ref. [48]	101	0.140625	0.421875	0.49633789	0.578125	0.46679688	0.49379185	0.51953125	0.03125
Ref. [49]	104	0.140625	0.421875	0.50390625	0.59375	0.4765625	0.50585938	0.5390625	0.0234375
Ref. [50]	100	0.140625	0.40625	0.50097656	0.609375	0.44726563	0.50634766	0.53320313	0.03125
Ref. [51]	106	0.140625	0.390625	0.49414063	0.609375	0.47070313	0.50132533	0.53320313	0.0234375
Ref. [52]	102	0.140625	0.421875	0.49804688	0.640625	0.4765625	0.50746373	0.53320313	0.0234375
Ref. [53]	104	0.0391	0.3906	-	0.6250	0.4707	-	0.53125	0.0391
Ref. [54]	104	0.0547000	0.4018	0.4946	0.5781	0.4667969	0.4988839	0.5332031	0.0391
Ref. [55]	108	0.1328	0.40625	0.4985352	0.59375	0.46484375	0.5020229	0.52734375	0.0234375

From Table 8 it follows that the NL of $\zeta_{E_{1607}}(1607, 182, 0)$ is greater than the S-boxes in [31,47–50,52–54], equal to that of [51] and less than the S-box developed in [55], which indicates that $\zeta_{E_{1607}}(1607, 182, 0)$ is highly nonlinear in comparison to the S-boxes in [31,47–50,52–54]. Additionally, the LAP of $\zeta_{E_{1607}}(1607, 182, 0)$ is comparable to all the S-boxes in Table 8. The SAC (average) value of $\zeta_{E_{1607}}(1607, 182, 0)$ is greater than the S-boxes in [51,54], and the SAC (max) value is less than or equal to the S-boxes in [31,47,50–53]. Similarly the BIC (min) value of $\zeta_{E_{1607}}(1607, 182, 0)$ is closer to the optimal value 0.5 than that of [31,47,48,50,51,53–55], and the BIC (max) value of the new S-box is better than that of the S-boxes in [31,49–55]. Thus the confusion/diffusion creation capability of $\zeta_{E_{1607}}(1607, 182, 0)$ is better than [31,50–53,55]. The DAP value of our suggested S-box $\zeta_{E_{1607}}(1607, 182, 0)$ is lower than the DAP of the S-boxes presented in [31,48,50,53,54] and equal to that of [47,49,51,52,55]. Thus from the above discussion it follows that the newly designed S-box shows high resistance to linear as well as differential attacks.

4.3. Evaluation of the Proposed Encryption Technique

In this section the current scheme is implemented on all gray images of the USC-SIPI Image Database [56]. The USC-SIPI database contains images of size $m \times m$, $m = 256,512,1024$. Furthermore, some security analyses that are explained one by one in the associated subsections are presented. To validate the quality of the proposed scheme, the experimental results are compared with some other encryption schemes. The parameters used for the experiments are $A_1 = A_2 = -1.0541$, $A_3 = 401$, $B_1 = B_2 = -0.8514$ and $B_3 = 691,3036,5071$ for $m = 256,512,1024$, respectively; $a_1 = 2$, $b_1 = 1000$, $a_2 = 19$, $b_2 = 1000$, $a_3 = 5$, $\delta = 1000$, $t = 2$, $p = 293$, $L = 90{,}000$ and S_P varies for each P. The experiments were performed using Matlab R2016a on a personal computer with a 1.8 GHz Processor and 6 GB RAM. All encrypted images of the database along with histograms are available at [57]. Some plain images, House$_{256 \times 256}$, Stream$_{512 \times 512}$, Boat$_{512 \times 512}$ and Male$_{1024 \times 1024}$ and their cipher images are displayed in Figure 1.

(a)   (b)   (c)   (d)

(e)   (f)   (g)   (h)

Figure 1. (a–d) Plain images House, Stream, Boat and Male; (e–h) cipher images of the plain images (a–d), respectively.

4.3.1. Statistical Attack

A cryptosystem is said to be secure if it has high resistance against statistical attacks. The strength of resistance against statistical attacks is measured by entropy, correlation and histogram tests. All of these tests are applied to evaluate the performance of the discussed scheme.

(1)   Histogram. A histogram is a graphical way to display the frequency distribution of pixel values of an image. A secure cryptosystem generates cipher images with uniform histograms. The histograms of the encrypted images using the proposed method are available at [57]. However, the respective histograms for the images in Figure 1 are shown in Figure 2. The histograms of the encrypted images are almost uniform. Moreover, the histogram of an encrypted image is totally different from that of the respective plain image, so that it does not allow useful information to the adversaries, and the proposed algorithm can resist any statistical attack.

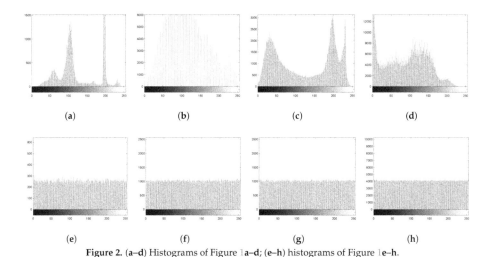

Figure 2. (a–d) Histograms of Figure 1a–d; (e–h) histograms of Figure 1e–h.

(2) Entropy. Entropy is a standout feature to measure the disorder. Let I be a source of information over a set of symbols N. Then the entropy of I is defined by:

$$H(I) = \sum_{i=1}^{\#N} p(I_i) \log_2 \frac{1}{p(I_i)},\tag{17}$$

where $p(I_i)$ is the probability of occurrence of symbol i. The ideal value of $H(I)$ is $\log_2(\#N)$, if all symbols of N occur in I with the same probability. Thus, an image I emanating 256 gray levels is highly random if $H(I)$ is close to 8 (notice, however, that this definition of entropy does not take into account pixel correlations). The entropy results for all images encrypted by the suggested technique are shown in Figure 3, where the minimum, average and maximum values are 7.9966, 7.9986 and 7.9999, respectively. These results are close to 8, and hence the developed mechanism is secure against entropy attacks.

(3) Pixel correlation. A meaningful image has strong correlation among the adjacent pixels. In fact, a good cryptosystem has the ability to break the pixel correlation and bring it close to zero. For any two gray values x and y, the pixel correlation can be computed as:

$$C_{xy} = \frac{E\big[(x - E[x])(y - E[y])\big]}{\sqrt{K[x]K[y]}},\tag{18}$$

where $E[x]$ and $K[x]$ denote expectation and variance of x, respectively. The range of C_{xy} is -1 to 1. The gray values x and y are in low correlation if C_{xy} is close to zero. As the pixels may be adjacent in horizontal, diagonal and vertical directions, the correlation coefficients of all encrypted images along all three directions are shown in Figure 3, where the respective ranges of C_{xy} are $[-0.0078, 0.0131]$, $[-0.0092, 0.0080]$ and $[-0.0100, 0.0513]$. These results show that the presented method is capable of reducing the pixel correlation near to zero.

In addition, 2000 pairs of adjacent pixels of the plain image and cipher image of Lena$_{512\times512}$ are randomly selected. Then correlation distributions of the adjacent pixels in all three directions are shown in Figure 4, which reveals the strong pixel correlation in the plain image but a weak pixel correlation in the cipher image generated by the current scheme.

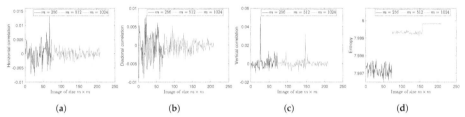

Figure 3. (**a–c**) The horizontal, diagonal and vertical correlations among pixels of each image in USC-SIPI database; (**d**) the entropy of each image in USC-SIPI database.

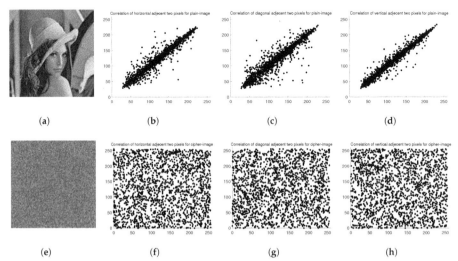

Figure 4. (**b–d**) The distribution of pixels of the plane image (**a**) in the horizontal, diagonal and vertical directions; (**f–h**) the distribution of pixels of the cipher image (**e**) in the horizontal, diagonal and vertical directions.

4.3.2. Differential Attack

In differential attacks the opponents try to get the secret keys by studying the relation between the plain image and cipher image. Normally attackers encrypt two images by applying a small change to these images, then compare the properties of the corresponding cipher images. If a minor change in the original image can cause a significant change in the encrypted image, then the cryptosystem has a high security level. The two tests NPCR (number of pixels change rate) and UACI (unified average changing intensity) are usually used to describe the security level against differential attacks. For two plain images P and P' different at only one pixel value, let C_P and $C_{P'}$ be the cipher images of P and P', respectively, then NPCR and UACI are calculated as:

$$\text{NPCR} = \sum_{u=1}^{m} \sum_{v=1}^{n} \frac{\tau(u,v)}{m \times n}, \tag{19}$$

$$\text{UACI} = \sum_{u=1}^{m} \sum_{v=1}^{n} \frac{|C_P(u,v) - C_{P'}(u,v)|}{255 \times m \times n}, \tag{20}$$

where $\tau(u,v) = 0$ if $C_P(u,v) = C_{P'}(u,v)$ and $\tau(u,v) = 1$, otherwise. The expected values of NPCR and UACI for 8-bit images are 0.996094 and 0.334635, respectively [13]. We applied the above two tests to each image of the database by randomly changing the pixel value of each image. The experimental results are shown in Figure 5, giving average values of NPCR and UACI of 0.9961 and 0.3334, respectively. It follows from the obtained results that our scheme is capable of resisting a differential attack.

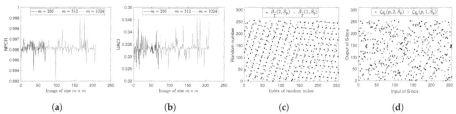

| (a) | (b) | (c) | (d) |

Figure 5. (**a,b**) The NPCR and UACI results for each image in the USC-SIPI database; (**c**) First 256 pseudo-random numbers and (**d**) two S-boxes generated for Lena$_{512\times512}$ with a small change in an input key t.

4.3.3. Key Analysis

For a secure cryptosystem it is essential to perform well against key attacks. A cryptosystem is highly secure against key attacks if it has key sensitivity and large key space and strongly opposes the known-plaintext/chosen-plaintext attack. The proposed scheme is analyzed against key attacks as follows.

(1) Key sensitivity. Attackers usually use slightly different keys to encrypt a plain image and then compare the obtained cipher image with the original cipher image to get the actual keys. Thus, high key sensitivity is essential for higher security. That is, cipher images of a plain image generated by two slightly different keys should be entirely different. The difference of the cipher images is quantified by Equations (19) and (20). In experiments we encrypted the whole database by changing only one key, while other keys remain unchanged. The key sensitivity results are shown in Table 9, where the average values of NPCR and UACI are 0.9960 and 0.3341, respectively, which specify the remarkable difference in the cipher images. Moreover, our cryptosytem is based on the pseudo-random numbers and S-boxes. The sensitivity of pseudo-random numbers sequences $\beta_{*\atop T}(2, S_P)$ and $\beta_{*\atop T}(1, S_P)$ and S-boxes $\zeta_{E_p}(p, 2, S_P)$ and $\zeta_{E_p}(p, 1, S_P)$ for Lena$_{512\times512}$ is shown in Figure 5.

Table 9. Difference between two encrypted images when key $t = 2$ is changed to $t = 1$. NPCR: number of pixels change rate; UACI: unified average changing intensity.

Image	NPCR(%)	UACI(%)	Image	NPCR(%)	UACI(%)	Image	NPCR(%)	UACI(%)
Female	99.62	33.39	House	99.62	33.23	Couple	99.56	33.30
Tree	99.59	33.35	Beans	99.64	33.23	Splash	99.60	33.97

(2) Key space. In order to resist a brute force attack, key space should be sufficiently large. For any cryptosystem, key space represents the set of all possible keys required for the encryption process. Generally, the size of the key space should be greater than 2^{128}. In the present scheme the parameters $a_1, b_1, a_2, b_2, a_3, \delta, L, S_P, t$ and p are used as secret keys, and we store each of them in 28 bits. Thus the key space of the proposed cryptosystem is 2^{280} which is larger than 2^{128} and hence capable to resist a brute force attack.

(3) Known-plaintext/chosen-plaintext attack. In a known-plaintext attack, the attacker has partial knowledge about the plain image and cipher image, and tries to break the cryptosystem, while in a chosen-plaintext attack the attacker encrypts an arbitrary image to get the encryption keys. An all-white/black image is usually encrypted to test the performance of a scheme against these powerful attacks [29,58]. We analyzed our scheme by encrypting an all-white/black image of size 256 × 256. The results are shown in Figure 6 and Table 10, revealing that the encrypted

images are significantly randomized. Thus the proposed system is capable of preventing the above mentioned attacks.

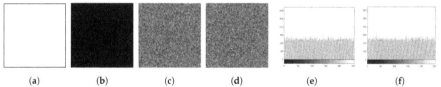

| (a) | (b) | (c) | (d) | (e) | (f) |

Figure 6. (a) All-white; (b) all-black; (c,d) cipher images of (a,b); (e,f) histograms of (c,d).

Table 10. Security analysis of all-white/black encrypted images by the proposed encryption technique.

Plain Image	Entropy	Correlation of Plain Image			NPCR (%)	UACI (%)
		Hori.	Diag.	Ver.		
All-white	7.9969	0.0027	0.0020	−0.0090	99.60	33.45
All-black	7.9969	−0.0080	0.0035	0.0057	99.62	33.41

4.4. Comparison and Discussion

Apart from security analyses, the proposed scheme is compared with some well-known image encryption techniques. The gray scale images of $Lena_{256 \times 256}$ and $Lena_{512 \times 512}$ are encrypted using the presented method, and experimental results are listed in Table 11.

Table 11. Comparison of the proposed encryption scheme with several existing cryptosystems for image $Lena_{m \times m}$, $m = 256,512$.

Size m	Algorithm	Entropy	Correlation			NPCR (%)	UACI(%)	# S-Boxes	Dynamic S-Boxes
			Hori.	Diag.	Ver.				
	Ours	7.9974	0.0001	−0.0007	−0.0001	99.91	33.27	1	Yes
	Ref. [31]	7.9993	0.0012	0.0003	0.0010	99.60	33.50	1	Yes
	Ref. [3]	7.9973	-	-	-	99.50	33.30	0	-
256	Ref. [27]	7.9046	0.0164	−0.0098	0.0324	98.92	32.79	>1<50	Yes
	Ref. [26]	7.9963	−0.0048	−0.0045	−0.0112	99.62	33.70	8	Yes
	Ref. [59]	7.9912	−0.0001	0.0091	0.0089	100	33.47	0	-
	Ref. [60]	7.9974	0.0020	0.0020	0.0105	99.59	33.52	0	-
	Ours	7.9993	0.0001	0.0042	0.0021	99.61	33.36	1	Yes
512	Ref. [25]	7.9992	0.0075	0.0016	0.0057	99.61	33.38	1	No
	Ref. [29]	7.9993	−0.0004	0.0001	−0.0018	99.60	33.48	1	No
	Ref. [61]	7.9970	−0.0029	0.0135	0.0126	99.60	33.48	0	-
-	Ref. [62]	7.9994	0.0018	−0.0012	0.0011	99.62	33.44	>1	Yes
	Ref. [2]	7.9993	0.0032	0.0011	−0.0002	99.60	33.47	>1	Yes

It is deduced that our scheme generates cipher images with comparable security. Furthermore, we remark that the scheme in [29] generates pseudo-random numbers using group law on EC, while the proposed method generates pseudo-random numbers by constructing triads using auxiliary parameters of elliptic surfaces. Group law consists of many operations, which makes the pseudo-random number generation process slower than the one we present here. The scheme in [26] decomposes an image to eight blocks and uses dynamic S-boxes for encryption purposes. The computation of multiple S-boxes takes more time than computing only one S-box. Similarly the techniques in [2,27] use a set of S-boxes and encrypt an image in blocks, while our newly developed scheme encrypts the whole image using only one dynamic S-box. Thus, our scheme is faster than the schemes in [2,27]. The security system in [61] uses a chaotic system to encrypt blocks of an image. The results in Table 11 reveal that our proposed system is cryptographically stronger than the scheme in [61]. The algorithms in [3,59] combine chaotic systems and different ECs to encrypt images. It follows from Table 11 that the security level of our scheme is comparable to that of the schemes in [3,59]. The technique in [60] uses double chaos along with DNA coding to get good results, as shown in Table 11, but the results obtained by the new scheme are better than that of [60]. Similarly the technique in [31] encrypts images using ECs but does not guarantee an S-box for each set of input parameters, thus making our scheme faster and more robust than the scheme developed in [31].

Furthermore, the following facts put our scheme in a favorable position:

(i) Our scheme uses a dynamic S-box for each input image while the S-box used in [29] is a static one, which is vulnerable [63] and less secure than a dynamic one [64].

(ii) The presented scheme guarantees an S-box for each image, which is not the case in [31].

(iii) To get random numbers, the described scheme generates triads for all images of the same size, while in [31] the computation of an EC for each input image is necessary, which is time consuming.

(iv) The scheme in [26] uses eight dynamic S-boxes for a plain image, while the current scheme uses only one dynamic S-box for each image to get the desired cryptographic security.

5. Conclusions

An image encryption scheme based on quasi-resonant triads and MECs was introduced. The proposed technique constructs triads to generate pseudo-random numbers and computes an MEC to construct an S-box for each input image. The pseudo-random numbers and S-box are then used for altering and scrambling the pixels of the plain image, respectively. As for the advantages of our proposed method, firstly triads are based on auxiliary parameters of elliptic surfaces, and thus pseudo-random numbers and S-boxes generated by our method are highly sensitive to the plain image, which prevents adversaries from initiating any successful attack. Secondly, generation of triads using auxiliary parameters of elliptic surfaces consumes less time than computing points on ECs (we find a 4x speed increase for a range of image resolutions $m \in [128, 512]$), which makes the new encryption system relatively faster. Thirdly, our algorithm generates the cipher images with an appropriate security level.

In summary, all of the above analyses imply that the presented scheme is able to resist all attacks. It has high encryption efficiency and less time complexity than some of the existing techniques. In the future, the current scheme will be further optimized by means of new ideas to construct the S-boxes using the constructed triads, so that we will not need to compute an MEC for each input image.

Author Contributions: All authors contributed equally to this work. All authors have read and agree to the published version of the manuscript.

Funding: This research is funded through the HEC project NRPU-7433.

Acknowledgments: We thank Gene Kopp for useful comments and suggestions.

Conflicts of Interest: The authors declare no conflict of interest. The funding sponsors had no role in the design of the study; in the collection, analyses, or interpretation of data; in the writing of the manuscript, and in the decision to publish the results.

Abbreviations

The following abbreviations are used in this manuscript:

MEC   Mordell elliptic curve
S-box   Substitution box
EC   Elliptic curves

References

1. Mahmud, M.; Lee, M.; Choi, J.Y. Evolutionary-Based Image Encryption using RNA Codons Truth Table. *Optics Laser Technol.* **2020**, *121*, 105818. [CrossRef]

2. Zhang, X.; Mao, Y.; Zhao, Z. An Efficient Chaotic Image Encryption Based on Alternate Circular S-boxes. *Nonlinear Dyn.* **2014**, *78*, 359–369. [CrossRef]

3. El-Latif, A.A.A.; Niu, X. A Hybrid Chaotic System and Cyclic Elliptic Curve for Image Encryption. *AEU-Int. J. Electron. Commun.* **2013**, *67*, 136–143. [CrossRef]

4. Yang, Y.G.; Pan, Q.X.; Sun, S.J.; Xu, P. Novel Image Encryption Based on Quantum Walks. *Sci. Rep.* **2015**, *5*, 1–9. [CrossRef] [PubMed]

5. Zhong, H.; Chen, X.; Tian, Q. An Improved Reversible Image Transformation using K-Means Clustering and Block Patching. *Information* **2019**, *10*, 17. [CrossRef]

6. Li, C.; Lin, D.; Lü, J. Cryptanalyzing an Image-Scrambling Encryption Algorithm of Pixel Bits. *IEEE MultiMedia* **2017**, *24*, 64–71. [CrossRef]

7. Hua, Z.; Yi, S.; Zhou, Y. Medical Image Encryption using High-Speed Scrambling and Pixel Adaptive Diffusion. *Signal Process.* **2018**, *144*, 134–144. [CrossRef]

8. Xie, E.Y.; Li, C.; Yu, S.; Lü, J. On the Cryptanalysis of Fridrich's Chaotic Image Encryption Scheme. *Signal Process.* **2017**, *132*, 150–154. [CrossRef]

9. Azam, N.A. A Novel Fuzzy Encryption Technique Based on Multiple Right Translated AES Gray S-Boxes and Phase Embedding. *Secur. Commun. Netw.* **2017**, *2017*, 5790189. [CrossRef]

10. Luo, Y.; Tang, S.; Qin, X.; Cao, L.; Jiang, F.; Liu, J. A Double-Image Encryption Scheme Based on Amplitude-Phase Encoding and Discrete Complex Random Transformation. *IEEE Access* **2018**, *6*, 77740–77753. [CrossRef]

11. Li, J.; Li, J.S.; Pan, Y.Y.; Li, R. Compressive Optical Image Encryption. *Sci. Rep.* **2015**, *5*, 10374. [CrossRef] [PubMed]

12. Hua, Z.; Xu, B.; Jin, F.; Huang, H. Image Encryption using Josephus Problem and Filtering Diffusion. *IEEE Access* **2019**, *7*, 8660–8674. [CrossRef]

13. Wu, J.; Cao, X.; Liu, X.; Ma, L.; Xiong, J. Image Encryption using the Random FrDCT and the Chaos-Based Game of Life. *J. Modern Opt.* **2019**, *66*, 764–775. [CrossRef]

14. Yousaf, A.; Alolaiyan, H.; Ahmad, M.; Dilbar, N.; Razaq, A. Comparison of Pre and Post-Action of a Finite Abelian Group Over Certain Nonlinear Schemes. *IEEE Access* **2020**, *8*, 39781–39792. [CrossRef]

15. Ismail, S.M.; Said, L.A.; Radwan, A.G.; Madian, A.H.; Abu-ElYazeed, M.F. A Novel Image Encryption System Merging Fractional-Order Edge Detection and Generalized Chaotic Maps. *Signal Process.* **2020**, *167*, 107280. [CrossRef]

16. Tang, Z.; Yang, Y.; Xu, S.; Yu, C.; Zhang, X. Image Encryption with Double Spiral Scans and Chaotic Maps. *Secur. Commun. Netw.* **2019**, *2019*, 8694678. [CrossRef]

17. Abdelfatah, R.I. Secure Image Transmission using Chaotic-Enhanced Elliptic Curve Cryptography. *IEEE Access* **2019**, *8*, 3875–3890. [CrossRef]

18. Yu, J.; Guo, S.; Song, X.; Xie, Y.; Wang, E. Image Parallel Encryption Technology Based on Sequence Generator and Chaotic Measurement Matrix. *Entropy* **2020**, *22*, 76. [CrossRef]

19. Zhu, S.; Zhu, C.; Wang, W. A Novel Image Compression-Encryption Scheme Based on Chaos and Compression Sensing. *IEEE Access* **2018**, *6*, 67095–67107. [CrossRef]

20. ElKamchouchi, D.H.; Mohamed, H.G.; Moussa, K.H. A Bijective Image Encryption System Based on Hybrid Chaotic Map Diffusion and DNA Confusion. *Entropy* **2020**, *22*, 180. [CrossRef]

21. Zhou, Y.; Bao, L.; Chen, C.P. Image Encryption using a New Parametric Switching Chaotic System. *Signal Process.* **2013**, *93*, 3039–3052. [CrossRef]

22. Hua, Z.; Zhou, Y.; Huang, H. Cosine-Transform-Based Chaotic System for Image Encryption. *Inf. Sci.* **2019**, *480*, 403–419. [CrossRef]

23. Xu, Y.; Wang, H.; Li, Y.; Pei, B. Image Encryption Based on Synchronization of Fractional Chaotic Systems. *Commun. Nonlinear Sci. Numer. Simul.* **2014**, *19*, 3735–3744. [CrossRef]

24. Ahmad, M.; Shamsi, U.; Khan, I.R. An Enhanced Image Encryption Algorithm using Fractional Chaotic Systems. *Procedia Comput. Sci.* **2015**, *57*, 852–859. [CrossRef]

25. Cheng, P.; Yang, H.; Wei, P.; Zhang, W. A Fast Image Encryption Algorithm Based on Chaotic Map and Lookup Table. *Nonlinear Dyn.* **2015**, *79*, 2121–2131. [CrossRef]

26. Belazi, A.; El-Latif, A.A.A.; Belghith, S. A Novel Image Encryption Scheme Based on Substitution-Permutation Network and Chaos. *Signal Process.* **2016**, *128*, 155–170. [CrossRef]

27. Rehman, A.U.; Khan, J.S.; Ahmad, J.; Hwang, S.O. A New Image Encryption Scheme Based on Dynamic S-boxes and Chaotic Maps. *3D Res.* **2016**, *7*, 7. [CrossRef]

28. Jia, N.; Liu, S.; Ding, Q.; Wu, S.; Pan, X. A New Method of Encryption Algorithm Based on Chaos and ECC. *J. Inf. Hiding Multimedia Signal Process.* **2016**, *7*, 637–643.
29. Toughi, S.; Fathi, M.H.; Sekhavat, Y.A. An Image Encryption Scheme Based on Elliptic Curve Pseudo Random and Advanced Encryption System. *Signal Process.* **2017**, *141*, 217–227. [CrossRef]
30. Liu, H.; Liu, Y. Cryptanalyzing an Image Encryption Scheme Based on Hybrid Chaotic System and Cyclic Elliptic Curve. *Opt. Laser Technol.* **2014**, *56*, 15–19. [CrossRef]
31. Hayat, U.; Azam, N.A. A Novel Image Encryption Scheme Based on an Elliptic Curve. *Signal Process.* **2019**, *155*, 391–402. [CrossRef]
32. Bustamante, M.D.; Hayat, U. Complete Classification of Discrete Resonant Rossby/Drift Wave Triads on Periodic Domains. *Commun. Nonlinear Sci. Numer Simul.* **2013**, *18*, 2402–2419. [CrossRef]
33. Hayat, U.; Amanullah, S.; Walsh, S.; Abdullah, M.; Bustamante, M.D. Discrete Resonant Rossby/Drift Wave Triads: Explicit Parameterisations and a Fast Direct Numerical Search Algorithm. *Commun. Nonlinear Sci. Numer. Simul.* **2019**, *79*, 104896. [CrossRef]
34. Azam, N.A.; Hayat, U.; Ullah, I. An Injective S-Box Design Scheme over an Ordered Isomorphic Elliptic Curve and Its Characterization. *Secur. Commun. Netw.* **2018**, *2018*, 3421725. [CrossRef]
35. Charney, J.G. On the scale of atmospheric motions. *Geophys. Public* 1948, *17*, 3–17.
36. Hasegawa, A.; Mima, K. Pseudo-three-dimensional turbulence in magnetized nonuniform plasma. *Phys. Fluids* **1978**, *21*, 87–92. [CrossRef]
37. Connaughton, C.P.; Nadiga, B.T.; Nazarenko, S.V.; Quinn, B.E. Modulational instability of Rossby and drift waves and generation of zonal jets. *J. Fluid Mech.* **2010**, *654*, 207–231. [CrossRef]
38. Harris, J.; Connaughton, C.; Bustamante, M.D. Percolation Transition in the Kinematics of Nonlinear Resonance Broadening in Charney–Hasegawa–Mima Model of Rossby Wave Turbulence. *New J. Phys.* **2013**, *15*, 083011. [CrossRef]
39. Galperin, B.; Read, P.L. (Eds.) *Zonal Jets: Phenomenology, Genesis, and Physics*; Cambridge University Press: Cambridge, UK, 2019.
40. Kopp, G.S. The Arithmetic Geometry of Resonant Rossby Wave Triads. *SIAM J. Appl. Algebra Geomet.* **2017**, *1*, 352–373. [CrossRef]
41. Washington, L.C. *Elliptic Curves Number Theory and Cryptography, Discrete Mathematics and Its Applications*, 2nd ed.; Chapman and Hall/CRC, University of Maryland College Park: College Park, MD, USA, 2003.
42. Azam, N.A.; Hayat, U.; Ullah, I. Efficient Construction of S-boxes Based on a Mordell Elliptic Curve Over a Finite Field. *Front. Inf. Technol. Electron. Eng.* **2019**, *20*, 1378–1389. [CrossRef]
43. Adams, C.; Tavares, S. The Structured Design of Cryptographically Good S-boxes. *J. Cryptol.* **1990**, *3*, 27–41. [CrossRef]
44. Matsui, M. Linear cryptanalysis method of DES cipher. In *Advances in Cryptology, Proceedings of the Workshop on the Theory and Application of of Cryptographic Techniques (EURO-CRYPT-93), Lofthus, Norway, 23–27 May 1993*; Springer: Berlin/Heidelberg, Germany, 1994; pp. 386–397.
45. Webster, A.; Tavares, S.E. On the design of S-boxes. In *Conference on the Theory and Application of Cryptographic Techniques*; Springer: Berlin/Heidelberg, Germany, 1985; pp. 523–534.
46. Biham, E.; Shamir, A. Differential Cryptanalysis of DES-like Cryptosystems. *J. Cryptol.* **1991**, *4*, 3–72. [CrossRef]
47. Ye, T.; Zhimao, L. Chaotic S-box: Six-Dimensional Fractional Lorenz–Duffing Chaotic System and O-shaped Path Scrambling. *Nonlinear Dyn.* **2018**, *94*, 2115–2126. [CrossRef]
48. Özkaynak, F.; Çelik, V.; Özer, A.B. A New S-box Construction Method Based on the Fractional-Order Chaotic Chen System. *Signal Image Video Process.* **2017**, *11*, 659–664. [CrossRef]
49. Çavuşoğlu, Ü.; Zengin, A.; Pehlivan, I.; Kaçar, S. A Novel Approach for Strong S-Box Generation Algorithm Design Based on Chaotic Scaled Zhongtang System. *Nonlinear Dyn.* **2017**, *87*, 1081–1094. [CrossRef]
50. Belazi, A.; El-Latif, A.A.A. A Simple yet Efficient S-box Method Based on Chaotic Sine Map. *Optik* **2017**, *130*, 1438–1444. [CrossRef]
51. Özkaynak, F. Construction of robust substitution boxes based on chaotic systems. *Neural Comput. Appl.* **2019**, *31*, 3317–3326. [CrossRef]
52. Liu, L.; Zhang, Y.; Wang, X. A Novel Method for Constructing the S-box Based on Spatiotemporal Chaotic Dynamics. *Appl. Sci.* **2018**, *8*, 2650. [CrossRef]

53. Hayat, U.; Azam, N.A.; Asif, M. A Method of Generating 8 × 8 Substitution Boxes Based on Elliptic Curves. *Wirel. Pers. Commun.* **2018**, *101*, 439–451. [CrossRef]

54. Wang, X.; Çavuşoğlu, Ü.; Kacar, S.; Akgul, A.; Pham, V.T.; Jafari, S.; Alsaadi, F.E.; Nguyen, X.Q. S-box Based Image Encryption Application using a Chaotic System without Equilibrium. *Appl. Sci.* **2019**, *9*, 781. [CrossRef]

55. Alzaidi, A.A.; Ahmad, M.; Ahmed, H.S.; Solami, E.A. Sine-Cosine Optimization-Based Bijective Substitution-Boxes Construction using Enhanced Dynamics of Chaotic Map. *Complexity* **2018**, *2018*, 9389065. [CrossRef]

56. USC-SIPI Image Database. Available online: http://sipi.usc.edu/database/database.php (accessed on 21 February 2020).

57. Available online: https://github.com/ikram702314/Results (accessed on 15 April 2020).

58. Wang, X.; Zhao, H.; Hou, Y.; Luo, C.; Zhang, Y.; Wang, C. Chaotic Image Encryption Algorithm Based on Pseudo-Random Bit Sequence and DNA Plane. *Modern Phys. Lett. B* **2019**, *33*, 1950263. [CrossRef]

59. Wu, J.; Liao, X.; Yang, B. Color Image Encryption Based on Chaotic Systems and Elliptic Curve ElGamal Scheme. *Signal Process.* **2017**, *141*, 109–124. [CrossRef]

60. Wan, Y.; Gu, S.; Du, B. A New Image Encryption Algorithm Based on Composite Chaos and Hyperchaos Combined with DNA Coding. *Entropy* **2020**, *22*, 171. [CrossRef]

61. Tong, X.J.; Zhang, M.; Wang, Z.; Ma, J. A Joint Color Image Encryption and Compression Scheme Based on Hyper-Chaotic System. *Nonlinear Dyn.* **2016**, *84*, 2333–2356. [CrossRef]

62. Zhang, Y.; Xiao, D. An Image Encryption Scheme Based on Rotation Matrix Bit-Level Permutation and Block Diffusion. *Commun. Nonlinear Sci. Numer. Simul.* **2014**, *19*, 74–82. [CrossRef]

63. Rosenthal, J. A Polynomial Description of the Rijndael Advanced Encryption Standard. *J. Algebra. Its Appl.* **2003**, *2*, 223–236. [CrossRef]

64. Kazlauskas, K.; Kazlauskas, J. Key-Dependent S-box Generation in AES Block Cipher system. *Informatica* **2009**, *20*, 23–34.

© 2020 by the authors. Licensee MDPI, Basel, Switzerland. This article is an open access article distributed under the terms and conditions of the Creative Commons Attribution (CC BY) license (http://creativecommons.org/licenses/by/4.0/).

Article

A New Algorithm for Medical Color Images Encryption Using Chaotic Systems

Seyed Shahabeddin Moafimadani *, Yucheng Chen and Chunming Tang *

School of Mathematics and Information Science, Guangzhou University, Guangzhou 510006, China;
yuchengchen@e.gzhu.edu.cn
* Correspondence: ctang@gzhu.edu.cn (C.T.); Shahabmadani@e.gzhu.cn (S.S.M.)

Received: 11 May 2019; Accepted: 4 June 2019; Published: 10 June 2019

Abstract: In this paper, we present a new algorithm based on chaotic systems to protect medical images against attacks. The proposed algorithm has two main parts: A high-speed permutation process and adaptive diffusion. After the implementation of the algorithm in the MATLAB software, it is observed that the algorithm is effective and appropriate. Also, to quantitatively evaluate the uniformity of the histogram, the chi-square test is done. Key sensitivity analysis demonstrates that images cannot be decrypted whenever a small change happens in the key, which indicates that the algorithm is suitable. Clearly, part of special images is selected to test the selected plain-text, like an all-white image and an all-black image. Entropy results obtained from the implementation of the algorithm on this type of images show that the proposed method is suitable for this particular type of images. In addition, the obtained results from noise and occlusion attacks analysis show that the proposed algorithm can withstand against these types of attacks. Moreover, it can be seen that the images after encryption and decryption are of good quality; the measures such as the correlation coefficients, the entropy, the number of pixel change rate (NPCR), and the uniform average change intensity (UACI) have suitable values; and the method is better than previous methods.

Keywords: image encryption; medical color images; RGB; chaotic system

1. Introduction

The confidentiality of patient information is one of the vital security aspects of electronic health services. For example, the confidentiality of patients' medical records is necessary. In addition, the methods of protection should be improved due to the rapid advancement of technologies for accessing the personal information of individuals. The security and privacy of medical image transferring is one of the acute subjects that should be seriously considered in telecare medical information systems (TMIS). In the past years, medical images were grayscale, but today, color images have entered the medical arena, and they can show more accurate information about body conditions. Color images that are acquired by new scanners using the Medipix3RX chip technology are very important in the medical arena. Image-data transferring from a position to another via an unsafe network are usually determined in qualifications of privacy, validity, totality, and confidentiality. Therefore, more significance should be given to the security of the sensitive data that are included in medical images by DICOM (digital imaging communication in medicine). In this textuality, many problems using various cryptographic techniques have been proposed in the literature to overcome this problem [1–4]. In their paper, Abdel-Nabi and Al-Haj proposed a hybrid encryption algorithm using watermarking that offers high embedding capabilities for medical images. The proposed algorithm is a combination of reversible data-caching techniques with standard encryption techniques for ensuring the security requirements for transferred and stored medical images [5]. In their paper, Abdmouleh et al. presented a partial cryptographic approach that was based on the digital wavelet transform (DWT) and was

JPEG2000 compliant to ensure the safe transfer and storage of medical images [6]. Lakshmi et al. presented a similar algorithm using a discrete wavelet transform (DWT), with the difference being that they used a fuzzy chaotic map for the watermarking [7]. In their paper, Cao et al. presented a medical image encryption algorithm using edge maps that were derived from a source image. The algorithm consists of three parts: Bit-plane decomposition, a random-sequence generator, and permutation [8]. Ismail et al. presented the double-humped (DH) logistic map to produce pseudorandom numbers keys (PRNG) in their paper. The generalized parameter that is added to the map provides more control on the map chaotic range [9]. Jeong et al. proposed a new medical image encrypting method using a 2D chaotic map and C-MLCA in their article. The 2D chaotic map is a construction with self-guarding attributes, which moves the location of the pixel and encrypts the image [10]. In their method, Ke et al. offered an encryption algorithm that was based on reversible data using the MSB-based prediction [11]. In their study, Nematzadeh et al. proposed an encryption method for medical images based on a hybrid pattern of the improved genetic algorithm (IGA) and paired map lattices. First, the assumed way employs a paired map lattice to produce a some of secure cipher-images as a primitive population of the IGA. Then, it exerts the IGA to both increase the entropy of the cipher images and reduce the algorithm's calculations time [12]. In their paper, Singh et al. proposed a medical image encryption scheme algorithm using the improved ElGamal encryption technique. Their proposed method made a new contribution since the necessity for separate calculations for encoding plain messages to elliptic curve coordinates was removed. The algorithm using the improved version of the ElGamal encryption scheme is designed to encrypt medical images [13]. Suganya and Amudha's proposed method uses two encryption algorithms, namely, RC4 and AES, which are the stream cipher and block cipher algorithms, respectively. The main objective of this method is to provide integrity control for medical images, although they are encrypted. Experimental security analysis is conducted using 8-bit ultrasound images and 16-bit positron emission tomography (PET) images [14]. The chaos systems that have become popular today have been used in many types of research. For example: Liang and Qi investigated mechanical analysis of generates the Chen chaotic system to the extensile Kolmogorov system. In Hu et al.'s research, the Chen chaotic system is designed as a pseudorandom sequence producer. In their research, Wang et al. used the memristor chaotic systems (MCSs). Gong et al. provided a method for image encryption based on hyper-chaotic system and discrete fractional random transform. Michail et al.'s research was based on chaotic systems and hash functions to implement totally self-checking (TSC). James et al., in their research based on chaotic systems and hash functions, discussed the performance of SHA-3 256- bit core. Ahmad and Das, based on chaos and hash algorithms, discussed Hardware performance analysis of SHA-256 and SHA-512 algorithms on FPGAs. In their research, Xu et al. improved chaotic cryptosystem based on circular bit shift and XOR operations. Pareek and Patidar, in their research, designed medical image protection based on genetic algorithm and chaotic system. Hua et al., in their study, designed medical image encryption using hash algorithm high-speed scrambling [15–24]. Also, Chai et al. designed a color image encryption method based on dynamic DNA encryption and chaotic system [25] and Ma et al. provided a new method of plaintext-related and chaos-based image encryption [26]. Niyat et al. offered an image encryption algorithm with the rule of cellular automata (CA). CA is a self- establishing construction with a group of cells in which any cell is updated based on certain regulation that are to depend on a limited number of neighboring cells [27]. Chen et al. designed an image encryption method based on hyperchaotic system in the turner transfigure domain. The RGB ingredients of the main color image are encrypted into 1D circulation. [28]. The method in [5] focused on the embedding capacity, but no results are given with respect to other criteria, such as the correlation coefficients entropy, the number of pixel change rate (NPCR), or the uniform average change intensity (UACI). The methods in [6,7] are based on the digital wavelet transform (DWT) and allow for safe transport, but they do not provide suitable results with respect to the correlation coefficients. The results that were obtained in [8] are only suitable for the NPCR, and other values are not suitable. In the method in [9], the values that are obtained are appropriate, but they are only suitable for grayscale images. The results of the methods in [10–12] are only suitable with respect to their correlation coefficients, but there are no improvements with respect other criteria. The results that are obtained in [13,14] are only suitable with respect to entropy. The methods in [27,28] perform well in color images encrypting and can be used to encrypt medical images. Nevertheless, our method is better in terms of the correlation coefficients, the entropy, the NPCR, and the UACI.

Our aim in this paper is to provide an algorithm that protects medical color images based on chaotic systems and SHA-256 systems. The algorithm is composed of two parts: A high-speed permutation process and adaptive diffusion. For this reason, in Section 2, we will present the basic concepts of chaotic systems and SHA-256 systems. In Section 3, we will describe the proposed algorithm's equations, and in Section 4, the empirical results from the implementation of the proposed algorithm that is simulated in the MATLAB software will be given. In Section 5, we compare the results of the proposed method with previous methods, and in Section 6, we will explain the quality and appropriateness of this method.

2. Preliminary Work

2.1. Chaotic Systems

Chaos theory is a chapter of mathematics centralization on the action of dynamical systems that are highly sensitive to initial situation. "Chaos" is a notion denoting that within the obvious accidentalness of chaotic systems, there are basically models, stable feedback rings, iteration, self-likeness, fractals, self- formation, and dependence on programming at the initial part, which is known to have sensitive to depend on initial situation.

Little differences in initial situation, like those owing to rounding errors in numerical calculations, output widely in different outcomes for dynamical systems, thus, generally interpretation of the long-term oracle of their action impossible. This action is known as certain chaos, or simply chaos. The theory was tabloid by Edward Lorenz as follows:

Chaos: When the design specifies the future, but the proximate present does not proximately specify the future.

In 1963, Lorenz studied chaotic systems using a nonlinear differential equation, which is one of the first examples of algebraic chaotic systems in dissipative systems. Chaotic systems are very abundant in nature and they are used in many branches of science, such as the physics of dynamics and photonics, medical sciences, chemistry, and demography. In recent years, much research has been done on chaotic systems by scientists and more practical systems have been introduced, such as Chen's system, Lu's system, and Qi's 3D four-wing chaotic system [15–17]. The chaotic system that is used in this algorithm is the Chen-based hyper-chaotic system. The Chen-based hyper-chaotic system in [18] is described as follows:

$$\begin{cases} \dot{s} = a(t-s) \\ \dot{t} = ds - su + ct - v \\ \dot{u} = st - bu \\ \dot{v} = s + w \end{cases} \tag{1}$$

where s, t, u, and v are the fixed variables and a, b, c, d, and w are the controlling parameters of the system. The dynamical cycle will be hyper-chaotic when a = 36, b = 3, c = 28, d = −16, and −0.7 < w < 0.7.

2.2. SHA-256 (Secure Hash Algorithm 256)

A cryptographic hash (sometimes called a "digest") is a type of 'signature' for a text or data file. The SHA-256 products an almost-unique 256-bit (32-byte) signature for a text. The SHA-256 is a type of the deputy hash functions to the SHA-1 (referential to as SHA-2) and is one of the existing powerfulness hash functions. The SHA-256 is not much more complicated to code than the SHA-1 and has not yet been agreement in any path. The 256-bit key makes it a good common-function for the

AES. It is explained in the NIST (National Institute of Standards and Technology) standard 'FIPS 180-4'. The NIST also prepared a number of test vectors to investigate the validity of its execution [19–21].

To the SHA-256, the message is decomposed to n blocks with 512 bits, and at the end of its the final block, bit '1' is added to be followed by k zero bits, where k is the least nonnegative the answer path of the equation l + 1 + k = 448mod512. Next, a 64-bit binary block that is equivalent to l is added. Clearly, a "1" followed by k "0" s that is followed by 64 bits are added at the end of M to generate a crooked message of length $512 * n$ bits. For instance, the 8-bit ASCII message "abc" has a length of $l = 8 \times 3 = 24$. Therefore, the message is padded with a one bit, then $448 - (24 + 1) = 423$ zero, and then the 64 bits of the length of message $(11000)_2 = (24)_{10}$. Then, one message plan carries out on the blocks of M, generating the W_t amount, any of which is to the corresponding t-th repetition of the transmutation. The transmutation takes W_t, a fixed value, K_t and the primary amounts $H^{(0)}$ (in the repetition one) or the values generated in the past repetition; carries out the transmutation procedure; and produces a series of hash values via a number of repetitions. The last produced hash value is considered as the message digest, h [19]. The SHA-256 needs 64 repetition to generate its message digest. The round contains additives and rational functions that are set to generate the round's output values. The included NLFs are shown in Equation (2):

$$Ch(x, \, y, z) = (x \bullet y) \oplus (\overline{x} \bullet y)$$
$$Maj(x, \, y, \, z) = (x \bullet y) \oplus (x \bullet z) \oplus (x \bullet z)$$
$$\sum_{0}^{256}(x) = ROTR^2(x) \oplus ROTR^{13}(x) \oplus ROTR^{22}(x) \tag{2}$$
$$\sum_{1}^{256}(x) = ROTR^6(x) \oplus ROTR^{11}(x) \oplus ROTR^{25}(x)$$

where $\oplus$, $\bullet$, and $\overline{}$ denote the XOR, AND, and NOT bitwise rational functions, respectively, and x, y, and z are 32-bit words. $ROTR^X$ Shows x right round bit spin. According to the 64 W_t values that are necessary, the first 16 are organized by the 512-bit input block whereas the remaining 48 W_t amounts are calculated using Equation (3). The functions σ_0 and σ_1 are computed using Equation (4).

$$W_t = \sigma_1^{\{256\}}(W_{t-2}) + W_{t-7} + \sigma_0^{\{256\}}(W_{t-15}) + W_{t-16} \, 16 \leq t \leq 63 \tag{3}$$

$$\sigma_0^{\{256\}} = ROTR^7(x) \oplus ROTR^{18}(x) \oplus SHR^3(x)$$
$$\sigma_1^{\{256\}} = ROTR^{17}(x) \oplus ROTR^{19}(x) \oplus SHR^{10}(x) \tag{4}$$

Here, SHR^x stands for the right bit shift. The SHA-256 base transformation rounds are shown in Figure 1.

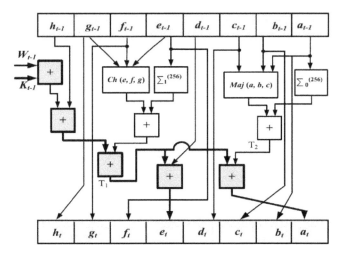

Figure 1. Secure hash algorithm 256 (SHA-256) base transformation rounds.

2.3. SHA-256 Architectures

The performance of the SHA-256 construction's transformation round is shown in Figure 1. It takes, as inputs, eight 32-bit characters, $(a_{t-1} - h_{t-1})$, the value W_{t-1}, and the stable value K_{t-1}, performs the calculations shown in Figure 1, and generates the values $(a_t - h_t)$ after 64 repetitions [19].

3. Proposed Algorithm

The proposed algorithm has two parts: A high-speed permutation process and adaptive diffusion.

3.1. High-Speed Permutation Process

Step 1. First, the plain image P of size $M \times N \times D$ is used as the input; it has the initial state values of $a0$, $b0$, $c0$, and $d0$; and uses the SHA-256 function, which is constructed according to the plain image. We consider $DM = D \times M$ and $s = sum(sha256)/(64 \times 256)$, which reshape matrix P into the 2D matrix P1. Then, the new values of the chaos system $a0$, $b0$, $c0$, and $d0$ are generated using Equation (4) as follows:

$$\begin{cases} a0 = s + a0 - s + a0 \\ b0 = s + b0 - s + b0 \\ c0 = s + c0 - s + c0 \\ d0 = s + d0 - s + d0 \end{cases}. \tag{5}$$

Step 2. Then, the initial values and parameters are used to iterate the chaotic systems to obtain the vectors $a1$, $a2$, $a3$, and $a4$ and quantize to generate four different vectors $PR1$, $PC1$, $PR2$, and $PC2$, which are as follows:

$$\begin{cases} PR1 = (|a1| - a1) \times 10^{15} \bmod N + 1 \\ PC1 = (|a2| - a2) \times 10^{15} \bmod DM + 1 \\ PR2 = (|a3| - a3) \times 10^{15} \bmod N + 1 \\ PC2 = (|a4| - a4) \times 10^{15} \bmod DM + 1 \end{cases}. \tag{6}$$

Here, we have to use the *circshift* (shift array circularly) rule and its definition is as follows: If A and B are matrixes, $B = circshift(A, shiftsize)$ circularly shifts the values in the array A using the shift size elements [22].

Step 3. We consider $PR, PC \in \mathbb{N}^{DM \times N}$ and for $i = 1$ to DM, if i is odd, then we get the following:

$$\begin{cases} PR(:,i) = circshift\ P1(i,:)\ by\ step\ PR1(i) \\ \qquad\qquad and\ else \\ PR(:,i) = circshift\ P1(i,:)by\ step - PR2(i) \end{cases} \tag{7}$$

Step 4. For $j = 1$ to N, if j is odd, then we get the following:

$$\begin{cases} PC(:,j) = circshift\ IR(:,j)\ by\ step\ IC1(j) \\ \qquad\qquad and\ else \\ PC(:,j) = circshift\ IR(:,j)by\ step - IC2(j) \end{cases} \tag{8}$$

Now, $P2 = PC$ is the permutated image.

This Process explained in Algorithm 1 with the title: High-speed Permutation Process.

Algorithm 1 High-speed Permutation Process

Input: Image P of size M × N × D, Initial state: a0, b0, c0, d0, and Sha256 value of P
Output: Permutated Image

1. Let DM = D × M, s = sum (sha256)/(64 × 256), and reshape P to 2-dimension matrix P1.
2. Generate a new initial value of the chaotic system: a0, b0, c0, d0;
3. Use the initial value and parameters to iterate the chaotic system to get the vectors: a1, a2, a3, a4, and quantize to generate four different vectors: PR1, PC1, PR2, PC2.
4. Set PR, PC $\in \mathbb{N}^{DM} \times N$
5. for i = 1 to DM do
6.   if i is odd then
7.   PR(:, i) = circshift P1(i,:) by step PR1(i)
8.   else
9.   PR(:, i) = circshift P1(i,:) by step −PR2(i)
10.   end if
11. end for
12. for j = 1 to N do
13.   if j is odd then
14.   PC(:,j) = circshift PR(:,j) by step PC1(j)
15.   else
16.   PC(:,j) = circshift PR(:,j) by step −PC2(j)
17.   end if
18. end for
19. Let PC be the permutated imagere
20. turn Permutated Image

3.2. Adaptive Diffusion

Step 1. First, the other initial values and parameters are used to iterate the chaotic systems to obtain the vectors $a110$, $b110$, $c110$, and $d110$, and they are quantized to generate four different vectors $a11$, $b11$, $c11$, and $d11$.

We set $N0$ as a random number, $N00 = N0 + 1$ and $nn = (DM \times N)/2$.

$$\begin{cases} a11 = |(a110 + b110) \times 10^{15}|mod2^3 + 1 \\ b11 = |(b110 - a110) \times 10^{15}|mod2^3 + 1 \\ c11 = |(c110 + d110) \times 10^{15}|mod2^8 \\ d11 = |(c110 + d110) \times 10^{15}|mod2^8 \end{cases} \tag{9}$$

Step 2. Set $A, B \in \mathbb{N}^{DM \times N}$ and $i = 1$ to DM. If $i \geq 1$ and $i \leq DM/2$, then we get the following:

$$\begin{cases} A(i,:) = a11(((i-1) \times N + 1) : (i \times N),:) \\ B(i,:) = c11(((i-1) \times N + 1) : (i \times N),:) \end{cases}.$$ (10)

Otherwise,

$$\begin{cases} A(i,:) = b11(((i-1-(DM/2)) \times N + 1) : ((i-(DM/2)) \times N),:) \\ B(i,:) = d11(((i-1-(DM/2)) \times N + 1) : ((i-(DM/2)) \times N),:) \end{cases}.$$ (11)

Step 3. Here, we have to apply *bitcircshift* rule, which is an action that is done on all of the bits of a binary amount, in which they are transformed by a determined number of locations to the left or right. *Bitcircshift* is used when the operand is being used as a series of bits relatively than generally. In other words, the operand behaves as single bits that show something and are not values [22].

Let $P2 = PC$, and $P3 \in \mathbb{N}^{DM \times N}$. If $i = 1$ to DM *and* $j = 1$ to N, then we get the following:

$$P3(i, j) = bitcircshift \; P2(i, j) \; by \; step \; A(i, j).$$ (12)

Step 4. Let $key_{r0} = (a110(1:N) + b110(1:N))'/2$ and $key_{c0} = (c110(1:DM) + d110(1:DM))/2$, and quantize them as key_r and key_c, respectively, in [0, 255].

$$\begin{cases} key_r = |key_{r0}| \times 10^{15} mod256 \\ key_c = |key_{c0}| \times 10^{15} mod256 \end{cases}$$ (13)

Step 5. Set $P4R, P4C \in \mathbb{N}^{DM \times N}$. If $i = 1$ to DM, $r1 = circshift(B(i,:), [0, i])$ and $i = 1$, then we get the following:

$$\begin{cases} P4R(i,:) = ((P3(i,:) + r1)mod256) \oplus key_r \\ \qquad\qquad and \; else \\ P4R(i,:) = ((P3(i,:) + r1)mod256) \oplus P4R((i-1),:) \end{cases}.$$ (14)

Step 6. If $j = 1$ to N, $c1 = circshift(B(j,:), [j, 0])$ and $j = 1$, then we get the following:

$$\begin{cases} P4C(:, j) = ((P4R(:, j) + c1)mod256) \oplus key_c \\ \qquad\qquad and \; else \\ P4C(:, j) = ((P4R(:, j) + c1)mod256) \oplus P4C(:, (j-1)) \end{cases}.$$ (15)

P4C, the final image, is encrypted.

This Process explained in Algorithm 2 with the title: Adaptive diffusion.

Our proposed algorithm is a symmetric algorithm. The decryption procedure is the opposite of the encryption method and decryption is done using the encryption method's formulas. This is shown in Figure 2.

Remark 1. *The proposed algorithm is suitable for all color images (RGB). Because medical images are very important in today's technological world, we decided to use the proposed algorithm for medical images.*

Algorithm 2 Adaptive diffusion

Input: Input data from permutation procession
Output: Encrypted Image

1:     Use other initial values and parameters to iterate the chaotic system again to get the vectors: a110, b110, c110, d110, and quantize them to generate four different vectors: a11, b11, c11, d11.

2:     Set A, B $\in N^{DM} \times N$

3:     for i = 1 to DM do

4:       if i $\geq$ 1 and i $\leq$ DM/2 then

5:         A(i,:) = a11(((i − 1) × N + 1) : (i × N),:)

6:         B(i,:) = c11(((i − 1) × N + 1) : (i × N),:)

7:       else

8:          A(i,:) = b11(((i − 1 − (DM/2)) × N + 1) : ((i − (DM/2)) × N),:) B(i,:) = d11(((i − 1 − (DM/2)) × N + 1) : ((i − (DM/2)) × N),:)

9:       end if

10:    end for

11:    Let P2 = PC, and set P3 $\in N^{DM} \times N$

12:    for i = 1 to DM do

13:    for j = 1 to N do

14:    P3(i,j) = bitcircshift P2(i,j) by step A(i,j)

15:    end for

16:    end for

17:    Let key_r0 = (a110(1 : N)+b110(1 : N))'/2, key_c0 = (c110(1 : DM)+d110(1 : DM))/2, and quantize them key_r, key_c, in [0, 255]

18:    Set P4R, P4C $\in N^{DM} \times N$

19:    for i = 1 to DM do

20:    r1 = circshift (B(i,:),[0,i])

21:    if i = 1 then

22:       P4R(i,:) = bitxor(mod((P3(i, :)+r1), 256), key_r)

23:    else

24:       P4R(i,:) = bitxor(mod((P3(i, :)+r1), 256), P4R((i-1), :))

25:         end if

26:    end for

27:    for j = 1 to N do

28:    c1 = circshift (B(j,:),[j,0])

29:    if j = 1 then

30:       P4C(:,j) = bitxor(mod((P4R(:, j)+c1), 256), key_c)

31:    else

32:       P4C(:,j) = bitxor(mod((P4R(:, j)+c1), 256), P4C(:, (j-1)))

33:    end if

34:    end for

35:    Let P4C be the final encrypted image

36:    return Encrypted Image

The decryption process is inverse encryption process.
Input: Input data from permutation procession
Output: Encrypted Image

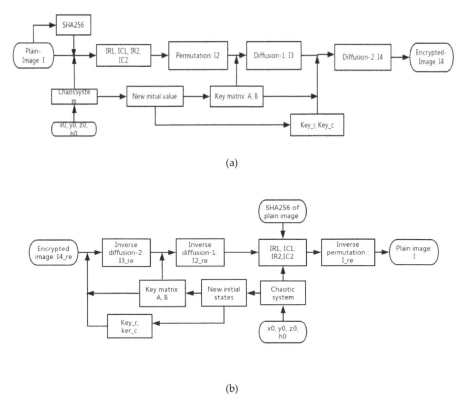

(a)

(b)

Figure 2. Schematic of the proposed encryption algorithm (**a**), and schematic of the proposed decryption algorithm (**b**).

4. Experiment Result and Security Analysis

In this section, we implemented the proposed algorithm on two medical color images using the MATLAB 2017a software environment (in personal computer with core i7, 3.4GHz, RAM 16GB). As we stated in the introduction, the medical color images are obtained using the Medipix3RX chip technology that is used in today's imaging devices [1–4].

For example, four color images 256 × 256 in size have been selected as the plain images. In Figure 3, images b, e, h, and k are images that are encrypted by the proposed algorithm for the plain images a, d, g, and j, respectively; and images c, f, i, and l are the decrypted images.

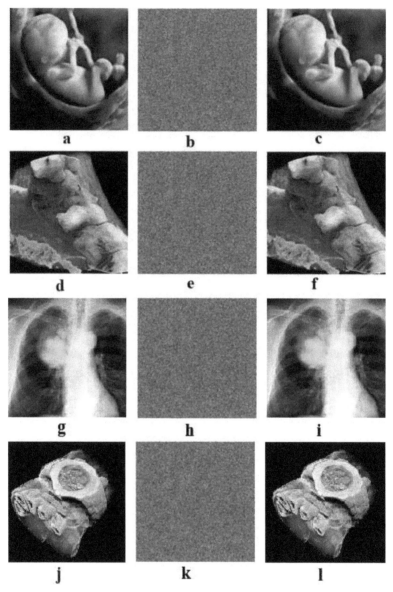

Figure 3. (**a,d,g,j**): Plain images. (**b,e,h,k**): Respective encrypted images. (**c,f,i,l**): Respective decrypted images. Initial values for all images: (a0 = 0.1314, b0 = 0.5214, c0 = 0.3698, and d0 = 0.8419). Values of the chaotic system: Image (**a**): sha256 = '9ADBBFB88CFD90C23CE114E4740 2054E6DDC4182510E80980EA7151CD11E6D18', image (**d**): sha256 = '8BF6A886E4B58D2B530749 EE9BAB54A3C360D406DC5B901CC169D7870FA3CA09', image (**g**): sha256 = '49A22186DB65786789 CD1391CDE4D9737039E758F39A45C59D8338DE05353337', and image (**j**): sha256 = '6EB1ADE45F27A 67E09A25265835F05BC11E057255DA81359299631F4724936C8'.

4.1. Security Analysis

As seen, it is not possible to visually compare the plain images with images that were obtained from the decryption process, and the measures such as the correlation coefficients of two adjacent pixels in the plain image and the cipher image, the entropy, the NPCR (number of pixel change rate), and the UACI (uniform average change intensity) should be mathematically examined. We consider an example of a baby's image.

4.2. Histogram Analysis

Color images include three main color channels (red, green, and blue), and these images are called RGB images. Figure 4 shows the histograms of these three channels that are observed for the baby's image.

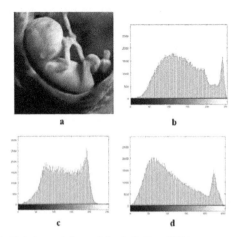

Figure 4. (**a**): Plain image baby, and (**b–d**): R, G, and B histograms, respectively.

In Figure 5, we can see the baby's decrypted image from the three-channel color histograms.

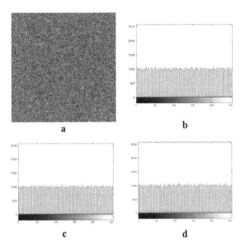

Figure 5. (**a**): Baby's decrypted image, and (**b–d**): R, G, and B histograms, respectively.

Entropy **2019**, *21*, 577

Chi-square Analysis

Statistical analysis is a type of the commonplace cryptology procedures. The monotony of the histogram of cipher demonstrates the strength of the encryption path to statistical analysis. The ocular effect of the histogram is not sufficient to verify the accident of a cipher image's pixel values [26]. To quantitatively measure the monotony of the histogram, we use the chi-square test as a metric. The description of the chi-square is as follows:

$$\chi^2_{exp} = \sum_{i=1}^{Q} \frac{(Q_i - e_i)^2}{e_i},$$
$$e_i = \frac{M \times N}{Q}, \tag{16}$$

where $Q = 256$ in our method, o_i is the observed incidence frequency of each rate on the histogram of the ciphered image, e_i is the envisage incidence frequency of the uniform distribution, and $M \times N$ is the length of an image trail. For an ideal image encryption system, the empirical chi-square value should be less than the theoretic amount. With the importance level of 0.05, the theoretic chi-square value is 293 [26]. The chi-square test conclusions and transition rates are listed in Tables 1 and 2. All the test images transition the test, which shows that our plan has a satisfying encryption effect.

Table 1. Chi-square test results (part 1).

Images		a			d	
Channels	R	G	B	R	G	B
χ^2_{test}	247.0762	204.9082	220.0742	251.8379	260.2695	256.4063
$\chi^2_{255.0.05}$	293	293	293	293	293	293
Pass or not	Yes	Yes	Yes	Yes	Yes	Yes

Table 2. Chi-square test results (part 2).

Images		g			j	
Channels	R	G	B	R	G	B
χ^2_{test}	285.8359	261.9980	247.2793	250.0836	210.7622	231.1039
$\chi^2_{255.0.05}$	293	293	293	293	293	293
Passor not	Yes	Yes	Yes	Yes	Yes	Yes

4.3. Correlation Analysis

The correlation coefficient of two adjacent pixels in the plain image and the cipher image is one of the important factors in determining the quality of image encryption algorithms [23]. In Figure 6, we can see the correlation histograms for the plain image and cipher image, and the correlation histograms are shown in three directions: Horizontal, vertical, and diagonal.

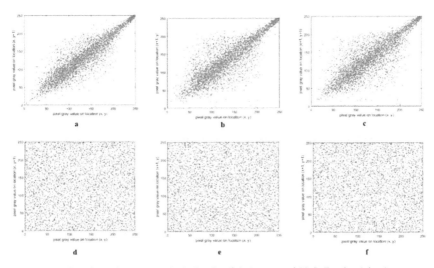

Figure 6. Correlation histograms. (**a–c**): For the plain image; and (**d–f**): For the cipher image.

In Table 3, the numerical values of the correlation for the plain image and the cipher image are given, and the values in the table are calculated for three directions: Horizontal, vertical, and diagonal. We can specifically see that the correlation coefficients of the plain image are near to 1, however the correlation coefficients of the cipher-image are about equal 0, which may explain why the designed encrypted algorithm has a powerful resistance to possible statistical attacks. The table specifies that the proposed algorithm has the required quality. The correlation coefficient of two adjacent pixels in the plain image and the cipher image is obtained as follows:

$$r_{xy} = \frac{E((x_i - E(x))(y_i - E(y))}{\sqrt{D(x)D(y)}} \tag{17}$$

where

$$E(x) = \frac{1}{N}\sum_{i=1}^{n} x_i,$$
$$D(x) = \frac{1}{N}\sum_{i=1}^{n}(x_i - E(x))^2 .$$

$E((x_i - E(x))(y_i - E(y)) = cov(x, y)$, $E(x) = \frac{1}{N}\sum_{i=1}^{N} x_i$ is the expected value, N is the number of image pixels, and $D(x) = \frac{1}{N}\sum_{i=1}^{N}(x_i - E(x))^2$ is the variance. x and y are the gray values of two adjacent pixels, and N is the total number of pixels that are chosen from the image.

Table 3. Correlation coefficients in the plain image and the cipher image.

Image	Channel	Plain-Text			Cipher-Text		
		H	**V**	**D**	**H**	**V**	**D**
	R	0.9952	0.9978	0.9897	−0.0115	0.0048	−0.0026
a	G	0.9825	0.9881	0.9688	0.0109	0.0097	−0.0161
	B	0.9833	0.9803	0.9615	−0.0224	−0.0091	0.0062
	R	0.9217	0.8673	0.8580	0.0097	−0.0091	−0.0094
d	G	0.8575	0.7647	0.7328	0.0015	0.0075	−0.0055
	B	0.9140	0.8966	0.8626	0.0137	0.0051	0.0065

Table 3. *Cont.*

Image	Channel	Plain-Text			Cipher-Text		
		H	**V**	**D**	**H**	**V**	**D**
	R	0.9942	0.9968	0.9887	−0.0125	0.0047	−0.0025
g	G	0.9815	0.9871	0.9678	0.0109	0.0095	−0.0160
	B	0.9823	0.9793	0.9605	−0.0214	−0.0093	0.0063
	R	0.9217	0.8663	0.8570	0.0095	−0.0092	−0.0093
j	G	0.8575	0.7637	0.7318	0.0014	0.0073	−0.0056
	B	0.9140	0.8956	0.8616	0.0135	0.0052	0.0066

4.4. Entropy Analysis

The entropy randomly measures the data sequence and is defined as follows [24]:

$$H(S) = \sum_{i=0}^{2^N-1} P(s_i) \log\left(\frac{1}{P(s_i)}\right) \tag{18}$$

where N is the number of grayscale levels in an image, and $P(s_i)$ is the incidence possibility of grayscale "I" in the image. The entropy amount will be 8 for images that are wholly accidentally generated. The nearer the entropy of an encryption method is to 8, the less foreseeable it is, and thus, the more secure the plan. The entropies for the designed encryption method have been measured for a sample image and the conclusions are shown in Table 4.

Table 4. Information entropies results for plain and cipher images.

Image	Channel	Image a	Image d	Image g	Image j
	R	6.9581	7.7047	6.9571	7.7067
Plain image	G	6.8945	7.4724	6.8955	7.4734
	B	6.1365	7.7502	6.1355	7.7512
	RGB	7.2528	7.7604	7.2548	7.7614
	R	7.9992	7.9991	7.9982	7.9993
Cipher image	G	7.9993	7.9991	7.9983	7.9995
	B	7.9993	7.9991	7.9994	7.9981
	RGB	7.9997	7.9996	7.9996	7.9994

4.5. NPCR (Number of Pixel Change Rate) and UACI (Uniform Average Change Intensity)

In a differential attack, a little variation is built to the plain image, and the designed algorithm is employed to encrypt the plain image before and after this variation. These two encrypted images have been evaluated to detect any possible connection between the plain image and the cipher image. The (UACI) and the (NPCR) are two indicator that are regularly used by researchers to test the differential attack resistor of any image encryption method [12].

Suppose that C_1 and C_2 are two cipher images that are encrypted from two plain images with only one-bit difference. The NPCR and UACI are defined as follows:

$$NPCR(C_1, C_2) = \sum_{i,j} \frac{H(i,j)}{M} \times 100\% \tag{19}$$

and

$$UACI(C_1, C_2) = \sum_{i,j} \frac{|C_1(i,j) - C_2(i,j)|}{(S-1) \times M} \times 100\% \tag{20}$$

where M shows the total number of pixels in any cipher-image, S illustrates the number of allowed pixels, and $H(i,j)$ demonstrates the difference between C_1 and C_2, which is specified as follows.

$$H(i,j) = \begin{cases} 0, & if \quad C_1(i,j) = C_2(i,j) \\ 1, & if \quad C_1(i,j) \neq C_2(i,j) \end{cases} . \tag{21}$$

The larger the NPCR and UACI are, the better the quality of the algorithm. For four randomly selected points, the NPCRs and UACIs are listed in Table 5.

Table 5. Number of pixel change rates (NPCRs) and uniform average change intensities (UACIs) of different positions (%).

Position	(12,34)	(34,56)	(56,78)	(78,90)
NPCR	99.6232	99.6215	99.6170	99.5971
UACI	33.4574	33.4952	33.5326	33.4755

4.6. Key Space

The key space for encryption algorithms should be large enough to withstand potential attacks. The minimum key space should be 2^{100}. The input values of $(x_0, y_0, z_0, h_0, SHA256)$ act as a secret key, and, based on this, the secret key space is $10^{14} \times 10^{14} \times 10^{14} \times 10^{14} \times 2^{128} = 10^{56} \times 2^{128}$. This indicates that the designed method has good key space.

4.7. Key Sensibility Analysis

A safe encryption system must be sensitive to the key; for example, the little change of encryption keys can lead to very different cipher image, and a small change of the decryption keys cannot decrypt the image. Several key sensitivity tests are performed. Figure 7 shows the encrypted images of the baby (plain image b). Figure 3a shows the encrypted image using user keys with a 1-bit difference. The plain encrypted image is indicated in Figure 3b. When the keys of the initial state x_0, y_0, z_0, h_0, and $SHA256$ are changed by one bit (i.e., 10^{-14} for x_0, y_0, z_0, and h_0 and 2^{-128} for SHA256), the five new encrypted images are obtained and shown in Figure 7a–e. We compare them with the image in Figure 7e, and the five differential images are shown in Figure 7f–j. This shows that there are very big differences between the images in Figure 7e,f–j.

In addition, to experience the capability of the designed method to resist the cipher text attack, the keys x_0, y_0, z_0, and h_0 will be modified by 10^{-14} and SHA256 will be changed by 2^{-128} to decrypt the plain encrypted image. The decrypted images are indicated in Figure 7, which are wholly different from the plain image. Therefore, it can be seen that the cipher text cannot be suitably decrypted without the correct keys, which shows that the proposed method can effectively hamper the cipher text merely attack.

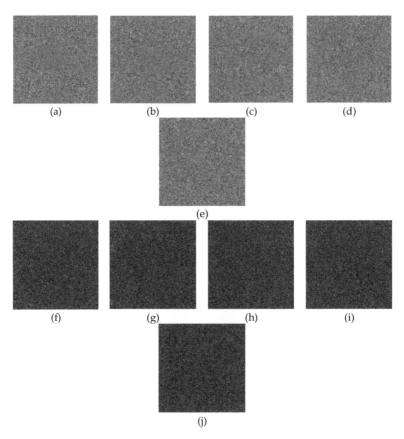

Figure 7. Cipher-images within authentic primary keys and the differences between them and the plain encrypted images: (**a–e**) Five new encrypted images with the keys $x_0 + 10^{-14}$, $y_0 + 10^{-14}$, $z_0 + 10^{-14}$, $h_0 + 10^{-14}$, and SHA256+2^{-128}, respectively; and (**f–j**) differences between the unauthentic encrypted images and the plain image.

4.8. Known-Plain Image and Chosen-Plain Image Analysis

Clearly, some specific images are selected to test the selected plain-text attack, like a full-white image in Figure 8a and a full-black image in Figure 8d. The results are shown in Figure 8, which indicate that the cryptology is appropriate for these specific images and can resist the chosen-plain-text attack. In Table 6, we can see the entropy values for all-white and all-black images. It can be seen that all the values that are obtained are close to 8, indicating the suitability of the proposed algorithm.

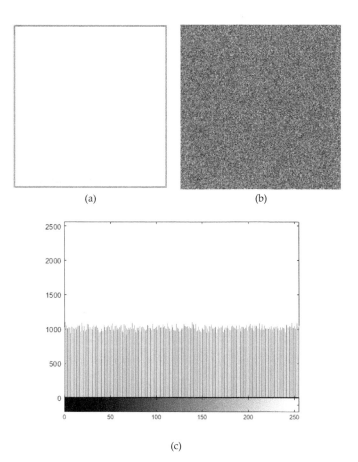

(a)

(b)

(c)

Figure 8. *Cont.*

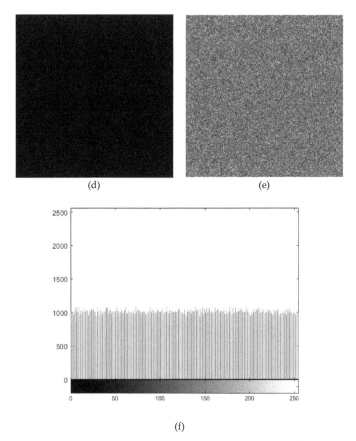

(d)   (e)

(f)

Figure 8. Selected plain-image test for white and black images, display (**a**) the full-white image, (**b**) the cipher image of panel (**a**), (**c**) the histogram of channel R (**b**), (**d**) the full-black image, (**e**) the cipher image of panel (**d**), and (**f**) the histogram of channel R.

Table 6. Information entropy of the full-black image and full-white image.

Image	R	G	B
Black	7.9993	7.9994	7.9993
White	7.9994	7.9994	7.9993

4.9. Noise Attack and Occlusion Attack

In the digital world, the images will unexpectedly experience noise and occlude attacks in the transition process, and an effective cryptology must be strong against them. The baby image is used as the test image. Figure 9 shows the noisy cipher images that are contaminated by Gaussian noise (GN), salt and pepper noise (SPN), and speckle noise (SN) with different noise compression and their decrypted images. As seen from Figure 10, the most information of the plain image can be intuitively identified from the decrypted image's presentation of the cipher images with different occlusion effects and their corresponding retrieved images. Specifically, the decrypted images still include most date of the baby image. The PSNR (peak signal-to-noise-ratio) is employed to compute the condition of the decrypted image after a possible attack. For a gray image, the PSNR and MSE can be computed as follows:

$$PSNR = 10 \times log_{10}\left(\frac{255 \times 255}{MSE}\right)(db) \tag{22}$$

$$MSE = \frac{1}{mn}\sum_{i=1}^{m}\sum_{j=1}^{n}\|I_1(i,j) - I_2(i,j)\|^2 MSE = \frac{1}{mn}\sum_{i=1}^{m}\sum_{j=1}^{n}\|I_1(i,j) - I_2(i,j)\|^2 \tag{23}$$

where MSE shows the mean square error between the cipher image $I_1(i,j)$ and the plain image $I_2(i,j)$, and m and n are the width and height, respectively [25]. The results are explained in Tables 7 and 8.

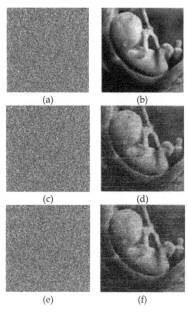

(a)　　(b)

(c)　　(d)

(e)　　(f)

Figure 9. Noise attack test results. (**a,b**) Gaussian noise (GN); (**c,d**) salt and pepper noise (SPN); (**e,f**) speckle noise (SN).

Table 7. Noise attack test results.

Item	PSNR		
	R	G	B
Salt and Pepper	34.2863	33.6124	33.3711
Gaussian	30.6104	29.9219	29.7102
Speckle	30.6023	29.8951	29.7059

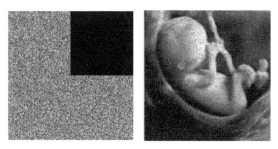

Figure 10. *Cont.*

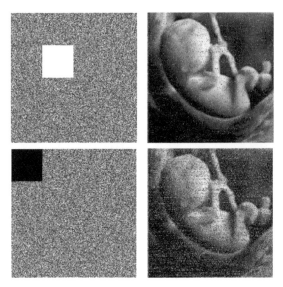

Figure 10. Occlude attack test results.

Table 8. Occlude attack test results.

Item	PSNR		
	R	G	B
Salt and pepper	34.0961	33.6237	33.5174
Gaussian	34.0106	33.6064	33.4199
Speckle	31.0996	30.6760	30.4795

5. Comparison

In this section, we will compare the results that were obtained from the proposed algorithm with previous algorithms. The results that were obtained from the designed method are measured with the methods in References [10,24] with respect to their correlation, entropy, NPCR, and UACI, and the key space was compared with references [9,12,24]. The results of the comparison are shown in Tables 9–14.

Table 9. Correlation coefficients results in the original image and the encrypted image for the proposed method and the two methods that were presented in existing methods.

Image	Methods	Channel	Plain Image			Cipher Image		
			H	V	D	H	V	D
		R	0.9952	0.9978	0.9897	−0.0115	0.0048	−0.0026
	proposed	G	0.9825	0.9881	0.9688	0.0109	0.0097	−0.0161
		B	0.9833	0.9803	0.9615	−0.0224	−0.0091	0.0062
		R	0.9952	0.9978	0.9897	−0.0122	−0.0117	−0.0238
Image a	Algorithm [10]	G	0.9825	0.9881	0.9688	−0.0113	0.0079	−0.0230
		B	0.9833	0.9803	0.9615	−0.0099	0.0149	0.0092
		R	0.9952	0.9978	0.9897	−0.0114	−0.0110	−0.0174
	Algorithm [24]	G	0.9825	0.9881	0.9688	−0.0206	−0.0071	0.0180
		B	0.9833	0.9803	0.9615	−0.0134	−0.0106	−0.00194

Table 9. *Cont.*

Image	Methods	Channel	Plain Image			Cipher Image		
			H	V	D	H	V	D
Image d	proposed	R	0.9217	0.8673	0.8580	0.0097	−0.0091	−0.0094
		G	0.8575	0.7647	0.7328	0.0015	0.0075	−0.0055
		B	0.9140	0.8966	0.8626	0.0137	0.0051	0.0065
	Algorithm [10]	R	0.9217	0.8673	0.8580	0.0133	−0.0095	−0.0070
		G	0.8575	0.7647	0.7328	−0.0182	0.0230	−0.0056
		B	0.9140	0.8966	0.8626	−0.0282	-7.3230×10^{-4}	−0.0073
	Algorithm [24]	R	0.9217	0.8673	0.8580	−0.0154	−0.0242	0.0094
		G	0.8575	0.7647	0.7328	−0.0059	0.0109	1.7711×10^{-4}
		B	0.9140	0.8966	0.8626	−0.0216	−0.0089	0.0077
Image g	proposed	R	0.9942	0.9968	0.9887	−0.0125	0.0047	−0.0025
		G	0.9815	0.9871	0.9678	0.0109	0.0095	−0.0160
		B	0.9823	0.9793	0.9605	−0.0214	−0.0093	0.0063
	Algorithm [10]	R	0.9942	0.9968	0.9887	−0.0136	0.0066	−0.0035
		G	0.9815	0.9871	0.9678	0.0113	0.0103	−0.0190
		B	0.9823	0.9793	0.9605	−0.0237	−0.0098	0.0073
	Algorithm [24]	R	0.9942	0.9968	0.9887	−0.0129	0.0049	−0.0076
		G	0.9815	0.9871	0.9678	0.0111	0.0101	−0.0171
		B	0.9823	0.9793	0.9605	−0.0231	−0.0156	0.0067
Image j	proposed	R	0.9217	0.8663	0.8570	0.0095	−0.0092	−0.0093
		G	0.8575	0.7637	0.7318	0.0014	0.0073	−0.0056
		B	0.9140	0.8956	0.8616	0.0135	0.0052	0.0066
	Algorithm [10]	R	0.9217	0.8663	0.8570	0.0115	−0.0102	−0.0105
		G	0.8575	0.7637	0.7318	0.0084	0.0094	−0.0083
		B	0.9140	0.8956	0.8616	0.0196	0.0067	0.0089
	Algorithm [24]	R	0.9217	0.8663	0.8570	0.0115	−0.0111	−0.0106
		G	0.8575	0.7637	0.7318	0.0082	0.0103	−0.0074
		B	0.9140	0.8956	0.8616	0.0161	0.0083	0.0090

Table 10. Information entropies of the cipher images and plain images, part 1.

Image	Channel	Proposed Algorithm Image a	Method [10] Image a	Method [24] Image a	Proposed Algorithm Image d	Method [10] Image d	Method [24] Image d
Plainimage	R	6.9581	6.9581	6.9581	7.7047	7.7047	7.7047
	G	6.8945	6.8945	6.8945	7.4724	7.4724	7.4724
	B	6.1365	6.1365	6.1365	7.7502	7.7502	7.7502
	RGB	7.2528	7.2528	7.2528	7.7604	7.7604	7.7604
Cipherimage	R	7.9992	7.9991	7.9992	7.9991	7.9991	7.9988
	G	7.9993	7.9989	7.9992	7.9991	7.9991	7.9992
	B	7.9993	7.9988	7.9993	7.9991	7.9990	7.9989
	RGB	7.9997	7.9996	7.9997	7.9996	7.9995	7.9996

Table 11. Information entropies of the cipher images and plain images, part 2.

Image	Channel	Proposed Algorithm Image g	Method [10] Image g	Method [24] Image g	Proposed Algorithm Image j	Method [10] Image j	Method [24] Image j
Plainimage	R	6.9571	6.9571	6.9571	7.7067	7.7067	7.7067
	G	6.8955	6.8955	6.8955	7.4734	7.4734	7.4734
	B	6.1355	6.1355	6.1355	7.7512	7.7512	7.7512
	RGB	7.2548	7.2548	7.2548	7.7614	7.7614	7.7614
Cipherimage	R	7.9982	7.9981	7.9982	7.9993	7.9993	7.9989
	G	7.9983	7.9979	7.9982	7.9995	7.9995	7.9994
	B	7.9994	7.9985	7.9994	7.9981	7.9980	7.9979
	RGB	7.9996	7.9995	7.9996	7.9994	7.9993	7.9994

Table 12. NPCRs and UACIs at different positions (%), Part 1.

Position	Proposed Algorithm Position (12, 34)	Method [10] Position (12, 34)	Method [24] Position (12, 34)	Proposed Algorithm Position (34, 56)	Method [10] Position (34, 56)	Method [24] Position (34, 56)
NPCR	99.6232	99.6222	99.6218	99.6215	99.5015	99.6187
UACI	33.4574	33.4504	33.4767	33.4952	33.3952	33.4850

Table 13. NPCRs and UACIs at different positions (%), Part 2.

Position	Proposed Algorithm Position (56, 78)	Method [10] Position (56, 78)	Method [24] Position (56, 78)	Proposed Algorithm Position (78, 90)	Method [10] Position (78, 90)	Method [24] Position (78, 90)
NPCR	99.6170	99.4170	99.6123	99.5971	99.1971	99.6223
UACI	33.5326	33.2326	33.3979	33.4755	33.3755	33.4759

Table 14. Comparison of the proposed algorithm's key space with other algorithms.

Algorithm	Proposed	[9]	[12]	[24]
Key space	$10^{56} \times 2^{128}$	2^{192}	2^{128}	2^{256}

6. Conclusions

In this paper, we present a new algorithm based on chaotic systems to protect these images against attacks. The proposed algorithm has two main parts: A high-speed permutation process and adaptive diffusion, which lead to a very efficient and reliable approach in this regard. By examining the results that were obtained from the implementation of the proposed algorithm in the MATLAB software environment and comparing these results with existing methods, it is observed that the designed method is better than those algorithms with respect to the important factors that are mentioned. Such that, to quantitatively evaluate the uniformity of the histogram, the chi-square test is done and the obtained results are desirable. Also, key sensitivity analysis shows that the image is not decrypted with a small change in the key, which indicates that the algorithm is suitable. Clearly, particular images are selected to experiment the selected plain-text, such as a full-white image and a full-black image. Entropy results obtained from the implementation of the algorithm on this type of images show that the proposed method is suitable for this particular type of images. In the real world, the images will inevitably experience noise and occlude attacks while shifting, and an effective cryptosystem must be powerful versus them. The obtained results from noise and occlusion attacks analyses show that the proposed algorithm can withstand against these types of attacks. It is also observed that the values that are obtained with respect to the entropy, NPCR, and UACI are better than those from the methods in existing papers. As we have already mentioned, the key space should be large enough (at least 2^100).

Compared to the old methods, we observe that the key space of our method is very large and more resistant than other methods in dealing with attacks.

Author Contributions: For research articles with several authors, a short paragraph specifying their individual contributions must be provided. The following statements should be used "conceptualization, S.S.M., Y.C. and C.T; methodology, S.S.M., Y.C. and C.T.; software, S.S.M., Y.C. and C.T.; validation, S.S.M., Y.C. and C.T.; formal analysis, S.S.M., Y.C. and C.T.; investigation, S.S.M., Y.C. and C.T.; resources, S.S.M., Y.C. and C.T.; data curation, S.S.M., Y.C. and C.T.; writing—original draft preparation, S.S.M.; writing—review and editing, S.S.M. and C.T.; visualization, S.S.M., Y.C. and C.T.; supervision, C.T.; project administration, C.T.; funding acquisition, C.T.

Funding: This research was funded by the Foundation of National Natural Science of China (No. 61772147), Guangdong Province Natural Science Foundation of major basic research and Cultivation project (No. 2015A030308016), Project of Ordinary University Innovation Team Construction of Guangdong Province (No. 2015KCXTD014), Collaborative Innovation Major Projects of Bureau of Education of Guangzhou City (No. 1201610005) and National Cryptography Development Fund (No. MMJJ20170117).

Conflicts of Interest: The authors declare no conflict of interest.

References

1. Benssalah, M.; Rhaskali, Y.; Azzaz, M.S. Medical Images Encryption Based on Elliptic Curve Cryptography and Chaos Theory. In Proceedings of the 2018 International Conference on Smart Communications in Network Technologies (SaCoNeT), El Oued, Algeria, 27–31 October 2018; pp. 222–226.
2. Mohamed Parvees, M.Y.; Abdul Samath, J.; Parameswaran Bose, B. Medical Images are Safe—An Enhanced Chaotic Scrambling Approach. *J. Med. Syst.* **2017**, *41*, 167. [CrossRef] [PubMed]
3. Rinkel, J.; Magalhães, D.; Wagner, F.; Frojdh, E.; Sune, R.B. Equalization method for Medipix3RX. *Nucl. Instrum. Methods Phys. Res. Sect. A Accel. Spectrom. Detect. Assoc. Equip.* **2015**, *801*, 1–6. [CrossRef]
4. Gimenez, E.N.; Astromskas, V.; Horswell, I.; Omar, D.; Spiers, J.; Tartoni, N. Development of a Schottky CdTe Medipix3RX hybrid photon counting detector with spatial and energy resolving capabilities. *Nucl. Instrum. Methods Phys. Res. Sect. A Accel. Spectrom. Detect. Assoc. Equip.* **2016**, *824*, 101–103. [CrossRef]
5. Abdel-Nabi, H.; Al-Haj, A. Medical imaging security using partial encryption and histogram shifting watermarking. In Proceedings of the 2017 8th International Conference on Information Technology (ICIT), Amman, Jordan, 17–18 May 2017; pp. 802–807.
6. Abdmouleh, M.K.; Khalfallah, A.; Bouhlel, M.S. A Novel Selective Encryption DWT-Based Algorithm for Medical Images. In Proceedings of the 2017 14th International Conference on Computer Graphics, Imaging and Visualization, Marrakesh, Morocco, 23–25 May 2017; pp. 79–84.
7. Lakshmi, C.; Thenmozhi, K.; Rayappan, J.B.B.; Amirtharajan, R. Encryption and watermark-treated medical image against hacking disease—An immune convention in spatial and frequency domains. *Comput. Methods Progr. Biomed.* **2018**, *159*, 11–21. [CrossRef] [PubMed]
8. Cao, W.; Zhou, Y.; Chen, C.L.P.; Xia, L. Medical image encryption using edge maps. *Signal Process.* **2017**, *132*, 96–109. [CrossRef]
9. Ismail, S.M.; Said, L.A.; Radwan, A.G.; Madian, A.H.; Abu-Elyazeed, M.F. Generalized double-humped logistic map-based medical image encryption. *J. Adv. Res.* **2018**, *10*, 85–98. [CrossRef] [PubMed]
10. Jeong, H.; Park, K.; Cho, S.; Kim, S. Color Medical Image Encryption Using Two-Dimensional Chaotic Map and C-MLCA. In Proceedings of the 2018 Tenth International Conference on Ubiquitous and Future Networks (ICUFN), Prague, Czech Republic, 3–6 July 2018; pp. 501–504.
11. Ke, G.; Wang, H.; Zhou, S.; Zhang, H. Encryption of medical image with most significant bit and high capacity in piecewise linear chaos graphics. *Measurement* **2019**, *135*, 385–391. [CrossRef]
12. Nematzadeh, H.; Enayatifar, R.; Motameni, H.; Guimarães, F.G.; Coelho, V.N. Medical image encryption using a hybrid model of modified genetic algorithm and coupled map lattices. *Opt. Lasers Eng.* **2018**, *110*, 24–32. [CrossRef]
13. Laiphrakpam, D.S.; Khumanthem, M.S. Medical image encryption based on improved ElGamal encryption technique. *Optik* **2017**, *147*, 88–102. [CrossRef]
14. Suganya, G.; Amudha, K. Medical image integrity control using joint encryption and watermarking techniques. In Proceedings of the 2014 International Conference on Green Computing Communication and Electrical Engineering (ICGCCEE), Coimbatore, India, 6–8 March 2014; pp. 1–5.

15. Liang, X.; Qi, G. Mechanical analysis of Chen chaotic system. *Chaos Solitons Fractals* **2017**, *98*, 173–177. [CrossRef]
16. Hu, H.; Liu, L.; Ding, N. Pseudorandom sequence generator based on the Chen chaotic system. *Comput. Phys. Commun.* **2013**, *184*, 765–768. [CrossRef]
17. Wang, L.; Dong, T.; Ge, M.-F. Finite-time synchronization of memristor chaotic systems and its application in image encryption. *Appl. Math. Comput.* **2019**, *347*, 293–305. [CrossRef]
18. Gong, L.; Deng, C.; Pan, S.; Zhou, N. Image compression-encryption algorithms by combining hyper-chaotic system with discrete fractional random transform. *Opt. Laser Technol.* **2018**, *103*, 48–58. [CrossRef]
19. Michail, H.E.; Athanasiou, G.S.; Theodoridis, G.; Gregoriades, A.; Goutis, C.E. Design and implementation of totally-self checking SHA-1 and SHA-256 hash functions' architectures. *Microprocess. Microsyst.* **2016**, *45*, 227–240. [CrossRef]
20. James, J.; Karthika, R.; Nandakumar, R. Design & Characterization of SHA 3- 256 Bit IP Core. *Procedia Technol.* **2016**, *24*, 918–924.
21. Ahmad, I.; Shoba Das, A. Hardware implementation analysis of SHA-256 and SHA-512 algorithms on FPGAs. *Comput. Electr. Eng.* **2005**, *31*, 345–360. [CrossRef]
22. Xu, S.-J.; Chen, X.-B.; Zhang, R.; Yang, Y.-X.; Guo, Y.-C. An improved chaotic cryptosystem based on circular bit shift and XOR operations. *Phys. Lett. A* **2012**, *376*, 1003–1010. [CrossRef]
23. Pareek, N.K.; Patidar, V. Medical image protection using genetic algorithm operations. *Soft Comput.* **2016**, *20*, 763–772. [CrossRef]
24. Hua, Z.; Yi, S.; Zhou, Y. Medical image encryption using high-speed scrambling and pixel adaptive diffusion. *Signal Process.* **2018**, *144*, 134–144. [CrossRef]
25. Chai, X.; Fu, X.; Gan, Z.; Lu, Y.; Chen, Y. A color image cryptosystem based on dynamic DNA encryption and chaos. *Signal Process.* **2019**, *155*, 44–62. [CrossRef]
26. Ma, S.; Zhang, Y.; Yang, Z.; Hu, J.; Lei, X. A New Plaintext-Related Image Encryption Scheme Based on Chaotic Sequence. *IEEE Access* **2019**, *7*, 30344–30360. [CrossRef]
27. Niyat, A.Y.; Moattar, M.H.; Torshiz, M.N. Color image encryption based on hybrid hyper-chaotic system and cellular automata. *Opt. Lasers Eng.* **2017**, *90*, 225–237. [CrossRef]
28. Chen, H.; Tanougast, C.; Liu, Z.; Hao, B. Securing color image by using hyperchaotic system in gyrator transform domains. *Opt. Quantum Electron.* **2016**, *48*, 396. [CrossRef]

© 2019 by the authors. Licensee MDPI, Basel, Switzerland. This article is an open access article distributed under the terms and conditions of the Creative Commons Attribution (CC BY) license (http://creativecommons.org/licenses/by/4.0/).

Article

A New Image Encryption Algorithm Based on Composite Chaos and Hyperchaos Combined with DNA Coding

Yujie Wan [1], Shuangquan Gu [1] and Baoxiang Du [1,2,*]

[1] Electronic Engineering College, Heilongjiang University, Harbin 150080, China;
2181237@s.hlju.edu.cn (Y.W.); 2181235@s.hlju.edu.cn (S.G.)
[2] Kunpad Communications(KunShan) Co., Ltd., Kunshan 215300, China
* Correspondence: dubaoxiang@hlju.edu.cn

Received: 18 December 2019; Accepted: 31 January 2020; Published: 2 February 2020

Abstract: In order to obtain chaos with a wider chaotic scope and better chaotic behavior, this paper combines the several existing one-dimensional chaos and forms a new one-dimensional chaotic map by using a modular operation which is named by LLS system and abbreviated as LLSS. To get a better encryption effect, a new image encryption method based on double chaos and DNA coding technology is proposed in this paper. A new one-dimensional chaotic map is combined with a hyperchaotic Qi system to encrypt by using DNA coding. The first stage involves three rounds of scrambling; a diffusion algorithm is applied to the plaintext image, and then the intermediate ciphertext image is partitioned. The final encrypted image is formed by using DNA operation. Experimental simulation and security analysis show that this algorithm increases the key space, has high sensitivity, and can resist several common attacks. At the same time, the algorithm in this paper can reduce the correlation between adjacent pixels, making it close to 0, and increase the information entropy, making it close to the ideal value and achieving a good encryption effect.

Keywords: chaotic systems; image encryption; DNA coding; security analysis

1. Introduction

With the rapid development of the Internet, more and more multimedia image information is transmitted online. Images are widely used because of their vivid and intuitive characteristics. People can easily access other people's information through the Internet with the help of an ordinary computer and network cable. Therefore, the question of how to transfer the information safely and ensure its security has become an urgent problem to be solved. Image encryption is the primary solution. Due to high redundancy and correlation between image pixels, large amounts of data, and fidelity, traditional text encryption technology cannot meet the needs of image encryption [1]. Therefore, the development of secure and effective image encryption algorithms is still the focus of the communication field [2].

Due to its high sensitivity to initial values and system parameters, excellent ergodicity, and good pseudo-randomicity, chaotic systems have become the primary choice of cryptographic systems [3,4]. Therefore, many image encryption schemes based on chaos have been proposed [5,6]. Among them, chaotic image encryption methods are divided into one-dimensional chaotic and multidimensional chaotic encryption methods. A one-dimensional chaotic system has a simple structure which is easy to implement. However, they also have some problems: the scope of chaotic behavior is small, and the Lyapunov index is low [7]. Some improved encryption schemes for one-dimensional chaotic maps have been proposed. Wu et al. improved the existing one-dimensional chaos and proposed a new image encryption method [8]. A new method was proposed by Chao et al. who took the output of tent mapping as the input of Chebyshev mapping, and then applied perturbations to generate excellent pseudo-random chaotic sequences for encryption [9]. Hua et al. proposed to combine two one-dimensional chaotic systems in parallel to form a new one-dimensional chaotic system through cosine transform to encrypt the image [10], which increased the scope of system chaotic mapping. C P et al. defined a new one-dimensional chaotic map with the difference of two chaotic output sequences [11]. These methods expand the scope of chaotic mapping and improve chaotic properties to some extent, but the system parameters are still limited. On the other hand, the multi-dimensional chaotic phase space is complex, the system parameters have more flexibility, and the dynamic behavior is difficult to predict. In particular, the hyperchaotic system has two or more positive Lyapunov exponents, and the characteristics of the chaos are better for this system. A multi-dimensional chaotic system can produce multiple chaotic sequences at the same time, which can be used in image scrambling and diffusion, respectively, with high security. Sun adopts a 5-D hyperchaotic system to generate pseudo-random sequences and decompose permutation images, which can resist statistical attacks and differential attacks and is suitable for practical application [12].

Since DNA molecules can be processed in parallel on a large scale, with huge storage and ultra-low power consumption, many image encryption methods are proposed by many researchers who combine chaotic mapping and DNA coding technology. In 1994, Aldeman proposed DNA computing for the first time, ushering in a new era of information processing [13]. In 2002, Gehani et al. proposed to encrypt images one by one with DNA strings [14]. In 2012, an image encryption method based on piecewise linear mapping of DNA and PWLCM was proposed, which increased the key space [15]. However, these encryption methods cannot resist selective plaintext attacks and known plaintext attacks. In 2019, Zhang et al. proposed a new image encryption method based on quantum chaos and DNA coding, which has high security and can resist brute force attacks and statistical attacks [16]. In 2019, a color image encryption algorithm based on dynamic DNA encryption and chaos was proposed, using the hash function and external parameters to calculate its initial value, which can effectively resist the selected plaintext attack with better security [17]. Guan et al. proposed a digital image encryption algorithm based on DNA and frequency domain hyperchaos, which improved security against differential attacks [18]. Yang et al. proposed an image compression and encryption scheme based on fractional-order hyperchaotic system combining 2D compressed sensing and DNA coding. The fractional order and initial value of the fractional hyperchaos system are used as the key of the encryption scheme, which greatly expands the key space and has a strong ability to resist multiple attacks [19].

In order to provide a better encryption effect, a new image encryption scheme based on double chaos (one-dimensional composite chaos and hyperchaos) and DNA coding technology is proposed. This algorithm has the following advantages: (1) First, Fibonacci transformation and diffusion operation of modularization are performed on the plaintext image, and the pixel position and value of the plaintext image are fully changed to reduce the image correlation. (2) The first-round scrambling-diffusion operation is repeated three times, so that the value of each encrypted pixel is affected by the previous one, which increases the sensitivity to its clear text. (3) A new one-dimensional complex chaos is proposed, which has no period window within the chaos scope, that is to say it is a full map, and is larger than the corresponding one-dimensional chaotic Lyapunov exponents. Combining the new chaotic sequence with DNA technology, the secondary encryption extends the complexity and improves its security. (4) Taking the pixel value of the plaintext image as the initial value of the chaotic system can resist the plaintext attack and increase the key space. In this paper, key space, statistical analysis, differential attack, and anti-noise attack are analyzed. Experimental results and security analysis also confirm that the algorithm proposed in this paper increases key space, has high sensitivity, can resist multiple attacks, and can effectively protect the security of image information.

The rest of this paper is arranged as follows. The second section mainly introduces the theoretical knowledge required in this paper, such as typical chaotic systems, newly constructed LLS system, and DNA coding technology. The third section introduces in detail the image encryption scheme based on double chaos and DNA coding technology. The fourth section is the experimental simulation and security analysis. Finally, the fifth section draws the conclusion of this paper.

2. The Basic Principle

2.1. One-Dimensional Chaotic Mapping

2.1.1. Logistic Chaotic Mapping

Logistic chaotic mapping is a classical one-dimensional chaotic mapping with a simple structure and few control parameters, which is convenient for implementation and generalization involving other chaos [20]. The expression of Logistic chaotic mapping is shown as Formula (1):

$$x_{n+1} = \mu x_n(1 - x_n), \ n = 0, 1, 2, 3 \cdots \tag{1}$$

where μ is the system control parameter, and x_0 is the initial value of the system $0 < x_0 < 1$. The bifurcation diagram and lyapunov exponent of logistic chaotic mapping are shown in Figures 1a and 2a. It can be seen that with the increase of μ and the number of bifurcations of the system, when μ varies from 3.5699456 to 4, the system enters a chaotic state.

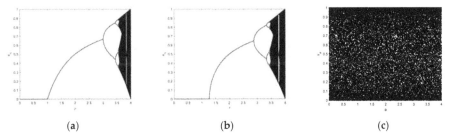

(a)   (b)   (c)

Figure 1. The Bifurcation diagrams of the (**a**) Logistic map, (**b**) Sine map, (**c**) LLSS map.

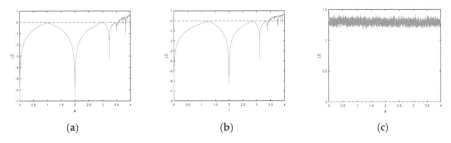

(a)   (b)   (c)

Figure 2. The Lyapunov Exponent of the (**a**) Logistic map, (**b**) Sine map, (**c**) LLSS map.

2.1.2. Sine Chaotic Mapping

Sine chaotic mapping is a mapping derived from the Sine function, which can convert the input angle in the range from 0 to $1/\pi$ to the output angle in a certain range [21]. The expression of Sine chaotic mapping is shown as Formula (2):

$$x_{n+1} = r\sin(\pi x_n)/4, \tag{2}$$

where x_n is the input and r is the system control parameter. The bifurcation diagram of Sine's chaotic mapping and lyapunov exponent are shown as Figures 1b and 2b.

2.2. LLSS Chaotic Mapping

A new one-dimensional chaotic system can be obtained by using the existing one-dimensional chaos as a seed map. In this paper, two logistic maps and Sine map were connected in parallel, and then a mod operation was used to form a new one-dimensional chaos algorithm named LLSS. The structure is shown as Figure 3. The system expression is defined by Formula (3):

$$x_{n+1} = mod(2ax(n)(1 - x(n)) + (8 - 2a)\sin(\pi x(n))/4, 1). \tag{3}$$

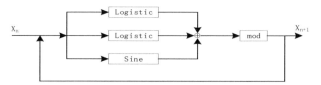

Figure 3. The new chaotic system of the LLS map.

The bifurcation diagram and Lyapunov exponent of LLSS are shown in Figures 1c and 2c. As can be seen from the figure, the LLSS is fully mapped within the range of [0,4] and has no period window. Compared with the classical one-dimensional chaotic map, the Lyapunov exponent also increases.

2.3. Qi Hyperchaotic System

In 2005, Qi et al. discovered and named a new chaos algorithm called the Qi chaotic system [22]. On the basis of the experience of increasing dimensions to obtain hyperchaos, Qi et al. further proposed the Qi hyperchaos system. In comparison, the dynamic characteristics are more complex and the motion trajectory traversal range in phase space is larger [23]. The Qi hyperchaotic system is a four-dimensional hyperchaotic system. The dynamic equation is shown as Formula (4) as follows:

$$\begin{cases} \dot{x} = a(y - x) + yzw \\ \dot{y} = b(x + y) - xzw \\ \dot{z} = -cz + exyw \\ \dot{w} = -dw + xyz \end{cases} \tag{4}$$

When the system parameters $a = 50$, $b = 4$, $c = 13$, $d = 20$, $e = 4$, the system is in a hyperchaotic state. When the initial value [1; 2; 3; 4] is selected, its attractor phase diagram develops as shown in Figure 4.

2.4. DNA Coding Technique

A DNA sequence is a string of molecules that represent the genetic information carried. The sequence consists of four deoxyribonucleic acids, which are A(adenine), T(thymine), C(cytosine), and G(guanine) [24]. A and T as well as C and G are complementary pairs. When applying DNA sequences to binary Numbers, 0 and 1 are complementary. Four deoxyribonucleic acids are represented by two binary Numbers, so 00 and 11 are complementary, and 01 and 10 are also complementary. There are eight combinations satisfying the principle of base complementary pairing, that is, there are eight combinations of coding rules [25].

Plaintext can be thought of as a matrix with a pixel value from 0 to 255, and each plaintext pixel can be represented by a DNA sequence with a length of 4. For example, this information with a pixel value of 182 is converted into a binary sequence [10110110], which is encoded according to coding rule 1 in Table 1. The binary sequence obtained is [10111001], and the corresponding DNA sequence is CTGC. According to coding rule 2, the binary sequence obtained is [01111001]. DNA operations include XOR, addition, and subtraction, represented by a ternary number, where 0 represents DNA

XOR, 1 represents DNA addition, and 2 represents DNA subtraction. These three operation rules between DNA sequences are set as shown in Table 2.

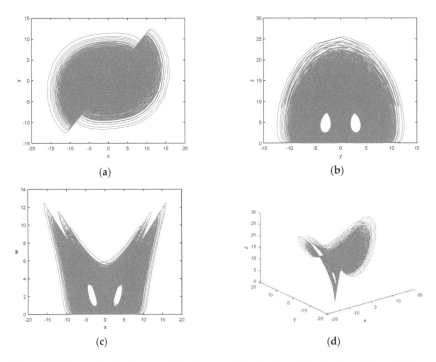

Figure 4. Qi Hyper-chaotic attractor: (**a**) (x-y) plane; (**b**) (y-z) plane; (**c**) (x-w) plane; (**d**) (x-y-z) plane.

Table 1. DNA coding rules.

Title 1	1	2	3	4	5	6	7	8
A	00	00	01	01	10	10	11	11
T	11	00	10	10	01	01	00	00
C	01	10	00	11	00	11	01	10
G	10	01	11	00	11	00	10	01

Table 2. DNA XOR, Addition and Subtraction.

XOR	A	G	C	T	+	A	G	C	T	-	C	A	T	G
A	A	G	C	T	A	A	G	C	T	C	C	A	T	G
G	G	A	T	C	G	G	C	T	A	A	G	C	A	T
C	C	T	A	G	C	C	T	A	G	T	T	G	C	A
T	T	C	G	A	T	T	A	G	C	G	A	T	G	C

3. Proposed Encryption Algorithm

The flow chart of the proposed encryption scheme is shown in Figure 5. Suppose that the size of the original image I is M × N, and the encryption process is as follows:

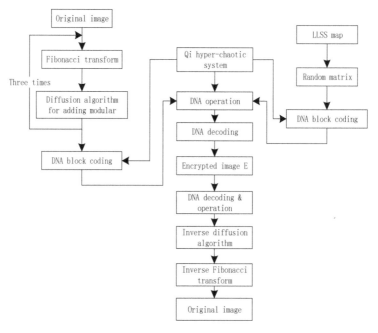

Figure 5. The flow chart of the proposed image encryption algorithm.

Step 1: read in the original image I and use the Fibonacci transform to produce scrambled image F.

Definition: Fibonacci is a scrambling algorithm based on two-dimensional chaotic mapping, which is a nonlinear transformation in modular form and reduces the correlation by changing the position relation of image pixels. Its definition is shown in Formula (5):

$$\begin{pmatrix} x' \\ y' \end{pmatrix} = \begin{pmatrix} 1 & 1 \\ 1 & 0 \end{pmatrix} \begin{pmatrix} x \\ y \end{pmatrix} \bmod N. \tag{5}$$

Step 2: The scrambled image F is diffused with the algorithm of adding and taking modules to obtain the diffusion image K. The main Formula is shown in Formula (6). This diffusion operation can make the scrambled image fully diffuse into the ciphertext,

$$K_i = (K_{i-1} + H_i + F_i) \bmod 256 \tag{6}$$

where K_i is the diffused image, F_i is the scrambled image, and H_i is the password pixel.

Step 3: Repeat step 1 and step 2 for the three times to fully obtain the middle ciphertext M.

Step 4: Generate an M × M random matrix using LLSS chaotic mapping denoted as R. Given the initial value and system parameters of LLSS, the chaotic sequence of LLSS is generated by iterating SUM + 999 times, and the first 1000 points are removed to obtain the sequence P, which is transformed into an integer from 0 to 255, and then transformed into a random matrix R of M rows and N columns.

Step 5: construct a control sequence with a hyperchaotic Qi system

(1) In order to resist the selective plaintext attack, the relationship between the initial value of the system and the plaintext is established, and the initial value of the hyperchaotic system X_0, Y_0, Z_0 and W_0 is obtained according to the Formula (7) to (10).

(2) In order to obtain better randomness, the first 1500 iterations is removed and four hyperchaotic sequences X, Y, Z and W are generated. To reconstruct the sequence, X and Y determine the

encoding mode of DNA, Z determines the operation of DNA, and W represents the decoding mode of DNA.

$$X_0 = sum(sum(bitand(I,3)))/(3 * SUM), \tag{7}$$

$$Y_0 = sum(sum(bitand(I,12)/4))/(3 * SUM), \tag{8}$$

$$Z_0 = sum(sum(bitand(I,48)/16))/(3 * SUM), \tag{9}$$

$$W_0 = sum(sum(bitand(I,192)/64))/(3 * SUM), \tag{10}$$

Step 6: The random matrix R and the middle image M are preprocessed and divided into four blocks. The middle image M is encoded according to the sequence number corresponding to X to get D1, and the random matrix R is encoded according to the sequence number corresponding to Y to get D2. Then the above two encoded blocks are calculated according to Z. Finally, the results of the operation are calculated with the results of the previous one again. Combine the split blocks to get the final encrypted image E.

Decryption is the reverse operation of encryption. Decryption is mainly divided into three modules: DNA decoding and operation, inverse diffusion operation, and inverse operation of Fibonacci transformation. These modules are shown in the lower part of Figure 5.

4. Simulation Results and Security Analysis

The five images size of are used as the test images 256 × 256 including Lena, Couple, Cameraman, Baboon, and Lake. Simultaneous, the MatlabR2015a is used as the platform. The original image, the encrypted image, and the corresponding decrypted image are shown in Figure 6. It can be seen from the comparison diagram that the encrypted image is a snowflake, in which there is no information of the original image, and the original image can also be decrypted from the encrypted image, indicating that the algorithm proposed in this paper has a good encryption effect. In this section, the proposed algorithm is analyzed for security.

4.1. Key Analysis

4.1.1. Key Space

A good encryption algorithm should have enough key space to resist exhaustive attacks. The key of the proposed algorithm consists of a total of seven keys: x_0, y_0, z_0, w_0, H_0, x_{01}, and μ_0. According to the international standard IEEE 754, the index portion is expressed as a positive value to simplify the comparison. The significant digit of a double-precision floating-point type is 52 bits, the size of the key space of the control parameter will be greater than $2^{52 \times 7} = 2^{364} > 2^{128}$. The results show that it is almost impossible to attack the algorithm correctly by brute force, so the encryption algorithm can resist brute force attacks.

4.1.2. Key Sensitivity

A small change in the decryption key makes a huge difference to the result, and the original image will not be decrypted correctly, indicating that the algorithm gas has a high sensitivity. First, set the initial values of the Qi hyperchaos system: $x_0 = 0.5001$, $y_0 = 0.5130$, $z_0 = 0.5170$, $w_0 = 0.3237$; and the initial values of the LLSS system: $x_{01} = 0.3711$, $\mu_0 = 3.9990$. Then, make a tiny change to the encryption key, select one of the key parameters, and add 10^{-10} so that the results can be compared as shown in Figure 7. It can be seen that only a slight change can have a huge effect. And the decryption diagram is completely different from the original image. Therefore, it can be concluded that it is impossible to decrypt by completely guessing the encryption key.

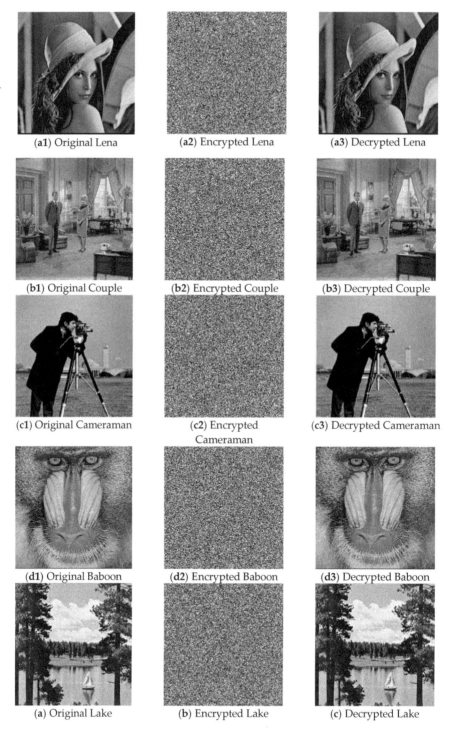

Figure 6. Original (**a1**)–(**e1**), encrypted (**a2**)–(**e2**), and decrypted of test image (**a3**)–(**e3**).

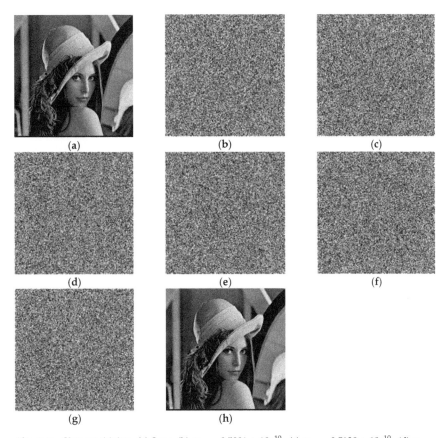

Figure 7. Key sensitivity: (a) Lena (b) $x_0 = 0.5001 + 10^{-10}$, (c) $y_0 = 0.5130 + 10^{-10}$, (d) $z_0 = 0.5170 + 10^{-10}$, (e) $w_0 = 0.3237 + 10^{-10}$, (f) $x_{01} = 0.3711 + 10^{-10}$, (g) $x_0 = 3.9990 + 10^{-10}$, (h) corrected decrypted image.

4.2. Statistic Analysis

4.2.1. Gray Histogram

Gray histogram is more intuitive, and the visibility is good. It can be intuitively seen from the figure that the frequency or probability of occurrence of the gray value. The more balanced the histogram, the better the encryption effect [26]. The comparison results are shown in Figure 8. The gray level histogram represents each gray level and the number of times that gray level occurs. The x-axis represents grayscale values of 0 to 255, and the y-axis represents the number of pixels in the corresponding grayscale in the figure. As can be seen from the figure, the histogram of the original image fluctuates greatly and is not uniform; Ciphertext images are roughly evenly distributed. The results show that the attacker cannot get information about the original image from the ciphertext, which indicates that the algorithm proposed in this paper has a good encryption effect.

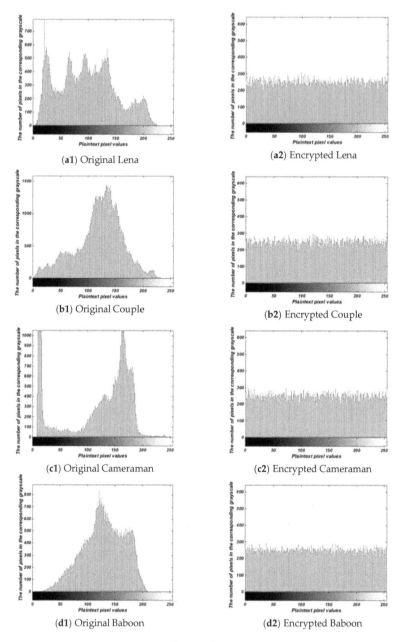

(**a1**) Original Lena

(**a2**) Encrypted Lena

(**b1**) Original Couple

(**b2**) Encrypted Couple

(**c1**) Original Cameraman

(**c2**) Encrypted Cameraman

(**d1**) Original Baboon

(**d2**) Encrypted Baboon

Figure 8. *Cont.*

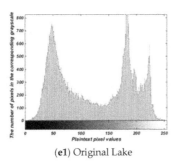

(e1) Original Lake

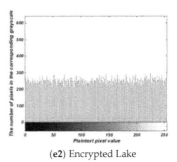

(e2) Encrypted Lake

Figure 8. Gray Histogram of original (a1)–(e1). Gray Histogram of decrypted image (a2)–(e2).

4.2.2. Correlation Analysis of Adjacent Pixels

Two thousand pairs of adjacent pixel values are randomly selected from the horizontal, vertical and diagonal directions of plaintext and ciphertext images. The following Formulas (11) to (14) are used to calculate the correlation coefficient of two adjacent pixel values:

$$\rho_{xy} = \frac{cov(x,y)}{\sqrt{D(x)}\sqrt{D(y)}}, \tag{11}$$

$$E(x) = \frac{1}{N}\sum_{i=1}^{N} x_i, \tag{12}$$

$$D(x) = \frac{1}{N}\sum_{i=1}^{N} (x_i - E(x))^2, \tag{13}$$

$$cov(x,y) = \frac{1}{N}\sum_{i=1}^{N} (x_i - E(x))(y_i - E(y)), \tag{14}$$

where x, y is the gray value of two adjacent pixels in the image, N is the total pixel value selected from the image, $E(x)$ and $E(y)$ are the mean value, $D(x)$ and $D(y)$ are the variance. The smaller the absolute value of the correlation coefficient is, the lower the correlation is. The correlation coefficient of plaintext and ciphertext is shown in Table 3. It can be seen from Table 3 that the absolute value of plaintext image correlation is close to 1, and the absolute value of ciphertext correlation is close to 0, which indicates that the image correlation after encryption is destroyed. The correlation diagram is shown in Figure 9, from which it can be seen that the pixels of the plaintext image are highly concentrated and distributed near the corners, while the pixels of the ciphertext image are evenly distributed.

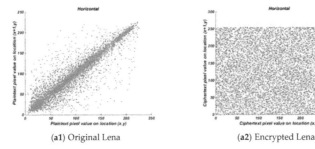

(a1) Original Lena

(a2) Encrypted Lena

Figure 9. *Cont.*

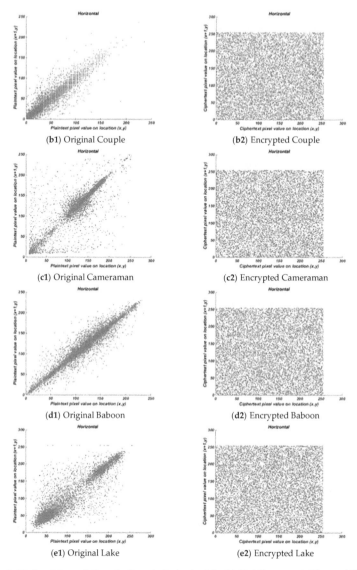

Figure 9. Horizontal correlation of adjacent pixels of original (**a1**)–(**e1**), encrypted image (**a2**)–(**e2**).

Table 3. Correlation coefficients of adjacent pixels for the test images.

Image	Scheme	Horizontal	Vertical	Diagonal
Lena	Original image	0.93767	0.97178	0.9104
	Cipher image	0.0020306	0.010543	0.0019857
Couple	Original image	0.9485	0.93625	0.89823
	Cipher image	0.0031994	0.0044791	−0.000148
Cameraman	Original image	0.92127	0.9633	0.89823
	Cipher image	0.0026387	0.010641	−0.000148
Baboon	Original image	0.90552	0.9228	0.8557
	Cipher image	−0.014249	0.0073645	0.0068203
Lake	Original image	0.93051	0.95735	0.89664
	Cipher image	0.0012594	−0.0014642	0.0020329

4.2.3. Information Entropy

The information entropy of the image is considered from the statistical characteristics and represents the overall characteristics of the image in the mean sense. It reflects the average amount of information in the image. The following Formula (15) is used to calculate the information entropy of the image:

$$H(x) = \sum_{i=0}^{2^n-1} p(m_i) log_2 \frac{1}{p(m_i)}, \tag{15}$$

where $p(m_i)$ represents the probability of signal m. For a 256 × 256 image, the ideal value of entropy is equal to 8, which means the image is uniform. The closer it gets to 8, the harder the cryptosystem leaves some information available. When the probability of each gray value is basically equal, the entropy reaches the maximum value. Table 4 is the information entropy of the algorithm proposed in this paper. It can be seen from Table 4 that the information entropy of this paper is close to 8, which indicates that the probability of accidental information leakage is very small.

Table 4. Information entropy.

Image	Lena	Couple	Cameraman	Baboon	Lake
Original image	7.5534	7.4601	7.0097	7.3649	7.5314
Cipher image	7.9974	7.9971	7.9970	7.9968	7.9973

4.3. Differential Attack

The difference between plaintext and ciphertext can be expressed by NPCR (the number of pixels change rate) and UACI (the number average changing intensity), where NPCR represents the ratio of different gray values of different ciphertext images at the same position, while UACI represents the average change density of different ciphertext images. UACI and NPCR can be used to test the ability of encryption algorithms to resist differential attacks. The Formulas (16) to (18) are to calculate NPCR and UACI.

$$NPCR = \frac{\sum_{i,j} D(i,j)}{M \times N} \times 100\%, \tag{16}$$

$$UACI = \frac{1}{M \times N} \times \frac{\sum_{i,j} |C_1(i,j) - C_2(i,j)|}{L} \times 100\%, \tag{17}$$

$$D(i,j) = \begin{cases} 0, & C_1(i,j) = C_2(i,j) \\ 1, & otherwise \end{cases}, \tag{18}$$

where $C_1(i,j)$ and $C_2(i,j)$ represent the ciphertext image corresponding to two plaintext images with only one pixel difference. For a 256-level image, the ideal values of UACI and NPCR are 33.4635% and 99.6094%. The test results are shown in Table 5. It can be seen from the table that the average UACI is 99.6130% and NPCR is 33.5211%, which is very close to the ideal value.

Table 5. UACI and NPCR.

Image	Lena	Couple	Cameraman	Baboon	Lake	Average
NPCR (%)	99.5987	99.6276	99.6002	99.6170	99.6216	99.6130
UACI (%)	33.5267	33.5208	33.3921	33.6318	33.5344	33.5211

4.4. Anti-Noise Ability

In order to test the anti-noise ability of the algorithm, add a different intensity of Salt and Pepper noise and Gaussian noise to the ciphertext image and decrypt it. Then use the peak signal to noise ratio (PSNR) to assess it, which is the most widely used image perception quality evaluation method, and defined by the mean square error (MSE):

$$MSE = -\frac{1}{m \times n} \sum_{i=1}^{m} \sum_{j=1}^{n} [I(i,j) - D(i,j)]^2,$$
(19)

$$PSNR = 10\lg\left(\frac{255^2}{MSE}\right),$$
(20)

where I is the original image and D is the decrypted image. The test results are shown in Table 6. First increase the noise of the density of 0.001, 0.005 and 0.01 to the cipher images. The noised cipher images are shown in the first column of Table 6, and then they can be decrypted. The decrypted images are shown in the third column of Figure 6. The corresponding PSNR is shown in the fourth column. It can be seen from the figure that in the case of noise, the algorithm in this paper can decrypt the noised cipher images and obtain the original image information. Even if the noise intensity reaches 0.01, the decrypted image can still be visually recognized. It can be seen that the encryption scheme can effectively resist a certain degree of noise attack.

Table 6. PSNR with different noises and intensities.

Noise	Noisy encrypted images	Noise intensities	Decrypted images	PSNR(dB)
		0.001		41.7268
Salt and Pepper noise		0.005		34.7189
		0.01		33.4257
		0.001		35.2165
Gaussian		0.005		33.8192
		0.01		32.483

4.5. Anti-Cropping Ability

To test the ability of the proposed algorithm to resist clipping attacks, set the gray values of some pixels of the encrypted image to 0, and then decrypt it with the correct key. As shown in Figure 10, it can be seen that after cutting off a pixel block, the original image can still be decrypted to a certain extent, indicating that the algorithm proposed in this paper has a certain degree of anti-cropping ability.

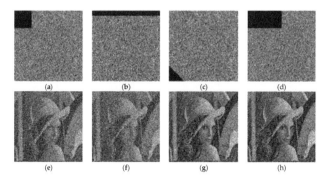

Figure 10. Cropping attacks with different areas. (**a–d**) Partially cut the encrypted image, (**e–h**) decrypted images.

4.6. Chosen-Plaintext Attack

In cryptanalysis, there are four typical attacks: ciphertext-only attack, known-plaintext attack, chosen-plaintext attack, and chosen-ciphertext attack. If it can resist a chosen-ciphertext attack, it has enough security to resist other attacks. In this paper, two kinds of images, all black and all white, are used for testing. The encryption diagram and its histogram are shown in Figure 11. At the same time, the correlation between information entropy and adjacent pixels can be analyzed, as shown in Table 7.

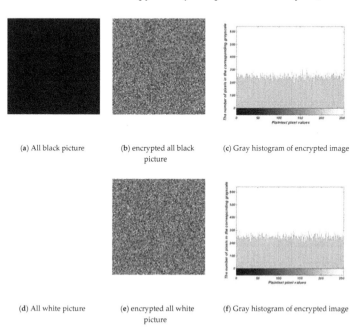

(**a**) All black picture   (**b**) encrypted all black picture   (**c**) Gray histogram of encrypted image

(**d**) All white picture   (**e**) encrypted all white picture   (**f**) Gray histogram of encrypted image

Figure 11. Test results with all black and all white.

Table 7. Information entropy and correlation coefficients of the test images.

	Entropy	Correlation Coefficients		
		Horizontal	Vertical	Diagonal
All black	0	—	—	—
Cipher with all black	7.9972	−0.0036	0.0261	0.0033
All white	0	—	—	—
Cipher with all white	7.9973	−0.0042	0.0187	−0.0021

4.7. Comparative Analysis with Other Literatures

The algorithm proposed in this paper is compared with other literatures in terms of key space, information entropy and differential attack. The results are shown in Table 8. It can be seen from the table that the algorithm proposed in this paper is close to the ideal value, and better than the algorithms discussed in other literatures in three ways, indicating that this algorithm has a good encryption effect.

Table 8. Comparative analysis.

Algorithm	Key space	Information entropy	UACI (%)	NPCR (%)
Ours	3.8×10^{109}	7.9971	33.5211	99.6130
Ref. [5]	1.2×10^{83}	7.9951	33.4624	99.4890
Ref. [15]	1.9×10^{126}	7.9973	30.2375	99.5950
Ref. [16]	1.6×10^{79}	7.9964	33.4694	99.6105
Ref. [19]	2.9×10^{138}	7.9845	28.6679	99.6101
Ref. [27]	6.5×10^{119}	7.9970	33.3443	99.7643

The algorithm proposed in this paper is compared with other literatures on related rows of adjacent pixels. The results are shown in Table 9. As can be seen from the table, the algorithm proposed in this paper reduces the pixel correlation from the three directions of horizontal, vertical, and diagonal, so that its absolute value is close to 0. Compared with other algorithms, the reduction effect of this algorithm is better.

Table 9. Comparative analysis of the correlation coefficients of adjacent pixels.

Algorithm	Vertical	Horizontal	Diagonal
Ours	0.0020	0.0105	0.0019
Ref. [5]	0.0298	−0.0359	0.0052
Ref. [15]	0.0021	0.0004	−0.0038
Ref. [16]	0.0054	−0.0011	−0.0038
Ref. [19]	0.0001	−0.0011	−0.0014
Ref. [27]	−0.0331	0.0125	−0.0236

4.8. Structural Similarity Index (SSIM)

SSIM is a measure of the similarity of two images. If the two images are before encryption and after decryption, then SSIM can be used to evaluate the quality of the encrypted image. The value is from 0 to 1. The larger the value, the smaller the image distortion. Calculated as follows:

$$\mu_X = \frac{1}{m \times n} \sum_{i=1}^{m} \sum_{j=1}^{n} X(i,j), \tag{21}$$

$$\sigma_X = \frac{1}{m \times n - 1} \sum_{i=1}^{m} \sum_{j=1}^{n} (X(i,j) - \mu_X)^2)^{1/2}, \tag{22}$$

$$\sigma_{XY} = \frac{1}{m \times n - 1} \sum_{i=1}^{m} \sum_{j=1}^{n} (X(i,j) - \mu_X)(Y(i,j) - \mu_Y), \tag{23}$$

$$SSIM = \frac{(2\mu_X\mu_Y + C_1)(2\sigma_{XY} + C_2)}{(\mu_X^2 + \mu_Y^2 + C_1)(\sigma_X^2 + \sigma_Y^2 + C_2)}, \tag{24}$$

where $C_1 = (0.01 \times 255)^2$, $C_2 = (0.03 \times 255)^2$. Calculate the SSIM value is 0.81085 according to the formula. It can be seen that it is within the range and the value is relatively high. This shows that the algorithm has less distortion.

4.9. Computational Complexity Analysis

The image encryption algorithm was implemented by Matlab on a personal computer with an Intel i5-4210U processor and 4.00G RAM. It takes time to record the encryption and decryption of different image sizes. The results are shown in Figure 12.

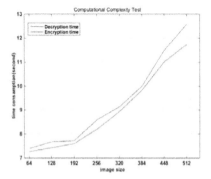

Figure 12. Image encryption algorithm computational complexity test.

5. Discussion

This paper proposes a new one-dimensional chaos, which is formed by parallel processing of Logistic and Sine chaos as seed maps and through modulo operation. The new chaos has the advantages of a simple one-dimensional chaotic structure, being easy to implement and full mapping in the chaos range. The algorithm in this paper is based on the combination of the double chaos, this new one-dimensional chaotic, and hyperchaos Qi, and uses DNA coding technology to achieve image encryption. In the fourth part of the experimental simulation and performance analysis, we can see that the algorithm proposed in this paper can increase the key space, have high sensitivity to the key, reduce the degree of correlation of the original image, and resist the advantages of multiple attacks. However, the efficiency of the algorithm discussed in this paper is not high, and the degree of anti-attack needs to be improved. This will be progressed in future research.

6. Conclusions

In this paper, a new image encryption scheme based on composite chaos and Qi hyperchaos combined with DNA coding is proposed. In this scheme, Fibonacci transformation and diffusion algorithm of adding modules are used for initial encryption. Then the intermediate ciphertext and the new compound chaos are calculated by DNA to form the final ciphertext. In order to resist chosen-plaintext attack, the algorithm takes the sum of original image pixels as the initial value of a chaotic sequence. Experimental simulation shows that this scheme can increase the key space and resist many common attacks. However, the efficiency of the scheme is not high, so the main work in the future will be to improve the efficiency of the algorithm.

Author Contributions: Conceptualization, Y.W.; methodology, Y.W.; software, Y.W. and S.G.; validation, Y.W., S.G. and B.D.; formal analysis, S.G.; investigation, Y.W. and S.G.; resources, Y.W. and S.G.; data curation, Y.W. and B.D.; writing—original draft preparation, Y.W.; writing—review and editing, Y.W., S.G. and B.D.; visualization, Y.W., S.G. and B.D.; supervision, Y.W. and S.G.; project administration, Y.W. and B.D.; funding acquisition, B.D. All authors have read and agreed to the published version of the manuscript.

Funding: This research was funded by [SCIENCE AND TECHNOLOGY INNOVATION SPECIAL PROJECT OF BASIC RESEARCH PROJECT OF BASIC SCIENTIFIC RESEARCH OPERATING EXPENSES OF COLLEGES AND UNIVERSITIES IN HEILONGJIANG PROVINCE IN 2019], grant number [KJCX201906].

Acknowledgments: The authors also gratefully acknowledge the helpful comments and suggestions of the reviewers, which have improved the presentation.

Conflicts of Interest: The authors declare no conflict of interest.

References

1. Huang, X.L.; Ye, G.D. An image encryption algorithm based on Time-Delay and random insertion. *Entropy* **2018**, *20*, 974. [CrossRef]
2. Ahmad, J.; Hwang, S.O. Chaos-based diffusion for highly autocorrelated data in encryption algorithms. *Nonlinear Dyn.* **2015**, *82*, 1839–1850. [CrossRef]
3. Zhou, H.M. Noise reduction multi-carrier differential chaos shift keying system. *J. Circuits Syst. Comput.* **2018**, *27*, 14. [CrossRef]
4. Budroni, M.A.; Calabrese, I.; Miele, Y. Control of chemical chaos through medium viscosity in a batch ferroin-catalysed Belousov-Zhabotinsky reaction. *Phys. Chem. Chem.* **2017**, *19*, 32235–32241. [CrossRef] [PubMed]
5. Zhou, N.; Pan, S.; Cheng, S. Image compression–encryption scheme based on hyper-chaotic system and 2D compressive sensing. *Opt. Laser Technol.* **2016**, *82*, 121–133. [CrossRef]
6. Hancerliogulları, A.; El Hadad, K.M.; Kurt, E. Implementation of a real time analog secure image communication system via a chaotic circuit. *Politek. Derg.* **2019**, *20*, 1083–1092. [CrossRef]
7. Zhou, Y.C.; Bao, L.; Chen, C.L.P. A new 1D chaotic system for image encryption. *Signal Process.* **2014**, *97*, 172–182. [CrossRef]
8. Wu, X.J.; Kan, H.B.; Kurths, J. A new color image encryption scheme based on DNA sequences and multiple improved 1D chaotic maps. *Appl. Soft Comput.* **2015**, *37*, 24–39. [CrossRef]
9. Cao, L.C.; Luo, Y.L.; Qiu, S.H. A perturbation method to the tent map based on Lyapunov exponent and its application. *Chin. Phys. B* **2015**, *24*, 78–85. [CrossRef]
10. Hua, Z.Y.; Zhou, Y.C.; Huang, H.J. Cosine-transform-based chaotic system for image encryption. *Inf. Sci.* **2019**, *480*, 403–419. [CrossRef]
11. Chanil, P.; Lilian, H. A new color image encryption using combination of the 1D chaotic map. *Signal Process.* **2017**, *138*, 129–137.
12. Sun, S.L. A Novel Hyperchaotic Image Encryption Scheme Based on DNA Encoding, Pixel-Level Scrambling and Bit-Level Scrambling. *IEEE Photonics J.* **2018**, *10*, 129–137. [CrossRef]
13. Adleman, L. Molecular computation of solutions to combinatorial problems. *Science* **1994**, *266*, 1021–1024. [CrossRef]
14. Gehani, A.; Labean, T.; Reif, J. DNA-based Cryptography. *Asp. Mol. Comput.* **2002**, *54*, 233–249.
15. Liu, H.J.; Wang, X.Y.; Kadir, A. Image encryption using DNA complementary rule and chaotic maps. *Appl. Soft Comput.* **2012**, *12*, 1457–1466. [CrossRef]
16. Zhang, J.; Huo, D. Image encryption algorithm based on quantum chaotic map and DNA coding. *Multimed. Tools Appl.* **2019**, *78*, 15605–15621. [CrossRef]
17. Chai, X.L.; Fu, X.L.; Gan, Z.H. A color image cryptosystem based on dynamic DNA encryption and chaos. *Signal Process.* **2019**, *155*, 44–62. [CrossRef]
18. Guan, M.M.; Yang, X.L.; Hu, W.S. Chaotic image encryption algorithm using frequency-domain DNA encoding. *IET Image Process.* **2019**, *119*, 105661. [CrossRef]
19. Yang, Y.G.; Guan, B.W.; Li, J. Image compression-encryption scheme based on fractional order hyperchaotic systems combined with 2D compressed sensing and DNA encoding. *Opt. Laser Technol.* **2019**, *13*, 1535–1539.
20. Huang, C.G.; Cheng, H.; Ding, Q. Logistic chaotic sequence generator based on physical unclonable function. *J. Commun.* **2019**, *40*, 186–193.

Entropy **2020**, *22*, 171

21. Bi, C.; Zhang, Q.; Xiang, Y.; Wang, J.M. Bifurcation and attractor of two-dimensional sinusoidal discrete map. *Acta Phys. Sin.* **2019**, *62*, 240503.

22. Qi, G.Y.; Chen, G.R.; Du, S. Analysis of a new chaotic system. *Phys. A* **2005**, *352*, 295–308. [CrossRef]

23. Qi, G.Y.; Chen, G.R. Analysis and circuit implementation of a new 4D chaotic system. *Phys. Lett. A* **2006**, *352*, 386–397. [CrossRef]

24. Xiao, G.Z.; Lu, M.X.; Qin, L.; Lai, X.J. New field of cryptography: DNA cryptography. *Chin. Sci. Bull.* **2006**, *51*, 1413–1420. [CrossRef]

25. Watson, J.D.; Crick, F.; Qin, L.; Lai, X.J. A structure for deoxyribose nucleic acid. *Nature* **1953**, *171*, 737–738. [CrossRef] [PubMed]

26. Ahmad, J.; Hwang, S.O. A secure image encryption scheme based on chaotic maps and affine transformation. *Multimed. Tools Appl.* **2016**, *75*, 13951–13976. [CrossRef]

27. Khan, F.A.; Ahmed, J.; Khan, J.S.; Ahmad, J.; Khan, M.A. A novel image encryption based on Lorenz equation, Gingerbreadman chaotic map and S8 permutation. *J. Intell. Fuzzy Syst.* **2017**. [CrossRef]

© 2020 by the authors. Licensee MDPI, Basel, Switzerland. This article is an open access article distributed under the terms and conditions of the Creative Commons Attribution (CC BY) license (http://creativecommons.org/licenses/by/4.0/).

Article

A Novel Image Encryption Approach Based on a Hyperchaotic System, Pixel-Level Filtering with Variable Kernels, and DNA-Level Diffusion

Jiang Wu, Jiayi Shi and Taiyong Li *

School of Economic Information Engineering, Southwestern University of Finance and Economics, Chengdu 611130, China; wuj_t@swufe.edu.cn (J.W.); 218081202002@smail.swufe.edu.cn (J.S.)
* Correspondence: litaiyong@gmail.com

Received: 15 October 2019; Accepted: 17 December 2019; Published: 19 December 2019

Abstract: With the rapid growth of image transmission and storage, image security has become a hot topic in the community of information security. Image encryption is a direct way to ensure image security. This paper presents a novel approach that uses a hyperchaotic system, Pixel-level Filtering with kernels of variable shapes and parameters, and DNA-level Diffusion, so-called PFDD, for image encryption. The PFDD totally consists of four stages. First, a hyperchaotic system is applied to generating hyperchaotic sequences for the purpose of subsequent operations. Second, dynamic filtering is performed on pixels to change the pixel values. To increase the diversity of filtering, kernels with variable shapes and parameters determined by the hyperchaotic sequences are used. Third, a global bit-level scrambling is conducted to change the values and positions of pixels simultaneously. The bit stream is then encoded into DNA-level data. Finally, a novel DNA-level diffusion scheme is proposed to further change the image values. We tested the proposed PFDD with 15 publicly accessible images with different sizes, and the results demonstrate that the PFDD is capable of achieving state-of-the-art results in terms of the evaluation criteria, indicating that the PFDD is very effective for image encryption.

Keywords: image encryption; hyperchaotic system; filtering; DNA computing; diffusion

1. Introduction

Images carry rich and direct information that is easy to perceive for the human visual system. In some specific fields, such as military, security, medical fields, and so on, it is very important to prevent image content from leaking. Therefore, image security has become a very hot research topic in the community of information security. Image encryption algorithms that change the values and/or the positions of pixels in images have been thought of as effective methods for image security. Although many popular encryption algorithms, such as DES (data encryption standard), advanced encryption standard (AES), and RSA (Rivest–Shamir–Adleman), were initially designed for block textual data, they can also be applied to encrypting images [1]. For example, AES with cipher block chaining (CBC) mode can achieve good performance in image encryption in spite of images having the apparent characteristics of bulky pixels, strong correlations, and high redundancy. Recently, chaos-based approaches have become another hot topic in the field of image encryption, since chaotic systems have many merits for encryption, such as ergodicity, unpredictability, pseudorandomness, and high sensitivity to parameters and initial values [2–5].

In chaos-based image encryption, chaotic systems are usually applied to generate chaotic sequences for changing the positions and/or values of pixels in images. Chen et al. generalized the 2D chaotic cat map to three dimensions and then applied the 3D cat map to conducting image encryption, and the results showed that the proposed scheme was fast and highly secure [2]. Pareek et al.

used two Logistic maps and eight different operations to encrypt the pixels in an image, and the experiments demonstrated that the proposed approach was a secure and efficient way for image encryption [6]. Borujeni and Eshghi used a logistic map to generate a bit sequence for pseudorandom number generation in Tompinks–Paige algorithm, and the results indicated that the proposed scheme could resist any brute-force and statistical attacks [7]. Sheela et al. proposed a novel 2D Henon map with broad chaotic regime, and then used this map and sine map to confuse and diffuse images. The experimental analysis revealed the proposed scheme was advantageous over some compared traditional ones [8]. Low-dimensional chaotic systems have many advantages, such as simple form, few parameters, and easy implementation, but they are vulnerable to attack. A simple but effective solution is to use high-dimensional chaotic systems instead of low-dimensional ones. Lyapunov exponent (LE) is a poplar way to measure chaos. When a chaotic system has two or more positive LEs, it is called a hyperchaotic system, which usually has a larger key space and higher security for encryption [9,10]. Norouzi and Mirzakuchaki used two hyperchaotic systems to modify the gray-level of each pixel and crack the strong correlation among neighboring pixels in an image at the same time [11]. Zhu et al. put forward an image encryption scheme using a compound homogeneous hyperchaotic system to permute the plain image twice and then to diffuse the permutated pixels with dynamic local binary pattern operations, and the experiments demonstrated its security and effectiveness [12]. Xue et al. used a hyperchaotic system owning three positive Lyapunov exponents to encrypt the region of interest (ROI) of a color image [13]. A recently-emerged and hot research topic is to use chaotic systems and compressive sensing to encrypt and compress images simultaneously [14–17]. Some other hyperchaotic systems were also applied to image encryption [18–22].

As far as operations of image encryption are concerned, permutation and diffusion are among the most important ones. The former changes the positions of the data in an image, while the latter changes the values of the data. An encryption operation may involve one block of pixels, one pixel, one DNA unit (two bits), or even one bit [10,23–25]. The work by Xu et al. indicated that a scheme with block permutation and dynamic index based diffusion was very effective for chaotic image encryption [23]. Chaos-based S-Boxes are very popular in block encryption methods [26–28]. Zhang et al. proposed an image fusion encryption with a hyperchaotic system and DNA-level operations [29]. Chai et al. integrated several types of chaotic systems and DNA computing to encrypt images, showing that the proposed schemes had high security and could resist different attacks [30,31]. Khan et al. proposed a novel image encryption approach that integrated DNA computing, the intertwining logistic map, and the affine transformation for medical image encryption. The experiments demonstrated that the proposed approach was robust, efficient, and secure for medical image encryption [32]. Zhan et al. proposed a scheme with a hyperchaotic system, global bit permuting, and DNA computing (HCDNA) to improve the security and robustness of encryption [33]. In order to improve the performance of diffusion, Zhu et al. used hyperchaotic systems and ciphertext diffusion in a crisscross pattern (CDCP) to encrypt pixel-level data, and the experiments revealed the CDCP had very promising performance regarding time and diffusion [34]. Sun put forward an image encryption algorithm that used a 5D hyperchaotic system for operations on pixel-level, DNA-level and bit-level data, and both the theoretical analysis and the experimental results demonstrated that the encryption approach was secure and could resist types of attacks [35]. Zhou et al. combined a hyperchaotic system and quantum operations for bit-level image encryption [36]. To eliminate the weakness of an image encryption scheme [37], Ahmad et al. integrated discrete cosine transformation (DCT), chaotic skew tent map, and XOR operations to encrypt images. The proposed cryptosystem was capable of resisting many types of attacks and achieved very promising results in terms of several tests [38]. Very recently, Hua and Zhou have proposed a novel image cipher algorithm using block-based scrambling and image filtering (IC-BSIF), which introduced filtering, a classic operation in digital image processing, into image encryption by designing a special filter [39]. In spite of the effectiveness for image encryption, the existing filtering-based schemes usually adopt a fixed shape of filters, lacking the diversity of the filters. Hence, they may have negative impacts on encryption performance [5,10].

Motivated by the merits of hyperchaotic systems for image encryption as well as the diffusion performance by filtering and pixel-level CDCP, this paper proposes a novel scheme integrating a hyperchaotic system, pixel-level filtering with filters of different shapes, and DNA-level CDCP-like diffusion, namely, PFDD, for image encryption. PFDD consists of four stages. First, we use a 4D hyperchaotic system to generate chaotic sequences for subsequent encryption operations. Second, each pixel is filtered by a specific kernel/filter, whose shape and weights are determined by the chaotic sequences. In other words, the kernels for the pixels in an image are totally different from each other, which helps to enhance the diversity of kernels. Third, the filtered image is transformed into a bit stream, and then a global bit-level permutation is conducted on the bit stream to change the position of each bit and naturally change the values of corresponding pixels. The bit stream is then encoded into DNA-level data by rules decided by the chaotic sequences. Finally, we propose a DNA-level diffusion scheme to improve encryption performance. The main novelty of the PFDD is two-fold: (1) we propose a novel filtering operation for image encryption, which uses variable kernel shapes and kernel parameters determined by hyperchaotic sequence; and (2) we also propose a DNA-level diffusion scheme to further change the values of images.

The main contributions of this paper are as follows: (1) we use a hyperchaotic system to generate sequences for all the encryption operations; (2) kernels with variable shapes and different parameters determined by hyperchaotic sequences are used to conduct filtering to change the pixel values; (3) novel DNA-level diffusion is proposed to expand any tiny changes in a plain image to the whole cipher image; (4) pixel-level, bit-level, and DNA-level operations are used to improve the encryption effectiveness; and (5) extensive experiments demonstrate the proposed PFDD is very promising for image encryption.

The main advantages of the PFDD are three aspects: (1) permutation or diffusion is conducted with different-levels of data (pixel-level, bit-level, and DNA-level), improving the effectiveness of the PFDD; (2) a novel pixel-level filtering strategy with different kernel types and parameters determined by hyperchaotic sequences increases the diversity of kernels and hence enhances the security of the PFDD; and (3) the DNA-level diffusion is able to expand a tiny change in a plain image to the whole cipher image to resist differential attacks very well.

The rest of this paper is organized as follows. First, we briefly describe a 4D hyperchaotic system with two positive LEs, filtering operations, and DNA computing in Section 2. Then the proposed image encryption scheme that integrates the hyperchaotic system, pixel-level filtering with variable kernels, and DNA-level diffusion, is proposed in detail in Section 3. In Section 4, we display our extensive experiments on 15 testing images; the results are reported and analyzed. Finally, the paper is concluded in Section 5.

2. Preliminaries

2.1. Hyperchaotic Systems

Hyperchaos, first reported by Rössler [40], is usually defined as a chaotic attractor which has more than one Lyapunov exponent. Due to its advantages in security, hyperchaos is becoming more and more popular in image encryption. Recently, Gu and Gao made a 4D hyperchaotic system by adding a general linear controller to the 3D autonomous Chen's chaotic system [41,42]. This system has two positive Lyapunov exponents and can be formulated by Equation (1):

$$\begin{cases} \dot{x} = a(y-x) \\ \dot{y} = dx - xz + cy - w \\ \dot{z} = xy - bz \\ \dot{w} = mx + k \end{cases}, \tag{1}$$

where x, y, z, and w are state variables; and a, b, c, d, m, and k are variable constants. In our work, we use the 4th-order Runge-Kutta method with a step size of $h = 0.001$ to solve the hyperchaotic system.

When the parameters $(a, b, c, d, m, k) = (36, 3, 28, -16, 0.5, 0.5)$ and initial values $(x^0, y^0, z^0, w^0) = (-1, -1, 0.3333, -5.9583)$, the attractors of this 4D hyperchaotic system are illustrated in Figure 1.

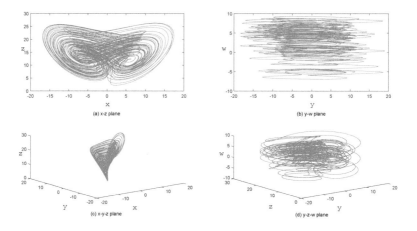

Figure 1. Hyperchaotic attractors of the 4D Hyperchaotic system.

2.2. Filtering

Filtering, also known as convolution, can be used for smoothing, denoising, and sharpening images, and thus becomes essential for image processing. By applying a convolution operation between a kernel/mask/filter and an image, the pixel values of the processed image will be changed. Thus, filtering can be used for diffusing an image. However, an image filtering operation is usually irreversible, making it impossible to decrypt images. As a result, filtering cannot be directly used for diffusion in image encryption. Fortunately, Hua and Zhou proposed a new method to solve this problem: by setting the value of right-bottom position of the filtering kernel to "1," the corresponding point in the encrypted image can be recovered [39]. In spite of the magic this technique is, there are limitations of this function, including using a kernel with a fixed shape and fixed parameters to do convolution. An ideal method should use dynamic kernel shape and variable kernel parameters for filtering.

2.3. DNA Computing

DNA computing could be used to solve a computational problem [43]. Different from the binary alphabets (0 and 1) in traditional computers, information is expressed by four-character genetic alphabets; i.e., A, C, G, and T for adenine, cytosine, guanine, and thymine in DNA computing, respectively. The crucial technologies of DNA computing in image encryption are encoding and decoding rules and algebraic operations. Considering the four characters of DNA alphabet, there ought to exist 4! = 24 combinations in DNA encoding. However, only eight categories of DNA combinations satisfy the DNA complementary rules, as shown in Table 1. A pixel of eight bits in a grayscale image can be encoded to four characters by using these encoding rules. For instance, a decimal value 180 can be converted to a binary value "10110100," and further be transformed into DNA sequences "GACT" and "ACTG" by Rule 3 and Rule 8, respectively. Obviously, different encoding rules lead the identical decimal or binary value to completely different DNA sequences.

Table 1. The encoding and decoding rules of DNA computing.

RULE	Rule 1	Rule 2	Rule 3	Rule 4	Rule 5	Rule 6	Rule 7	Rule 8
00	A	T	T	A	C	G	C	G
01	C	G	C	G	A	A	T	T
10	G	C	G	C	T	T	A	A
11	T	A	A	T	G	C	G	C

Like binary algebraic operations, DNA has its own algebraic operations, such as addition ($\oplus$), subtraction ($\ominus$), and XOR ($\otimes$). Different from traditional binary operations, different DNA encoding rules can produce different results. In other words, once the encoding rule is decided, the results of DNA algebraic operations are fixed. For example, with encoding Rule 1, the results of DNA addition, subtraction, and XOR operations are listed in Table 2. These operations are usually used to change the values of DNA characters.

Table 2. DNA algebraic addition ($\oplus$), subtraction ($\ominus$), and XOR ($\otimes$) operations.

$\oplus$	A	C	G	T	$\ominus$	A	C	G	T	$\otimes$	A	C	G	T
A	A	C	G	T	A	A	T	G	C	A	A	C	G	T
C	C	G	T	A	C	C	A	T	G	C	C	A	T	G
G	G	T	A	C	G	G	C	A	T	G	G	T	A	C
T	T	A	C	G	T	T	G	C	A	T	T	G	C	A

3. The Proposed Image Encryption Scheme

3.1. Generating Hyperchaotic Sequences

In this paper, we use the 4D hyperchaotic system described in Section 2.1 to generate the hyperchaotic sequences for encryption. Generally, the procedure is divided into three steps:

Step 1: The 4D hyperchaotic system begins to iterate to generate long enough sequences for image encryption. In the i-th iteration, we can obtain four state values denoted as $s^i = \{x^i, y^i, z^i, w^i\}$.

Step 2: The sequences generated by the first n_0 iterations are discarded to eliminate the adverse effects.

Step 3: When the iteration completes, a hyperchaotic sequence S can be obtained by concatenating all the $s^j (j = 1, 2, \cdots, N)$ as in Equation (2):

$$
\begin{aligned}
S = \{s^1, s^2, \cdots, s^N\} &= \{x^1, y^1, z^1, w^1, \cdots, x^N, y^N, z^N, w^N\} \\
&= \{s_1, s_2, s_3, s_4, \cdots, s_{4N-3}, s_{4N-2}, s_{4N-1}, s_{4N}\}.
\end{aligned}
\tag{2}
$$

Then the generated sequence S is further cast to an integral sequence by Equation (3):

$$
s_i = \lfloor (|s_i| - \lfloor |s_i| \rfloor) \times 10^{14} \rfloor \% 256,
\tag{3}
$$

where $|\cdot|$, $\lfloor \cdot \rfloor$, and % denote the operations of absolute value, flooring, and modulo, respectively [5,9].

3.2. Pixel-Level Filtering with Variable Kernels

Having generated hyperchaotic sequence via Section 3.1, two sub-sequences, S_h and S_w, can be obtained. S_h is a $1 \times h$ vector while S_w is a $w \times 1$ vector, so a parameter p can be computed by Equation (4):

$$p = S_h \cdot I \cdot S_w \%256, \tag{4}$$

where I represents the plain image; h and w denote its height and width, respectively; and $\cdot$ is the operation of matrix multiplication. It is clear that p is associated with the plain image and it can be further used to change filtering kernels. In this way, different plain images will be diffused by different kernels when conducting filtering.

According to the work of IC-BSIF, filtering can be used for image encryption [39]. However, it employs convolution operation to images with a kernel with a fixed shape and fixed kernel parameters values, lacking the diversity of the kernel. Very recently, Li et al. used a 1×3 or 3×1 variable kernel with different parameters to implement convolution on an image; in other words, the kernels associated with each pixel in an image for convolution are different in so-called dynamic filtering [5]. The experimental results have shown the effectiveness of dynamic filtering. Nevertheless, there still remains some room for improvement with dynamic filtering. An ideal method is to conduct filtering with variable kernel shapes and parameters, which may lead to better performance. To this end, we can use the hyperchaotic sequence to determine the shapes and parameters of the kernels. For a 3×3 kernel, since the value at the right-bottom corner is fixed to "1," it only requires $3 \times 3 - 1 = 8$ bits to determine the kernel shapes. Fortunately, a single value in the hyperchaotic sequence is exactly an 8 bit integer, which can determine $2^8 = 256$ types of kernel shapes. For example, an 8 bit integer "0" denotes a kernel of all "0," which means all the contents in the kernel are "0" and the shape of the kernel is blank, and hence the kernel is independent of the filtering. In contrast, an 8 bit integer "256" denotes a kernel of all "1," implying all the values in the kernel are involved in filtering. A detailed example is shown in Figure 2. The integer "17" ("00010001" in binary) in the hyperchaotic sequence determines the shape of a 3×3 kernel, which has only three non-zero cells with blue background including the "1" in the right-bottom cell, as shown in a red border. The next eight integers first conduct bit, the XOR operation with the parameter p defined in Equation (4), and a new sequence containing eight integers can be obtained. Then, the new sequence is used to fill the red kernel, and we can get the kernel k_1. After that, filtering can be conducted on the 3×3 part with a red border in the plain image P, and then the pixel value "211" in P can be encrypted to "125" in the cipher image C. Likewise, the next nine integers in the hyperchaotic sequence can generate another kernel k_2. With this kernel and the part with a green border in P, the pixel "137" in P can be encrypted to "183" in C.

From this example, we can see that both the shapes and the parameters of the kernels are completely determined by the hyperchaotic sequence. The filtering operation can be applied to diffusing an image.

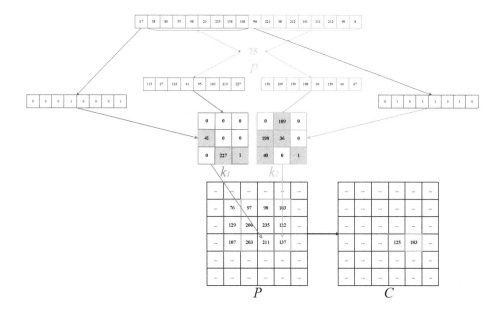

Figure 2. An example of multi-shape dynamic filtering.

3.3. Global Bit-Level Permutation

Permutation is usually used to change the positions of pixels, and it can be further applied to permuting bit-streams. The main procedure of such an operation is as the following. First, generate a hyperchaotic sequence that has the same length as a bit-stream. Then, sort the hyperchaotic sequence to get the sorting index. Finally, rearrange the bit-stream according to the sorting index [9,33].

3.4. DNA-Level Diffusion

Diffusion is a frequently used way to change the pixels in images. The existing diffusion schemes are usually associated with pixel-level data or bit-level data only. Motivated by the effectiveness of CDCP [34], a pixel-level diffusion scheme, this paper proposes a DNA-level diffusion approach. In this approach, DNA addition and XOR are used to further diffuse the image since DNA algebraic operations have a property of changing the values of nucleic acids. The main idea of such DNA-level diffusion is to expand the changes in one DNA character to the whole DNA sequence. Given the length of the DNA sequence S, $L = h \times w \times d/2$, where h, w, and d denote the height, width, and depth of a plain image, respectively, and half of the L, $H = L/2$. The pseudocode of such diffusion is described as follows:

Step 1: $C(1) = S(1) \otimes (C0 \oplus K(1))$; $C(H + 1) = S(H + 1) \otimes (C(1) \oplus K(H + 1))$
Step 2: for i = 2 → H

$$C(i) = S(i) \otimes (C(H + i - 1) \oplus K(i))$$
$$C(H + i) = S(H + i) \otimes (C(i) \oplus K(H + i))$$

end for

Step 3: $D(1) = C(1) \otimes (C(DL) \oplus K(1))$; $D(H + 1) = C(H + 1) \otimes (D(1) \oplus K(H + 1))$
Step 4: for i = 2 → H

$$D(i) = C(i) \otimes (D(H + i - 1) \oplus K(i))$$

$$D(H + i) = C(H + i) \otimes (D(i) \oplus K(H + i))$$

end for

where $\oplus$ and $\otimes$ are DNA addition and XOR, respectively; $C0$ is a user-defined parameter; K is an auxiliary DNA-level sequence generated from the hyperchaotic system; and D is the obtained diffused image.

3.5. PFDD: The Proposed Image Encryption Approach Using a Hyperchaotic System, Pixel-Level Filtering with Variable Kernels, and DNA-Level Diffusion

Due to the effectiveness of hyperchaotic systems in image encryption, permutation power of bit-level scrambling, and diffusion power of filtering and CDCP, this paper proposes a novel image encryption scheme by integrating such advantages. The proposed scheme conducts encryption on various levels, including pixel-level data, bit-level data, and DNA-level data. First, it uses a 4D hyperchaotic system with two positive LEs to generate chaotic sequences for encryption. Second, dynamic filtering operations with kernels with different shapes and parameters are conducted on pixels to diffuse the image. Third, the image is transformed into a bit stream and the global bit permutation is conducted twice. Then, the bit stream is transformed into DNA-level data. Finally, DNA-level diffusion is operated with DNA-level data, and then the DNA-level data is transformed into a pixel-level cipher image. The flowchart of the PFDD is shown in Figure 3 and the steps are described in detail as the following.

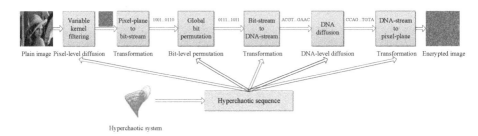

Figure 3. The framework of the proposed PFDD (Pixel-level Filtering with kernels of variable shapes and parameters and DNA-level Diffusion).

Step 1: Use initial values to generate a hyperchaotic sequence via Equations (1)–(3);

Step 2: For each pixel in the plain image, create a kernel whose shape and parameters are determined by the hyperchaotic sequence. Then, conduct a filtering operation on the pixel with the kernel. This is named pixel-level filtering with variable kernels, which results in a diffused image, as described in Section 3.2;

Step 3: Transform the diffused image into a bit stream;

Step 4: Perform the global bit permutation twice;

Step 5: Encode the bit stream into a DNA-stream. Every pair of two adjacent bits is encoded into a DNA symbol through a DNA encoding rule determined by the hyperchaotic sequence;

Step 6: Conduct DNA-level diffusion on the DNA-stream as described in Section 3.4;

Step 7: Transform the DNA-level diffused plane into a pixel plane, i.e., the cipher image.

The core of the PFDD consists of pixel-level filtering with variable kernels (Step 2), global bit permutation (Step 4), and DNA-level diffusion (Step 6). The PFDD conducts encryption in the pixel-level, bit-level, and DNA-level data, and hence it has the potential to improve encryption. The PFDD is a typical strategy of "divide and conquer"; that is, the task of image encryption is divided into several sub-tasks of encrypting different level data [44–46]. The decryption is the reverse of the encryption.

4. Experimental Results

4.1. Experimental Settings

In order to evaluate the performance of the proposed PFDD, some state-of-the-art encryption schemes were used for comparison, such as image encryption using pixel-level diffusion with dynamic filtering and DNA-level permutation with 3D Latin cubes (DFDLC) [10], image encryption with a hyperchaotic system and DNA computing (HCDNA) [33], CDCP [34], and IC-BSIF [39]. We set the parameters for the PFDD as follows. For the 4D hyperchaotic system, we set $(x^0, y^0, z^0, w^0) = (-1, -1, 0.3333, -5.9583)$ and 1200 as the discard iterating time, respectively. For these comparison encryption methods' parameters, we generally set their parameters according to the corresponding original references. We used 15 publicly-accessed, 256-level grayscale images with different sizes to test the proposed PFDD, and the sizes and names of the images are listed in Table 3. Note that Lena1024, Male2048, and Airport2048 were generated from corresponding test images with sizes of 512×512, 1024×1024, and 1024×1024 via interpolation, respectively.

Table 3. Testing images.

Image	Size ($h \times w$)	Image	Size ($h \times w$)	Image	Size ($h \times w$)
Lena256	256×256	Airplane256	256×256	Aerial512	512×512
Finger512	512×512	Clown512	512×512	Martha512	512×512
Crowd512	512×512	Reagan512	512×512	Trucks512	512×512
Woman512	512×512	Lena512	512×512	Lena1024	1024×1024
Male1024	1024×1024	Male2048	2048×2048	Airport2048	2048×2048

All the experiments were conducted with Matlab R2017a on a PC with 64-bit Windows 10 Ultimate, 16 GB memory, and a 3.60 GHz I7 CPU.

4.2. Security Key Analysiss

Security keys are essential for image encryption. A large key space and high sensitivity of keys enhance the security of encryption and are capable of resisting brute-force attacks. In this subsection, we analyze those two attributes of the proposed PFDD.

4.2.1. Key Space

According to the existing research, if a cryptographic system has a key space greater than 2^{100}, it is able to resist brute-force attacks [14,47]. The initial values (x^0, y^0, z^0, w^0) for the hyperchaotic system can be used as a part of the keys of the PFDD. If the precision of each value is 10^{-15}, the key space will be $(10^{-15})^4 = 10^{-60} \approx 2^{199}$. Besides, the number of discarded iterations in generation of chaotic sequence, n_0, and the value by multiplying a chaotic sequence and the plain image (p in Equation (4)) can also be used as keys, enhancing the key space. Since the key space of the PFDD is much greater than 2^{100}, it can resist brute-force attacks.

4.2.2. Sensitivity to Security Keys

A good and practical image encryption system should be extremely sensitive to the security keys. In other words, a tiny change with keys will lead to a completely different recovered image from the plain image. It is one of the natural characteristics of hyperchaotic systems. To verify it, we used the right security key, K_1, and a tiny change key, K_2, to decrypt some cipher images. Specifically, K_1 is $(x^0, y^0, z^0, w^0) = (-1, -1, 0.3333, -5.9583)$, and then we added 10^{-15} to one of the initial value, x^0, and kept the other values were unchanged to obtain K_2; i.e., $K_2 = (x^0 + 10^{-15}, y^0, z^0, w^0) = (-1 + 10^{-15}, -1, 0.3333, -5.9583)$. The decrypted images with K_1 and K_2 are shown in the first and the second row in Figure 4, respectively.

Figure 4. Sensitivity to security keys. From left to right, the images are Lena256, Airplane256, Aerial512, Finger512, Lena1024, Male1024, Male2048, and Airport2048.

It is clear that K_1 can decrypt the cipher images correctly, whereas K_2 cannot do it at all, so that the results decrypted by K_2 are random-like. The experimental results demonstrate that the sensitivity of the key of the PFDD is extremely high, which is a good attribute of an ideal image encryption system.

4.3. Statistical Analysis

Typical statistical analysis includes information entropy (IE) analysis, histogram analysis, and correlation analysis. The cipher images with a well-designed encryption algorithm should have evenly distributed histograms and very high entropies, and the neighboring pixels should have very weak correlations.

4.3.1. Information Entropy Analysis

Information entropy, a key concept in information theory, exists to measure the degree of randomness or uncertainty in a given complex system. Typically, for a 256-level grayscale image I, the IE can be computed by Equation (5) [10].

$$IE(I) = -\sum_{i=0}^{255} p(i) log_2(p(i)), \qquad (5)$$

where $p(i)$ indicates the probability of occurrence of the i-th gray level. For an image that only contains one type gray level, e.g., an all white image, the IE obtains the minimum, 0, while if all gray levels appear with the same probability, i.e., $\frac{1}{256}$, the image can achieve the highest IE, 8. A well-designed image encryption algorithm will result in an IE as close as possible to 8. The IEs of the images with the proposed PFDD and the compared algorithms are listed in Table 4.

Entropy **2020**, *22*, 5

Table 4. The information entropies (IEs) of the test images.

Image	Input	Cipher Images				
		PFDD	**DFDLC** [10]	**HCDNA** [33]	**CDCP** [34]	**IC-BSIF** [39]
Lena256	7.5954	**7.9973**	7.9971	7.9965	7.9966	7.9972
Airplane256	6.4523	7.9972	7.9969	7.9962	**7.9973**	**7.9973**
Aerial512	6.9940	**7.9993**	**7.9993**	7.9985	**7.9993**	**7.9993**
Finger512	6.7279	**7.9993**	**7.9993**	7.9990	7.9992	7.9992
Clown512	5.3684	7.9992	**7.9993**	7.9892	7.9992	**7.9994**
Martha512	7.5222	**7.9993**	**7.9993**	7.9991	**7.9993**	**7.9993**
Crowd512	7.4842	7.9992	**7.9993**	7.9946	**7.9994**	**7.9993**
Reagan512	7.1923	**7.9993**	**7.9993**	**7.9993**	**7.9993**	7.9992
Trucks512	6.5632	**7.9994**	**7.9994**	**7.9994**	7.9993	7.9993
Woman512	6.9542	7.9992	7.9992	**7.9993**	7.9992	**7.9993**
Lena512	7.4455	7.9993	7.9993	**7.9994**	7.9993	7.9993
Lena1024	7.4439	**7.9998**	**7.9998**	7.9991	**7.9998**	**7.9998**
Male1024	7.5237	**7.9998**	**7.9998**	7.9940	**7.9998**	**7.9998**
Male2048	7.5369	**8.0000**	**8.0000**	7.9935	**8.0000**	**8.0000**
Airport2048	6.8106	**8.0000**	**8.0000**	7.9994	**8.0000**	**8.0000**

From this table, we can see that the IEs of all plain images fall in the range of $[5.3648, 7.5954]$—far lower than 8. In contrast, the IEs by all the encryption methods are very close or even equal to the theoretical maximum 8. More specifically, PFDD, DFDLC, HCDNA, CDCP, and IC-BSIF achieve the highest IEs with 10, 9, 4, 9, and 9 out of 15 cases, respectively. The PFDD achieved the highest IE 10 times, which is superior to the other models, indicating that the PFDD can effectively resist entropy attacks. It is worth pointing out that some entropies of the last two images are equal to 8, which shows the pixel distributions in the last two cipher images are very uniform.

4.3.2. Histogram Analysis

A histogram is a graph that can directly reflect the distribution of pixel values in an image. The histogram of a natural image usually shows some shapes with mountains and valleys, whereas that of a cipher image by an ideal encryption algorithm should be nearly uniformly distributed to avoid histogram attacks. The images and the corresponding histograms are shown in Figure 5.

It can be found that all the histograms of plain images are very different. For example, Lena with different sizes, Finger512, Martha512, Crowd512, Male1024, and Male2048 have a wide range of grayscale values, while Airplane256, Trucks512, Woman512, and Airport2048 have a narrow one. At the same time, the different shapes of the histograms mean that the distributions of the plain images are totally different. However, when we investigate the cipher images, we can find that they are all random-like, even for the plain images with narrow pixel ranges. The histograms of all the encrypted images are so flat that they are very close to uniform distributions, showing that the proposed PFDD exhibits ideal performance regarding the histogram distribution. In particular, the tops of all bars in the histograms of cipher images with large size (the last five images) seem like horizontal lines, indicating the pixels distributes more uniformly in the cipher images with large sizes than those with small sizes.

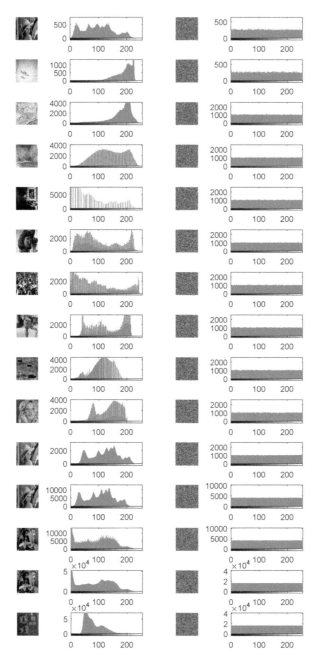

Figure 5. Images and histograms. From left to right, the images are plain images, the histograms of the plain images, the cipher images, and the histograms of the cipher images. In each histogram, the *x*-axis and the *y*-axis represent the pixel values and the total times the corresponding pixel occurs, respectively. From top to bottom, the names of the involved images are the same as the first column in Table 4.

4.3.3. Correlation Analysis

Correlation reflects the relevance between two neighboring pixels in an image. Generally speaking, the correlation in a natural image is high because any two neighboring pixels are very similar, which is probably utilized to crack the image. Therefore, a practical encryption scheme should decrease such a correlation to a very low level. The correlation coefficient γ between a sequence of pixels x and the sequence of its neighboring pixels y in an image can be formulated by Equation (6) [10].

$$E(x) = \frac{1}{L} \sum_{i=1}^{L} x_i,$$

$$D(x) = \frac{1}{L} \sum_{i=1}^{L} (x_i - E(x))^2,$$

$$\rho(x, y) = \frac{1}{L} \sum_{i=1}^{L} (x_i - D(x))(y_i - D(y)),$$

$$\gamma = \frac{\rho(x, y)}{\sqrt{D(x)D(y)}},$$

(6)

where L is the length of the sequence of x; $E(x)$ and $D(x)$ denote the mathematical expectation and the standard deviation of x, respectively; and $\rho(x, y)$ is the covariance of the two given sequences: x and y.

For each plain image and each cipher image, we calculate the correlation coefficients in the horizontal, vertical, and diagonal directions, represented by γ_h, γ_v, and γ_d, respectively. Since all the pixels in an image are involved, we can think of the correlation from a global perspective. The results are listed in Table 5. We can see that the correlation coefficients of the plain images are in the range of $[0.8003, 0.9899]$, which is close to 1, confirming the strong correlation existing in natural images. However, such a strong correlation is destroyed drastically by the encryption methods. We can also find that all the correlation coefficients of the cipher images are very close to 0, showing that there is almost no correlation in the encrypted images. A typical example is the image of Martha. It has the highest correlation in the vertical direction, i.e., $\gamma_v = 0.9899$. With the encryption schemes, however, the absolute values of γ_v are less than 0.002, demonstrating the strong correlation in plain Martha has almost been completely broken. The PFDD achieves the lowest correlation coefficients in 11 out of 45 cases and all the correlation coefficients are close to 0. The results indicate that the PFDD can obtain correlation coefficients comparable to the competitive methods.

Table 5. The correlation coefficients γ of the test images.

Image	γ	Input	Cipher Images				
			PFDD	DFDLC [10]	HCDNA [33]	CDCP [34]	IC-BSIF [39]
Lena256	γ_h	0.9144	−0.0014	0.0045	−0.0042	0.0041	−0.0004
	γ_v	0.9545	0.0028	0.0012	−0.0011	**0.0004**	−0.0020
	γ_d	0.9098	0.0066	0.0001	0.0029	**−0.0000**	0.0028
Airplane256	γ_h	0.9562	0.0080	−0.0038	−0.0040	−0.0027	**0.0009**
	γ_v	0.8742	−0.0104	0.0004	−0.0007	**0.0001**	−0.0036
	γ_d	0.8995	**−0.0000**	−0.0019	0.0003	0.0022	−0.0022
Aerial512	γ_h	0.8993	−0.0024	−0.0009	**0.0007**	0.0009	−0.0014
	γ_v	0.8549	−0.0011	0.0021	−0.0011	**−0.0009**	0.0014
	γ_d	0.8003	**0.0003**	0.0005	0.0021	0.0010	−0.0011
Finger512	γ_h	0.9343	0.0002	**−0.0001**	0.0007	−0.0023	−0.0026
	γ_v	0.9168	0.0013	**0.0002**	0.0029	−0.0032	−0.0030
	γ_d	0.8664	**−0.0007**	0.0017	−0.0022	−0.0010	0.0011

Table 5. *Cont.*

Image	γ	Input	Cipher Images				
			PFDD	**DFDLC** [10]	**HCDNA** [33]	**CDCP** [34]	**IC-BSIF** [39]
Clown512	γ_h	0.9763	−0.0018	−0.0026	**0.0001**	0.0019	0.0022
	γ_v	0.9888	0.0009	**−0.0004**	0.0020	−0.0033	0.0012
	γ_d	0.9699	**−0.0002**	0.0002	0.0010	−0.0008	−0.0015
Martha512	γ_h	0.9864	0.0014	0.0020	0.0002	−0.0009	**−0.0001**
	γ_v	0.9899	−0.0017	0.0008	**−0.0003**	0.0003	−0.0013
	γ_d	0.9815	−0.0015	−0.0004	0.0014	−0.0003	−0.0030
Crowd512	γ_h	0.9059	0.0021	**−0.0003**	−0.0004	0.0019	−0.0013
	γ_v	0.9047	**0.0001**	0.0014	−0.0029	−0.0005	0.0003
	γ_d	0.8525	−0.0018	−0.0022	0.0017	**−0.0007**	0.0012
Reagan512	γ_h	0.9668	−0.0031	0.0003	−0.0017	0.0003	0.0015
	γ_v	0.9757	0.0010	0.0003	−0.0007	0.0035	**−0.0002**
	γ_d	0.9573	**0.0005**	0.0008	0.0013	0.0022	0.0023
Trucks512	γ_h	0.9408	−0.0016	−0.0034	**−0.0013**	0.0014	0.0028
	γ_v	0.9110	0.0017	−0.0021	**−0.0003**	−0.0023	−0.0019
	γ_d	0.8906	−0.0008	**0.0000**	0.0001	−0.0029	−0.0007
Woman512	γ_h	0.9250	0.0028	0.0002	−0.0032	0.0008	−0.0004
	γ_v	0.9570	−0.0013	−0.0015	0.0008	−0.0020	0.0003
	γ_d	0.9217	0.0011	0.0014	0.0030	0.0003	−0.0003
Lena512	γ_h	0.9691	0.0013	0.0023	−0.0015	−0.0004	0.0023
	γ_v	0.9841	0.0021	**0.0009**	−0.0020	0.0028	**0.0009**
	γ_d	0.9639	0.0013	**0.0008**	0.0024	0.0016	**0.0008**
Lena1024	γ_h	0.9918	**0.0007**	0.0008	−0.0012	0.0015	0.0008
	γ_v	0.9962	−0.0003	−0.0003	−0.0020	−0.0012	−0.0003
	γ_d	0.9902	−0.0004	0.0001	0.0001	−0.0005	0.0001
Male1024	γ_h	0.9769	−0.0012	−0.0001	−0.0003	−0.0005	−0.0001
	γ_v	0.9804	**0.0008**	0.0014	0.0011	0.0009	0.0014
	γ_d	0.9669	0.0009	0.0008	−0.0002	0.0006	0.0008
Male2048	γ_h	0.9942	0.0014	**0.0001**	**0.0001**	0.0002	**0.0001**
	γ_v	0.9950	−0.0002	0.0002	−0.0004	−0.0002	0.0002
	γ_d	0.9905	0.0002	0.0004	−0.0003	0.0002	0.0004
Airport2048	γ_h	0.9781	0.0009	0.0010	−0.0003	0.0001	0.0010
	γ_v	0.9764	−0.0004	0.0001	−0.0007	−0.0003	0.0001
	γ_d	0.9581	−0.0002	0.0002	−0.0006	−0.0002	0.0002

All the information entropy analysis, histogram analysis, and the correlation analysis demonstrate that the proposed PFDD can effectively resist statistical attacks.

4.4. Analysis of Resisting Differential Attacks

As a type of cryptanalysis, differential attacks aim to analyze how a tiny change in a plain image affects the corresponding cipher image. To defend differential attacks, a good encryption scheme should ensure that any tiny changes in the plain image are able to produce a completely different cipher image.

To measure the ability of resisting differential attacks of encryption schemes, the unified average changing intensity, UACI for short, and the number of pixels change rate, NPCR for short, are two very popular indices, as defined by Equations (7) and (8), respectively [48].

$$UACI = \frac{\sum_{i=1}^{W}\sum_{j=1}^{H}|C_1(i,j) - C_2(i,j)|}{255WH} \times 100\%,\qquad(7)$$

$$NPCR = \frac{\sum_{i=1}^{W}\sum_{j=1}^{H}\delta(i,j)}{WH} \times 100\%,\qquad(8)$$

where C_1 and C_2 are two cipher images, whose width and height are W and H, respectively, and $\delta(i,j)$ is an indicator to judge whether the two pixel values at the position of (i,j) in C_1 and C_2 are identical, which is defined as Equation (9).

$$\delta(i,j) = \begin{cases} 0, & C_1(i,j) = C_2(i,j) \\ 1, & C_1(i,j) \neq C_2(i,j) \end{cases}. \tag{9}$$

According to [48], for a given 256×256 8 bit gray image and a significance level $\alpha = 0.05$, if the UACI falls into the interval of $\left(\mathcal{U}_{0.05}^{*l1}, \mathcal{U}_{0.05}^{*u1}\right) = (33.2824\%, 33.6447\%)$, and the NPCR is greater than $\mathcal{N}_{0.05}^{*1} = 99.5693\%$, it is said that the corresponding method passes the UACI and the NPCR test at $\alpha = 0.05$, respectively. Likewise, if the UACI falls into $\left(\mathcal{U}_{0.05}^{*l2}, \mathcal{U}_{0.05}^{*u2}\right) = (33.3730\%, 33.5541\%)$, $\left(\mathcal{U}_{0.05}^{*l3}, \mathcal{U}_{0.05}^{*u3}\right) = (33.4183\%, 33.5088\%)$, and $\left(\mathcal{U}_{0.05}^{*l4}, \mathcal{U}_{0.05}^{*u4}\right) = (33.4409\%, 33.4862\%)$ for an 8 bit gray image of 512×512, 1024×1024, and 2048×2048, respectively, the encryption scheme also passes the UACI test. If the NPCR is greater than $\mathcal{N}_{0.05}^{*2} = 99.5893\%$, and $\mathcal{N}_{0.05}^{*3} = 99.5994\%$, $\mathcal{N}_{0.05}^{*4} = 99.6044\%$ for an 8-bit gray image of these sizes, the encryption scheme is said to pass the NPCR test.

To compute the UACI and the NPCR once, we add one to a randomly selected pixel. The computation is repeated 10 times, and the mean UACI and NPCR are listed in Tables 6 and 7, respectively. The values that passed corresponding tests are shown in bold. From Table 6, we can see that all the UACI by PFDD, DFDLC, CDCP, and IC-BSIF fell in the specified intervals $\left(\mathcal{U}_{0.05}^{*l1}, \mathcal{U}_{0.05}^{*u1}\right)$, $\left(\mathcal{U}_{0.05}^{*l2}, \mathcal{U}_{0.05}^{*u2}\right)$, $\left(\mathcal{U}_{0.05}^{*l3}, \mathcal{U}_{0.05}^{*u3}\right)$, and $\left(\mathcal{U}_{0.05}^{*l4}, \mathcal{U}_{0.05}^{*u4}\right)$, showing they can pass the UACI test for images with all sizes of the testing images. It is worth pointing out that the PFDD achieved the highest UACI values in seven out of 15 cases. The HCDNA obtains so poor UACI that none of the image with HCDNA can pass the UACI test. As far as the NPCR is concerned, we found that PFDD, DFDLC, and IC-BSIF can pass the test. In contrast, CDCP passes the test in eight out of 15 cases, and once again, none of the images with HCDNA can pass it. The possible reason is that the encryption schemes with filtering operations (PFDD, DFDLC, and IC-BSIF) are capable of improving the performance of diffusion.

Table 6. The average unified average changing intensities (UACI, in precentages) of running the schemes 10 times.

Image	PFDD	DFDLC [10]	HCDNA [33]	CDCP [34]	IC-BSIF [39]
Lena256	**33.4440**	**33.4741**	18.7430	**33.4862**	**33.4200**
Airplane256	**33.4620**	**33.4367**	20.3208	**33.5691**	**33.4330**
Aerial512	**33.4745**	**33.4471**	22.1490	**33.4430**	**33.4575**
Finger512	**33.4711**	**33.4095**	13.0616	**33.4836**	**33.4601**
Clown512	**33.4742**	**33.4437**	26.4164	**33.4142**	**33.4787**
Martha512	**33.4810**	**33.4748**	22.0456	**33.4501**	**33.4810**
Crowd512	**33.4718**	**33.4624**	21.2259	**33.4466**	**33.4612**
Reagan512	**33.4267**	**33.4657**	13.9140	**33.4909**	**33.5007**
Trucks512	**33.4885**	**33.4700**	25.9466	**33.4382**	**33.4385**
Woman512	**33.5120**	**33.4505**	21.6499	**33.4779**	**33.4719**
Lena512	**33.4840**	**33.4363**	26.4423	**33.4275**	**33.4568**
Lena1024	**33.4776**	**33.4674**	31.1754	**33.4320**	**33.4630**
Male1024	**33.4459**	**33.4536**	30.3316	**33.4876**	**33.4475**
Male2048	**33.4587**	**33.4641**	23.7265	**33.4629**	**33.4683**
Airport2048	**33.4556**	**33.4550**	29.0287	**33.4661**	**33.4590**

Table 7. The average number of pixels change rates (NPCRs (%)) of running the schemes 10 times.

Image	PFDD	DFDLC [10]	HCDNA [33]	CDCP [34]	IC-BSIF [39]
Lena256	**99.6124**	**99.6202**	46.0794	**100.0000**	99.6045
Airplane256	**99.6260**	**99.6155**	47.1913	**100.0000**	99.5866
Aerial512	**99.6101**	**99.6130**	55.1017	99.5516	99.6142
Finger512	**99.5956**	**99.6077**	30.8046	**99.6445**	99.6118
Clown512	**99.6141**	**99.6107**	60.8291	99.4683	99.6124
Martha512	**99.6112**	**99.6056**	54.7043	**99.6180**	99.6130
Crowd512	**99.6112**	**99.6066**	59.4704	99.5816	99.6156
Reagan512	**99.6111**	**99.6054**	35.9236	**99.5967**	99.6089
Trucks512	**99.6112**	**99.6121**	67.8079	**99.6015**	99.6055
Woman512	**99.6120**	**99.6133**	58.5091	99.5684	99.6168
Lena512	**99.6062**	**99.6140**	94.1631	99.2096	99.6173
Lena1024	**99.6075**	**99.6100**	78.9105	99.2248	99.6100
Male1024	**99.6113**	**99.6107**	78.9105	99.2470	99.6084
Male2048	**99.6104**	**99.6092**	87.1085	**100.0000**	99.6099
Airport2048	**99.6077**	**99.6089**	87.1085	**100.0000**	99.6088

The analysis indicates that the PFDD can pass the UACI and the NPCR tests for all the experimental images, and hence it can resist differential attacks.

4.5. Plaintext and Ciphertext Attack Analysis

For a system of image encryption, there are four typical types of attacks; i.e., ciphertext only, chosen ciphertext, known plaintext, and chosen plaintext attacks. Among these attacks, the chosen plaintext attack is known as the most powerful one. If a cryptosystem can withstand it, it is said to have the ability to resist against other types of attacks [49].

From the aforementioned analysis, it is known that any tiny changes (even a bit) in the plain image will produce a totally different cipher image, so the proposed PFDD can resist differential attacks, which is a typically chosen plain text attack. Besides, the security keys include a value (p in Equation (4)) which is related to the plain image. Therefore, different plain images can generate different security keys and then obtain different results of permutation and diffusion. The cipher images by the proposed PFDD are all noise-like and all the corresponding histograms are very close to uniform distributions, further enhancing the security. In a word, the proposed PFDD highly depends on the content of the plain image, and it can resist against plaintext and ciphertext attacks.

4.6. Running Time and Results on Large Images

Encryption speed is another index to evaluate approaches of image encryption. Since the speed is not related to the content but to the sizes of images, we report the running time of the proposed PFDD and the compared approaches with four different types of sizes, as shown in Table 8. It can be seen that with the increase of image sizes, the running times of all the encryption approaches increase. Among the approaches, CDCP ranks first in all cases because of the simplicity of its operations, and is followed by IC-BSIF. The proposed PFDD ranks third in all cases. Since the main operations of PFDD include filtering and DNA-diffusion, its speed slightly underperforms against IC-BSIF, which conducts encryption mainly via filtering operations. DFDLC and HCDNA rank fourth and last regarding running time, respectively. Note that the running time of HCDNA is extremely high, and the possible reason is that it uses encoding/decoding rules and DNA algebraic rules directly for each operation. In the proposed PFDD, we use lookup tables instead of the rules directly for DNA encoding and DNA operations, so the running time of the PFDD is much less than that of HCDNA. Another interesting finding from this table is that the running time of all the encryption is linear with the image size. Therefore, for an encryption approach, we may estimate the running time for an image with a specific size.

Table 8. Running time (in seconds).

Image Size	PFDD	DFDLC [10]	HCDNA [33]	CDCP [34]	IC-BSIF [39]
256 × 256	0.9802	2.4491	8.2463	**0.2274**	0.6879
512 × 512	3.8264	7.6971	30.4855	**0.7035**	3.1478
1024 × 1024	14.4212	36.2733	123.9747	**2.7289**	9.9495
2048 × 2048	56.4122	127.8365	494.1457	**10.6423**	39.881

A good image encryption approach should process images of different sizes well. Since the PFDD treats each unit of images (bit, DNA, and pixel) equally, there is no obvious relationship between the effectiveness of encryption and image size. In other words, the PFDD can handle images of different sizes very well. This has been demonstrated by the aforementioned analysis and discussion in terms of entropy, correlation, histogram, UACI, and NPCR. Just like a coin has two sides, the processing strategy of the PFDD limits the speed because it has to conduct filtering on the pixels one by one. Therefore, although the PFDD can achieve good encryption results for large images, it will take a lot of running time to encrypt them, and hence the time efficiency is at an intermediate level. This might be a limitation of the proposed PFDD.

5. Conclusions

Image encryption is very important for information security. This paper proposed a novel and effective image encryption scheme integrating a 4D hyperchaotic system, pixel-level filtering with variable kernels, and DNA-level diffusion, namely, PFDD, for image encryption. In addition, a global bit-level scrambling operation was introduced to change the position of each single bit. The advantages of the PFDD come from three aspects: (1) it performs encryption with not only pixel-data and DNA-level data, but also bit-level data; (2) the filtering kernels with different shapes and different parameters are used to enhance the diversity of the kernels, and hence improve the performance of diffusion; and (3) a DNA-level diffusion algorithm is proposed to further enhance the diffusion. We conducted extensive experiments to verify the proposed PFDD, and the results showed that the PFDD has reliable security keys and is capable of resisting types of attacks. In the future, we will extend the PFDD to color image encryption. Besides that, we will study how to improve the efficiency of the PFDD.

Author Contributions: Formal analysis, J.W.; investigation, J.S.; methodology, T.L.; software, J.S. and T.L.; supervision, T.L.; writing—original draft, J.W., J.S., and T.L.; writing—review and editing, J.W. and T.L. All authors have read and agreed to the published version of the manuscript.

Funding: This research was funded by the Fundamental Research Funds for the Central Universities (grant number JBK1902029), the Ministry of Education of Humanities and Social Science Project (grant number 19YJAZH047), and the Scientific Research Fund of Sichuan Provincial Education Department (grant number 17ZB0433).

Acknowledgments: This work was supported by the Fundamental Research Funds for the Central Universities (grant number JBK1902029), the Ministry of Education of Humanities and Social Science Project (grant number 19YJAZH047), and the Scientific Research Fund of Sichuan Provincial Education Department (grant number 17ZB0433).

Conflicts of Interest: The authors declare no conflict of interest.

References

1. Singh, G. A study of encryption algorithms (RSA, DES, 3DES and AES) for information security. *Int. J. Comput. Appl.* **2013**, *67*. [CrossRef]
2. Chen, G.; Mao, Y.; Chui, C.K. A symmetric image encryption scheme based on 3D chaotic cat maps. *Chaos Solitons Fractals* **2004**, *21*, 749–761. [CrossRef]
3. Li, X.; Li, T.; Wu, J.; Xie, Z.; Shi, J. Joint image compression and encryption based on sparse Bayesian learning and bit-level 3D Arnold cat maps. *PLoS ONE* **2019**, *14*, e0224382. [CrossRef]

4. Zhou, S.; Zhang, Q.; Wei, X.; Zhou, C. A Summarization on Image Encryption. *IETE Tech. Rev.* **2010**, *27*, 503–510. [CrossRef]
5. Li, X.; Xie, Z.; Wu, J.; Li, T. Image Encryption Based on Dynamic Filtering and Bit Cuboid Operations. *Complexity* **2019**, *2019*, 7485621. [CrossRef]
6. Pareek, N.K.; Patidar, V.; Sud, K.K. Image encryption using chaotic logistic map. *Image Vis. Comput.* **2006**, *24*, 926–934. [CrossRef]
7. Borujeni, S.E.; Eshghi, M. Chaotic Image Encryption Design Using Tompkins-Paige Algorithm. *Math. Probl. Eng.* **2009**, *2009*, 762652. [CrossRef]
8. Sheela, S.J.; Suresh, K.V.; Tandur, D. Image encryption based on modified Henon map using hybrid chaotic shift transform. *Multimed. Tools Appl.* **2018**, *77*, 25223–25251. [CrossRef]
9. Li, T.; Yang, M.; Wu, J.; Jing, X. A Novel Image Encryption Algorithm Based on a Fractional-Order Hyperchaotic System and DNA Computing. *Complexity* **2017**, *2017*, 9010251. [CrossRef]
10. Li, T.; Shi, J.; Li, X.; Wu, J.; Pan, F. Image Encryption Based on Pixel-Level Diffusion with Dynamic Filtering and DNA-Level Permutation with 3D Latin Cubes. *Entropy* **2019**, *21*, 319. [CrossRef]
11. Norouzi, B.; Mirzakuchaki, S. A fast color image encryption algorithm based on hyper-chaotic systems. *Nonlinear Dyn.* **2014**, *78*, 995–1015. [CrossRef]
12. Zhu, H.; Zhang, X.; Yu, H.; Zhao, C.; Zhu, Z. An image encryption algorithm based on compound homogeneous hyper-chaotic system. *Nonlinear Dyn.* **2017**, *89*, 61–79. [CrossRef]
13. Xue, H.W.; Du, J.; Li, S.L.; Ma, W.J. Region of interest encryption for color images based on a hyperchaotic system with three positive Lyapunov exponets. *Opt. Laser Technol.* **2018**, *106*, 506–516. [CrossRef]
14. Chai, X.; Zheng, X.; Gan, Z.; Han, D.; Chen, Y. An image encryption algorithm based on chaotic system and compressive sensing. *Signal Process.* **2018**, *148*, 124–144. [CrossRef]
15. Gong, L.; Qiu, K.; Deng, C.; Zhou, N. An optical image compression and encryption scheme based on compressive sensing and RSA algorithm. *Opt. Lasers Eng.* **2019**, *121*, 169–180. [CrossRef]
16. Zhou, N.; Jiang, H.; Gong, L.; Xie, X. Double-image compression and encryption algorithm based on co-sparse representation and random pixel exchanging. *Opt. Lasers Eng.* **2018**, *110*, 72–79. [CrossRef]
17. Zhu, S.; Zhu, C. A new image compression-encryption scheme based on compressive sensing and cyclic shift. *Multimed. Tools Appl.* **2019**, *78*, 20855–20875. [CrossRef]
18. Tong, X.; Liu, Y.; Zhang, M.; Xu, H.; Wang, Z. An Image Encryption Scheme Based on Hyperchaotic Rabinovich and Exponential Chaos Maps. *Entropy* **2015**, *17*, 181–196. [CrossRef]
19. Wang, Z.; Min, F.; Wang, E. A new hyperchaotic circuit with two memristors and its application in image encryption. *AIP Adv.* **2016**, *6*, 095316. [CrossRef]
20. Zhang, J.; Hou, D.; Ren, H. Image Encryption Algorithm Based on Dynamic DNA Coding and Chen's Hyperchaotic System. *Math. Probl. Eng.* **2016**, *2016*, 6408741. [CrossRef]
21. Yu, S.; Zhou, N.; Gong, L.; Nie, Z. Optical image encryption algorithm based on phase-truncated short-time fractional Fourier transform and hyper-chaotic system. *Opt. Lasers Eng.* **2020**, *124*, 105816. [CrossRef]
22. Sun, S.; Guo, Y.; Wu, R. A Novel Image Encryption Scheme Based on 7D Hyperchaotic System and Row-column Simultaneous Swapping. *IEEE Access* **2019**, *7*, 28539–28547. [CrossRef]
23. Xu, L.; Gou, X.; Li, Z.; Li, J. A novel chaotic image encryption algorithm using block scrambling and dynamic index based diffusion. *Opt. Lasers Eng.* **2017**, *91*, 41–52. [CrossRef]
24. Gayathri, J.; Subashini, S. A spatiotemporal chaotic image encryption scheme based on self adaptive model and dynamic keystream fetching technique. *Multimed. Tools Appl.* **2018**, *77*, 24751–24787. [CrossRef]
25. Wu, X.; Wang, K.; Wang, X.; Kan, H.; Kurths, J. Color image DNA encryption using NCA map-based CML and one-time keys. *Signal Process.* **2018**, *148*, 272–287. [CrossRef]
26. Zhu, C.; Wang, G.; Sun, K. Cryptanalysis and Improvement on an Image Encryption Algorithm Design Using a Novel Chaos Based S-Box. *Symmetry* **2018**, *10*, 399. [CrossRef]
27. Zhu, S.; Wang, G.; Zhu, C. A Secure and Fast Image Encryption Scheme Based on Double Chaotic S-Boxes. *Entropy* **2019**, *21*, 790. [CrossRef]
28. Liu, H.; Zhao, B.; Huang, L. Quantum image encryption scheme using Arnold transform and S-box scrambling. *Entropy* **2019**, *21*, 343. [CrossRef]
29. Zhang, Q.; Guo, L.; Wei, X. A novel image fusion encryption algorithm based on DNA sequence operation and hyper-chaotic system. *Optik* **2013**, *124*, 3596–3600. [CrossRef]

30. Chai, X.; Chen, Y.; Broyde, L. A novel chaos-based image encryption algorithm using DNA sequence operations. *Opt. Lasers Eng.* **2017**, *88*, 197–213. [CrossRef]
31. Chai, X.; Fu, X.; Gan, Z.; Lu, Y.; Chen, Y. A color image cryptosystem based on dynamic DNA encryption and chaos. *Signal Process.* **2019**, *155*, 44–62. [CrossRef]
32. Khan, J.S.; Ahmad, J.; Abbasi, S.F.; Kayhan, S.K. DNA Sequence Based Medical Image Encryption Scheme. In Proceedings of the 10th Computer Science and Electronic Engineering (CEEC), Colchester, UK, 19–21 September 2018; pp. 24–29.
33. Zhan, K.; Wei, D.; Shi, J.; Yu, J. Cross-utilizing hyperchaotic and DNA sequences for image encryption. *J. Electron. Imaging* **2017**, *26*, 013021. [CrossRef]
34. Zhu, C.; Hu, Y.; Sun, K. New image encryption algorithm based on hyperchaotic system and ciphertext diffusion in crisscross pattern. *J. Electron. Inf. Tech.* **2012**, *34*, 1735–1743. [CrossRef]
35. Sun, S. A Novel Hyperchaotic Image Encryption Scheme Based on DNA Encoding, Pixel-Level Scrambling and Bit-Level Scrambling. *IEEE Photonics J.* **2018**, *10*, 1–14. [CrossRef]
36. Zhou, N.; Chen, W.; Yan, X.; Wang, Y. Bit-level quantum color image encryption scheme with quantum cross-exchange operation and hyper-chaotic system. *Quantum Inf. Process.* **2018**, *17*, 137. [CrossRef]
37. Ahmed F.; Siyal M.; Abbas, V. A perceptually scalable and jpeg compression tolerant image encryption scheme. In Proceedings of the 4th Pacific-RIM Symposium on Image and Video Technology (PSIVT), Singapore, 14–17 November 2010; pp. 232–238.
38. Ahmad, J.; Khan, M.A.; Ahmed, F.; Khan, J.S. A novel image encryption scheme based on orthogonal matrix, skew tent map, and XOR operation. *Neural Comput. Appl.* **2018**, *30*, 3847–3857. [CrossRef]
39. Hua, Z.; Zhou, Y. Design of image cipher using block-based scrambling and image filtering. *Inf. Sci.* **2017**, *396*, 97–113. [CrossRef]
40. Rossler, O.E. An equation for hyperchaos. *Phys. Lett. A.* **1979**, *71*, 155–157. [CrossRef]
41. Gu, Q.; Gao, T. Analysis of transition between chaos and hyper-chaos of an improved hyper-chaotic system. *Chin. Phys. B* **2009**, *18*, 84–90.
42. Chen, G.; Ueta, T. Yet another chaotic attractor. *Int. J. Bifurc. Chaos* **1999**, *9*, 1465–1466. [CrossRef]
43. Adleman, L.M. Molecular computation of solutions to combinatorial problems. *Nature* **1994**, *369*, 40. [CrossRef]
44. Li, T.; Hu, Z.; Jia, Y.; Wu, J.; Zhou, Y. Forecasting Crude Oil Prices Using Ensemble Empirical Mode Decomposition and Sparse Bayesian Learning. *Energies* **2018**, *11*, 1882. [CrossRef]
45. Zhao H.; Zheng J.; Xu, J.; Deng W. Fault diagnosis method based on principal component analysis and broad learning system. *IEEE ACCESS* **2019**, *7*, 99263–99272. [CrossRef]
46. Li, T.; Zhou, Y.; Li, X; Wu, J.; He, T. Forecasting Daily Crude Oil Prices Using Improved CEEMDAN and Ridge Regression-Based Predictors. *Energies* **2019**, *12*, 3063.
47. Alvarez, G.; Li, S. Some basic cryptographic requirements for chaos-based cryptosystems. *Int. J. Bifurc. Chaos* **2006**, *16*, 2129–2151. [CrossRef]
48. Yue, W.; Joseph, P.N.; Sos, A. NPCR and UACI Randomness Tests for Image Encryption. *J. Sel. Areas Telecommun.* **2011**, *1*, 31–38.
49. Chai, X.; Gan, Z.; Lu, Y.; Chen, Y.; Han, D. A novel image encryption algorithm based on the chaotic system and DNA computing. *Int. J. Mod. Phys. C* **2017**, *28*, 1750069. [CrossRef]

 © 2019 by the authors. Licensee MDPI, Basel, Switzerland. This article is an open access article distributed under the terms and conditions of the Creative Commons Attribution (CC BY) license (http://creativecommons.org/licenses/by/4.0/).

Article

On the Security of a Latin-Bit Cube-Based Image Chaotic Encryption Algorithm

Zeqing Zhang * and Simin Yu

School of Automation, Guangdong University of Technology, Guangzhou 510006, China; siminyu@163.com
* Correspondence: 2111704029@mail2.gdut.edu.cn

Received: 15 August 2019; Accepted: 9 September 2019; Published: 12 September 2019

Abstract: In this paper, the security analysis of an image chaotic encryption algorithm based on Latin cubes and bit cubes is given. The proposed algorithm adopts a first-scrambling-diffusion-second-scrambling three-stage encryption scheme. First, a finite field is constructed using chaotic sequences. Then, the Latin cubes are generated from finite field operation and used for image chaotic encryption. In addition, according to the statistical characteristics of the diffusion image in the diffusion stage, the algorithm also uses different Latin cube combinations to scramble the diffusion image for the second time. However, the generation of Latin cubes in this algorithm is independent of plain image, while, in the diffusion stage, when any one bit in the plain image changes, the corresponding number of bits in the cipher image follows the change with obvious regularity. Thus, the equivalent secret keys can be obtained by chosen plaintext attack. Theoretical analysis and experimental results indicate that only a maximum of $2.5 \times \sqrt[3]{w \times h} + 6$ plain images are needed to crack the cipher image with $w \times h$ resolution. The size of equivalent keys deciphered by the method proposed in this paper are much smaller than other general methods of cryptanalysis for similar encryption schemes.

Keywords: image chaotic encryption; cryptography; Latin cube; bit cube; chosen plaintext attack

1. Introduction

Image chaotic encryption algorithms have attracted some special attention in the field of information security [1–7]. In recent years, many image chaotic encryption schemes combined chaos theories with other technologies, such as one-time keys [8], bit-level permutation [9], DNA operations [10–13], parallel computing system [14], matrix semi-tensor product theory [15], cellular automata [16,17], neural network [18,19], Latin square or Latin cube [20–22], and so on, have been proposed. However, the security issues of image chaotic encryption algorithms have also attracted much attention. As a basic requirement of security, the ciphertext image of the image chaotic encryption algorithm must have good uniformity. In addition, the algorithm must have a large enough key space to resist brute force attacks. For instance, in order to show the security of the image chaotic encryption algorithm in the statistical sense, the key space analysis, statistical analysis, and differential analysis of the chaos encryption algorithm proposed in [23] and its corresponding extended algorithm are given in Sections 4 and 5 of [23], respectively. However, the high uniformity of ciphertext does not mean that the encryption algorithm has high security performance. For example, in [24], the security analysis of an image chaotic encryption algorithm proposed in [16] is given, and it is found that the generation of key stream is related to the sum of pixel values of plain images. Under the premise of satisfying the sum of pixel values of a plain image unchanged, only two pixel values of cipher image are changed corresponding to the variation of two pixel values of a plain image, which is vulnerable to differential attack. Therefore, the equivalent secret keys can be obtained by selecting 512 plain images. In [25], the cryptanalysis of a DNA encoding-based image scrambling and diffusion encryption algorithm proposed in [10] is reported to find that the scrambling algorithm is also independent of plain image, so that it can be deciphered by chosen plaintext attack. In addition, by choosing some specific plain images, the original image chaotic encryption algorithm can be simplified into scrambling-only encryption algorithm, which has been proven to be insecure [26,27]. In [28], the security analysis of an image encryption algorithm based on a compound chaotic system proposed in [29] is given, and it is pointed out that there are a large number of equivalent secret keys in the image chaotic encryption algorithm. In [30], an 8D self-synchronous and feedback-based chaotic stream cipher using the lower 8 bits of one state variable for encryption is proposed. However, in [31], most of the secret keys are successfully acquired by means of a divide and conquer attack, known plaintext attack, and a chosen ciphertext attack, respectively. In [32], the security analysis of a Latin square based image chaotic encryption algorithm proposed in [22] is given to find the security vulnerabilities both in the diffusion stage and in the scrambling stage through chosen text attack. In [33], the chosen plaintext attack is adopted for the safety performance assessment of a 1D combinatorial chaotic encryption algorithm proposed in [34]. In addition, in [35], the chosen plaintext attack is also utilized for analyzing the security of a bit cube-based image chaotic encryption algorithm proposed in [36]. In addition, some chaotic cipher designers have also discovered the importance of cryptanalysis. For example, in Section 3 of [37], the resistance to the four classic attack methods is analyzed in detail. The analysis shows that the proposed encryption algorithm has resistance to the chosen plaintext attack because it is sensitive to the initial parameters.

In 2019, an image chaotic encryption algorithm based on orthogonal Latin cubes and bit cubes is given in [20]. First, a chaotic sequence is generated by logistic mapping, and it is further arranged in ascending order to obtain its corresponding chaotic index sequence. Next, a finite field is constructed by the chaotic index sequence, and three orthogonal Latin cubes are also generated. Then, the generated three orthogonal Latin cubes are used for the first-scrambling-diffusion- second-scrambling three-stage encryption. Although the designer claims that the algorithm has passed various statistical tests, the analysis results in this paper demonstrate that the algorithm has at least two security vulnerabilities as follows:

(1) The generation of Latin cubes in this algorithm is independent of plain image.
(2) When any one bit in the plain image changes, the corresponding number of bits in the cipher image follows the change with obvious regularity.

Based on the above-mentioned security vulnerabilities, this paper adopts both chosen plaintext attack and differential attack for analyzing the safety performance for the image chaotic encryption algorithm proposed in [20]. First, a full zero plain image and multiple non-full zero plain images are selected, and the differential operation is performed between the cipher image corresponding to this full zero plain image and the cipher image corresponding to those non-full zero plain images. On the premise that the sum of bit 1 in each differential operation is even, the chaotic index sequence lx can be deciphered. Next, based on the obtained lx, and on the condition that there exists an intersection in the solutions of unary quadratic equation on finite field $GF(q)$, the secret keys α, β, γ can be further deciphered.

The rest of the paper is organized as follows: Section 2 briefly introduces the image chaotic encryption algorithm. Section 3 presents the security analysis. Section 4 gives the steps for deciphering image chaotic encryption algorithm. Section 5 demonstrates the numerical simulation experiments. Section 6 gives some improvement suggestions for the image chaotic encryption algorithm. Finally, Section 7 concludes the paper.

2. Description of an Image Chaotic Encryption Algorithm

2.1. A Brief View of an Image Chaotic Encryption Algorithm

In [20], the image chaotic encryption algorithm consists of secret keys selection, Latin cube generation, scrambling encryption, and diffusion encryption, as shown in Figure 1, where key_0, μ_0,

α, β, γ are the secret keys, x_n ($n = 0, 1, 2, \cdots$) is a chaotic sequence generated by Logistic mapping, lx is a chaotic index sequence, L_1, L_2, L_3 are three Latin cubes, P is a 2D plain gray image, M is a bit cube representation of P, S_1 is a first-scrambling image of M, D is a diffusion image of S_1, S_2 is a second-scrambling image of D, E is a 2D cipher gray image of S_2, and B is generated by L_1. When the size of the image is $w \times h$, the length of x_n and lx is $q = \sqrt[3]{8 \times w \times h}$, the side length of Latin cubes and bit cubes is $q = \sqrt[3]{8 \times w \times h}$, and the secret keys α, β, $\gamma \in \{0, 1, 2, \cdots, q-1\}$. Note that an appropriate image size $w \times h$ should be selected to ensure that $q = \sqrt[3]{8 \times w \times h} = 2 \times \sqrt[3]{w \times h}$ is an even number. In Figure 1, L_1, L_2, $L_3 \in \{0, 1, 2, \cdots, q-1\}$ are Latin cubes, M, S_1, D, S_2, $B \in \{0, 1\}$ are bit cubes, P is a 2D plain gray image, E is a 2D cipher gray image, p_k, p_t, s_1, b, d, $e_k \in \{0, 1\}$ are 1D bit sequences corresponding to P, S_1, B, D, E, and $t = T(k)$ is a position scrambling rule corresponding to the first-scrambling stage.

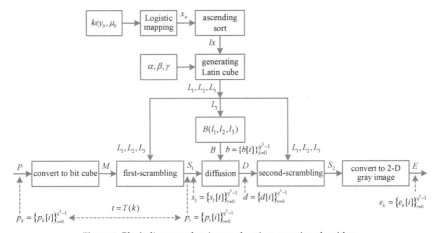

Figure 1. Block diagram of an image chaotic encryption algorithm.

2.2. Logistic Map

According to Figure 1, the chaotic sequence is generated through logistic mapping, given by

$$x_{n+1} = \mu x_n (1 - x_n),$$

(1)

where $n = 0, 1, 2, \cdots$, $x_n \in (0, 1)$, $0 \le \mu \le 4$. When $\mu > 3.573815$, Equation (1) is chaotic.

2.3. Generation of Latin Cubes

Let the side length of L_1, L_2, L_3 be $q = \sqrt[3]{8 \times w \times h}$, where q is an even number. For a given (l_1, l_2, l_3), one gets $L_1(l_1, l_2, l_3) = \psi_1$, $L_2(l_1, l_2, l_3) = \psi_2$, $L_3(l_1, l_2, l_3) = \psi_3$, $0 \le \psi_1, \psi_2, \psi_3 \le q-1$. If $(l_1, l_2, l_3) \neq (l_1', l_2', l_3')$, $(\psi_1, \psi_2, \psi_3) \neq (\psi_1', \psi_2', \psi_3')$, then L_1, L_2, L_3 are orthogonal to each other [38]. When $q = 3$, one gets three orthogonal Latin cubes, as shown in Figure 2a, and the corresponding triple tuple is shown in Figure 2b, respectively.

	$l_1 = 0$ $0 \le l_2, l_3 \le 2$			$l_1 = 1$ $0 \le l_2, l_3 \le 2$			$l_1 = 2$ $0 \le l_2, l_3 \le 2$		
L_1	0	1	2	1	2	0	2	0	1
	1	2	0	2	0	1	0	1	2
	2	0	1	0	1	2	1	2	0
L_2	0	1	2	1	2	0	2	0	1
	2	0	1	0	1	2	1	2	0
	1	2	0	2	0	1	0	1	2
L_3	0	2	1	1	0	2	2	1	0
	2	1	0	0	2	1	1	0	2
	1	0	2	2	1	0	0	2	1

(a)

	$l_1 = 0$, $0 \le l_2, l_3 \le 2$			$l_1 = 1$, $0 \le l_2, l_3 \le 2$			$l_1 = 2$, $0 \le l_2, l_3 \le 2$		
	0,0,0	1,1,2	2,2,1	1,1,1	2,2,0	0,0,2	2,2,2	0,0,1	1,1,0
(L_1, L_2, L_3)	1,2,2	2,0,1	0,1,0	2,0,0	0,1,2	1,2,1	0,1,1	1,2,0	2,0,2
	2,1,1	0,2,0	1,0,2	0,2,2	1,0,1	2,1,0	1,0,0	2,1,2	0,2,1

(b)

Figure 2. Three orthogonal Latin cubes and the corresponding triple tuple when $q = 3$. (a) three orthogonal Latin cubes; (b) the corresponding triple tuple.

The algorithm for generating Latin cubes proposed in [20] is implemented by replacing the ordered set $\{0, 1, 2, ..., q\}$ in the generation method proposed in [38] with the chaotic index sequence lx. The detailed steps for generating three orthogonal Latin cubes by means of a finite field are in Algorithm 1.

Algorithm 1 Steps for Generation of Latin Cubes.

Input: Secret keys key_0, μ_0, α, β, γ; Side length $q = \sqrt[3]{8 \times w \times h}$;
Output: Three orthogonal Latin cubes L_1, L_2 and L_3;
1: Generate the chaotic sequence $x = \{x_0, x_1, ..., x_{q-1}\}$ by using Logistic mapping.
2: Obtain the corresponding chaotic index sequence $lx = \{c_0, c_1, \cdots, c_i, \cdots, c_{q-1}\}$ by arranging $x = \{x_0, x_1, ..., x_{q-1}\}$ in ascending order, where $0 \le c_i$, $i \le q - 1$, satisfying $lx[i] = c_i$. Note that the chaotic index sequence lx can only be determined after the sequence value c_i and the sequence number i are simultaneously obtained. When the sequence value c_i is obtained, but the sequence number i is uncertain, the general form of the chaotic index sequence lx is in the form of

$$lx = \{c_{i_0}, c_{i_1}, \cdots, c_{i_k}, \cdots, c_{i_{q-1}}\}, \tag{2}$$

where $0 \le c_{i_k}$ $i_k \le q - 1$, $i_0 \ne i_1 \ne \cdots \ne i_k \ne \cdots \ne i_{q-1}$, $lx[i_k] = c_{i_k}$. In the following, ξ or ξ' denotes the sequence value and i_ξ or i'_ξ denotes the sequence number in Equation (2), respectively.
3: Construct a finite field by using chaotic index sequence lx, and then one gets the orthogonal Latin cubes on the finite field, given by

$$\begin{cases} L_1 (l_1, l_2, l_3) = \alpha^2 \times c_{l_1} + \alpha \times c_{l_2} + c_{l_3}, \\ L_2 (l_1, l_2, l_3) = \beta^2 \times c_{l_1} + \beta \times c_{l_2} + c_{l_3}, \\ L_3 (l_1, l_2, l_3) = \gamma^2 \times c_{l_1} + \gamma \times c_{l_2} + c_{l_3}, \end{cases} \tag{3}$$

where "+" denotes addition operation on the finite field, "×" denotes multiplication operation on the finite field, α, β, $\gamma \in lx$, c_{l_1}, c_{l_2}, c_{l_3} are sequence values of lx.
4: **return** L_1, L_2, L_3.

2.4. Steps for Image Chaotic Encryption

According to Figure 1, and taking a plain gray image with 512×512 resolution as an example, one has $q = \sqrt[3]{512 \times 512 \times 8} = 128$. The steps for image chaotic encryption are in Algorithm 2.

Algorithm 2 Steps for Image Chaotic Encryption.

Input: Secret keys key_0, μ_0, α, β, γ; Plaintext image P;
Output: Ciphertxet image E;
1: Convert the 2D plain gray image P into the bit cube M;
2: Obtain three orthogonal Latin cubes L_1, L_2, L_3 by Algorithm 1;
3: Scramble bit cube M by using three orthogonal Latin cubes L_1, L_2, L_3, and get the corresponding first-scrambling image S_1 in the form of bit cube, such that

$$S_1 (l_1, l_2, l_3) = M (L_1 (l_1, l_2, l_3), L_2 (l_1, l_2, l_3), L_3 (l_1, l_2, l_3)). \tag{4}$$

4: Obtain the diffusion bit cube $B (l_1, l_2, l_3)$ by using Latin cube L_1, given by

$$B (l_1, l_2, l_3) = \begin{cases} 0, & \text{if } L_1 (l_1, l_2, l_3) \geq 64, \\ 1, & \text{if } L_1 (l_1, l_2, l_3) < 64. \end{cases} \tag{5}$$

Then, get the diffusion 1D bit sequence $b[t]$ corresponding to diffusion bit cube $B(l_1, l_2, l_3)$ as

$$b[t] = B \left(\left\lfloor t/128^2 \right\rfloor, \ \lfloor t/128 \rfloor \%128, \ t\%128 \right), \tag{6}$$

where $t \in \{0, 1, 2, \cdots, q^3 - 1\}$, $\lfloor \cdot \rfloor$ is a round down operation, and "%" is a modulo operation.
5: Convert $S_1(l_1, l_2, l_3)$ into the 1D bit sequence $s_1[t]$ as

$$s_1[t] = S_1 \left(\left\lfloor t/128^2 \right\rfloor, \ \lfloor t/128 \rfloor \%128, \ t\%128 \right). \tag{7}$$

Then, get the 1D bit sequence $d[t]$ by using $s_1[t]$ and $b[t]$ as

$$d[t] = s_1[t] \oplus d[t-1] \oplus b[t], \tag{8}$$

where $0 \leq t \leq 128^3 - 1$, $d[-1] = 0$, "$\oplus$" denotes bitwise exclusive or operation.
6: Calculate $G(d) = \left(\sum_{i=0}^{q^3-1} d[i] \right)$, and convert the 1D bit sequence $d[t]$ into the bit cube $D(l_1, l_2, l_3)$.
Then, get the bit cube $S_2 (l_1, l_2, l_3)$ by utilizing $D(l_1, l_2, l_3)$, such that

$$S_2 (l_1, l_2, l_3) = \begin{cases} D (L_2 (l_1, l_2, l_3), L_3 (l_1, l_2, l_3), L_1 (l_1, l_2, l_3)), & (G(d)\%2 = 0), \\ D (L_3 (l_1, l_2, l_3), L_1 (l_1, l_2, l_3), L_2 (l_1, l_2, l_3)), & (G(d)\%2 = 1), \end{cases} \tag{9}$$

where $G(d)\%2 \in \{0, 1\}$ denotes the modular 2 operation on $G(d)$.
7: Convert the bit cube $S_2(l_1, l_2, l_3)$ into the 2D cipher gray image E with 512×512 resolution.
8: **return** E.

An example of encrypting a gray image with 2×4 resolution using the original encryption algorithm is shown in Figure 3. Figure 3a shows the three Latin cubes and the corresponding bit cubes L used for encryption. Figure 3b shows the encryption process. The numbers in the cells of P and E represent pixel values, and the bit values are represented in the cells of S_1, B, and S_2. The red cells in M indicate that they are bit representations of the red cell corresponding to P, i.e., the binary representation of 166 is $(10100110)_2$.

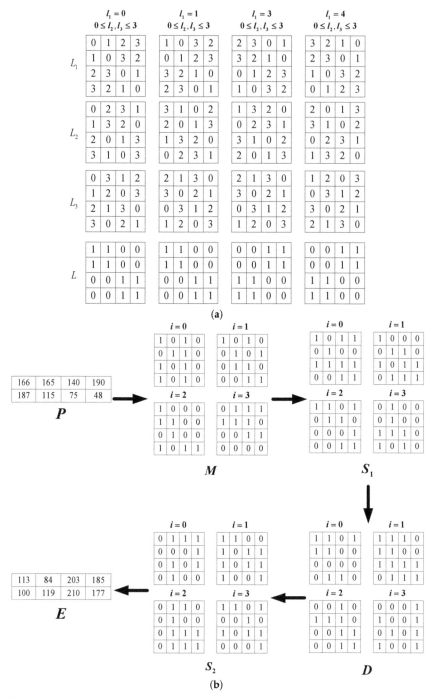

Figure 3. An example of encrypting a gray image with 2 × 4 resolution. (**a**) three orthogonal Latin cubes and the corresponding bit cubes *L* used for encryption; (**b**) the encryption process.

3. Security Analysis

According to Figure 1, it is found that the generation of three orthogonal Latin cubes L_1, L_2, L_3 is not related to the plain image. When the secret keys are given, the three orthogonal Latin cubes L_1, L_2, L_3 remain unchanged for different input plain images, which are provided a prerequisite for chosen plaintext attack. Therefore, one can decipher the equivalent secret keys lx, α, β, γ corresponding to the original secret keys key_0, μ_0, α, β, γ.

3.1. Analysis of Chaotic Index Sequence lx

3.1.1. Relation between the First-Scrambling Image S_1 and the Plain Image M

Proposition 1. *Suppose that M is the bit cube representation of P; S_1 is the first-scrambling image of M. The relationship between M and S_1 satisfies $S_1(i_0, i_0, i_\xi) = M(\xi, \xi, \xi)$, where $lx[i_0] = 0$, $lx[i_\xi] = \xi$, i_0, $\xi \in \{0, 1, 2, \cdots, q-1\}$, i_0 denotes the sequence number corresponding to the sequence value 0, and i_ξ denotes the sequence number corresponding to the sequence value ξ.*

Proof. Let $l_1 = l_2 = i_0$, $l_3 = i_\xi$, and substitute them into Equation (4), then, one gets

$$S_1(i_0, i_0, i_\xi) = M(L_1(i_0, i_0, i_\xi), L_2(i_0, i_0, i_\xi), L_3(i_0, i_0, i_\xi)). \tag{10}$$

In addition, let $l_1 = l_2 = i_0$, $l_3 = i_\xi$, and substitute them into Equation (3), then, one gets

$$\begin{cases} L_1(i_0, i_0, i_\xi) = \alpha^2 \times c_{i_0} + \alpha \times c_{i_0} + c_{i_\xi}, \\ L_2(i_0, i_0, i_\xi) = \beta^2 \times c_{i_0} + \beta \times c_{i_0} + c_{i_\xi}, \\ L_3(i_0, i_0, i_\xi) = \gamma^2 \times c_{i_0} + \gamma \times c_{i_0} + c_{i_\xi}. \end{cases} \tag{11}$$

Since $lx[i_0] = 0$, $lx[i_\xi] = \xi$, one has $lx[i_0] = c_{i_0} = 0$, $lx[i_\xi] = c_{i_\xi} = \xi$. In addition, substituting $c_{i_0} = 0$ and $c_{i_\xi} = \xi$ into Equation (11), one gets

$$L_1(i_0, i_0, i_\xi) = L_2(i_0, i_0, i_\xi) = L_3(i_0, i_0, i_\xi) = \xi. \tag{12}$$

In addition, substituting Equation (12) into Equation (10), it follows that $S_1(i_0, i_0, i_\xi) = M(\xi, \xi, \xi)$ holds. The proof is finished. □

3.1.2. The First Case for Analysis of Chaotic Index Sequence lx

Suppose that the 1D bit sequence corresponding to plain image P_0 is $\{p_0[i]\}_{i=0}^{q^3-1} = \{0\}_{i=0}^{q^3-1}$, the cipher image corresponding to plain image P_0 is E_0, the 1D bit sequence corresponding to cipher image E_0 is $\{e_0[i]\}_{i=0}^{q^3-1}$, and the 1D bit sequence corresponding to plain image P_k is $\{p_k[i]\}_{i=0}^{q^3-1}$, where $p_k[i]$ is given by

$$p_k[i] = \begin{cases} 1, & \text{if } i = k, \\ 0, & \text{if } i \neq k. \end{cases} \tag{13}$$

In addition, suppose that the cipher image corresponding to plain image P_k is E_k, the 1D bit sequence corresponding to cipher image E_k is $\{e_k[i]\}_{i=0}^{q^3-1}$, the 1D bit sequence corresponding to plain image $P_{k_1 k_2} = P_{k_1} \oplus P_{k_2}$ is $\{p_{k_1 k_2}[i]\}_{i=0}^{q^3-1} = \{p_{k_1}[i] \oplus p_{k_2}[i]\}_{i=0}^{q^3-1}$, the cipher image corresponding to plain image $P_{k_1 k_2}$ is $E_{k_1 k_2}$, the 1D bit sequence corresponding to cipher image $E_{k_1 k_2}$ is $\{e_{k_1 k_2}[i]\}_{i=0}^{q^3-1}$, the 1D bit sequence corresponding to plain image $P_{k_1 k_2 k_3} = P_{k_1} \oplus P_{k_2} \oplus P_{k_3}$ is $\{p_{k_1 k_2 k_3}[i]\}_{i=0}^{q^3-1} = \{p_{k_1}[i] \oplus p_{k_2}[i] \oplus p_{k_3}[i]\}_{i=0}^{q^3-1}$, the cipher image corresponding to $P_{k_1 k_2 k_3}$ is $E_{k_1 k_2 k_3}$, and the 1D bit sequence corresponding to $E_{k_1 k_2 k_3}$ is $\{e_{k_1 k_2 k_3}[i]\}_{i=0}^{q^3-1}$.

Proposition 2. *Suppose that the cipher image corresponding to plain image P_k is E_k, the 1D bit sequence corresponding to cipher image E_k is $\{e_k[i]\}_{i=0}^{q^3-1}$, the cipher image corresponding to plain image P_0 is E_0, and the 1D bit sequence corresponding to cipher image E_0 is $\{e_0[i]\}_{i=0}^{q^3-1}$. A differential operation is performed in the form of $\sum_{i=0}^{q^3-1}(e_0[i] \oplus e_k[i]) = q^3 - m_{k,0}$, in which $e_k[i_l] = e_0[i_l]$ ($l = 1, 2, \cdots, m_{k,0}$; $i_l \in \{0, 1, 2, \cdots, q^3 - 1\}$), $q^3$is an even number. If $(q^3 - m_{k,0})\%2 = q^3\%2 - m_{k,0}\%2 = m_{k,0}\%2 = 0$, then $T(k) = m_{k,0}$ holds, where $T(k)$ denotes the position scrambling rule in the first-scrambling stage, k denotes the position of the k-th bit before the first-scrambling of plain image, and $T(k)$ denotes the position of k-th bit after the first-scrambling of plain image.*

Proof. According to Equation (6), the relationship between the coordinates (l_1, l_2, l_3) of bit cube $B(l_1, l_2, l_3)$ and the position t of 1D bit sequence $b[t]$ corresponding to $B(l_1, l_2, l_3)$ is given by

$$\begin{cases} l_1 = \lfloor t/q^2 \rfloor = \lfloor t/128^2 \rfloor, \\ l_2 = \lfloor t/q \rfloor \%q = \lfloor t/128 \rfloor \%128, \\ l_3 = t\%q = t\%128. \end{cases} \tag{14}$$

On the other hand, the relationship between the coordinates (ξ, ξ, ξ) of bit cube $M(\xi, \xi, \xi)$ and the position k of 1D bit sequence $p_k[i]$ in Equation (13) is given by

$$k = \xi(q^2 + q + 1). \tag{15}$$

Thus, the relationship between the position of t-th bit after the first-scrambling of plain image and the position of k-th bit before the first-scrambling of plain image is given by

$$t = T(k) = T(\xi(q^2 + q + 1)). \tag{16}$$

(1) Consider the first-scrambling stage. In the first-scrambling stage, only change the bit position, but the bit value should remain unchanged. Suppose that the input 1D bit sequence corresponding to plain image P_k is p_k, after the first-scrambling of plain image, the corresponding output 1D bit sequence is p_t. According to Equation (16), the relationship between position t and k satisfies $t = T(k)$. In particular, if the input 1D bit sequence corresponding to plain image P_0 is $p_0 = \{p_0[i]\}_{i=0}^{q^3-1} = \{0\}_{i=0}^{q^3-1}$, after the first-scrambling of plain image, the corresponding output 1D bit sequence is $p_t = \{p_t[i]\}_{i=0}^{q^3-1}$, then one has $p_t = p_0 = \{0\}_{i=0}^{q^3-1}$. (2) Consider the diffusion stage. Take the output 1D bit encryption sequence $\{p_0[i]\}_{i=0}^{q^3-1}$ in the first-scrambling stage as the input 1D bit sequence in the diffusion stage. According to Equation (8), diffuse $\{p_0[i]\}_{i=0}^{q^3-1}$ by using the diffusion 1D bit sequence $\{b[i]\}_{i=0}^{q^3-1}$, obtain the corresponding output $\{d_0[i]\}_{i=0}^{128^3-1}$ in the diffusion stage. By substituting $s_1[i] = p_0[i] = 0$ into Equation (8), one has

$$\begin{cases} d_0[0] = p_0[0] \oplus d_0[-1] \oplus b[0] = 0 \oplus 0 \oplus b[0] = b[0], \\ d_0[1] = p_0[1] \oplus d_0[0] \oplus b[1] = 0 \oplus d_0[0] \oplus b[1] = b[0] \oplus b[1], \\ d_0[2] = p_0[2] \oplus d_0[1] \oplus b[2] = 0 \oplus d_0[1] \oplus b[2] = b[0] \oplus b[1] \oplus b[2], \\ \cdots \\ d_0[i] = b[0] \oplus b[1] \oplus b[2] \cdots \oplus b[i], \end{cases} \tag{17}$$

where $i = 0, 1, 2, \cdots, q^3 - 1$, $d_0[-1] = 0$. Similarly, take the output 1D bit encryption sequence $\{p_t[i]\}_{i=0}^{q^3-1}$ in the first-scrambling stage as the input 1D bit sequence in the diffusion stage. According to Equation (8), diffuse $\{p_t[i]\}_{i=0}^{q^3-1}$ by using the diffusion 1D bit sequence $\{b[i]\}_{i=0}^{q^3-1}$ and obtain the corresponding output $\{d_t[i]\}_{i=0}^{128^3-1}$ in the diffusion stage. By substituting $s_1[i] = p_t[i]$ into Equation (8), and also by utilizing Equation (17), one has

$$
\begin{cases}
d_t[0] = p_t[0] \oplus d_t[-1] \oplus b[0] = 0 \oplus 0 \oplus b[0] = d_o[0], \\
d_t[1] = p_t[1] \oplus d_t[0] \oplus b[1] = 0 \oplus d_o[0] \oplus b[1] = 0 \oplus b[0] \oplus b[1] = d_o[1], \\
\cdots \\
d_t[t] = p_t[t] \oplus d_t[t-1] \oplus b[t] = 1 \oplus d_o[t-1] \oplus b[t] = 1 \oplus d_o[t] = \overline{d_o[t]}, \\
d_t[t+1] = p_t[t+1] \oplus d_t[t] \oplus b[t+1] = 0 \oplus d_t[t] \oplus b[t+1] = \overline{d_o[t]} \oplus b[t+1] = \overline{d_o[t+1]}, \\
\cdots \\
d_t[i] = p_t[i] \oplus d_t[i-1] \oplus b[i] = 1 \oplus d_o[i-1] \oplus b[i] = \overline{d_o[i]},
\end{cases}
\tag{18}
$$

where $d_t[-1] = 0$. According to Equation (18), one has

$$
\begin{cases}
d_t[i] = d_o[i] & (0 \le i < t), \\
d_t[i] = \overline{d_o[i]} & (t \le i \le (q^3 - 1)),
\end{cases}
\tag{19}
$$

where $\overline{d_o[i]}$ denotes the bitwise NOT of $d_o[i]$. (3) Consider the second-scrambling stage. Take the output 1D bit encryption sequences $\{d_o[i]\}_{i=0}^{q^3-1}$ and $\{d_t[i]\}_{i=0}^{q^3-1}$ in the diffusion stage as the input 1D bit sequences in the second-scrambling stage, calculate $G(d_0) = \left(\sum_{i=0}^{q^3-1} d_0[i]\right)$, $G(d_t) = \left(\sum_{i=0}^{q^3-1} d_t[i]\right)$, respectively. If $t\%2 = 0$ in Equation (19) holds, then it follows that

$$
G(d_t)\%2 = G(d_0)\%2.
\tag{20}
$$

According to Equation (9) with Equation (20), it is noted that the same scrambling rule for $\{d_o[i]\}_{i=0}^{q^3-1}$ and $\{d_t[i]\}_{i=0}^{q^3-1}$ is used in the second-scrambling stage. By comparing the first equation $d_t[i] = d_o[i]$ $(0 \le i < t)$ of Equation (19) with $e_k[i_l] = e_0[i_l]$ $(l = 1, 2, \cdots, m_{k,0}; i_l \in \{0, 1, 2, \cdots, q^3 - 1\})$, it follows that $t = m_{k,0}$. Then, according to Equation (16), $T(k) = m_{k,0}$ holds. The proof is finished. $\qquad\square$

Based on Proposition 1, one has $S_1(i_0, i_0, i_\xi) = M(\xi, \xi, \xi)$, where $\xi \in \{0, 1, 2, \cdots, q - 1\}$ is the sequence value of chaotic index sequence lx, i_ξ is the sequence number of lx. However, even though ξ is given, since $S_1(i_0, i_0, i_\xi)$ is the first-scrambling result of bit cube $M(\xi, \xi, \xi)$, but the scrambling rule $T(\cdot)$ is unknown beforehand, the sequence numbers i_0 and i_ξ cannot be directly available. Thus, Proposition 2 is needed to obtain the specific numbers i_0 and i_ξ.

Based on Proposition 2, suppose that the input plain image $M(l_1, l_2, l_3)$ is given by

$$
M(l_1, l_2, l_3) =
\begin{cases}
1, & \text{if } l_1 = l_2 = l_3 = \xi, \\
0, & \text{otherwise},
\end{cases}
\tag{21}
$$

where $\xi \in \{0, 1, \cdots, q - 1\}$. Based on Equation (15) with Equation (21), one has $k = \xi \cdot (q^2 + q + 1)$. Next, one obtains $m_{k,0}$ by a chosen plaintext attack. If $m_{k,0}\%2 = 0$ holds, then the same scrambling rule is used for d_0 and d_t in the second-scrambling stage, such that $T(k) = m_{k,0} = t$. Finally, according to Equation (14), it follows that

$$
\begin{cases}
i_0 = \lfloor t/q^2 \rfloor = \lfloor T(\xi \cdot (q^2 + q + 1))/q^2 \rfloor = \lfloor T(k)/q^2 \rfloor = \lfloor m_{k,0}/q^2 \rfloor, \\
i_\xi = t\%q = T(\xi \cdot (q^2 + q + 1))\%q = T(k)\%q = m_{k,0}\%q.
\end{cases}
\tag{22}
$$

An example of Proposition 2 is as in Figure 4. Figure 4a shows the ciphertext corresponding to the grayscale image lena. Figure 4b shows the corresponding ciphertext image after changing the bit at the bit-cube coordinates (6, 6, 6) of lena. Figure 4c is a bitwise exclusive or result between Figure 4a,b. Figure 4d is a bit statistical histogram of Figure 4c.

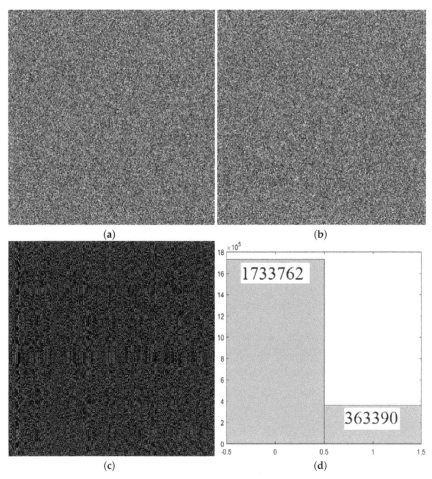

(a)

(b)

(c)

(d)

Figure 4. An example of Proposition 2. (**a**) the ciphertext corresponding to the grayscale image lena; (**b**) the corresponding ciphertext image after changing the bit at the bit-cube coordinates (6, 6, 6) of lena; (**c**) the bitwise exclusive or result between Figure 4a,b; (**d**) the bit statistical histogram of Figure 4c.

The difference between the two plaintexts is only 1 bit. It can be found from Figure 4d that the number of identical bits between their corresponding ciphertexts is 1,733,762, which is an even number. Substituting $m_{k,0} = 1,733,762$, $\zeta = 6$, and $q = 128$ into Equation (22) yields $i_0 = 105$ and $i_6 = 2$.

3.1.3. The Second Case for Analysis of Chaotic Index Sequence Ix

If $m_{k,0}\%2 \neq 0$, the above-mentioned method is no longer available, which needs to be further consideration.

Corollary 1. *Supposing that the cipher image corresponding to plain image* $P_{k_1 k_2} = P_{k_1} \oplus P_{k_2}$ $(k_1 \neq k_2)$ *is* $E_{k_1 k_2}$, *the 1D bit sequence corresponding to* $E_{k_1 k_2}$ *is* $\{e_{k_1 k_2}[i]\}_{i=0}^{q^3-1}$, *the cipher image corresponding to plain image* P_0 *is* E_0, *the 1D bit sequence corresponding to* E_0 *is* $\{e_0[i]\}_{i=0}^{q^3-1}$. *A differential operation is performed in the form* of $\sum_{i=0}^{q^3-1}\left(e_0[i]\oplus e_{k_1k_2}[i]\right)=m_{k_1k_2,0}$, in which $e_k[i_l]\neq e_0[i_l]$ $(l=1,2,\cdots,m_{k_1k_2,0};\ i_l\in\{0,1,2,\cdots,q^3-1\})$. If $m_{k_1k_2,0}\%2=0$, then $|T(k_1)-T(k_2)|=m_{k_1k_2,0}$ holds. In addition, if $|T(k_1)-T(k_2)|\%2=0$, then $m_{k_1k_2,0}=|T(k_1)-T(k_2)|$ also holds.

Corollary 2. *Suppose that the cipher image corresponding to plain image $P_{k_1k_2k_3}=P_{k_1}\oplus P_{k_2}\oplus P_{k_3}$ $(k_1\neq k_2\neq k_3)$ is $E_{k_1k_2k_3}$, the 1D bit sequence corresponding to $E_{k_1k_2k_3}$ is $\{e_{k_1k_2k_3}[i]\}_{i=0}^{q^3-1}$, the cipher image corresponding to plain image P_0 is E_0, the 1D bit sequence corresponding to E_0 is $\{e_0[i]\}_{i=0}^{q^3-1}$. A differential operation is performed in the form of $\sum_{i=0}^{q^3-1}\left(e_0[i]\oplus e_{k_1k_2k_3}[i]\right)=q^3-m_{k_1k_2k_3,0}$, in which $e_{k_1k_2k_3}[i_l]=e_0[i_l]$ $(l=1,2,\cdots,m_{k_1k_2k_3,0};\ i_l\in\{0,1,2,\cdots,q^3-1\})$, q^3 is an even number. If $(q^3-m_{k_1k_2k_3,0})\%2=q^3\%2-m_{k_1k_2k_3,0}\%2=m_{k_1k_2k_3,0}\%2=0$, then $T(k_1)+T(k_2)-T(k_3)=m_{k_1k_2k_3,0}$ holds, where $T(k_1)<T(k_3)<T(k_2)$ or $T(k_1)>T(k_3)>T(k_2)$. In addition, if $[T(k_1)+T(k_2)-T(k_3)]\%2=0$, then $m_{k_1k_2k_3,0}=T(k_1)+T(k_2)-T(k_3)$ also holds.*

Suppose that the set of all sequence values corresponding to the chaotic index sequence lx is $\Omega=\{\xi_{i_1},\xi_{i_2},\cdots,\xi_{i_{q/2}},\xi'_{i'_1},\xi'_{i'_2},\cdots,\xi'_{i'_{q/2}}\}$. Let $\Psi=\{\xi_{i_1},\xi_{i_2},\cdots,\xi_{i_{q/2}}\}$ be the set of sequence value ξ corresponding to sequence number i_ξ, where i_ξ is obtained by using Equation (22). The relationship among ξ, k, t is $k=\xi(q^2+q+1)$ and $t=T(k)=T(\xi\cdot(q^2+q+1))$. For $\forall\xi\in\Psi$, $m_{k,0}\%2=0$ and $t=m_{k,0}$ hold. Similarly, let $\Psi'=\{\xi'_{i'_1},\xi'_{i'_2},\cdots,\xi'_{i'_{q/2}}\}$ be the set of sequence value ξ' corresponding to sequence number i'_ξ. The relationship among ξ', k', t' is $k'=\xi'\cdot(q^2+q+1)$ and $t'=T(k')=T(\xi'\cdot(q^2+q+1))$. For $\forall\xi'\in\Psi'$, $m_{k',0}\%2=0$ and $t'=m_{k',0}$ do not hold.

When $\xi\in\Psi$, one has $k=\xi(q^2+q+1)$ and $m_{k,0}\%2=0$, based on the Proposition 2, $t=m_{k,0}$ holds. According to Equation (22), the sequence number i_ξ corresponding to sequence value ξ is given by $i_\xi=t\%q$. However, when $\xi'\in\Psi'$, one has $k'=\xi'(q^2+q+1)$ and $m_{k',0}\%2\neq0$, the Proposition 2 is not available, $t'=m_{k',0}$ does not hold. Therefore, the sequence number $i'_{\xi'}$ corresponding to sequence value $\xi'\in\Psi'$ cannot be determined by using Equation (22).

To further solve the above-mentioned problem, by selecting k'_1, k'_2 $(k'_1\neq k'_2)$, one can obtain $m_{k'_1,0}$ corresponding to k'_1, and $m_{k'_2,0}$ corresponding to k'_2 by using chosen plaintext attack, which satisfies $m_{k'_1,0}\%2=1$ and $m_{k'_2,0}\%2=1$. Under this circumstance, although $T(k'_1)$ and $T(k'_2)$ are unknown, but according to the Proposition 2, $\forall k$ corresponding to $T(k)\%2=0$ can be found, so that the remained $\forall k'$ satisfies $T(k'_1)\%2=1$ and $T(k'_2)\%2=1$, $|T(k'_1)-T(k'_2)|\%2=0$. According to the Corollary 1, it follows that

$$m_{k'_1k'_2,0}=|T(k'_1)-T(k'_2)|=|t'_1-t'_2|. \tag{23}$$

According to the chosen plaintext attack, $m_{k'_1k'_2,0}$ in Equation (23) can be obtained from the given ξ'_1, $\xi'_2\in\Psi'$, where ξ'_1 corresponding to t'_1 satisfies $t'_1=T(\xi'_1(q^2+q+1))$, and ξ'_2 corresponding to t'_2 satisfies $t'_2=T(\xi'_2(q^2+q+1))$, respectively.

For the same k'_1, k'_2, by selecting a suitable k such that $k=\xi(q^2+q+1)$, $m_{k,0}\%2=0$, one gets $[T(k'_1)+T(k'_2)-T(k)]\%2=0$. Then, according to the Corollary 2, it follows that

$$m_{k'_1k'_2k,0}=T(k'_1)+T(k'_2)-T(k)=t'_1+t'_2-t, \tag{24}$$

where $T(k'_1)<T(k)<T(k'_2)$ or $T(k'_1)>T(k)>T(k'_2)$, $t'_1<t<t'_2$ or $t'_1>t>t'_2$.

According to the chosen plaintext attack, $m_{k'_1k'_2k,0}$ in Equation (24) can be obtained from the given ξ'_1, $\xi'_2\in\Psi'$ and $\xi\in\Psi$, where ξ'_1 corresponding to t'_1 satisfies $t'_1=T(\xi'_1(q^2+q+1))$, ξ'_2 corresponding to t'_2 satisfies $t'_2=T(\xi'_2(q^2+q+1))$, ξ corresponding to t satisfies $t=T(\xi(q^2+q+1))=m_{k,0}$, in which $m_{k,0}$ is known by a chosen plaintext attack as well.

Note that one can also select t'_i,t'_{i+1} $(i=2,3,\cdots,(q/2-1))$ in the same way, which is omitted here due to the limited length of the article.

According to Equations (23) and (24), four cases are given as follows:

(1)   If $t'_1 < t < t'_2$, then one has

$$\begin{cases} t'_2 = (m_{k'_1 k'_2,0} + m_{k'_1 k'_2 k,0} + t)/2 = A_1, \\ t'_1 = (-m_{k'_1 k'_2,0} + m_{k'_1 k'_2 k,0} + t)/2 = B_1. \end{cases} \tag{25}$$

(2)   If $t'_1 > t > t'_2$, then one has

$$\begin{cases} t'_1 = (m_{k'_1 k'_2,0} + m_{k'_1 k'_2 k,0} + t)/2 = A_1, \\ t'_2 = (-m_{k'_1 k'_2,0} + m_{k'_1 k'_2 k,0} + t)/2 = B_1. \end{cases} \tag{26}$$

(3)   If $t'_2 < t < t'_3$, then one has

$$\begin{cases} t'_3 = (m_{k'_2 k'_3,0} + m_{k'_2 k'_3 k,0} + t)/2 = A_2, \\ t'_2 = (-m_{k'_2 k'_3,0} + m_{k'_2 k'_3 k,0} + t)/2 = B_2. \end{cases} \tag{27}$$

(4)   If $t'_2 > t > t'_3$, then one has

$$\begin{cases} t'_2 = (m_{k'_2 k'_3,0} + m_{k'_2 k'_3 k,0} + t)/2 = A_2, \\ t'_3 = (-m_{k'_2 k'_3,0} + m_{k'_2 k'_3 k,0} + t)/2 = B_2. \end{cases} \tag{28}$$

Based on Equations (25)–(28), it follows that

$$\begin{cases} \{t'_1, t'_2\} = \{A_1, B_1\}, \\ \{t'_2, t'_3\} = \{A_2, B_2\}. \end{cases} \tag{29}$$

Then, according to Equation (29), it follows that

$$\begin{cases} t'_2 = \{A_1, B_1\} \bigcap \{A_2, B_2\}, \\ t'_1 = \{A_1, B_1\} - \{t'_2\}, \\ t'_3 = \{A_2, B_2\} - \{t'_2\}. \end{cases} \tag{30}$$

Similarly, for t'_{i-1}, t'_i, t and t'_i, t'_{i+1}, t, one has

$$\begin{cases} t'_i = \{A_{i-1}, B_{i-1}\} \bigcap \{A_i, B_i\}, \\ t'_{i-1} = \{A_{i-1}, B_{i-1}\} - \{t'_i\}, \\ t'_{i+1} = \{A_i, B_i\} - \{t'_i\}, \end{cases} \tag{31}$$

where $i = 2, 3, \cdots, (q/2 - 1)$.

For any given $\xi'_l \in \Psi'$ and $\xi \in \Psi$, according to Equation (31), first, one can get the corresponding t'_l. Then, the sequence number $i'_{\xi'_l}$ corresponding to the sequence value ξ'_l can be further obtained by using t'_l, such that

$$\begin{cases} i'_{\xi'_l} = t'_l \% q, \\ lx[i'_{\xi'_l}] = \xi'_l, \end{cases} \tag{32}$$

where $l \in \{1, 2, \cdots, q/2\}$.

Finally, according to the Equations (22) and (32), one can determine all the sequence values $\xi \in \Psi$, $\xi'_l \in \Psi'$ and all the corresponding sequence numbers $i_\xi, i'_{\xi'}$ in Equation (2), so that the chaotic index sequence lx can be completely deciphered.

3.2. Analysis of Secret Keys α, β, γ

Proposition 3. *Under the condition that the chaotic index sequence lx is obtained, for any $(l_1, l_2, l_3) \neq (l_1', l_2', l_3')$, where $l_i, l_i' \in \{0, 1, 2, \cdots, q-1\}$ $(i = 1, 2, 3)$, if $L_1(l_1, l_2, l_3) = L_2(l_1, l_2, l_3) \neq 0$ and $L_2(l_1', l_2', l_3') = L_3(l_1', l_2', l_3') \neq 0$, then the secret keys α, β, γ can be uniquely determined.*

Proof. According to Equation (3), if $L_1(l_1, l_2, l_3) = L_2(l_1, l_2, l_3) \neq 0$ and $L_2(l_1', l_2', l_3') = L_3(l_1', l_2', l_3') \neq 0$ for any $(l_1, l_2, l_3) \neq (l_1', l_2', l_3')$, then it follows that

$$\begin{cases} L_1(l_1, l_2, l_3) = L_2(l_1, l_2, l_3) = c_{l_1} \times \chi_1{}^2 + c_{l_2} \times \chi_1 + c_{l_3} \neq 0, \\ L_2(l_1', l_2', l_3') = L_3(l_1', l_2', l_3') = c_{l_1'} \times \chi_2{}^2 + c_{l_2'} \times \chi_2 + c_{l_3'} \neq 0, \end{cases} \tag{33}$$

where $c_{l_1}, c_{l_2}, c_{l_3}$ are sequence values of chaotic index sequence lx, $\chi_1 \in \{\alpha, \beta\}$, $\chi_2 \in \{\beta, \gamma\}$.

According to the first equation of Equation (33), one gets two solutions $\chi_1^{(1)}, \chi_1^{(2)}$ for χ_1. Similarly, according to the second equation of Equation (33), one gets two solutions $\chi_2^{(1)}, \chi_2^{(2)}$ for χ_2. Thus, there exists an intersection for the first equation and the second equation of Equation (33), given by $\beta = \{\chi_1^{(1)}, \chi_1^{(2)}\} \cap \{\chi_2^{(1)}, \chi_2^{(2)}\}$. Based on the deciphered secret key β, the remaining two secret keys $\alpha = \{\chi_1^{(1)}, \chi_1^{(2)}\} - \{\beta\}$ and $\gamma = \{\chi_2^{(1)}, \chi_2^{(2)}\} - \{\beta\}$ can further be deciphered as well.

If $L_1(l_1, l_2, l_3) = L_2(l_1, l_2, l_3) = 0$ and $L_2(l_1', l_2', l_3') = L_3(l_1', l_2', l_3') = 0$, then, an intersection for the first equation and the second equation of Equation (33) does not exist, so the secret keys α, β, γ cannot be obtained [39]. The proof is finished. □

3.3. Flowchart of Security Analysis

The flowchart of security analysis is shown in Figure 5.

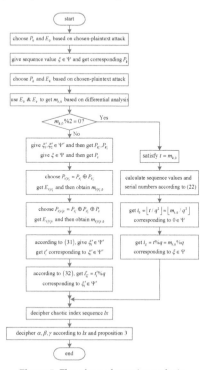

Figure 5. Flowchart of security analysis.

4. Steps for Deciphering the Image Chaotic Encryption Algorithm

The steps for deciphering image chaotic encryption algorithm are as Algorithm 3.

Algorithm 3 Steps for Deciphering Image Chaotic Encryption Algorithm.

Output: The equivalent secret keys lx, α, β, γ;
1: According to the chosen plaintext attack, choose the plain image as P_0, the corresponding cipher image is E_0, the 1D bit sequence corresponding to E_0 is $\{e_0[i]\}_{i=0}^{q^3-1}$.
2: According to the chosen plaintext attack, choose the plain image as P_k, the corresponding cipher image is E_k, the 1D bit sequence corresponding to E_k is $\{e_k[i]\}_{i=0}^{q^3-1}$. where $k = \xi \cdot (q^2 + q + 1)$, $\xi \in \Psi$.
3: According to the differential attack, calculate $m_{k,0}$ by using $\{e_0[i]\}_{i=0}^{q^3-1}$ and $\{e_k[i]\}_{i=0}^{q^3-1}$ obtained in step 1 and step 2.
4: If $m_{k,0}\%2 = 0$, then $t = m_{k,0}$ holds. According to Equation (22), the sequence number corresponding to sequence value 0 is $i_0 = \lfloor t/q^2 \rfloor = \lfloor m_{k,0}/q^2 \rfloor$, the sequence number corresponding to sequence value $\xi \in \Psi$ is $i_\xi = t\%q = m_{k,0}\%q$.
5: If $m_k\%2 = 1$, then $t \neq m_{k,0}$ holds, Equation (22) is not available. According to the chosen plaintext attack, choose the plain image as $P_{k_1'k_2'} = P_{k_1'} \oplus P_{k_2'}$, the corresponding cipher image is $E_{k_1'k_2'}$, the 1D bit sequence corresponding to $E_{k_1'k_2'}$ is $\{e_{k_1'k_2'}[i]\}_{i=0}^{q^3-1}$. In addition, choose the plain image as $P_{k_1'k_2'k} = P_{k_1'} \oplus P_{k_2'} \oplus P_k$, the corresponding cipher image is $E_{k_1'k_2'k}$, the 1D bit sequence corresponding to $E_{k_1'k_2'k}$ is $\{e_{k_1'k_2'k}[i]\}_{i=0}^{q^3-1}$.
6: According to the differential attack, first calculate $m_{k_1'k_2',0}$ by using $\{e_0[i]\}_{i=0}^{q^3-1}$ and $\{e_{k_1'k_2'}[i]\}_{i=0}^{q^3-1}$ obtained in step 1 and step 5. Then, calculate $m_{k_1'k_2'k,0}$ by using $\{e_0[i]\}_{i=0}^{q^3-1}$ and $\{e_{k_1'k_2'k}[i]\}_{i=0}^{q^3-1}$ obtained in step 1 and step 5.
7: According to Equation (32), calculate the sequence number $i'_{\xi_i'} = t_i'\%q$ corresponding to sequence value $\xi_i' \in \Psi'$.
8: Decipher the chaotic index sequence lx by using Equation (22) and Equation (32). Then, decipher the secret keys α, β, γ according to the Proposition 3.
9: **return** lx, α, β, γ;

Theoretical analysis and experimental results indicate that only a maximum of $2.5 \times \sqrt[3]{w \times h}$ plain images are needed to decipher the chaotic index sequence lx, and only a maximum of six plain images are needed to decipher secret keys α, β, γ. Therefore, only a maximum of $2.5 \times \sqrt[3]{w \times h} + 6$ is needed to crack the cipher image with $w \times h$ resolution.

5. Numerical Simulation Experiments

In the numerical simulation experiments, the secret keys are set as $key_0 = 0.34$, $\mu_0 = 3.9$, $\alpha = 20$, $\beta = 37$, $\gamma = 46$, the image is with 512×512 resolution. According to the steps for deciphering the image chaotic encryption algorithm given in Section 4, the deciphering algorithm of the origin cipher is implemented by the C program language. Simulations are operated under a laptop computer with Intel Core i7-8550U CPU (Santa Clara, CA, USA) 1.80 GHz, 8 GB RAM, the operating system is Microsoft Windows 10 (Redmond, WA, USA). Using the original algorithm to encrypt and use the algorithm proposed in this paper to crack an image with size of 512×512 takes about 0.115 s and 10.702 s, respectively. Since the encryption process of the algorithm is independent of plaintext and ciphertext, the equivalent key obtained by deciphering any ciphertext image can be used to decipher all ciphertext images of the same resolution. Taking the standard 2D plain gray image Lena, Cameraman, Livingroom as three examples, the plain images, the cipher images, and the deciphered images are shown in Figure 6, respectively.

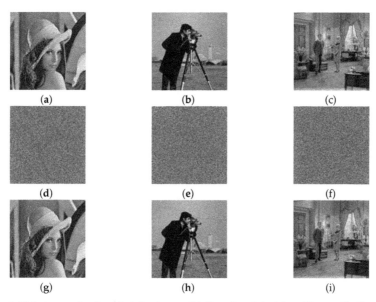

Figure 6. Plain images ((**a–c**) row), cipher images ((**d–f**) row), and deciphered images ((**g–i**) row) of Lena ((**a–g**) column), cameraman ((**b–h**) column), and living room ((**c–i**) column).

Although the previous analysis is for grayscale images, the original encryption algorithm can be easily extended to encrypt color images by encrypting each of the three channels of the color image as a separate grayscale image. In this case, the attack method proposed in this paper is still valid. Take a real-life image with 1024×2048 resolution as an example. Encrypting this image using the original encryption algorithm, it takes about 0.53 s to encrypt the three color channels with the same key, and it takes about 107.36 s to decipher the corresponding ciphertext using the attack method proposed in this paper. Encrypting three color channels with three different sets of keys takes about 1.42 s, and it takes about 318.45 s to decipher the corresponding ciphertext. The results are shown in Figure 7.

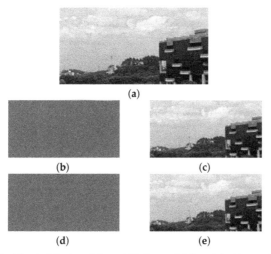

Figure 7. The result of the deciphering of the real-life image. (**a**) the original image; (**b**) encrypting the three color channels with the same key; (**c**) the deciphered image corresponding to (**b**); (**d**) encrypting the three color channels with three different sets of keys; (**e**) the deciphered image corresponding to (**d**).

6. Suggestions for Improvement

According to the analysis in Section 3, the original algorithm is insecure and cannot resist the choice of plaintext attack, and the complexity of the attack method is relatively low. To deal with its security defects, the corresponding suggestions for improvement to enhance the security are as follows:

(1) Enhance the sensitivity of the encryption algorithm to plaintext and ciphertext. According to the analysis in Section 3, the original algorithm has a universal equivalent key lx, α, β, γ. The original algorithm is not sensitive to both plaintext and ciphertext. The root cause of this defect is that the generation of Latin cubes is independent of plaintext image. This vulnerability can be solved by introducing some statistical properties of plaintext, such as the sum of all pixel values, into the generation phase of the Latin cubes.

(2) The mechanism used in the diffusion phase is too simple to achieve the avalanche effect of cryptography, which makes the encryption algorithm vulnerable to differential attacks. To fulfill this demand, increasing the number of encryption rounds or exploiting some complex diffusion mechanisms are worthy options.

7. Conclusions

This paper investigates the security of a Latin-bit cube-based image chaotic encryption algorithm. The algorithm adopts a first-scrambling-diffusion-second-scrambling three-stage encryption scheme. Although the designer claims that the algorithm has passed various statistical tests, the security analysis results in this paper demonstrate that the algorithm has some security vulnerabilities. In particular, the generation of Latin cubes is independent of plain image, and the change in the number of bits in the cipher image follows the change of any one bit in the plain image with obvious regularity. Thus, the equivalent secret keys lx, α, β, γ can be cracked by a chosen plaintext attack and differential attack. Only a maximum of $2.5 \times \sqrt[3]{w \times h} + 6$ plain images are needed to decipher the equivalent secret keys. Theoretical analysis and numerical simulation experiment results verify the effectiveness of the analytical method.

Author Contributions: Methodology, Z.Z.; Project administration, S.Y.; Software, Z.Z.; Supervision, S.Y.; Validation, S.Y.

Funding: This research was funded by the National Key Research and Development Program of China (No. 2016YFB0800401) and the National Natural Science Foundation of China (No. 61532020, 61671161).

Conflicts of Interest: The authors declare no conflict and interest.

References

1. Özkaynak, F. Brief review on application of nonlinear dynamics in image encryption. *Nonlinear Dyn.* **2018**, *92*, 305–313. [CrossRef]
2. Abdallah, E.E.; Ben Hamza, A.; Bhattacharya, P. Video watermarking using wavelet transform and tensor algebra. *Signal Image Video Process.* **2010**, *4*, 233–245. [CrossRef]
3. Abdallah, E.E.; Hamza, A.B.; Bhattacharya, P. MPEG Video Watermarking Using Tensor Singular Value Decomposition. In *Image Analysis and Recognition*; Kamel, M., Campilho, A., Eds.; Springer: Berlin/Heidelberg, Germany, 2007; pp. 772–783.
4. Wang, J.; Ding, Q. Dynamic Rounds Chaotic Block Cipher Based on Keyword Abstract Extraction. *Entropy* **2018**, *20*, 693. [CrossRef]
5. Wang, X.; Yang, L.; Liu, R.; Kadir, A. A chaotic image encryption algorithm based on perceptron model. *Nonlinear Dyn.* **2010**, *62*, 615–621. [CrossRef]
6. Zhang, Y.; Wang, X. A new image encryption algorithm based on non-adjacent coupled map lattices. *Appl. Soft. Comput.* **2015**, *26*, 10–20. [CrossRef]
7. Wang, X.; Liu, L.; Zhang, Y. A novel chaotic block image encryption algorithm based on dynamic random growth technique. *Opt. Lasers Eng.* **2015**, *66*, 10–18. [CrossRef]

8. Liu, H.; Wang, X. Color image encryption based on one-time keys and robust chaotic maps. *Comput. Math. Appl.* **2010**, *59*, 3320–3327. [CrossRef]

9. Liu, H.; Wang, X. Color image encryption using spatial bit-level permutation and high-dimension chaotic system. *Opt. Commun.* **2011**, *284*, 3895–3903. [CrossRef]

10. Song, C.; Qiao, Y. A Novel Image Encryption Algorithm Based on DNA Encoding and Spatiotemporal Chaos. *Entropy* **2015**, *17*, 6954–6968. [CrossRef]

11. Chai, X.; Fu, X.; Gan, Z.; Lu, Y.; Chen, Y. A color image cryptosystem based on dynamic DNA encryption and chaos. *Signal Process.* **2019**, *155*, 44–62. [CrossRef]

12. Liu, H.; Wang, X.; kadir, A. Image encryption using DNA complementary rule and chaotic maps. *Appl. Soft. Comput.* **2012**, *12*, 1457–1466. [CrossRef]

13. Wang, X.; Zhang, Y.; Bao, X. A novel chaotic image encryption scheme using DNA sequence operations. *Opt. Lasers Eng.* **2015**, *73*, 53–61. [CrossRef]

14. Wang, X.; Feng, L.; Zhao, H. Fast image encryption algorithm based on parallel computing system. *Inf. Sci.* **2019**, *486*, 340–358. [CrossRef]

15. Wang, X.; Gao, S. Image encryption algorithm for synchronously updating Boolean networks based on matrix semi-tensor product theory. *Inf. Sci.* **2019**, *507*, 16–36. [CrossRef]

16. Yaghouti Niyat, A.; Moattar, M.H.; Niazi Torshiz, M. Color image encryption based on hybrid hyper-chaotic system and cellular automata. *Opt. Lasers Eng.* **2017**, *90*, 225–237. [CrossRef]

17. Chai, X.; Gan, Z.; Yang, K.; Chen, Y.; Liu, X. An image encryption algorithm based on the memristive hyperchaotic system, cellular automata and DNA sequence operations. *Signal Process.-Image Commun.* **2017**, *52*, 6–19. [CrossRef]

18. Wang, X.; Li, Z. A color image encryption algorithm based on Hopfield chaotic neural network. *Opt. Lasers Eng.* **2019**, *115*, 107–118. [CrossRef]

19. Bigdeli, N.; Farid, Y.; Afshar, K. A robust hybrid method for image encryption based on Hopfield neural network. *Comput. Electr. Eng.* **2012**, *38*, 356–369. [CrossRef]

20. Xu, M.; Tian, Z. A novel image cipher based on 3D bit matrix and latin cubes. *Inf. Sci.* **2019**, *478*, 1–14. [CrossRef]

21. Xu, M.; Tian, Z. A novel image encryption algorithm based on self-orthogonal Latin squares. *Optik* **2018**, *171*, 891–903. [CrossRef]

22. Chen, J.; Zhu, Z.; Fu, C.; Zhang, L.; Zhang, Y. An efficient image encryption scheme using lookup table-based confusion and diffusion. *Nonlinear Dyn.* **2015**, *81*, 1151–1166. [CrossRef]

23. Zhang, Y.; Wang, X. A symmetric image encryption algorithm based on mixed linear–nonlinear coupled map lattice. *Inf. Sci.* **2014**, *273*, 329–351. [CrossRef]

24. Li, M.; Lu, D.; Wen, W.; Ren, H.; Zhang, Y. Cryptanalyzing a Color Image Encryption Scheme Based on Hybrid Hyper-Chaotic System and Cellular Automata. *IEEE Access* **2018**, *6*, 47102–47111. [CrossRef]

25. Wen, H.; Yu, S.; Lü, J. Breaking an Image Encryption Algorithm Based on DNA Encoding and Spatiotemporal Chaos. *Entropy* **2019**, *21*, 246. [CrossRef]

26. Li, C.; Lo, K. Optimal quantitative cryptanalysis of permutation-only multimedia ciphers against plaintext attacks. *Signal Process.* **2011**, *91*, 949–954. [CrossRef]

27. Jolfaei, A.; Wu, X.; Muthukkumarasamy, V. On the Security of Permutation-Only Image Encryption Schemes. *IEEE Trans. Inf. Forensic Secur.* **2016**, *11*, 235–246. [CrossRef]

28. Feng, W.; He, Y.; Li, H.; Li, C. Cryptanalysis and Improvement of the Image Encryption Scheme Based on 2D Logistic-Adjusted-Sine Map. *IEEE Access* **2019**, *7*, 12584–12597. [CrossRef]

29. Hua, Z.; Zhou, Y. Image encryption using 2D Logistic-adjusted-Sine map. *Inf. Sci.* **2016**, *339*, 237–253. [CrossRef]

30. Lin, Z.; Yu, S.; Lü, J.; Cai, S.; Chen, G. Design and ARM-Embedded Implementation of a Chaotic Map-Based Real-Time Secure Video Communication System. *IEEE Trans. Circuits Syst. Video Technol.* **2015**, *25*, 1203–1216.

31. Lin, Z.; Yu, S.; Feng, X.; Lü, J. Cryptanalysis of a Chaotic Stream Cipher and Its Improved Scheme. *Int. J. Bifurc. Chaos* **2018**, *28*, 1850086. [CrossRef]

32. Hu, G.; Xiao, D.; Wang, Y.; Li, X. Cryptanalysis of a chaotic image cipher using Latin square-based confusion and diffusion. *Nonlinear Dyn.* **2017**, *88*, 1305–1316. [CrossRef]

33. Wang, H.; Xiao, D.; Chen, X.; Huang, H. Cryptanalysis and enhancements of image encryption using combination of the 1D chaotic map. *Signal Process.* **2018**, *144*, 444–452. [CrossRef]

34. Pak, C. A new color image encryption using combination of the 1D chaotic map. *Signal Process.* **2017**, *138*, 129–137. [CrossRef]
35. Wu, J. Cryptanalysis and enhancements of image encryption based on three-dimensional bit matrix permutation. *Signal Process.* **2018**, *142*, 292–300. [CrossRef]
36. Zhang, W.; Yu, H.; Zhao, Y.; Zhu, Z. Image encryption based on three-dimensional bit matrix permutation. *Signal Process.* **2016**, *118*, 36–50. [CrossRef]
37. Wang, X.; Teng, L.; Qin, X. A novel colour image encryption algorithm based on chaos. *Signal Process.* **2012**, *92*, 1101–1108. [CrossRef]
38. Arkin, J.; Straus, E.G. Latin k-cubes. *Fibonacci Q.* **1974**, *12*, 288–292.
39. Berlekamp, E.; Rumsey, H.; Solomon, G. On the solution of algebraic equations over finite fields. *Inf. Comput.* **1967**, *10*, 553–564. [CrossRef]

© 2019 by the authors. Licensee MDPI, Basel, Switzerland. This article is an open access article distributed under the terms and conditions of the Creative Commons Attribution (CC BY) license (http://creativecommons.org/licenses/by/4.0/).

Article

A Secure and Fast Image Encryption Scheme Based on Double Chaotic S-Boxes

Shenli Zhu [1], Guojun Wang [2] and Congxu Zhu [3,*]

[1] School of Computer Science, University of South China, Hengyang 421001, China
[2] School of Computer Science, Guangzhou University, Guangzhou 510006, China
[3] School of Computer Science and Engineering, Central South University, Changsha 410083, China
* Correspondence: zhucx@csu.edu.cn; Tel.: +86-0731-8882-7601

Received: 22 July 2019; Accepted: 12 August 2019; Published: 13 August 2019

Abstract: In order to improve the security and efficiency of image encryption systems comprehensively, a novel chaotic S-box based image encryption scheme is proposed. Firstly, a new compound chaotic system, Sine-Tent map, is proposed to widen the chaotic range and improve the chaotic performance of 1D discrete chaotic maps. As a result, the new compound chaotic system is more suitable for cryptosystem. Secondly, an efficient and simple method for generating S-boxes is proposed, which can greatly improve the efficiency of S-box production. Thirdly, a novel double S-box based image encryption algorithm is proposed. By introducing equivalent key sequences {$\mathbf{r}$, $\mathbf{t}$} related with image ciphertext, the proposed cryptosystem can resist the four classical types of attacks, which is an advantage over other S-box based encryption schemes. Furthermore, it enhanced the resistance of the system to differential analysis attack by two rounds of forward and backward confusion-diffusion operation with double S-boxes. The simulation results and security analysis verify the effectiveness of the proposed scheme. The new scheme has obvious efficiency advantages, which means that it has better application potential in real-time image encryption.

Keywords: image encryption; compound chaotic system; S-box; image information entropy

---

1. Introduction

With the rapid development of network communication, image encryption has become a research hotspot in the field of image processing and information security. Since image information has the characteristics of large amounts of data, strong redundancy and high correlation between adjacent pixels, image encryption algorithms need not only high security, but also fast encryption speed. If the speed of encryption is low, the time consumed will be too long because of the large amount of image data. To encrypt multimedia information with large amounts of data, security and efficiency should be considered comprehensively [1–5]. Chaos-based cryptosystem just meets the need of image encryption, which leads to the research of chaos-based image encryption technology has been widely concerned by scholars. As for chaotic cryptography, a new chaotic system with better cryptographic performance deserves to be established. Some representative studies have contributed to this aspect [6–9]. How to generate key stream or encryption component with good performance is very important to the security of the image Cryptosystem [10–12]. How to design encryption algorithm is the core research content of the image Cryptosystem [13]. Cryptanalysis [14–16] is another important research direction of cryptography, which can help cryptographic designers improve the security of cryptographic algorithms.

Among many chaos-based image encryption algorithms, the permutation and diffusion (PD) pattern encryption algorithm proposed by Fridrich [17] is the most popular one. This image encryption algorithm structure consists of shuffling pixel positions and changing pixel values. The permutation (or shuffling, scrambling) process plays a role in confusing the relationship between the cipher image and plain image. The function of the diffusion process is to spread the change of one pixel value in the plain image to the whole range of the cipher image. Based on the basic confusion-diffusion architecture, researchers have proposed many novel concrete encryption strategies. In Ref. [18–24], authors proposed some different permutation strategies for image scrambling aiming at the confusion process. In Ref. [22,25–29], authors put forward some novel image diffusion algorithm. In Ref. [30–36], authors adopt new chaotic systems to improve the complexity and randomness of chaotic key streams. Some other cryptographic methods have also been tried by many researchers. For example, some cryptographic algorithms are based on bit-level permutation and diffusion [30], and some algorithms introduce the DNA coding mechanism [37], and some algorithms mainly use S-box to encrypt images [38–40]. However, some image encryption schemes exist as obvious security vulnerabilities. Thus, these image encryption schemes cannot resist some attacks, such as the chosen/known plaintext. In addition, some image encryption algorithms are inefficient, such as using bit-level image scrambling, DNA encoding mechanism, key related to plaintext Hash value [41,42], and the high-dimensional chaotic system [43,44]. Encryption algorithms with low efficiency are not suitable for some resource-constrained environments, such as mobile social network [45], sensor network communication environment [46] and searchable encryption [47]. Compared with high-dimensional continuous-time chaotic systems, low-dimensional discrete chaotic systems can generate chaotic sequences with higher efficiency. Moreover, some studies show that the complexity of discrete systems is higher than that of continuous systems [48–50].

Substitution-boxes (abbreviated as S-boxes) are important non-linear components in the block cipher system, which play an important role in the security of cryptosystems. Therefore, some image encryption systems based on chaos also use S-box. Majid Khan [51] employed multi-parameters chaotic systems in the construction of S-boxes that are applied to the encryption of images. The multi-parameters chaotic systems are hyper-chaotic systems. Moreover, the output trajectory points of the system need to be sampled, so the time cost of generating S-boxes in the encryption scheme is bound to be long. In addition, the S-box in the scheme is equivalent to the original key and is independent of the image content. Therefore, it is vulnerable to the chosen-plaintext attack. In order to resist the selective plaintext attack, some image encryption algorithms based on chaos introduce the mechanism of the key and plaintext association. Wang et al. [52] proposed a novel image encryption algorithm based on dynamic S-boxes constructed by chaos, in which a system up to 50 S-boxes need to be generated. It is time-consuming and unsuitable for real-time encryption. M.A. Murillo-Escobar et al. [53] proposed a color image encryption algorithm based on total plain image characteristics and 1D logistic map with optimized distribution. They have a diffusion process optimized by the modified chaotic sequence. In addition, the pseudorandom sequence for the encryption process is based on the total plain image characteristic and a 128 bits secret key, so the encryption algorithm can resist the powerful chosen-plaintext attack. Zhang et al. [54] proposed a plaintext-related image encryption algorithm based on chaos. The Zhang's system can also fight against the chosen-plaintext attacks due to using a plaintext-related key sequence. However, in order to make the final key related to the plaintext, the process of generating the final key in the above algorithms is complex. So far, most image and video encryption algorithms based on chaos mainly rely on the empirical security analysis. However, the recent study [55] has shown that the empirical safety analysis is not enough. A encryption algorithm passing the empirical safety tests is merely a necessary condition for security, but is not a sufficient criterion.

In order to improve the security and real-time performance of the image encryption algorithm, this paper presents a simple yet security image encryption algorithm based on chaotic S-boxes. The main goal of this paper is to improve the encryption efficiency of the encryption system on the premise of ensuring a certain level of security. The main innovations of this paper are as follows: (1) A new compound chaotic system, the Sine-Tent system (STS), is proposed. The compound system has wider chaotic range and better chaotic performance than any of the original systems, so it is more suitable for cryptographic applications. (2) A simple and effective S-box construction method based on the new compound chaotic system is proposed, which can speed up the generation of S-boxes. (3) A double S-boxes based image encryption algorithm is designed. Double S-boxes can not only meet the security requirements of the system, but also make the time cost much lower than multiple S-boxes. The algorithm makes the parameters of the permutation and diffusion process interrelated and related with image ciphertext so that the encryption algorithm can resist chosen-ciphertext attack. Additionally, two rounds of forward and backward confusion-diffusion operation enhances the resistance of the system to the differential analysis attack.

The rest of this paper is organized as follows. Section 2 introduces the new Sine-Tent system (STS) model. Section 3 describes the simple and effective S-box construction method based on the Sine-Tent system. Section 4 describes the new double S-boxes based image encryption algorithm. Section 5 presents the results of experiments and analysis of the proposed scheme. Finally, some concluding remarks are given in Section 6.

2. The Proposed New Chaotic System

1D discrete chaotic systems have many advantages in image encryption because of their simple structures. In this section, we firstly review two 1D chaotic maps: The Sine and Tent maps. They will be used for constructing our new chaotic system. Then, a new discrete compound chaotic system is proposed to solve the problems existing in the Sine and Tent maps.

2.1. Sine Chaotic Map

The Sine map is one of the famous 1D chaotic maps. It is a simple dynamical system with complex chaotic behavior similar to the Logistic map. The mathematical model of the Sine map can be expressed as

$$x(n+1) = \mu/4 \times \sin(\pi \times x(n)) \tag{1}$$

where μ is the system parameter in the range of $(0, 4]$, $x(0)$ is the initial state value of the system and $\{x(n), n = 1, 2, \dots \}$ is the output sequence of state values. To observe the chaotic behaviors of the Sine map, its Lyapunov Exponent and bifurcation diagram are presented in Figure 1a,b.

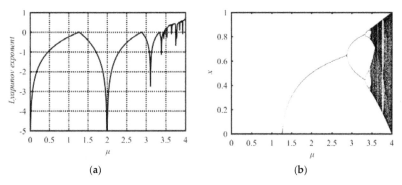

Figure 1. Lyapunov Exponent and bifurcation diagram of the Sine map. (**a**) Lyapunov Exponent diagram; (**b**) bifurcation diagram.

As is well known, for a dynamical system, a positive Lyapunov Exponent means chaotic behavior occurs in the dynamical system. So, from Figure 1a, one can see that only when the parameter $\mu \geq 3.57$ can chaotic behavior occur in the Sine map. The bifurcation diagram depicts the possible state values of the system under each parameter. Corresponding to a value of system parameter, if there are infinite state values, the system with the parameter has chaotic behavior. Corresponding to a value of system parameter, if only one or a limited number of state values output, the system with the parameter does not have chaotic behavior. In the bifurcation diagram shown in Figure 1b, the areas of μ with dense points shows its good chaotic behavior and the areas of μ with the solid line represents its non-chaotic property. There are two problems in the Sine map. First, the range of system parameters corresponding to chaotic phenomena is limited only within the range of [3.57, 4]. Even within this range, there are some parameters which make the Sine map have no chaotic behaviors. This is verified by its Lyapunov Exponent diagram and the blank zone in its bifurcation diagram. Second, when the system parameter value is less than four, the state values of the system output sequence are distributed in a narrower range than the [0, 1] interval. Only when the system parameter value is four, the state values of the system output sequence are distributed in the whole [0, 1] range. It shows the nonuniform distribution in the range of [0, 1]. These two problems reduce the application value of the Sine map.

2.2. Tent Chaotic Map

The name "Tent map" comes from its bifurcation diagram, which has the tent-like shape. Its mathematical model can be expressed as

$$x(n+1) = \begin{cases} \mu/2 \times x(n) & x(n) < 0.5 \\ \mu/2 \times (1 - x(n)) & x(n) \geq 0.5 \end{cases} \tag{2}$$

where μ is the system parameter in the range of (0, 4].

Its chaotic property is shown in the Lyapunov Exponent analysis in Figure 2a and bifurcation analysis in Figure 2b. Both analysis results indicate that its parameter value range with chaotic behavior is $2 \leq \mu \leq 4$. The Tent map has the same problems as the Sine map: The small parameter value range with chaotic behavior and the nonuniform distribution of the output state values.

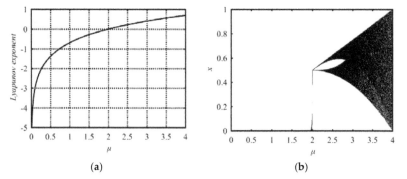

(a) (b)

Figure 2. Lyapunov Exponent and bifurcation diagram of the Tent map. (a) Lyapunov Exponent diagram; (b) bifurcation diagram.

2.3. The Sine-Tent System

We put forward a new compound system by combining the Sine and Tent maps and called the new system the Sine-Tent system (STS). Its mathematical model is as follows:

$$x(n+1) = \begin{cases} (4-\mu)/4 \times \sin(\pi \times x(n)) + \mu/2 \times x(n) & x(n) < 0.5 \\ (4-\mu)/4 \times \sin(\pi \times x(n)) + \mu/2 \times (1 - x(n)) & x(n) \geq 0.5 \end{cases} \tag{3}$$

where μ is the system parameter in the range of [0, 4]. When $\mu = 0$, Equation (3) degenerates to the Sine map, while $\mu = 4$, Equation (3) degenerates to the Tent map. Therefore, both the Sine map and Tent map can be regarded as special cases of the Sine-Tent system.

The Lyapunov Exponent and bifurcation diagram of the STS are shown in Figure 3a,b, respectively. From Figure 3 one can see that its parameter value range with chaotic behavior is $\mu \in [0, 4]$, which is much larger than those of the Sine or Tent maps. Its output sequences uniformly distribute within [0, 1] (see Figure 3b). Hence, the STS has better chaotic performance than the Sine and Tent maps.

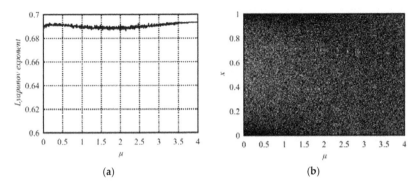

(a)   (b)

Figure 3. Lyapunov Exponent and bifurcation diagram of the Sine-Tent map. (a) Lyapunov Exponent diagram; (b) bifurcation diagram.

The new compound system has at least three advantages compared with the Sine and Tent maps. First, the output sequences of the new compound system spread out in the entire value range between zero and one. Second, the proposed Sine-Tent system has a wider chaotic range. The Lyapunov Exponents of the Sine-Tent system is positive in the entire range of $0 \leq \mu \leq 4$. However, the Sine map and Tent map have positive values of Lyapunov Exponents only within much smaller ranges. Thirdly, we know that a larger Lyapunov Exponent means stronger chaotic properties. From the Lyapunov Exponent diagrams, one can see that the new system has larger Lyapunov Exponents (Lyapunov Exponents is always close to 0.7) in the whole parameter range of [0, 4], while the Sine and Tent maps have large Lyapunov Exponents only when the parameter is close to four. Therefore, the chaotic characteristic of the new system is stronger, and it always maintains the invariable excellent chaotic performance in the entire parameter range of $0 \leq \mu \leq 4$. These advantages guarantee that the proposed Sine-Tent system is more suitable for information security applications such as image encryption.

3. An Efficient New Method for Generating S-Boxes

In Ref. [56], Belazi et al. proposed a simple yet efficient S-box generating method based on the chaotic sine map, in which a prime number p and a one to one map from the real number interval (0, 1) to the integer set {0, 1, 2, ... , 255} need to be found. In this section, we present a simpler approach for designing S-boxes using the chaotic Sine-Tent map. The new method takes advantage of the excellent chaotic characteristics of the Sine-Tent map. The detailed steps of generating S-boxes are given below.

Step 1: Set parameter d as an odd positive integer and $d > 0$, d can be used as a secret key.

Step 2: Let **T1** = 1:256, then we obtain an array **T1** which contains 256 distinct integers in the range of [1, 256].

Step 3: Based on **T1** and d to obtain a new array **T** by Equation (4)

$$T(i) = \mathrm{mod}(d \times T1(i), 256), \ i = 1, 2, \ldots, 256 \tag{4}$$

The new array $\mathbf{T}_{1 \times 256}$ will contain 256 distinct integers in the range of [0, 255]. As long as d is a finite odd integer and $T1(i) \neq T1(j)$ if $i \neq j$, then $T(i) \neq T(j)$ if $i \neq j$. This conclusion is true and can be proved by experimental tests.

Step 4: Set the parameters μ, initial state value x_0 of the Sine-Tent map, and an integer $N_0 > 0$. Iterate Sine-Tent map (N_0 + 256) times to generate a chaotic sequence of length (N_0 + 256). Discard the first N_0 elements of the original chaotic sequence, then we can obtain a new chaotic sequence of length 256, which is represented by **X**.

Step 5: Sort the chaotic sequence **X**, then we can get a position index array $J = \{J(1), J(2), \dots , J(256)\}$, $J(i) \in \{1, 2, \dots , 256\}$. As a result of the non-periodicity of the chaotic sequence, it will inevitably lead to that $J(i) \neq J(j)$ as long as $i \neq j$.

Step 6: Calculate the 1D array **S** as follows:

$$S(i) = T(J(i)), \ i = 1, 2, \dots , 256 \tag{5}$$

Step 7: Transform the 1D array $S_{1 \times 256}$ into a 2D matrix $S_{16 \times 16}$, and this is the proposed S-box.

By the above method, the length of chaotic sequences to be used in constructing a 16×16 sized S-box is only 256. Therefore, the time cost of this method is very low. In our experiments, double S-boxes are generated by the above S-box generation algorithm. The initial condition x_0, system parameter μ of the Sine-Tent map and the parameters $\{d, N_0\}$ for the S-box generation are set as $\{x_{10} = 0.21, \mu_1 = 0.399, d_1 = 43, N_0 = 500\}$ and $\{x_{20} = 0.27, \mu_2 = 3.999, d_2 = 241, N_0 = 500\}$ for S-box **S1** and **S2**, respectively. The generated double S-boxes are shown in Tables 1 and 2, which are used in our proposed image encryption algorithm.

Table 1. The chaotic S-box **S1** generated with parameters $\{x_{10} = 0.21, \mu_1 = 0.399, d_1 = 43, N_0 = 500\}$.

S-box	c1	c2	c3	c4	c5	c6	c7	c8	c9	c10	c11	c12	c13	c14	c15	c16
r1	27	4	47	58	146	86	137	215	61	68	129	80	131	214	97	119
r2	168	210	253	91	219	30	112	63	52	188	73	139	55	16	158	204
r3	124	71	21	45	169	32	208	121	198	179	246	8	175	194	35	5
r4	70	3	114	42	205	89	101	159	173	127	75	235	118	243	143	141
r5	147	13	196	163	11	62	134	76	191	133	132	145	33	43	120	31
r6	17	156	245	186	25	237	88	161	0	83	87	72	116	150	255	226
r7	138	74	46	34	136	99	12	218	110	195	105	57	172	65	2	216
r8	211	184	19	20	84	242	85	98	189	22	24	185	166	109	15	217
r9	167	48	56	78	90	59	36	244	6	107	142	180	23	238	106	7
r10	28	247	199	201	40	250	206	183	223	200	29	67	128	126	10	241
r11	113	233	207	140	152	135	122	174	228	151	102	148	79	176	49	95
r12	190	103	92	39	64	1	171	220	212	51	221	130	249	170	164	230
r13	60	162	117	154	157	160	229	187	100	26	37	155	225	222	232	104
r14	181	224	53	18	108	96	66	38	248	182	178	251	165	231	202	81
r15	50	93	149	9	239	192	209	82	115	236	44	144	69	111	153	125
r16	254	41	227	213	193	14	77	197	54	123	203	177	94	252	234	240

Table 2. The chaotic S-box S2 generated with parameters $\{x_{20} = 0.27, \mu_2 = 3.999, d_2 = 241, N_0 = 500\}$.

S-box	c1	c2	c3	c4	c5	c6	c7	c8	c9	c10	c11	c12	c13	c14	c15	c16
r1	75	140	59	156	233	234	149	214	126	105	134	228	101	84	111	35
r2	113	241	53	202	17	96	93	168	172	82	78	203	159	182	249	118
r3	115	68	195	107	189	104	165	80	39	94	150	254	199	183	157	74
r4	52	210	55	200	229	48	132	163	219	201	117	146	153	43	71	230
r5	60	70	103	211	95	92	36	12	81	133	46	176	209	251	237	186
r6	98	136	20	44	178	185	177	19	137	50	21	206	65	192	129	79
r7	240	7	121	38	27	196	25	167	89	72	162	221	148	147	24	223
r8	100	47	248	164	34	29	73	69	245	1	10	191	216	26	204	18
r9	37	15	32	108	9	160	139	220	238	232	58	161	109	6	169	62
r10	45	3	0	180	114	120	246	250	33	194	198	13	158	31	66	155
r11	83	125	244	51	212	97	91	99	77	138	173	243	253	102	123	166
r12	225	208	110	40	222	87	218	197	170	184	124	131	4	112	179	255
r13	85	64	193	88	56	16	236	207	181	144	231	239	152	135	122	67
r14	151	171	42	154	142	247	28	41	14	252	224	188	54	175	217	130
r15	22	215	49	5	141	11	2	127	145	86	116	213	205	63	242	128
r16	30	226	227	106	187	23	174	190	143	8	76	61	235	119	57	90

In the first row of Table 1, c1, c2, ... , c16 denotes the column numbers of the S-box. Additionally, in the first column of Table 1, r1, r2, ... , r16 denotes the row numbers of the S-box.

To determine the randomness of proposed S-box method, the statistical test suite (version 2.1.1), proposed by the National Institute of Standards and Technology (NIST) NIST-800-22 is introduced. The NIST-800-22 test results are listed in Table 3. We find that the 12 tests successfully passed. Moreover, the Random Excursions Test, Random Excursions Variant Test, and Universal Statistical Test were not applicable for the proposed S-box. This is because the sequence generated by an S-box only consists of 2048 bits. However, the Random Excursions Test and Random Excursions Variant Test require a long sequence consisting of a minimum of 1,000,000 bits, and the Universal Statistical Test also requires a long sequence consisting of a minimum of 387,840 bits.

Table 3. NIST-800-22 test results of the obtained S-box.

NIST-800-22 Tests	*p*-Value	Result
Frequency Test	1.00000	SUCCESS
Block Frequency Test	0.320250	SUCCESS
Cumulative Sums Test	0.536610	SUCCESS
Runs Test	0.894524	SUCCESS
Longest Run of Ones Test	1.0000	SUCCESS
Rank Test	0.481248	SUCCESS
Discrete Fourier Transform Test	0.807748	SUCCESS
Nonperiodic Template Matchings Test	0.861831	SUCCESS
Overlapping Template Matchings Test	0.282761	SUCCESS
Approximate Entropy Test	0.011732	SUCCESS
Serial Test	0.239176	SUCCESS
Linear Complexity Test	0.203697	SUCCESS
Random Excursions Test	\	TESTNOTAPPLICABLE
Random Excursions Variant Test	\	TESTNOTAPPLICABLE
Universal Statistical Test	\	TESTNOTAPPLICABLE

4. The Proposed S-Box based Encryption Scheme

4.1. Cryptanalysis of an S-Box Based Encryption Algorithm

In Ref. [57], Çavuşoğlu et al. proposed an image encryption scheme by using the S-box generated with a novel hyper-chaotic system. The sketch of the encryption scheme is shown in Figure 4.

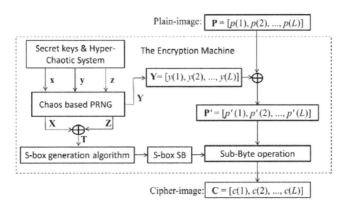

Figure 4. Sketch of the original encryption algorithm.

Suppose the input pixel value array of the plain image is $\mathbf{P} = [p(1), p(2), \ldots, p(L)]$. The output pixel value array of the cipher image is $\mathbf{C} = [c(1), c(2), \ldots, c(L)]$. The encryption steps can be described in detail below.

Step 1: Generate three real value chaotic sequences **x**, **y**, and **z** by using a hyper-chaotic system with given parameters and initial state values as secret keys.

Step 2: Transform the three real value sequences **x**, **y** and **z** into three integer sequences **X**, **Y** and **Z** by the chaos-based pseudo random number generator (PRNG). Each element in **X**, **Y** and **Z** is an 8-bit integer and its decimal number is in the range of [0, 255].

Step 3: The S-box, denoted as $S = [s(j, k)]$, is created by using sequences **X**, **Z** and a novel S-box generation algorithm. Where, $s(j, k) \in \{0, 1, \dots, 255\}$, $j = 1, 2, \dots, 16$, $k = 1, 2, \dots, 16$.

Step 4: The intermediate cipher image array $\mathbf{P'} = [p'(1), p'(2), \dots, p'(L)]$ is generated by using sequences $\mathbf{Y} = [y(1), y(2), \dots, y(L)]$ as

$$p'(i) = y(i) \oplus p(i), i = 1, 2, \dots, L \tag{6a}$$

where $\oplus$ denotes bitwise XOR. The decryption operation corresponding to Equation (6a) can be expressed as Equation (6b):

$$p(i) = y(i) \oplus p'(i), i = 1, 2, \dots, L \tag{6b}$$

Step 5: Perform sub-byte operation on **P'** with the 16×16 sized S-box **S**, and obtain the cipher image array $\mathbf{C} = [c(1), c(2), \dots, c(L)]$.

Here, the sub-byte operation is a process in which each pixel value in the image is substituted with an element value in the S-box. The sub-byte operation can be implemented by defining a function. For example, the function sub_byte[**S**, p] can find a substitute to p from the S-box **S**. Let $q = \text{sub\_byte}[\mathbf{S}, p]$, the algorithm of the function sub_byte[**S**, p] can be described as Algorithm 1. For example, if $p = 55 = (0011\ 0111)_2$, then $j = (0011)_2 + 1 = 4$, $k = (0111)_2 + 1 = 8$. Consequently, $q = \text{sub\_byte}[\mathbf{S}, p] = \text{sub\_byte}[\mathbf{S}, 55] = s(j, k) = s(4,8)$.

Algorithm 1 The algorithm pseudo code of function $q = \text{sub\_byte}[S, p]$.

Input:	$S = [s(j, k)], p; (j = 1, 2, \dots, 16, k = 1, 2, \dots, 16.)$
Output:	$q = \text{sub\_byte}[S, p];$
1:	Convert p to a binary number $(b_8b_7 \dots b_2b_1)_2;$
2:	Let $j = (b_8b_7b_6b_5)_2 = 8 \times b_8 + 4 \times b_7 + 2 \times b_6 + 1 \times b_5;$ $k = (b_4b_3b_2b_1)_2 = 8 \times b_4 + 4 \times b_3 + 2 \times b_2 + 1 \times b_1;$
3:	Let $j = j + 1; k = k + 1;$
4:	Let $q = s(j, k);$

Therefore, Step 5 can be expressed by the following general form:

$$c(i) = \text{sub\_byte}[\mathbf{S}, p'(i)], i = 1, 2, \dots, L \tag{7a}$$

The decryption operation corresponding to Equation (7a) can be expressed as Equation (7b):

$$p'(i) = \text{sub\_byte\_1}[\mathbf{S}, c(i)], i = 1, 2, \dots, L \tag{7b}$$

where, function sub_byte_1[·, ·] is the inverse operation of the function sub_byte[·, ·].

The above S-box based encryption algorithm has the following potential defects:

(1) The chaotic sequence **Y** and S-box is actually the equivalent of the secret keys, which are not related with the image to be encrypted.

(2) The algorithm has no diffusion effect. While one pixel is changed in the plain image, there is only one changed pixel in the cipher image.

(3) The sequence **Y** and S-box are separated in the bitwise XOR process and Sub-Byte process, and the bitwise XOR process unrelated to the Sub-Byte process.

Based on the above analysis, we find that the above encryption scheme cannot resist the chosen-plaintext attack. Suppose the target cipher image to be recovered is $\mathbf{C} = [c(1), c(2), \dots, c(L)]$, we can launch chosen-plaintext attack on the above encryption scheme to recover its corresponding plain image $\mathbf{P} = [p(1), p(2), \ldots, p(L)]$. The simplest attacking algorithm can be described as Algorithm 2.

Algorithm 2 The simplest attacking algorithm pseudo code.

1: $n = 0$;
2: while $(n < 256)$ do
3:      Choose the n-th plain image $\mathbf{Pn} = [n, n, \ldots, n]$;
4:      Get its corresponding cipher image $\mathbf{Cn} = [cn(1), cn(2), \ldots, cn(L)]$ by using the encryption machine of Figure 4;
5:      for $i = 1, 2, \ldots, L$, do
            if $c(i) == cn(i)$, then we can get $p(i) = n$;
6:      end for
7:      $n = n + 1$;
8: end while

This simplest attack method with Algorithm 2 requires 256 selected plaintext images. However, a more efficient chosen-plaintext method only needs to select two plain images. For details, readers can refer to Ref. [58].

4.2. The Novel Double S-Boxes Based Image Encryption Algorithm

To eliminate the security defects that exist in some S-box based encryption algorithms, a novel double S-boxes based image encryption algorithm is proposed. The main innovations of the new scheme lie in the following three aspects: Firstly, the new Sine-Tent compound chaotic system is used to generate double S-boxes, which are used in the two rounds of the encryption process of the new scheme. Secondly, the first S-box is used to realize pixel confusion and substitution simultaneously. Thirdly, two rounds of the encryption process are correlated and the diffusion mechanism is introduced. The main steps of the novel double S-boxes based image encryption algorithm is described as follows:

Step 1: Input the secret parameters $\{x_{10}, \mu_1, d_1, x_{20}, \mu_2, d_2, r_0, t_0, m\}$ and the plain image $\mathbf{PI}$ with the size of $M \times N$. $\mathbf{PI}$ is reshaped to a 1D pixel array $\mathbf{P} = [p(1), p(2), \ldots, p(L)]$, where $L = M \times N$.

Step 2: Generate the first S-box $\mathbf{S1}$ by using the new S-box generation algorithm with parameters $\{x_{10}, \mu_1, d_1\}$.

Step 3: Generate the second S-box $\mathbf{S2}$ by using the new S-box generation algorithm with parameters $\{x_{20}, \mu_2, d_2\}$.

Step 4: Perform the first round of encryption on array $\mathbf{P}$ with the first S-box $\mathbf{S1}$, and obtain the temporary cipher image pixel array $\mathbf{B} = [b(1), b(2), \ldots, b(L)]$ as

$$\begin{cases} j = \mathrm{mod}(1 + m, L) + 1; \\ r = r_0; \\ b(1) = \mathrm{mod}(\mathrm{sub\_byte}[\mathbf{S1}, p(j)] + r, 256). \end{cases} \quad \text{for } i = 1 \qquad (8)$$

$$\begin{cases} j = \mathrm{mod}(i + m, L) + 1; \\ r = \mathrm{mod}(b(i-1) + r, 256); \\ b(i) = \mathrm{mod}(\mathrm{sub\_byte}[\mathbf{S1}, p(j)] + r + b(i-1), 256). \end{cases} \quad \text{for } i = 2, 3, \ldots L \qquad (9)$$

where, $\mathrm{sub\_byte}[\mathbf{S1}, x]$ denotes byte substitution for x using S-box $\mathbf{S1}$. The first round of encryption is the forward confusion-diffusion operation, in which permutation and diffusion are implemented simultaneously by introducing the location index j.

Step 5: Perform the second round of encryption on array $\mathbf{B}$ with the second S-box $\mathbf{S2}$, and obtain the final cipher image pixel array $\mathbf{C} = [c(1), c(2), \ldots, c(L)]$ as

$$\begin{cases} t = t_0; \\ c(L) = \mathrm{sub\_byte}[\mathbf{S2}, \mathrm{mod}(b(L) + t, 256)]. \end{cases} \quad \text{for } i = L \qquad (10)$$

$$\begin{cases} t = \mod(c(i+1) + t, 256); \\ c(i) = \text{sub\_byte}[\mathbf{S2}, \mod(b(i) + c(i+1) + t, 256)]. \end{cases} \quad \text{for } i = L-1, L-2, \ldots, 1 \qquad (11)$$

where, sub_byte[$\mathbf{S2}$, x] denotes byte substitution for x using S-box $\mathbf{S2}$. The second round of encryption is the backward diffusion operation.

Step 6: Transform the 1D vector $\mathbf{C}$ into a 2D matrix with size of $M \times N$, then the cipher image $\mathbf{CI}$ is obtained.

The decryption process is the inverse operation of the encryption process. To recover the plain image $\mathbf{P}$ from the cipher image $\mathbf{CI}$, the operating steps are as follows.

Step 1: Input the secret parameters $\{x_{10}, \mu_1, d_1, x_{20}, \mu_2, d_2, r_0, t_0, m\}$ and the cipher image $\mathbf{CI}$ with the size of $M \times N$, and $\mathbf{CI}$ is reshaped to a 1D pixel array $\mathbf{C} = [c(1), c(2), \ldots, c(L)]$, where $L = M \times N$.

Step 2: Generate the first S-box $\mathbf{S1}$. The operation is exactly the same as Step 2 of the encryption process.

Step 3: Generate the second S-box $\mathbf{S2}$. The operation is exactly the same as Step 3 of the encryption process.

Step 4: Recover the intermediate cipher image pixel array $\mathbf{B} = [b(1), b(2), \ldots, b(L)]$ as

$$\begin{cases} t = t_0; \\ b(L) = \mod(\text{sub\_byte\_1}(\mathbf{S2}, c(L)) - t + 256, 256). \end{cases} \quad \text{for } i = L. \qquad (12)$$

$$\begin{cases} t = \mod(c(i+1) + t, 256) \\ b(i) = \mod(\text{sub\_byte\_1}(\mathbf{S2}, c(i)) - t - c(i+1) + 256, 256) \end{cases} \quad \text{for } i = L-1, L-2, \ldots, 1 \qquad (13)$$

where, sub_byte_1[$\mathbf{S2}$, $\cdot$] denotes the inverse operation of sub_byte[$\mathbf{S2}$, $\cdot$] using S-box $\mathbf{S2}$.

Step 5: Recover the original plain image pixel array $\mathbf{P} = [p(1), p(2), \ldots, p(L)]$ as

$$\begin{cases} j = \mod(1 + m, L) + 1; \\ r = r0; \\ p(j) = \text{sub\_byte\_1}(\mathbf{S1}, \mod(b(1) - r + 256, 256)). \end{cases} \quad \text{for } i = 1. \qquad (14)$$

$$\begin{cases} j = \mod(i + m, L) + 1; \\ r = \mod(b(i-1) + r, 256); \\ p(j) = \text{sub\_byte\_1}(\mathbf{S1}, \mod(b(i) - b(i-1) - r + 256, 256)). \end{cases} \quad \text{for } i = 2, 3, \ldots, L \qquad (15)$$

where, sub_byte_1[$\mathbf{S1}$, $\cdot$] denotes the inverse operation of sub_byte[$\mathbf{S1}$, $\cdot$] using S-box $\mathbf{S1}$.

Step 6: Transform $\mathbf{P}$ into an $M \times N$ matrix, then the decrypted image $\mathbf{PI}$ is obtained.

5. Experimental Results and Security Analyses

To examine the security and efficiency of the proposed cryptosystem, we carry out some simulation experiments. All the algorithms are implemented with MATLAB R2016b run on a Microsoft Windows 7 operating system. The hardware environment is a PC with 3.3 GHz CPU, and 4 GB memory. Without losing generality, we adopted the public test images come from the USC-SIPI Image Database. Test images are 8-bit grayscale images with a size of 256×256, such as Lena, Baboon, Pepper. The all-black and all-white images are also used in the simulation experiments. The secret keys $\{x_{10}, \mu_1, d_1, x_{20}, \mu_2, d_2, r_0, t_0, m\}$ are set as $\{0.21, 0.399, 43, 0.27, 3.999, 241, 98, 200, 129\}$.

5.1. Experimental Results

The original plain images and their corresponding cipher-images are shown in Figures 5 and 6, respectively. While the decrypted images are identical to the corresponding original ones. As can be seen, the cipher-images are completely disordered and unrecognizable. Therefore, our proposed algorithm has a good encryption effect.

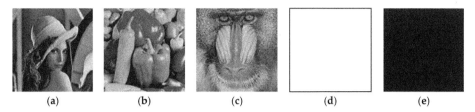

Figure 5. Original plain images. (**a**) The Lena plain image; (**b**) the Peppers plain image; (**c**) the Baboon plain image; (**d**) the all-white image; (**e**) the all-black image.

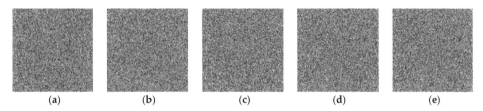

Figure 6. Encrypted cipher images. (**a**) The Lena cipher image; (**b**) the Peppers cipher image; (**c**) the Baboon cipher image; (**d**) the all-white cipher-image; (**e**) the all-black cipher image.

5.2. Key Space Analyses

A secure encryption scheme should have a large key space so as to resist brute-force attack. In our proposed encryption scheme, the secret keys include $\{x_{10}, \mu_1, d_1, x_{20}, \mu_2, d_2, r_0, t_0, m\}$. Among them, $\{x_{10}, \mu_1, x_{20}, \mu_2\}$ are four double-precision real numbers, each of them can reach the accuracy of 15 decimal places. d_1 and d_2 are two odd integers, each of them can have 10^4 different values. r_0 and t_0 are two integers, each of them has 255 different values. m is an integer range from 1 to L, where $L = 65536$. So, the key space of our proposed encryption scheme is $(10^{15\times4+4\times2}) \times 255 \times 255 \times 65536 \approx 2^{258}$, which is a key equivalent to 258 bits in length. Therefore, the key space is large enough to resist brute-force attack.

5.3. Statistical Analysis

5.3.1. Histogram Analysis

A histogram of an image demonstrates the distribution of the image pixel values, and it exposes the pixel distribution characteristics of the image. The more uniform the distribution of the pixel values, the closer the image is to the random signal image. Figure 7 shows the histograms of the above test plain images and cipher images encrypted by our proposed algorithm (the histograms of the all-white and all-black plain images are omitted). It can be seen from Figure 7 that the distributions of pixel values in plain images are clearly not uniform but in cipher images are very uniform.

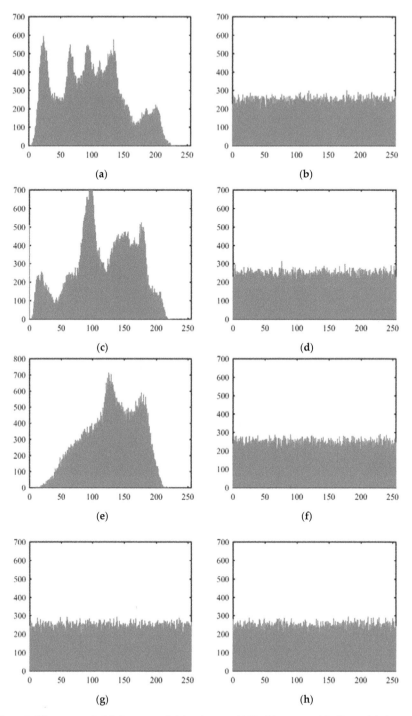

Figure 7. Histograms of plain images and cipher images. (**a**) The histogram of the Lena plain image; (**b**) the histogram of the Lena cipher image; (**c**) the histogram of the Peppers plain image; (**d**) the histogram of the Peppers cipher image; (**e**) the histogram of the Baboon plain image; (**f**) the histogram of the Baboon cipher image; (**g**) the histogram of the all-white cipher image; (**h**) the histogram of the all-black cipher image.

The distribution characteristics of a histogram can also be described quantitatively with the variance of a histogram, which is calculated by [16]

$$\text{var}(\mathbf{Z}) = \frac{1}{n^2} \sum_{i=1}^{n} \sum_{j=1}^{n} \frac{1}{2} (z_i - z_j)^2 \qquad (16)$$

where, n is the number of gray levels of an image, and $n = 256$ for 8-bit gray images. $\mathbf{Z}$ is a vector and $\mathbf{Z} = \{z_1, z_2, \ldots, z_n\}$, z_i and z_j are the numbers of pixels with gray values equal to $(i - 1)$ and $(j - 1)$ respectively. The lower value of variance indicates the higher uniformity of an image. In order to detect the variance values of the above test images and their cipher images, the variances of histograms of the plain images (size of 256 × 256) and their cipher images are calculated by using Equation (16). The results are listed in Table 4. Table 4 also lists the results obtained by the algorithm in References [39] and [40]. The average variance of five cipher images obtained with our proposed algorithm is 256.7125, which is much less than that of Zhang's algorithm [39], Wang's algorithm [40], and Çavuşoğlu's algorithm [57]. Thus, our proposed image encryption algorithm has better performance in resisting statistical attacks.

Table 4. Variances of histograms of the test images.

Images	Plain Image	Cipher Image	Cipher Image [39]	Cipher Image [40]	Cipher Image [57]
Lena	30,665.703	221.195	284.578	283.156	381.688
Peppers	36,379.133	224.234	269.727	227.898	332.898
Baboon	47,799.055	288.664	268.211	277.297	297.625
All-white image	16,711,680	293.039	544.234	41,725.063	1214.484
All-black image cipher image	16,711,680	256.430	1396.765	43,233.188	1214.484
Average	6,707,640.778	256.713	552.703	17,149.320	688.236

5.3.2. Correlation Analysis

Natural images usually have a strong correlation with adjacent pixels. An efficient encryption algorithm should reduce the correlation in cipher images. In order to exhibit the correlation strength intuitively, we randomly selected 2000 pairs of pixel along a certain direction (horizontal or vertical or diagonal) from an image to draw the correlation distribution diagram. Figure 8 shows the correlation distribution diagrams of the Lena plain and cipher image encrypted by our encryption algorithm. The abscissa and ordinate values at any point in the graph represent the values of a pair of neighbor pixels, respectively. For plaintext images, most of the points in the graph are distributed near a straight line with an inclination of 45 degrees. That is to say, the abscissa and ordinate coordinates of most points are basically equal, indicating that the pixel values of neighboring points in plaintext images are basically equal. However, the pixel values of each group of neighbor points in ciphertext images are not equal. The results confirm that the correlation among the adjacent pixels is reduced greatly by our proposed encryption algorithm.

To illustrate quantitatively the correlation of adjacent pixels in an image, we can calculate the correlation coefficient r_{XY} by using N pairs of an adjacent pixel. r_{XY} is defined as

$$r_{XY} = \text{cov}(X, Y) / \sqrt{D(X)} \sqrt{D(Y)} \qquad (17)$$

where, $X = \{x_1, x_2, \ldots, x_N\}$ and $Y = \{y_1, y_2, \ldots, y_N\}$, (x_i, y_i) is the i-th pairs of the adjacent pixel gray-scale values, and

$$D(X) = \frac{1}{N} \sum_{i=1}^{N} (x_i - \overline{X})^2, D(Y) = \frac{1}{N} \sum_{i=1}^{N} (x_i - \overline{Y})^2 \qquad (18)$$

$$\text{cov}(X, Y) = \frac{1}{N}\sum_{i=1}^{N}\left(x_i - \overline{X}\right)\left(y_i - \overline{Y}\right) \tag{19}$$

$$\overline{X} = \frac{1}{N}\sum_{i=1}^{N}x_i, \overline{Y} = \frac{1}{N}\sum_{i=1}^{N}y_i \tag{20}$$

Three types of correlation coefficients of adjacent pixels in the Lena plain and cipher image are calculated, respectively. Correlation coefficients of the Lena plain images are as: 0.9567 (horizontal direction), 0.9239 (vertical direction), 0.8888 (diagonal direction), showing that correlation coefficients of adjacent pixels in the Lena plain image are very high (all close to one). Results of the Lena cipher image are listed in Table 5. From Table 5, we can see that the correlation coefficients of adjacent pixels in the Lena cipher image are very low (all close to zero). Table 5 also lists the correlation coefficients of the Lena cipher image encrypted with Zhang's algorithm, Wang's algorithm and Çavuşoğlu's algorithm. The experimental results show that our proposed algorithm has the smallest absolute values of the correlation coefficient among the three algorithms, showing the best scrambling effect.

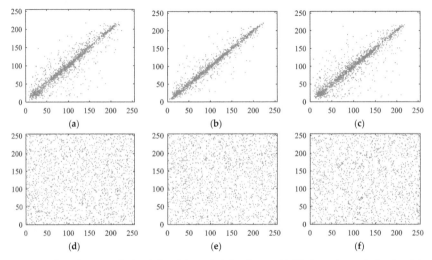

Figure 8. Correlation analysis of the plain and cipher Lena. (**a**) Horizontal correlation in plain image Lena; (**b**) vertical correlation in plain image Lena; (**c**) diagonal correlation in plain image Lena; (**d**) horizontal correlation in cipher image Lena; (**e**) vertical correlation in cipher image Lena; (**f**) diagonal correlation in cipher image Lena.

Table 5. Correlation coefficients of the Lena cipher images encrypted by different algorithms.

Algorithms	Horizontal	Vertical	Diagonal
The proposed algorithm	−0.002088	0.000312	0.001444
Zhang's algorithm [39]	−0.000582	0.001336	−0.004690
Wang's algorithm [40]	0.006057	0.012468	−0.006030
Çavuşoğlu's algorithm [57]	0.001640	0.031372	−0.000626

5.3.3. Information Entropy Analysis

Information entropy can be used to describe the degree of randomness or uncertainty of signals. The information entropy $H(m)$ of an image is calculated by

$$H(m) = -\sum_{i=0}^{2^n-1} P(m_i) \log_2[P(m_i)] \tag{21}$$

where $P(m_i)$ denotes the occurrence probability of the gray level i, and $i = 0, 1, 2, \ldots, 2^n$. Here, 2^n is the number of grayscale levels of an image. If each m_i has the same occurrence probability in an image, then $P(m_i) = 1/2^n$, then the image is completely random with $H(m) = n$. For an image with 256 gray-scale levels, $n = 8$, so, the information entropy of a completely random 8-bit gray image is eight. A good encryption algorithm should make the information entropy of its cipher image close to eight. We calculated the information entropy values of several cipher images obtained by four different encryption algorithms. The results are listed in Table 6. All the images have the same size of 256×256. From Table 6, one can see that all the entropy values are significantly closer to eight, so the randomness is satisfactory. Among these four algorithms, our proposed algorithm has the largest average entropy value, showing the best randomness of the cipher image encrypted by our proposed algorithm.

Table 6. Information entropy values of several cipher images obtained by different algorithms.

Test Images	Ref. [39]	Ref. [40]	Ref. [57]	Ours
Lena cipher image	7.9969	7.9969	7.9958	7.9976
Peppers cipher image	7.9970	7.9975	7.9963	7.9975
Baboon cipher image	7.9970	7.9969	7.9967	7.9968
All-black cipher image	7.9846	7.3901	7.9871	7.9972
All-white cipher image	7.9940	7.3998	7.9871	7.9968

5.3.4. Sensitivity Analysis

(1) Sensitivity to plain images

A secure encryption algorithm should be sensitive to the change of the plain image so as to resist the differential attack. To measure the sensitivity of an algorithm to tiny changes in a plain image, the number of pixels changing rate (NPCR) and the unified average changing intensity (UACI) are introduced. The NPCR and UACI are calculated by Equations (22)–(24).

$$\text{NPCR} = \frac{1}{M \times N} \sum_{i=1}^{M} \sum_{j=1}^{N} \delta(i,j) \times 100\% \tag{22}$$

$$\text{UACI} = \frac{1}{M \times N} \left(\sum_{i=1}^{M} \sum_{j=1}^{N} \frac{|c_1(i,j) - c_2(i,j)|}{255} \right) \times 100\% \tag{23}$$

where, M, N represent the number of rows and columns of an image, respectively. $C_1 = [c_1(i,j)]$ and $C_2 = [c_2(i,j)]$ express two encrypted images corresponding to two plain images with a tiny difference, and $\delta(i,j)$ is computed by

$$\delta(i,j) = \begin{cases} 1, & if \; c_1(i,j) \neq c_2(i,j), \\ 0, & if \; c_1(i,j) = c_2(i,j). \end{cases} \tag{24}$$

The larger the values of NPCR and UACI, the stronger the sensitivity of the algorithm to plaintext. For the best case, the ideal average value of NPCR is about 99.61%, and the ideal average value of UACI is about 33.46% [16].

To measure the sensitivity of our improved algorithm to the plain image, the original Lena gray image (size of 256×256) is adopted as the first plain image, and the second plain image is obtained by changing only one pixel of the first plain image. To obtain two cipher images C_1 and C_2 by executing the proposed encryption algorithm with the same secret keys, respectively. Then NPCR and UACI are computed with two cipher images, and the results are listed in Table 7. Table 7 also lists the results obtained by using the Zhang's, Wang's and Çavuşoğlu's algorithm. The results indicate that our proposed encryption algorithm is very sensitive to the plain image, and its sensitivity is better than those of Zhang's and Wang's algorithm.

Table 7. Values of number of pixels changing rate (NPCR) and unified average changing intensity (UACI) of Lena cipher images.

Position i	Values	Zhang's [39]	Wang's [40]	Çavuşoğlu's [57]	Ours
1	NPCR(%)	49.81	1.53×10^{-3}	1.53×10^{-3}	99.64
1	UACI(%)	16.86	1.14×10^{-3}	2.75×10^{-4}	33.55
L/4	NPCR(%)	74.69	1.53×10^{-3}	1.53×10^{-3}	99.59
L/4	UACI(%)	25.08	1.68×10^{-4}	8.26×10^{-4}	33.25
L/2	NPCR(%)	99.64	1.53×10^{-3}	1.53×10^{-3}	99.57
L/2	UACI(%)	33.54	6.10×10^{-4}	4.13×10^{-4}	33.41
L	NPCR(%)	49.84	1.53×10^{-3}	1.53×10^{-3}	99.62
L	UACI(%)	16.72	8.80×10^{-4}	8.62×10^{-4}	33.46

(2) Sensitivity to Secret Keys

A secure encryption algorithm should also be sensitive to the change of secret keys. That is to say, when secret keys change slightly, the cipher image should change dramatically. NPCR and UACI can also be used to measure the sensitivity of an encryption algorithm to secret keys. In our simulation tests, two groups of secret keys with a tiny difference are used to encrypt the same plain image Lena and two cipher images, C_1 and C_2, are obtained. The tiny change (to a float number is 10^{-15}, or to an integer number is one) is introduced to one of the secret keys ($x_{10}, \mu_1, d_1, x_{20}, \mu_2, d_2, r_0, t_0, m$) while keeping all the others unchanged. The NPCR and UACI of the cipher images C_1 and C_2 are calculated and listed in Table 8. The experimental results indicate that our proposed algorithm is very sensitive to a slight change in any secret key.

Table 8. NPCR and UACI of the proposed algorithm with a tiny difference in one of the secret keys.

Values	$\Delta x_{10} = 10^{-15}$	$\Delta \mu_1 = 10^{-15}$	$\Delta x_{20} = 10^{-15}$	$\Delta \mu_2 = 10^{-15}$	$\Delta d_1 = 1$	$\Delta d_2 = 1$	$\Delta r_0 = 1$	$\Delta t_0 = 1$	$\Delta m = 1$
NPCR(%)	99.63	99.62	99.56	99.62	99.61	99.58	99.63	99.61	99.61
UACI(%)	33.53	33.34	33.50	33.41	33.38	33.53	33.46	33.41	33.37

5.4. Analysis of Anti-Attack Performance

5.4.1. Classical Types of Attacks

According to Kerchoff's hypothesis, it is usually assumed that the cryptanalysts or opponents know the cryptosystem, and the security entirely depends on the secret key. A secure cryptosystem should resist all kinds of attacks; otherwise, the cryptosystem is insecure. Generally speaking, there are four classical types of attacks to break a cryptosystem, and their orders from the hardest types to the easiest types are listed as follows.

(1) Ciphertext-only attack: The cryptanalyst possesses one or more ciphertexts.

(2) Known-plaintext attack: The cryptanalyst has some plaintexts and the corresponding ciphertexts.

(3) Chosen-plaintext attack: The cryptanalyst has the opportunity to use the encryption machinery, so he or she can choose some plaintext and generate ciphertext.

(4) Chosen-ciphertext attack: The cryptanalyst has the opportunity to use the cryptograph, so he or she can choose some ciphertexts and generate plaintexts.

Among the four classical attack types mentioned above, the chosen-ciphertext attack is the most powerful attack. If a cryptosystem can resist this attack, it can resist other types of attacks.

In our proposed scheme, {**S1**, **S2**, **r**, **t**} become the equivalent keys to the original keys. It is not difficult to understand the following conclusions from the encryption formulas of Equations (8)–(11). First, it is difficult for an attacker to decipher the above equivalent keys even if he or she obtains known plaintext-ciphertext pairs $(p(i), c(i))$. Second, the equivalent keys **r** and **t** are updated before encrypting the i-th pixel and they are related with the intermediate ciphertext $b(i-1)$ or the final ciphertext $c(i+1)$. It means that a different cipher image will yield different sequences of {**r**, **t**}. Even if the attacker cracked the key sequences of {**r**, **t**} with some specially chosen-ciphertext, the key streams of {**r**, **t**} cannot be used to decrypt the target cipher image due to the key streams of the target cipher image that are different from the cracked key streams. Moreover, it is difficult to decipher the key streams {**r**, **t**} directly by using the chosen-ciphertext attack. Therefore, the proposed scheme can well resist the chosen-ciphertext attack and can resist the four classical types of attacks.

5.4.2. Analysis of Robustness against Noise and Occlusion

In order to resist the differential cryptanalysis attack brought by the opponent, a strong diffusion mechanism is introduced into the proposed encryption algorithm. As a result, the ciphertext is sensitive to the noise of the transmission channel, so the algorithm lacks robustness to noise and occlusion. However, the lack of such robustness also makes it impossible for the opponent to decipher the plaintext accurately, which can ensure that the confidentiality of the image content is protected. As for how to make the encrypted image not only resist differential attack, but also withstand a certain degree of noise, we consider introducing an error correction mechanism in channel coding and decoding. This is worthy of further study in the future.

5.5. Analysis of Speed

In addition to security performance, a practical cryptosystem should also have faster encryption speed. To evaluate the encryption efficiency of the proposed algorithm, the 8-bit greyscale images with a size of 256 × 256 and 512 × 512 are encrypted. And the same type of S-box based image encryption algorithms proposed by Zhang [39], Wang [40], and Çavuşoğlu [57] are also implemented on the same hardware and software platform mentioned at the beginning of Section 5. The average values of the encryption/decryption time taken by Zhang's algorithm, Wang's algorithm, Çavuşoğlu's algorithm and our proposed algorithm are shown in Table 9, respectively. The experimental results show the advantages of the proposed algorithm in time efficiency.

Table 9. The time cost tests (unit: s).

Image Size	Ref. [39]	Ref. [40]	Ref. [57]	This Paper
256 × 256	1.205	1.256	0.823	0.464
512 × 512	4.750	4.828	3.253	1.708

Our proposed algorithm has an execution time that includes: Two S-boxes generated by a novel simple method using the 1D discrete chaotic map, 2*L* times of byte substitution and 2*L* times mod 256 addition operations. Zhang's algorithm execution time include: Two S-boxes generated by an ordinary method using the 1D discrete chaotic map, 2*L* times of byte substitution, *L* times mod 256 addition operations and *L* times bitXor operations. Wang's algorithm has an execution time that includes: Three S-boxes generated by an ordinary method using the 3D continuous-time chaotic system, *L* times of byte substitution, *L* times mod 3 addition operations and *L* times bitXor operations. Çavuşoğlu's algorithm has an execution time that includes: One S-box generated by an ordinary method using the 3D continuous-time chaotic system, *L* times of byte substitution and *L* times bitXor operations. The mod addition operation has a less execution time than the bitXor operation, and the bitXor operation has a less execution time than the byte substitution operation. Our algorithm to generate the S-box has the least execution time among the four algorithms. As the result, the total execution time of our algorithm is the smallest one among the four algorithms.

6. Conclusions

In this paper, an efficient and secure image encryption scheme is presented. The main contributions of this paper are as follows: First, a new compound chaotic system, the Sine-Tent map, is proposed, which has wider chaotic range and better chaotic performance than any of the old one. And the new compound chaotic system is more suitable for cryptosystem. Second, an efficient and secure method for generating S-boxes is proposed, which has less execution time than the other ones. Third, a novel double S-boxes based image encryption algorithm is proposed. By introducing equivalent key sequences {r, t} related with image ciphertext, the proposed cryptosystem can resist the four classical types of attacks, which is an advantage over other S-box based encryption schemes. It overcomes the security defects of some old S-box based encryption algorithms. In addition, two rounds of forward and backward confusion-diffusion operation enhance the sensitivity of the algorithm. The simulation results and security analysis verify the effectiveness of the proposed scheme. The new scheme has obvious efficiency advantages, which means that it has better application potential in real-time image encryption. The proposed scheme is also suitable to color images by connecting three color channels of color image into gray image.

As for the research of the chaotic image encryption, there are two aspects worthy of further study in the future. First, we need to explore new security evaluation criteria to make up for the shortcomings of empirical security standards. Second, in order to ensure that the encryption system is not only resistant to differential cryptanalysis attacks, but also robust to noise, it may be an effective solution to introduce error-correcting codes in the process of cryptography and decoding.

Author Contributions: Conceptualization, C.Z. and S.Z.; Methodology, G.W.; Software, S.Z.; Validation, C.Z., S.Z. and G.W.; Formal analysis, C.Z.; Investigation, C.Z.; Resources, C.Z.; Data curation, S.Z.; Writing—Original draft preparation, S.Z.; Writing—Review and editing, C.Z. and G.W.; Visualization, C.Z.; Supervision, G.W.; Project administration, C.Z.; Funding acquisition, G.W.

Funding: This research was funded by [the National Natural Science Foundation of China] grant number [No. 61632009] and [Guangdong Provincial Natural Science Foundation] grant number [No. 2017A030308006].

Acknowledgments: The authors are thankful to the reviewers for their comments and suggestions to improve the quality of the manuscript.

Conflicts of Interest: The authors declare no conflict of interest.

References

1. Wang, J.; Ding, Q. Dynamic rounds chaotic block cipher based on keyword abstract extraction. *Entropy* **2018**, *20*, 693. [CrossRef]
2. Abdallah, E.E.; Ben Hamza, A.; Bhattacharya, P. Mpeg video watermarking using tensor singular value decomposition. In *Lecture Notes in Computer Science, Proceedings of the Image Analysis and Recognition (ICIAR 2007), Montreal, QC, Canada, 5–7 July 2017*; Kamel, M., Campilho, A., Eds.; Springer: Berlin/Heidelberg, Germany, 2007; Volume 4633, pp. 772–783.
3. Abdallah, E.E.; Ben Hamza, A.; Bhattacharya, P. Video watermarking using wavelet transform and tensor algebra. *Signal Image Video Process.* **2009**, *4*, 233–245. [CrossRef]
4. Zhang, S.; Li, X.; Tan, Z.; Peng, T.; Wang, G. A caching and spatial k-anonymity driven privacy enhancement scheme in continuous location-based services. *Future Gener. Comput. Syst.* **2019**, *94*, 40–50. [CrossRef]
5. Zhang, S.; Wang, G.; Bhuiyan, M.Z.A.; Liu, Q. A dual privacy preserving scheme in continuous location-based services. *IEEE Internet Things J.* **2018**, *5*, 4191–4200. [CrossRef]
6. Wang, X.; Pham, V.-T.; Jafari, S.; Volos, C.; Munoz-Pacheco, J.M.; Tlelo-Cuautle, E. A new chaotic system with stable equilibrium: From theoretical model to circuit implementation. *IEEE Access* **2017**, *5*, 8851–8858. [CrossRef]
7. Zhou, Y.; Bao, L.; Chen, C.L.P. A new 1D chaotic system for image encryption. *Signal Process.* **2014**, *97*, 172–182. [CrossRef]

8. Chen, E.; Min, L.; Chen, G. Discrete chaotic systems with one-line equilibria and their application to image encryption. *Int. J. Bifurc. Chaos* **2017**, *27*, 1750046. [CrossRef]
9. Zhu, S.; Zhu, C.; Cui, H.; Wang, W. A class of quadratic polynomial chaotic maps and its application in cryptography. *IEEE Access* **2019**, *7*, 34141–34152. [CrossRef]
10. Sahari, M.L.; Boukemara, I. A pseudo-random numbers generator based on a novel 3d chaotic map with an application to color image encryption. *Nonlinear Dyn.* **2018**, *94*, 723–744. [CrossRef]
11. Murillo-Escobar, M.A.; Cruz-Hernandez, C.; Cardoza-Avendano, L.; Mendez-Ramirez, R. A novel pseudorandom number generator based on pseudorandomly enhanced logistic map. *Nonlinear Dyn.* **2017**, *87*, 407–425. [CrossRef]
12. Islam, F.u.; Liu, G. Designing s-box based on 4D-4wing hyperchaotic system. *3D Res.* **2017**, *8*, 9. [CrossRef]
13. Alvarez, G.; Li, S. Some basic cryptographic requirements for chaos-based cryptosystems. *Int. J. Bifurc. Chaos* **2006**, *16*, 2129–2151. [CrossRef]
14. Li, C.; Lin, D.; Feng, B.; Lu, J.; Hao, F. Cryptanalysis of a chaotic image encryption algorithm based on information entropy. *IEEE Access* **2018**, *6*, 75834–75842. [CrossRef]
15. Zhu, C.; Wang, G.; Sun, K. Improved cryptanalysis and enhancements of an image encryption scheme using combined 1d chaotic maps. *Entropy* **2018**, *20*, 843. [CrossRef]
16. Zhu, C.; Sun, K. Cryptanalyzing and improving a novel color image encryption algorithm using rt-enhanced chaotic tent maps. *IEEE Access* **2018**, *6*, 18759–18770. [CrossRef]
17. Fridrich, J. Symmetric ciphers based on two-dimensional chaotic maps. *Int. J. Bifurc. Chaos* **1998**, *8*, 1259–1284. [CrossRef]
18. Zhang, X.; Fan, X.; Wang, J.; Zhao, Z. A chaos-based image encryption scheme using 2D rectangular transform and dependent substitution. *Multimed. Tools Appl.* **2014**, *75*, 1745–1763. [CrossRef]
19. Zhang, Y.; Xiao, D. Double optical image encryption using discrete chirikov standard map and chaos-based fractional random transform. *Opt. Lasers Eng.* **2013**, *51*, 472–480. [CrossRef]
20. Gan, Z.-h.; Chai, X.-l.; Han, D.-j.; Chen, Y.-r. A chaotic image encryption algorithm based on 3-d bit-plane permutation. *Neural Comput. Appl.* **2018**, *2018*, 1–20. [CrossRef]
21. Hu, G.; Xiao, D.; Zhang, Y.; Xiang, T. An efficient chaotic image cipher with dynamic lookup table driven bit-level permutation strategy. *Nonlinear Dyn.* **2016**, *87*, 1359–1375. [CrossRef]
22. Ye, G.; Zhao, H.; Chai, H. Chaotic image encryption algorithm using wave-line permutation and block diffusion. *Nonlinear Dyn.* **2016**, *83*, 2067–2077. [CrossRef]
23. Abd-El-Hafiz, S.K.; AbdElHaleem, S.H.; Radwan, A.G. Novel permutation measures for image encryption algorithms. *Opt. Lasers Eng.* **2016**, *85*, 72–83. [CrossRef]
24. Li, Y.; Wang, C.; Chen, H. A hyper-chaos-based image encryption algorithm using pixel-level permutation and bit-level permutation. *Opt. Lasers Eng.* **2017**, *90*, 238–246. [CrossRef]
25. Zhang, Y.; Xiao, D.; Shu, Y.; Li, J. A novel image encryption scheme based on a linear hyperbolic chaotic system of partial differential equations. *Signal Process. Image Commun.* **2013**, *28*, 292–300. [CrossRef]
26. Wang, X.; Liu, C.; Zhang, H. An effective and fast image encryption algorithm based on chaos and interweaving of ranks. *Nonlinear Dyn.* **2016**, *84*, 1595–1607. [CrossRef]
27. Xu, L.; Gou, X.; Li, Z.; Li, J. A novel chaotic image encryption algorithm using block scrambling and dynamic index based diffusion. *Opt. Lasers Eng.* **2017**, *91*, 41–52. [CrossRef]
28. Hua, Z.; Yi, S.; Zhou, Y. Medical image encryption using high-speed scrambling and pixel adaptive diffusion. *Signal Process.* **2018**, *144*, 134–144. [CrossRef]
29. Huang, H.; He, X.; Xiang, Y.; Wen, W.; Zhang, Y. A compression-diffusion-permutation strategy for securing image. *Signal Process.* **2018**, *150*, 183–190. [CrossRef]
30. Cao, C.; Sun, K.; Liu, W. A novel bit-level image encryption algorithm based on 2D-LICM hyperchaotic map. *Signal Process.* **2018**, *143*, 122–133. [CrossRef]
31. Chai, X. An image encryption algorithm based on bit level brownian motion and new chaotic systems. *Multimed. Tools Appl.* **2017**, *76*, 1159–1175. [CrossRef]
32. Hua, Z.; Jin, F.; Xu, B.; Huang, H. 2D Logistic-Sine-coupling map for image encryption. *Signal Process.* **2018**, *149*, 148–161. [CrossRef]
33. Hua, Z.; Zhou, Y. Image encryption using 2D Logistic-adjusted-Sine map. *Inf. Sci.* **2016**, *339*, 237–253. [CrossRef]
34. Kaur, M.; Kumar, V. Efficient image encryption method based on improved Lorenz chaotic system. *Electron. Lett.* **2018**, *54*, 562–564. [CrossRef]

35. Liu, J.; Yang, D.; Zhou, H.; Chen, S. A digital image encryption algorithm based on bit-planes and an improved Logistic map. *Multimed. Tools Appl.* **2018**, *77*, 10217–10233. [CrossRef]
36. Zhu, C. A novel image encryption scheme based on improved hyperchaotic sequences. *Opt. Commun.* **2012**, *285*, 29–37. [CrossRef]
37. Zhang, Y. The image encryption algorithm based on chaos and DNA computing. *Multimed. Tools Appl.* **2018**, *77*, 21589–21615. [CrossRef]
38. Farwa, S.; Shahy, T.; Muhammad, N.; Bibiz, N.; Jahangir, A.; Arshad, S. An image encryption technique based on chaotic S-box and Arnold transform. *Int. J. Adv. Comput. Sci. Appl.* **2017**, *8*, 360–364. [CrossRef]
39. Zhang, X.-P.; Guo, R.; Chen, H.-W.; Zhao, Z.-M.; Wang, J.-Y. Efficient image encryption scheme with synchronous substitution and diffusion based on double S-boxes. *Chin. Phys. B* **2018**, *27*, 080701. [CrossRef]
40. Wang, X.; Çavuşoğlu, Ü.; Kacar, S.; Akgul, A.; Pham, V.-T.; Jafari, S.; Alsaadi, F.; Nguyen, X. S-box based image encryption application using a chaotic system without equilibrium. *Appl. Sci.* **2019**, *9*, 781. [CrossRef]
41. Zhu, S.; Zhu, C.; Wang, W. A new image encryption algorithm based on chaos and secure hash SHA-256. *Entropy* **2018**, *20*, 716. [CrossRef]
42. Zhu, S.; Zhu, C.; Wang, W. A novel image compression-encryption scheme based on chaos and compression sensing. *IEEE Access* **2018**, *6*, 67095–67107. [CrossRef]
43. Zhu, S.; Zhu, C. Image encryption algorithm with an avalanche effect based on a six-dimensional discrete chaotic system. *Multimed. Tools Appl.* **2018**, *77*, 29119–29142. [CrossRef]
44. Sun, S.; Guo, Y.; Wu, R. A novel image encryption scheme based on 7D hyperchaotic system and row-column simultaneous swapping. *IEEE Access* **2019**, *7*, 28539–28547. [CrossRef]
45. Zhang, S.; Wang, G.; Liu, Q.; Abawajy, J.H. A trajectory privacy-preserving scheme based on query exchange in mobile social networks. *Soft Comput.* **2018**, *22*, 6121–6133. [CrossRef]
46. Bhuiyan, M.Z.A.; Wang, G.; Wu, J.; Cao, J.; Liu, X.; Wang, T. Dependable structural health monitoring using wireless sensor networks. *IEEE Trans. Dependable Secur. Comput.* **2017**, *14*, 363–376. [CrossRef]
47. Zhang, Q.; Liu, Q.; Wang, G. PRMS: A personalized mobile search over encrypted outsourced data. *IEEE Access* **2018**, *6*, 31541–31552. [CrossRef]
48. Sun, K.-H.; He, S.-B.; Yin, L.-Z.; Li-Kun, A.D.-L.D. Application of fuzzyen algorithm to the analysis of complexity of chaotic sequence. *Acta Phys. Sin.* **2012**, *61*, 130507.
49. Sun, K.-H.; He, S.-B.; He, Y.; Yin, L.-Z. Complexity analysis of chaotic pseudo-random sequences based on spectral entropy algorithm. *Acta Phys. Sin.* **2013**, *62*, 010501.
50. He, S.-B.; Sun, K.-H.; Zhu, C.-X. Complexity analyses of multi-wing chaotic systems. *Chin. Phys. B* **2013**, *22*, 050506. [CrossRef]
51. Khan, M. A novel image encryption scheme based on multiple chaotic S-boxes. *Nonlinear Dyn.* **2015**, *82*, 527–533. [CrossRef]
52. Wang, X.; Wang, Q. A novel image encryption algorithm based on dynamic S-boxes constructed by chaos. *Nonlinear Dyn.* **2013**, *75*, 567–576. [CrossRef]
53. Murillo-Escobar, M.A.; Cruz-Hernández, C.; Abundiz-Pérez, F.; López-Gutiérrez, R.M.; Acosta Del Campo, O.R. A rgb image encryption algorithm based on total plain image characteristics and chaos. *Signal Process.* **2015**, *109*, 119–131. [CrossRef]
54. Zhang, Y.; Tang, Y. A plaintext-related image encryption algorithm based on chaos. *Multimed. Tools Appl.* **2017**, *77*, 6647–6669. [CrossRef]
55. Preishuber, M.; Hutter, T.; Katzenbeisser, S.; Uhl, A. Depreciating motivation and empirical security analysis of chaos-based image and video encryption. *IEEE Trans. Inf. Forensics Secur.* **2018**, *13*, 2137–2150. [CrossRef]
56. Belazi, A.; El-Latif, A.A.A. A simple yet efficient S-box method based on chaotic Sine map. *Optik* **2017**, *130*, 1438–1444. [CrossRef]
57. Çavuşoğlu, Ü.; Kaçar, S.; Pehlivan, I.; Zengin, A. Secure image encryption algorithm design using a novel chaos based S-box. *Chaos Solitons Fractals* **2017**, *95*, 92–101. [CrossRef]
58. Zhu, C.X.; Wang, G.J.; Sun, K.H. Cryptanalysis and improvement on an image encryption algorithm design using a novel chaos based S-box. *Symmetry* **2018**, *10*, 399. [CrossRef]

© 2019 by the authors. Licensee MDPI, Basel, Switzerland. This article is an open access article distributed under the terms and conditions of the Creative Commons Attribution (CC BY) license (http://creativecommons.org/licenses/by/4.0/).

Article

High-Payload Data-Hiding Method for AMBTC Decompressed Images

Jung-Yao Yeh [1], Chih-Cheng Chen [2], Po-Liang Liu [1,*] and Ying-Hsuan Huang [3]

1 Graduate Institute of Precision Engineering, National Chung Hsing University, No. 250, Kuo-Kuang Road, Taichung City 402, Taiwan; honor0425@gmail.com
2 Department of Computer Science and Engineering, National Chung Hsing University, No. 250, Kuo-Kuang Road, Taichung City 402, Taiwan; salu.chen@gmail.com
3 Aeronautical Systems Research Division, National Chung-Shan Institute of Science and Technology, Taichung 40722, Taiwan; ying.hsuan0909@gmail.com
* Correspondence: pliu@dragon.nchu.edu.tw

Received: 19 December 2019; Accepted: 23 January 2020; Published: 25 January 2020

Abstract: Data hiding is the art of embedding data into a cover image without any perceptual distortion of the cover image. Moreover, data hiding is a very crucial research topic in information security because it can be used for various applications. In this study, we proposed a high-capacity data-hiding scheme for absolute moment block truncation coding (AMBTC) decompressed images. We statistically analyzed the composition of the secret data string and developed a unique encoding and decoding dictionary search for adjusting pixel values. The dictionary was used in the embedding and extraction stages. The dictionary provides high data-hiding capacity because the secret data was compressed using dictionary-based coding. The experimental results of this study reveal that the proposed scheme is better than the existing schemes, with respect to the data-hiding capacity and visual quality.

Keywords: data hiding; AMBTC; steganography; stego image; dictionary-based coding; pixel value adjusting

---

1. Introduction

The concealment of information within media files is commonly used in various applications. This process originates from the hieroglyphs used in the Egyptian civilization. Other cultures, such as the Chinese culture, adopted a more physical approach to hide messages by writing them on silk or paper, rolling the material into a ball, and covering the material with wax to communicate political or military secrets. Data hiding is nearly indispensable for every aspect in our daily lives whether for good or evil intentions.

Due to its rapid growth, the Internet has recently become far more popular than traditional media. Data is accessible by everyone due to the popularity of the Internet. Therefore, possessing the capabilities of detecting copyright violations, forgery, and fraud is crucial. Many techniques, such as steganography and cryptography, have been designed to secure digital data. The difference between steganography and cryptography is as follows: In cryptography (e.g., chaos-based encrypted systems, secure pseudo-random number generator, etc. [1]) users are aware that there is an encrypted image, but they cannot efficiently decode the encrypted image unless they know the proper key. In steganography, users can easily decode the encrypted message, but most people do not notice that there is an encrypted message. In this study, we focused on the techniques used for hiding data in images.

The schemes present for hiding data in an image can be broadly classified into two categories, irreversible data-hiding schemes [2–4] and reversible data-hiding schemes [5–7]. In the irreversible data-hiding schemes, a recipient can extract the secret information. However, the original image cannot be recovered after extracting the secret information. In the reversible data-hiding schemes, the hidden data can be extracted from the image, and the original image can be retrieved from a stego image without any distortion. Two factors affect a data-hiding scheme, i.e., visual quality and embedding payload. A high-quality data-hiding scheme should not raise any suspicions of adversaries. Therefore, this type of scheme should provide low image distortion and high payload.

To decrease the size of a digital image file or accelerate the transmission, a data-hiding scheme that employs a compressed image should be developed. Many compressed file formats have been proposed, such as JPEG and JPEG2000. Wang et al. [8] proposed a lossless data-hiding method for JPEG images by using adaptive embedding. Lee et al. [9] proposed a scheme in which a secret image was compressed using JPEG2000 and then, embedded in the cover image by using tri-way pixel value differencing. Nevertheless, both JPEG and JPEG2000 need complicated computation for image compression and decompression.

Another popular technique used for image compression is block truncation coding (BTC) [10]. Compared with the methods using JPEG and JPEG2000, BTC is a simple and efficient encoding technique that is used for image compression. Therefore, the computation cost is relatively low when a data-hiding scheme is based on BTC.

Lema and Mitchell [11] proposed the absolute moment BTC (AMBTC) technique to improve the compression performance of BTC. When AMBTC is used, the first absolute moment is maintained with the mean. To exploit the advantages of AMBTC compression, we proposed an AMBTC decompressed image-based data-hiding scheme by using a pixel adjusting strategy.

The basic idea of the proposed study is to preliminarily calculate the probability of secret data and then select the best codebook for embedding the secret data. The secret data are embedded into the AMBTC compression image by modifying the pixel value according to the codebook. Experimental results reveal that the proposed scheme is almost better than the current state of the art method in terms of the hiding capacity.

The remainder of the paper is organized as follows: Section 2 describes the relevant approaches such as the BTC and AMBTC techniques for data hiding; Section 3 describes the implementation flow of the scheme proposed for data hiding; Section 4 discusses several experimental results are presented, and some issues; and finally, Section 5 specifies the conclusions and future work.

2. Related Works

Before describing the high data-hiding capacity of the proposed scheme, we review the AMBTC technique and some recently developed AMBTC-based data-hiding methods.

2.1. Absolute Moment Block Truncation Coding (AMBTC)

BTC, a simple and efficient block-based lossy image compression method, is used for grayscale images. Although the BTC method provides a low compression ratio, it is a popular image compression method because of its low complexity with respect to both computation and implementation. In the BTC algorithm, an image X, with $M \times N$ pixels, is divided into nonoverlapping blocks. Each block has $n \times n$ pixels, and the pixel values can be different. The mean and standard deviation of each pixel value are calculated before conducting BTC. In general, two statistical characteristics change from one block to another.

The hardware implementation of BTC is challenging because the square and square root functions are involved. To resolve this problem, AMBTC [11] was proposed as a type of BTC. The AMBTC uses the first absolute moment and mean values instead of using the standard deviation value. The main difference between AMBTC and BTC is that the mean and standard deviation values of a block are preserved in BTC. However, in AMBTC, the high mean and low mean values of a block are preserved.

As in BTC, an image X is divided into nonoverlapping blocks with $n \times n$ pixels also in the AMBTC encoding phase. For each block, the mean $\bar{x}$ and the absolute moment a of the pixel values are calculated using

$$\bar{x} = \frac{1}{m} \sum_{i=1}^{m} x_i, \tag{1}$$

$$a = \frac{1}{m} \sum_{i=1}^{m} |x_i - \bar{x}|. \tag{2}$$

Note that $m = n \times n$.

The pixel value x_i is compared with the mean $\bar{x}$ for composing a bit plane for each pixel in the block. If the pixel value x_i is greater than the mean $\bar{x}$, then x_i is denoted as 1. Otherwise, the pixel value is denoted as 0. The equation of bit representation is

$$P_i = \begin{cases} 1, & \text{if } x_i > \bar{x}, \\ 0, & \text{otherwise.} \end{cases} \tag{3}$$

In the AMBTC-compressed block reconstruction phase, the block reconstruction is conducted using two values L_m and H_m. The values of L_m and H_m are computed using

$$L_m = \bar{x} - \frac{ma}{2(m-q)}, \tag{4}$$

$$H_m = \bar{x} + \frac{ma}{2q}. \tag{5}$$

In Equations (4) and (5), q represents the number of pixels with pixel values greater than $\bar{x}$. Thus, a compressed block has two values L_m and H_m, where L_m is the low mean value and H_m is the high mean value. To reconstruct a block, the pixels that are assigned the value of 0 in the bit plane are replaced with the L_m value, and the pixels assigned the value of 1 in the bit plane are replaced with the H_m value by

$$x'_i = \begin{cases} H_m, & \text{if } p_i = 1, \\ L_m, & \text{if } p_i = 0. \end{cases} \tag{6}$$

2.2. Related Work of BTC and AMBTC Based Data Hiding Schemes

BTC has significantly low complexity and requires less memory. Therefore, BTC is a good scheme for data hiding. Chuang and Chang proposed a data-hiding scheme for BTC-compressed images for embedding data in the bitmaps of smooth blocks to obtain an improved image quality. There are two steps in in the embedding process of the scheme proposed by Chuang and Chang. Initially, a cover image is compressed into blocks by using BTC for calculating two quantized data and the bit plane corresponding to each block. Finally, the secret data is embedded into the bitmaps of the predefined smooth blocks that satisfy the following equation: $H_m - L_m < Threshold$. The smooth blocks were selected because bit replacement in these bit planes causes a slight distortion in the BTC image. In the extraction process of the scheme proposed by Chuang and Chang [12], the difference $H_m - L_m$ has to be first calculated. If $H_m - L_m < Threshold$, then the secret bit in the bit plane p'_i is extracted. However, in this scheme, the stego image quality degrades significantly as the threshold values increases.

Hong et al. [13] proposed a reversible data-hiding scheme based on bit plane flipping according to the corresponding secret bit. In the embedding process, each image block was compressed using AMBTC-compressed codes to determine whether the block is embeddable or not. If $L_m < H_m$, then the block is considered embeddable. Otherwise, the block is considered non-embeddable. For each embeddable block, if the secret bit is 1, then the bit plane p_i is flipped to $\bar{p}_i$, where $\bar{p}_i$ is not an operator. If the secret bit is 0, then no operation is required. In the extraction process, if $L_m > H_m$ in p'_i, then the secret in p'_i is 1. Otherwise, the secret bit in p'_i is 0. The scheme presented by Hong et al. does not hide data in blocks with $L_m = H_m$. Therefore, Chen et al. [14] proposed a reversible data-hiding method to improve the scheme by Hong et al. The AMBTC-compressed block that has $L_m = H_m$ is a smooth area, which is considered unnecessary bit plane information. Thus, the secret bit can be embedded all bits in the bit plane block to improve the scheme by Hong et al.

Li et al. [15] introduced a data-hiding scheme by using the histogram shifting technique on BTC-compressed mean tables for further improving the hiding capacity, while maintaining the quality of the BTC-compressed image. The hiding scheme comprises two main steps. The first step is based on the bit plane flipping method that hides secret bits by swapping the high mean and low mean values. In the second step, histogram shifting is conducted on the resulting mean tables after swapping. This scheme requires no additional data in the stego code stream. Therefore, very low distortion is observed in this scheme after data embedding, and the security of the embedded data is enhanced. However, this technique cannot provide a sufficient data-hiding capacity and requires overhead information to record a histogram.

Lin et al. [16] proposed a technique to explore the redundancy in a block of AMBTC-compressed images to determine whether the block is embeddable. If the secret bits and bit plane combined in the block has more than three different cases, the block is marked as an embeddable block. Four disjoint sets were created using this technique of embeddable blocks for embedding data using different combinations of the mean value and its standard deviation.

Ou and Sun [17] proposed a data-hiding scheme with minimum distortion based on AMBTC. In this scheme, a predefined threshold is used to determine if a block of the AMBTC-compressed codes is a smooth or complex block in which data are embedded. If an AMBTC-compressed block $H_m - L_m < Threshold$, then the block is considered a smooth block. All bit planes in smooth blocks are used to embed data by replacing the bits of the block with secret data bits. The two quantization levels in the smooth block are then recalculated to reduce distortion in the image. In the complex blocks, a proportion of secret bits were concealed by exchanging the order of two quantization levels and toggling the bit plane. By performing this method, the payload can be increased without any distortion. Both smooth and complex blocks can be used to embed data in an AMBTC-compressed block. Therefore, the payload of this scheme was obviously enhanced.

Malik et al. [18] modified the AMBTC compression technique for embedding secret data. In their method, one-bit plane is converted to two-bit planes that can attain better image quality and high capacity. Although this scheme has high visual quality and high payload, it causes permanent distortion to the original AMBTC code and requires overhead information. Malik et al. [19] proposed an AMBTC compression-based data-hiding scheme by using the pixel value adjusting strategy. In this technique, the stream of secret bits was converted to digits with a base of three. Then, the pixel values of the AMBTC-compressed block are modified, at the most by one, to hide secret data. This scheme could maintain a balance between the hiding capacity and quality of a stego image.

As discussed above, data hiding by using the AMBTC technique is an issue worthy of more research. In this study, we extended the work of Malik et al. [19] to embed a larger amount of secret data. In the next section, the proposed scheme is discussed.

3. Proposed Scheme

Figure 1 shows the flowchart of our application. First, one monitoring image on the unmanned aerial vehicle was compressed because the transmitting volume of wireless network is limited. When the command post or chief's car receives the compression codes, they are decoded as the decompressed image. In addition, they embed secret data into the reconstructed image, thereby cheating hackers and avoiding attacks. Finally, the headquarters can extract secret data and recover the decompressed image.

The main aim of the study is to present a data-hiding scheme with high data-hiding capacity and high image quality. In the scheme, secret data is hidden in an AMBTC decompressed image. The AMBTC decompressed image is losslessly reconstructed and the secret data, then, is losslessly revealed from the reconstructed image. The AMBTC encoding procedures are described in Section 2.

Before embedding the secret data, the cover image must be compressed using the AMBTC algorithm. In other words, the proposed scheme uses the AMBTC decompressed image to embed the secret data.

The proposed scheme involves three stages: In the first stage, an appropriate encoding and decoding dictionary is found. The dictionary is used in the second stage to embed the data. In the third stage, the secret data is extracted. The details of the proposed scheme are presented in Figure 1.

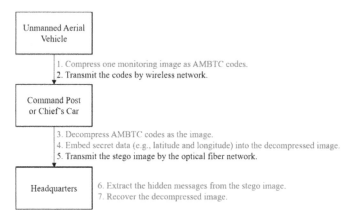

Figure 1. Flowchart of our applications.

3.1. Finding a Unique Decodable Dictionary

A binary secret sequence S comprises 0 and 1 values and is denoted as $S = \{s_1, s_2, \ldots, s_N\}$, where $s_i \in \{0, 1\}$ for $i = 1 \sim N$. Consider the dictionaries D_1, D_2, D_3, and D_4 formed using K subsets of S, that is, $S_{p1}, S_{p2}, \ldots, S_{pK}$. Different image quality is obtained due to the different dictionaries. Thus, we can calculate each probability of symbol S_p in S. The amount of information in each symbol I_a can be represented by

$$I_a = -log_2\big(pr\big(S_{pk}\big)\big). \tag{7}$$

Then, the average information per symbol interval is $H\big(S_{pk}\big)$ and can be represented by

$$H\big(S_{pk}\big) = -\sum_{k=1}^{n} pr(S_{pk})log_2\big(pr(S_{pk})\big). \tag{8}$$

The average information $H\big(S_{pk}\big)$ is referred to as the entropy. The dictionary with the smallest entropy H should be selected because it can achieve the best encoding benefit. The following explains why the dictionary of the smallest entropy is used: Assume that there is only one symbol's type in the whole secret sequence. In other words, the other types never occur. In this case, the entropy is equal to 0, i.e., $H\big(S_{pk}\big) = 0$. Afterwards, the specific symbols are replaced by the absolute minimum value "0", thereby controlling the distortion level in the data embedding phase. Consequently, the proposed method selects the dictionary of the smallest dictionary.

An example is used to explain the above procedure. Assume the secret sequence $S = \{0011101111001101100100000010011010\}$. In dictionary D_1 listed in Table 1, the secret sequence is represented as $S = \{001, 11, 01, 11, 10, 01, 10, 11, 001, 000, 001, 001, 10, 10\}$ for easy readability. According to D_1, the total number of information is 12.4670 and the average information $H\big(S_{pk}\big)$ per symbol at S is 2.1570. In dictionary D_2, which is listed in Table 2, the secret sequence can be represented as $S = \{00, 11, 10, 11, 11, 00, 11, 011, 00, 10, 00, 00, 10, 011, 010\}$. According to D_2, the total number of information is 12.1451 and the average information $H\big(S_{pk}\big)$ per symbol at S is 2.2264. The third and fourth dictionaries are constructed in the same manner, and their entropies values are listed in Tables 3 and 4, respectively. Obviously, the entropy of D_1 is the smallest among all the dictionaries. Therefore, we used D_1 to encode the secret sequence.

Table 1. Total number of data was 12.4670 with an entropy H of 2.1570 in the first dictionary D_1.

Symbol (S_p)	Freq. (S_p Count)	Amount of Information
000	1	3.4594
001	4	1.4594
01	2	3.0444
10	4	2.0444
11	3	2.4594

Table 2. Total number of data was 12.1451 with an entropy H of 2.2264 in the second dictionary D_2.

Symbol (S_p)	Freq. (S_p Count)	Amount of Information
00	5	1.7225
010	1	3.4594
011	2	2.4594
10	3	2.4594
11	4	2.0444

Table 3. Total number of data was 12.0751 with an entropy H of 2.2405 in the third dictionary D_3.

Symbol (S_p)	Freq. (S_p Count)	Amount of Information
00	3	2.4150
01	3	2.4150
100	3	1.8301
101	1	3.4150
11	4	2

Table 4. Total number of data was 12.5602 with an entropy H of 2.1726 in the fourth dictionary D_4.

Symbol (S_p)	Freq. (S_p Count)	Amount of Information
00	5	1.7225
01	3	2.4594
10	1	4.0444
110	3	1.8745
111	2	2.4594

Subsequently, the symbols in the selected dictionary are encoded further to obtaining the embedded digits. According to the rule of thumb of data encoding, S_p with the maximum occurrence frequency was encoded as the absolute minimum value. By contrast, S_p with the lowest occurrence frequency was encoded as the absolute maximum value. Consequently, S_p was sorted based on the occurrence frequency, and then, its sorted index was encoded to obtain the adjusting pixel values P_v, i.e.,

$$p_v = \begin{cases} -\left\lfloor \frac{Sort\ index}{2} \right\rfloor, & \text{if Sort index is an odd number,} \\ \left\lfloor \frac{Sort\ index}{2} \right\rfloor, & \text{otherwise.} \end{cases} \tag{9}$$

The following example is used to explain how to encode most symbols as smaller digits, as listed in Table 5. The occurrence frequencies of two symbols, "001" and "10", are 4, which are higher than those of other symbols. According to Equation (9), the symbol "001" is encoded as the absolute minimum value "0". Moreover, the symbol "10" is encoded as the second smallest value "1". The remaining symbols are encoded in the same manner.

Table 5. Dictionary example presenting the pixel value adjusting method.

Symbol (S_p)	Freq. (S_p Count)	Sorted Index	Adjusting Pixel Values (P_v)
000	1	5	−2
001	4	1	0
01	2	4	2
10	4	2	1
11	3	3	−1

3.2. Embedding Stage

The AMBTC decompressed blocks b_i in the original AMBTC decompressed image T are sequentially scanned. If the difference between H_m and L_m is smaller than 4, then the block is considered a non-embeddable block. Otherwise, the block is an embeddable block. In the first embeddable block, the binary representation of the ID number of the selected dictionary is embedded into the least significant bits (LSBs) of the second Hm and the second L_m. Note that the number of dictionaries is four, thus the two LSBS can effectively represent the ID number. The other blocks are then used to embed the secret data by using the pixel value adjusting strategy.

In each embeddable block, the first H_m and the first L_m are defined as non-embeddable pixels, which are used as the reference information of data extraction and image recovery. For the embeddable block b_i, each pixel x'_i except the first H_m and the first L_m is increased by the adjusting pixel values P_v, that is, $x''_i = x'_i + P_v$. The difference between maximum P_v and minimum P_v in the difference D is equal to 4. It implies that the distortion of pixels is low. The embedding pseudocode is shown in Algorithm 1 as follows:

Algorithm 1: Embedding pseudocode

---
       foreach *AMBTC − compressed block b_i in T* **do**
           if $H_m − L_m \leq 4$ **then**   /* non-embeddable block */
             Do nothing;
           else if b_i *is first embeddable block* **then**
             embedding dictionary D number;
           else
             foreach *pixel x'_i in b_i* **do**
               find adjusting pixel values P_v in D;
               $x''_i = x'_i + P_v$;
             end
           end
       end

Figure 2 displays the embedding example in which = {0011101111001101100 10000010011010}. Figure 2a presents the appropriate dictionary D found in Section 3.1. This dictionary was used to encode the secret sequence. After looking up the dictionary D, S is divided into many subsets S_p, as shown in Figure 2b. These subsets are mapped using the adjusting pixel values P_v, which are just the embedded value.

Symbol (S_p)	Adjusting Pixel Values (P_v)
000	-2
001	0
01	2
10	1
11	-1

$S =\{0011101111001101100100000010011010\}$

Concatenated S_p={001,11,01,11,10,01,10,11,001,000,001,001,10,10}

Concatenated P_v={ 0, -1, 2, -1, 1, 2, 1,-1, 0, -2, 0, 0, 1, 1}

(a) Unique decodable dictionary   (b) Lookup dictionary for adjusting pixel values

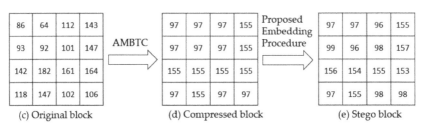

(c) Original block   (d) Compressed block   (e) Stego block

Figure 2. Example illustrating the proposed embedding stage.

To embed these values, the original block must be compressed and decompressed using the AMBTC algorithm, as shown in Figure 2c. After using the AMBTC algorithm, the AMBTC decompressed block can be reconstructed using a low mean value L_m of 97 and a high mean value H_m of 155, as shown in Figure 2d. Both the first L_m and first H_m are non-embeddable pixels and are marked with yellow color for easy readability. They are used as reference information of data extraction and image recovery. For the AMBTC decompressed block, the pixel, except the first H_m and the first L_m, is increased by the adjusting pixel values P_v to obtain the stego pixel. Figure 2e shows the stego block.

If the overflow or underflow problem occurs in any altered pixel of the block, then all of the pixels in the corresponding block remain unchanged. In other words, the block cannot be used to embed any secret bit. In addition, the proposed method records the ID number of the non-embeddable block to discriminate between the embeddable block and the non-embeddable block.

3.3. Extraction Stage

In the extraction stage, the secret data is extracted from the stego image T'. Moreover, T' can be used to recover the original AMBTC decompressed image T. The details of the procedures are listed as follows:

1. Scan the stego AMBTC decompressed block b'_i in T' sequentially. If the difference between H_m and L_m is smaller than 4, then this block is considered a non-embeddable block. Otherwise, it is an embeddable block.

2. Retrieve the ID number of the selected dictionary D from the first embedded block. In the first embedded block, both the LSBs of the second Hm and the second Lm are extracted, i.e., binary representation of the ID number of the selected dictionary. Therefore, the proposed method can reconstruct the selected dictionary. In addition, both the LSBs are replaced by the first Hm and the first Lm, thereby recovering the original decompressed pixel.

3. Calculate the adjusting pixel values by using $P_v = x''_i - H_m$ or $P_v = x''_i - L_m$ for each embeddable block b'_i. After obtaining P_v, we can look up the dictionary D to obtain the symbol S_p. After concatenating all S_p, we obtain the secret sequence S and recover the original AMBTC decompressed image T. The extraction and recovery pseudocode are shown in Algorithm 2.

Algorithm 2: Extraction and recovery pseudocode

> **foreach** *block b'_i in T'* **do**
>> **if** $H_m - L_m \leq 4$ **then**   /* non-embeddable block */
>>> Do nothing;
>>
>> **else if** b_i *is first embeddable block* **then**
>>> get the dictionary D number;
>>
>> **else**
>>> **foreach** *pixel x''_i in b'_i* **do**
>>>> get the first H_m and L_m;
>>>> **if** $x''_i = H_m$ **then**
>>>>> $P_v = x''_i - H_m$;
>>>>
>>>> **else**
>>>>> $P_v = x''_i - L_m$;
>>>>
>>>> **end**
>>>> find the symbol S_p in D;
>>>> $S = S + S_p$;
>>>
>>> **end**
>>
>> **end**
>
> **end**

Figure 3 illustrates the extraction and recovery example. First, the dictionary is retrieved from the first embeddable block. Second, the adjusting pixel values are calculated as $P_v = x''_i - H_m$ or $P_v = x''_i - L_m$. Third, P_v is mapped with the dictionary values to obtain S_p. Finally, S_p is concatenated for obtaining the secret sequence S and the AMBTC decompressed block.

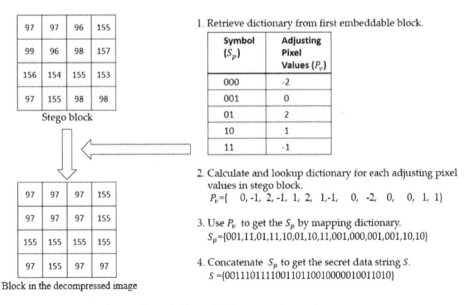

97	97	96	155
99	96	98	157
156	154	155	153
97	155	98	98

Stego block

1. Retrieve dictionary from first embeddable block.

Symbol (S_p)	Adjusting Pixel Values (P_v)
000	-2
001	0
01	2
10	1
11	-1

2. Calculate and lookup dictionary for each adjusting pixel values in stego block.
P_v={ 0, -1, 2, -1, 1, 2, 1,-1, 0, -2, 0, 0, 1, 1}

3. Use P_v to get the S_p by mapping dictionary.
S_p={001,11,01,11,10,01,10,11,001,000,001,001,10,10}

97	97	97	155
97	97	97	155
155	155	155	155
97	155	97	97

Block in the decompressed image

4. Concatenate S_p to get the secret data string S.
S ={00111011111001101100100000010011010}

Figure 3. Example illustrating the proposed extraction stage.

4. Experimental Results and Discussion

Some experimental cover images were tested to demonstrate the efficiency of the proposed scheme. In the experiments, the proposed scheme was verified using the following six test cover images: airplane, boat, lena, mandrill, peppers, and sailboat. As shown in Figure 4, all the images had the same size of 512×512 pixels with 256 grayscales, and the features of the images were diverse. The block size of the image presented in the AMBTC format was 4 × 4 pixels. A random binary sequence generated using a MATLAB (R2018a) function was used in the experiments as the secret sequence, where our secret data are the same as the secret data of the related works [15–17,19]. Note that each bit in the sequence has equal probability of being 0 or 1.

(a) Airplane   (b) Boat   (c) Lena

(d) Mandrill   (e) Peppers   (f) Sailboat

Figure 4. Test cover images.

The proposed scheme was evaluated and compared with the aforementioned schemes in terms of two performance measures, i.e., hiding capacity and peak signal-to-noise ratio (PSNR). The hiding capacity can be defined as the number of secret data bits that can be hidden into a cover image. The PSNR is an objective measure used for determining the visual quality of an image. The higher the PSNR of a stego image, the better its visual quality is. The rule of thumb is that when the PSNR is higher than 30 dB, the human eyes cannot easily perceive the difference between the cover image and the stego image. PSNR is defined by

$$PSNR = 10 \log_{10} \frac{255^2}{MSE},$$ (10)

$$MSE = \frac{1}{M \times N} \sum_{i=1}^{M} \sum_{j=1}^{N} \left(x_{ij} - x'_{ij} \right)^2,$$ (11)

where x_{ij} and x'_{ij} are the original and stego grayscale pixel values located at (i, j), respectively.

To present the superiority of the proposed scheme, we compared our scheme with the schemes presented by Li et al. [15], Lin et al. [16], Ou and Sun [17], and Malik et al. [19], as shown in Table 6. The proposed scheme achieved the highest data-hiding capacity for all five images except for the airplane image. The data-hiding capacity of the pixel value adjusting strategy was determined using the number of smooth blocks. If there are many smooth zones in a cover image, non-embeddable blocks are observed in abundance in the image. Moreover, the pixel value adjusting strategy used in the proposed scheme is modified at the most by 2, whereas the strategy used in the scheme proposed by Malik et al. is modified at the most by 1. Therefore, compared with the scheme presented by Malik et al., our scheme has a higher number of non-embeddable blocks in the airplane image. Non-embeddable blocks can be observed in black in Figure 5a,b. This is the main factor that causes the hiding capacity of the scheme proposed by Malik et al. to be better than that of the proposed scheme for the airplane image. For the other five images, the hiding capacity of our scheme is better than that of the scheme presented by Malik et al. by an enhancement value in the range of 10.13% to 29.89%. The hiding capacity of our scheme is better than the schemes proposed by Li et al., Lin et al., and Ou and Sun. Thus, we conclude that our scheme is better than the existing AMBTC- and BTC-based data-hiding schemes, in terms of the hiding capacity.

Table 6. Comparison between hiding capacity and PSNR for different images for the proposed scheme and other AMBTC- and BTC-based schemes.

Method	Performance	Airplane	Boat	Lena	Mandrill	Peppers	Sailboat
Proposed	Hiding capacity (bits)	338,836	495,582	437,577	515,836	478,591	479,407
	PSNR (dB)	32.0908	31.0397	33.0562	26.9348	33.0087	29.7857
Malik et al. (2018)	Hiding capacity (bits)	397,147	397,380	397,348	397,105	397,057	397,466
	PSNR (dB)	31.9018	31.0926	33.102	26.9474	33.304	29.8081
Ou and Sun (2015)	Hiding capacity (bits)	223,039	217,264	234,004	141,919	238,969	219,169
	PSNR (dB)	30.71	29.54	30.87	26.02	31.59	28.61
Lin et al. (2013)	Hiding capacity (bits)	261,984	262,096	262,112	262,144	261,984	262,064
	PSNR (dB)	31.64	30.32	33.05	26.0047	32.2021	28.8049
Li et al. (2011)	Hiding capacity (bits)	17,659	16,580	16,789	16,880	17,264	16,990
	PSNR (dB)	30.18	31.83	31.05	27	32.35	28.43

 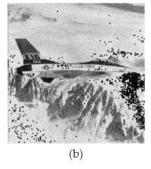

(a)   (b)

Figure 5. Non-embeddable blocks in "airplane.": (**a**) Proposed scheme and (**b**) scheme proposed by Malik et al.

Table 7 lists the comparison between the method by Malik et al. and the proposed method in terms of structural similarity index (SSIM). As mentioned above, the SSIM value of the method by Malik et al. is greater than that of the proposed method because the proposed method embeds more secret data. In other words, the maximum hiding capacity of the proposed method is higher than that of the method by Malik et al.

Table 7. Comparison between hiding capacity and SSIM for different images for the proposed scheme and the other method by Malik et al.

Image	Malik et al.'s Method		Proposed Method	
	Hiding Capacity (bits)	SSIM	Hiding Capacity (bits)	SSIM
Airplane	397,147	0.947	338,836	0.9447
Boat	397,380	0.918	495,582	0.915
Lena	397,348	0.937	437,577	0.933
Mandrill	397,105	0.886	515,836	0.885
Peppers	397,057	0.931	478,591	0.927
Sailboat	397,466	0.915	479,407	0.912

The PSNR is the other factor for evaluating performance of a hiding scheme. Table 6 presents that the PSNR of our scheme is better than the schemes proposed by Li et al., Lin et al., and Ou and Sun. For the airplane stego image, the proposed scheme has a better PSNR but a weaker hiding capacity than the scheme proposed by Malik et al., because our scheme has a higher number of non-embeddable blocks than the scheme by Malik et al. Note that a non-embeddable block maintains the image quality but decreases the hiding capacity. For the other five stego images, the PSNR obtained using the proposed scheme is weaker than that obtained using the scheme proposed by Malik et al. However, because the PSNR difference is less than 0.29 dB for the five stego images, they would not be distinguishable by human vision due to such negligible differences. By contrast, the hiding capacities are significantly increased by a value in the range of 10.13% to 29.89% for the other five stego images. This implies that a tradeoff exists between the PSNR and hiding capacity when the pixel value adjusting strategy is used. The visual quality of the proposed scheme is observed to be above the average value of that of the baseline schemes.

5. Conclusions

A high-capacity data-hiding scheme was proposed in this study for an AMBTC-compressed image. The proposed scheme has many more properties than only high capacity. In this scheme, the dictionary-based coding scheme and the pixel value adjusting strategy were combined to increase the hiding capacity and attain a satisfactory visual quality. Experimental results reveal that the proposed scheme is better than the existing AMBTC-based data-hiding schemes in terms of the hiding capacity. Moreover, the visual quality of the proposed scheme is better than that of baseline schemes. In the future, we should combine the method by Liao et al. [20] with the proposed method to discriminate the image smoothness, thereby enhancing the hiding capacity. In addition, we should try to add the concept of partition strategy [21] into the proposed method to embed more secret data into the color images.

Author Contributions: Conceptualization, C-C.C.; data curation, J-Y.Y.; funding acquisition, P-L.L.; investigation, J-Y.Y.; methodology, J-Y.Y. and C-C.C.; supervision, P-L.L.; validation, J-Y.Y. and Y-H.H.; writing—original draft, J-Y.Y.; writing—review & editing, P-L.L.

Funding: This research was funded by the Ministry of Science and Technology (Taiwan), No. MOST 108-2221-E-005-001.

Conflicts of Interest: The authors declare no conflict of interest.

References

1. Li, C.; Zhang, Y.; Xie, E.Y. When an Attacker Meets a Cipher-image in 2018, A Year in Review. *J. Inf. Secur. Appl.* **2019**, *48*, 1–9. [CrossRef]
2. Mielikainen, J. LSB Matching Revisited. *IEEE Signal Process. Lett.* **2006**, *13*, 285–287. [CrossRef]
3. Zhang, X.; Wang, S. Efficient Steganographic Embedding by Exploiting Modification Direction. *IEEE Commun. Lett.* **2006**, *10*, 781–783. [CrossRef]
4. Hong, W.; Chen, T. A Novel Data Embedding Method Using Adaptive Pixel Pair Matching. *IEEE Trans. Inf. Forensics Secur.* **2012**, *7*, 176–184. [CrossRef]
5. Tian, J. Reversible Data Embedding Using a Difference Expansion. *IEEE Trans. Circuits Syst. Video Technol.* **2003**, *13*, 890–896. [CrossRef]
6. Peng, F.; Li, X.; Yang, B. Adaptive Reversible Data Hiding Scheme Based on Integer Transform. *Signal Process.* **2012**, *92*, 54–62. [CrossRef]
7. Tai, W.L.; Yeh, C.M.; Chang, C.C. Reversible Data Hiding Based on Histogram Modification of Pixel Differences. *IEEE Trans. Circuits Syst. Video Technol.* **2009**, *19*, 906–910.
8. Wang, K.; Lu, Z.M.; Hu, Y.J. A High Capacity Lossless Data Hiding Scheme for JPEG Images. *J. Syst. Softw.* **2013**, *86*, 1965–1975. [CrossRef]
9. Lee, Y.P.; Lee, J.C.; Chen, W.K.; Chang, K.C.; Su, I.J.; Chang, C.P. High-Payload Image Hiding with Quality Recovery Using Tri-Way Pixel-Value Differencing. *Inf. Sci. (Ny).* **2012**, *191*, 214–225. [CrossRef]

10. Delp, E.; Mitchell, O. Image Compression Using Block Truncation Coding. *IEEE Trans. Commun.* **1979**, *27*, 1335–1342. [CrossRef]
11. Lema, M.; Mitchell, O. Absolute Moment Block Truncation Coding and Its Application to Color Images. *IEEE Trans. Commun.* **1984**, *32*, 1148–1157. [CrossRef]
12. Chuang, J.C.; Chang, C.C. Using A Simple and Fast Image Compression Algorithm to Hide Secret Information. *Int. J. Comput. Appl.* **2006**, *28*, 329–333.
13. Hong, W.; Chen, T.S.; Shiu, C.W. Lossless Steganography for AMBTC-Compressed Images. In Proceedings of the 2008 Congress on Image and Signal Processing, Sanya, Hainan, China, 27–30 May 2008; pp. 13–17.
14. Chen, J.; Hong, W.; Chen, T.S.; Shiu, C.W. Steganography for BTC Compressed Images Using no Distortion Technique. *Imaging Sci. J.* **2010**, *58*, 177–185. [CrossRef]
15. Li, C.H.; Lu, Z.M.; Su, Y.X. Reversible Data Hiding for BTC-Compressed Images Based on Bitplane Flipping and Histogram Shifting of Mean Tables. *Inf. Technol. J.* **2011**, *10*, 1421–1426. [CrossRef]
16. Lin, C.C.; Liu, X.L.; Tai, W.L.; Yuan, S.M. A Novel Reversible Data Hiding Scheme Based on AMBTC Compression Technique. *Multimed. Tools Appl.* **2015**, *74*, 3823–3842. [CrossRef]
17. Ou, D.; Sun, W. High Payload Image Steganography with Minimum Distortion Based on Absolute Moment Block Truncation Coding. *Multimed. Tools Appl.* **2015**, *74*, 9117–9139. [CrossRef]
18. Malik, A.; Sikka, G.; Verma, H.K. A High Payload Data Hiding Scheme Based on Modified AMBTC Technique. *Multimed. Tools Appl.* **2017**, *76*, 14151–14167. [CrossRef]
19. Malik, A.; Sikka, G.; Verma, H.K. An AMBTC Compression Based Data Hiding Scheme Using Pixel Value Adjusting Strategy. *Multidimens. Syst. Signal Process.* **2018**, *29*, 1801–1818. [CrossRef]
20. Liao, X.; Qin, Z.; Ding, L. Data Embedding in Digital Images Using Critical Functions. *Signal Process. Image Commun.* **2017**, *58*, 146–156. [CrossRef]
21. Liao, X.; Yu, Y.; Li, B.; Li, Z.; Qin, Z. A New Payload Partition Strategy in Color Image Steganography. *IEEE Trans. Circuits Syst. Video Technol.* **2019**, in press. [CrossRef]

© 2020 by the authors. Licensee MDPI, Basel, Switzerland. This article is an open access article distributed under the terms and conditions of the Creative Commons Attribution (CC BY) license (http://creativecommons.org/licenses/by/4.0/).

Article

Fusing Feature Distribution Entropy with R-MAC Features in Image Retrieval

Pingping Liu [1,2,3,*], Guixia Gou [1], Huili Guo [1], Danyang Zhang [1], Hongwei Zhao [1] and Qiuzhan Zhou [4]

[1] College of Computer Science and Technology, Jilin University, Changchun 130012, China; gougx18@mails.jlu.edu.cn (G.G.); huili6@staff.weibo.com (H.G.); zhangdy19@mails.jlu.edu.cn (D.Z.); zhaohw@jlu.edu.cn (H.Z.)
[2] Key Laboratory of Symbolic Computation and Knowledge Engineering of Ministry of Education, Jilin University, Changchun 130012, China
[3] School of Mechanical Science and Engineering, Jilin University, Changchun 130025, China
[4] College of Communication Engineering, Jilin University, Changchun 130012, China; tongxin@jlu.edu.cn
[*] Correspondence: liupp@jlu.edu.cn; Tel.: +86-138-4498-2003

Received: 22 September 2019; Accepted: 23 October 2019; Published: 25 October 2019

Abstract: Image retrieval based on a convolutional neural network (CNN) has attracted great attention among researchers because of the high performance. The pooling method has become a research hotpot in the task of image retrieval in recent years. In this paper, we propose the feature distribution entropy (FDE) to measure the difference of regional distribution information in the feature maps from CNNs. We propose a novel pooling method, which fuses our proposed FDE with region maximum activations of convolutions (R-MAC) features to improve the performance of image retrieval, as it takes the advantage of regional distribution information in the feature maps. Compared with the descriptors computed by R-MAC pooling, our proposed method considers not only the most significant feature values of each region in feature map, but also the distribution difference in different regions. We utilize the histogram of feature values to calculate regional distribution entropy and concatenate the regional distribution entropy into FDE, which is further normalized and fused with R-MAC feature vectors by weighted summation to generate the final feature descriptors. We have conducted experiments on public datasets and the results demonstrate that our proposed method could produce better retrieval performances than existing state-of-the-art algorithms. Further, higher performance could be achieved by performing these post-processing on the improved feature descriptors.

Keywords: image retrieval; pooling method; convolutional neural network; feature distribution entropy

1. Introduction

Content-based image retrieval (CBIR) has achieved appreciable performance over its long-standing development and has attracted more and more attention among researchers in recent years [1–3]. It aims to search the images with the same object, instance, and architecture from an image database and rank the images from the database to certain query images according to the similarities. The global features extracted from visual clues like texture and color were utilized to realize image retrieval in early times [1,2,4]. However, the global descriptors might change with the illumination, occlusion, and translation, and it is hard to keep invariance, which would reduce robustness and affect the performance of image retrieval. Later, the local descriptor of scale-invariant feature transform (SIFT) was proposed to meet invariance exception [5]. The appearance of SIFT has spawned a heavy load of excellent algorithms, which have achieved effective performance in image retrieval. At the beginning, most of the SIFT-based image retrieval methods relied on the bag-of-visual-words (BoW) model to obtain a compact vector of images [6]. Later, the vector of locally aggregated descriptors (VLAD) was proposed to consider all of cluster centers and the distance of local features to its nearest cluster center [7]. Fisher vector (FV) calculates the distances of local features to all cluster centers [8]. Then, spatial pyramid matching (SPM) was proposed, which is based on BoW, with spatial location information added to the feature descriptors [9]. All these methods fail to extract high-level semantic features.

Recently, convolutional neural network (CNN) has made a huge development with the success of AlexNet in the task of image classification [10], and has been widely applied in the tasks of image retrieval [11–15], object recognition [9,16,17], and target detection [18,19]. There are a number of CNN-based methods that have achieved acceptable performance in these tasks, as the deep features extracted from fully connected or convolutional layer contain richer high-level semantic information compared with traditional manual features. Especially for the task of image retrieval, a mass of methods based on the pre-trained network have achieved effective performance [3,12,17,20–24]. Recently, researchers have tended to use the feature maps generated from the last convolutional layer to achieve higher performance in image retrieval [3,12,16,20–26]. However, the high dimension of feature maps makes the descriptors hard to use directly. In early times, traditional aggregating methods were used to encode the feature map's output from the CNNs into deep feature descriptors [12,16,26]. With further development of deep learning, more and more methods have been proposed to utilize the feature maps from the CNNs to generate compact feature vectors by using pooling operation [12,20–25]. The key challenge for the pooling operation is how to extract the most pivotal features from feature maps and eliminate the effect of the irrelevant information noise. The region maximum activations of convolutions (R-MAC) pooling aim to consider the most prominent points for multiscale regions and has achieved outperforming results in image retrieval [24]. However, this method ignores the difference of distribution in different regions, which could be important to extract more effective feature descriptors for the task of image retrieval. In general, there are some effective post-process methods, like re-ranking [24,25,27] and query expansion (QE) [25,28,29], used in the model image retrieval. These operations could be significantly helpful to further increase the performance of image retrieval.

To solve the key challenge mentioned above, we tend to take the distribution information of feature maps into consideration to generate deep feature representations with richer information. In this paper, we propose a novel method to measure the distribution differences of multiple regions in feature maps called feature distribution entropy (FDE). We combine the proposed FDE with R-MAC to generate more effective features to improve the effectiveness of our image retrieval. To be specific, we make four contributions, as follows.

Firstly, we propose an effective scheme to compute FDE, which could be used to fully reflect the distribution differences of different regions. It would be helpful to focus on the more noteworthy regions and weaken the influence of irrelevant noise.

Secondly, we employ a superior strategy to combine our proposed FDE with R-MAC features to generate more discriminative features. The fused features could tend to extract compact feature representations with more information and are significant to eliminate influence of irrelevant information, especially the noise of background. The compact feature representations are more distinctive and could be more effective in improving the performance of image retrieval.

Thirdly, we perform the operation of re-ranking and QE on the deep-fused features produced by our proposed method. This helps us to obtain better retrieval results.

Fourth, we utilize the fine-tuned network [25] to perform our experiments on different datasets to verify the effectiveness of our proposed method.

To verify the superiority of our proposed method, we perform the experiments on the benchmarks with state-of-the-art re-ranking and QE approaches with the pre-trained and the fine-tuned network. The results of our experiments, which are described in detail in Section 4.3, show that our proposed method outperforms the existing state-of-the-art methods.

We organize the rest of our paper as follow. Section 2 is to illustrate the related work. The calculation of our proposed FDE and the fused features are represented in Section 3. We represent the results and analysis of our experiments in Section 4. Lastly, we make a conclusion for our paper in Section 5.

2. Related Work

Deep learning methods based on CNNs have made a great breakthrough in many tasks of computer vision. A general architecture of CNN usually consists of several convolutional layers followed by fully connected layers. The network is usually trained with a softmax layer. Recent works prefer to utilize the activations from the intermediate layer to realize some special tasks like target detection, semantic segmentation, target recognition, and so on, and have obtained effective performance [9,16–19,30,31]. Particularly for the task of image retrieval, Babenko et al. proposed to use the global features from the fully connected layers in image retrieval [22]. Gong et al. proposed to employ VLAD to aggregate the feature descriptors from the fully connected layers [20]. Recently, more and more works tend to use the feature representations generated by applying pooling on each channel feature maps output from the convolutional layers [3,12,16,20–26]. These feature representations usually contain richer high-level semantic information than fully connected ones and are significant to promote the effectiveness of image retrieval. More and more works show that better performance of image retrieval could be obtained when the deep feature descriptors are whitened [24,25]. Also, abundant works have shown that some post-process methods, like re-ranking and QE, would be significant to improve the performance of image retrieval [24,25,27–29]. In the rest of this section, we describe the related work for the methods we utilize in this paper in detail, which contains the pooling method, normalization, PCA, re-ranking, and QE.

2.1. Pooling Approaches

The methods based on CNNs have achieved superior performance in image retrieval. The early works using global features output from the fully connected layers are replaced by the local feature representations derived from the convolutional layers as it has more discriminative descriptive power. In early times, there were some popular encoding methods used in generating compact representations. Gong et al. proposed multi-orderless pooling CNN (MOP-CNN), which aim to extract multi-scale feature maps and utilize VLAD to encode them into the final feature descriptors [20]. Arandjelovic et al. later proposed to apply the VLAD to aggregate local features and design an end-to-end network for image retrieval [16]. Then, Mohedano et al. proposed a novel method, which applies the BOW into deep features [26]. It aggregates the features from CNN into compact representations. Multi-scale feature representation (MFC) was proposed by Hao et al. to extract features from three different scales and fuse the extracted features to generate the final feature vector [32]. However, all these methods mainly use traditional aggregation methods to encode the features from CNN into compact feature descriptors, which are always accompanied by huge consumption of computing.

There is another way to generate compact representations, which is derived from the pooling layer in CNN. The main idea is to utilize pooling on the activations of convolutional layers to produce more compact deep features [12,20–25]. The dimensionality of the deep features is consistent with the channels of feature maps from the corresponding convolutional layer. Babenko et al. propose sum pooling (SPoC), which computes the sum of values in the feature maps [22]. It has shown effective performance in image retrieval. The SPoC feature is calculated as following equation:

$$f^{(sum)} = \left[f_1^{sum}, \ldots, f_c^{(sum)}, \ldots, f_C^{(sum)} \right]^T, f_c^{(sum)} = \frac{1}{|X_c|} \sum_{x \in X_c} x \tag{1}$$

where C is the number of feature maps, c means the channel of features, $f_c^{(sum)}$ denotes the SPoC feature of c-th channel, $|X_c|$ is the amount of feature values in c-th channel feature map, X_c is c-th feature map, and x is the feature value in a certain feature map.

Later, Razavian et al. proposed max pooling (MAC) to select the maximum of each feature map [12]. The MAC pooling feature is computed as follows:

$$f^{(max)} = \left[f_1^{(max)}, \ldots, f_c^{(max)}, \ldots, f_C^{(max)} \right]^T, f_c^{(max)} = \max_{x \in X_c} x \qquad (2)$$

where C is the number of feature maps, c means the channel of feature, $f_c^{(max)}$ denotes the MAC feature of c-th channel, X_c is c-th feature map, and x is the feature values in a certain feature map.

According to the former work of MAC pooling, Giorgos et al. proposed R-MAC pooling, which uses sliding windows strategy to obtain a set of regions with different scale [24]. Each region performs MAC pooling in order to obtain the regional feature vectors. The computing equation is shown as follows:

$$f_R^{(r-max)} = \left[f_{R,1}^{(r-max)}, \ldots, f_{R,c}^{(r-max)}, \ldots, f_{R,C}^{(r-max)} \right]^T, f_{R,c}^{(r-max)} = \max_{x \in R} X_c(r) \qquad (3)$$

where $f_R^{(r-max)}$ denotes the maximum value of the given region, R denotes the regions extracted from the c-th feature map, and $X_c(r)$ means the feature value in the region r on c-th feature map

These methods have made great progress in improving the performance of image retrieval. However, the algorithms described above fail to take the regional distribution information of the feature maps into account. In order to make full use of distribution information, we introduce the concept of entropy to measure the distribution of feature maps. As the R-MAC pooling has obtained superior effectiveness and utilizes multi-scale strategy to extract regions, it is easy for us to analyze their difference. We design an effective scheme to calculate FDE, and then we fuse FDE with R-MAC features. Our experimental results show that our algorithm is better than many existing state-of-the-art algorithms.

2.2. Compact Features with Distribution Information

For image retrieval, many traditional algorithms ignore the distribution information of feature maps during generating the deep descriptors. To solve this issue, spatial distribution information is introduced as a supplement to feature descriptors to improve the retrieval performance. Based on BoW, Mehmood et al. combined histograms of local features with global features and constructed local feature maps in local regions [33]. Krapac et al. used Fisher kernel to calculate the spatial mean and cluster changes. Then, they encoded the BoW into a spatial map and combined them with Gaussian mixture model [34]. Koniusze et al. used spatial coordinate coding to simplify spatial pyramid representation [35]. Sancheset et al. improved the performance of FV-based object classification and prompted the spatial position of descriptors [36]. Liu et al. introduced the concept of spatial distribution entropy and applied spatial distribution entropy to the original VLAD algorithm [37]. These methods have achieved great performance in image retrieval, but these improvements were merely applied on traditional algorithms, which are no longer superior to the popular CNN-based algorithms.

Due to the rapid development of neural networks, CNN-based methods have shown excellent retrieval performance in mage retrieval. However, many retrieval algorithms do not make full use of regional distribution information. It is very important to preserve the regional distribution information of images to promote retrieval performance. In R-MAC, the local features of each region are directly concatenated to obtain global features. The contribution of each region is simply the biggest value in each region, which does not consider the difference of regional distribution information and fail to generate more informative feature descriptors. Entropy is an effective measurement to reflect the distribution information of regions, which is proposed by Shannon in 1948 [38]. We take the advantage of the regional distribution information to compute FDE in multiple-scale regions and fuse with the deep feature descriptors as supplement information to solve the disadvantages of R-MAC feature representations. We use FDE to measure the regional distribution information of feature maps and use it as a supplement to combine with the R-MAC features to enrich the deep feature representations.

Compared with R-MAC, the results the performance of our algorithm are improved, which indicates that the proposed algorithm is effective.

2.3. Normalization and PCA

Normalization plays an important role in image retrieval and has been largely used in image retrieval [13]. This operation aims to transform the data into a uniform scope to make a comparison among them. We would like to discuss two types of normalization, one being L2 normalization and the other being power normalization.

L2 normalization [13] aims to balance the impact of different values, as the values output from the convolutional layer are usually discrete and very different from each other. There would be a mass of extreme values, which would affect the performance of image retrieval. We utilize L2 normalization to narrow the difference of values without changing the proportional difference of values. Specifically, L2 normalization is to limit the values within the range from 0 to 1, and the formulation is defined as follows:

$$X_{L2} = \frac{X}{\|X\|} \tag{4}$$

where $\|X\|$ denotes the values in a certain vector and X means the magnitude of this vector.

Power normalization [13] functions the same as L2 normalization to eliminate the gap among extreme values. Power normalization is the reduction of the values in the vector in the form of power exponent, and we give the formulation as follows:

$$X_p = \mathrm{sgn}(X) \times X^p \tag{5}$$

where X is the values of a certain vector and $sig(X)$ denotes a symbolic function to prevent the sign of the value from changing after power normalization; the value would be 1 if X is larger than 0 and -1 if smaller than 0. p is a hyper-parameter.

The feature vectors generated from the pooling layer tend to have higher dimensions, which could cause large calculation consuming. The features would be accompanied by large noise, which always reduces the performance of image retrieval. To achieve better performance, whitening the feature descriptors is a common and essential stage used in image retrieval as described in the work of Chum et al. [39]. They focus on jointly down-weighting co-occurrences and aim to handle the problem of over-counting. Their work is further migrated in feature descriptors based on CNN. The principal component analysis (PCA) trained on an independent set is always used for whitening and dimensionality reduction [25,40,41]. It aims to project the original vector onto the direction in which the most original information can be retained. The values after PCA are expected to be as scattered as possible with high variance. Mokolajczyk et al. [41] used the training data to whiten local feature representations. Gordo et al. preferred to learn the whitening in an end-to-end manner based on CNN [40], and Filip et al. proposed a new method named learning whitening by taking advantage of training data provided by their 3D models and using liner discriminant projections to perform whitening on features [25]. In our paper, we prefer to utilize PCA to realize dimensionality reduction. It could reduce the computing consuming and eliminate the mutual influence between the original data components to promote the performance of image retrieval.

2.4. Re-ranking and Query Expansion

In image retrieval, the results of the first search are often not expressive enough, so reordering the first output will give better results. In image retrieval, re-ranking [27] is often followed by QE [28,29]. The operations of re-ranking and QE are generally helpful in achieving a better performance compared with the results retrieved by raw representations.

Giorgos et al. use approximate max-pooling localization (AML) to coarsely locate the local features of top N images by using the raw representations and then re-rank them [24]. The following

QE operation further improves the retrieval performance. In recent years, some new QE methods have been proposed. The most widely used is average query expansion (AQE), which extracts the features of the images ranking top K and averages them with the features of the query images, and then re-retrieving images to obtain a more accurate result. Inspired by AQE, Filip et al. added a weight to the features of the i-th image and named it αQE [25].

3. Proposed Method

In this section, we give some details of our proposed FDE and introduce how to combine FDE with R-MAC feature vectors to produce more discriminative feature representations, which is significant to improve the effectiveness of image retrieval. Furthermore, we perform re-ranking and QE, which have become standard post-processing used in improving the performance of image retrieval on our proposed method to obtain better performance. We would like to illustrate the process of our method in Section 3.3.

3.1. The Algorithm Background

For an image I, we use the pre-trained network without the fully connected layers to output the activations with 3D tensor of $W \times H \times C$ dimensions, where C means the number of feature maps output from the last convolution layer. For each feature map with size of $W \times H$, which could be represented as X_c and $k \in \{1, \ldots, C\}$, herein c denotes the feature channel, and each channel feature map with a certain region r, is represented as $X_{c(r)}$; we denote region location of each feature map as r. To ensure all these elements in the activations are non-negative, we apply the rectified linear units (ReLU) to the last layer.

As mentioned in Section 2.2, the R-MAC pooling method produces R different regions for each feature map and then calculates MAC features for each region. We represent the region as $r \in \{1, \ldots, R\}$. R-MAC uses multiple scale region extraction strategy to take full advantage of the convolutional layer activation information, which is different from MAC. The R-MAC performs MAC pooling on each R to produce an R-dimensional vector for each feature map. Then, all these feature vectors are encoded into a matrix of $R \times C$ for the activations. After the operation of normalization, the matrix is concatenated to obtain a C-dimensional feature vector, which could be denoted as $f = \{f_1, \ldots, f_C\}$. However, the feature vector does not make full use of the information of each region, because the distribution differs in feature regions. We propose to calculate FDE for feature maps that can reflect the difference in the distribution of pixel values of feature maps in different regions. Then, we combine our proposed FDE with R-MAC feature descriptors. To fuse the two parts better, we apply the operation of L2 and power normalization on FDE vector before fusing. Then, the L2 and power normalizations and PCA are performed on the fused feature descriptors to generate the final features, which would be used for retrieval. The distinctiveness of the final features can be enhanced, and this is further used to improve the performance of image retrieval.

3.2. Calculation of FDE

In this section, we represent our idea of the proposed FDE to take the difference of regions into consideration. Our proposed FDE could be helpful to generate more discriminative features that contain richer semantic information and simultaneously eliminate the effects of irrelevant background noise. To be specific, we calculate the proposed FDE and then combine FDE with R-MAC features to produce our final discriminative features. Herein, we design an effective scheme to calculate FDE. This scheme is proposed to focus on the different feature values in regions. We show the details of the processes of FDE calculation, as follows.

Herein, we would like to describe our proposed scheme in detail by taking one feature map as example. At the first step, we analyze statistical information of the feature values in each region. We build a histogram for each region of feature map. For each region of a certain feature map, there is a range of different feature values. We set the number of blocks in the histogram to B. Then, the value range s of each block is computed according to the maximum and minimum value of the current region. The information of distribution histogram on each region is calculated as the following equation:

$$h = \left\{ h(i,s) \middle| 1 \le i \le B; X_{r(min)} + (i-1)S \le s \le X_{r(min)} + is \right\}, S = \left(X_{r(max)} - X_{r(min)} \right)/B \tag{6}$$

where $X_{r(max)}$ and $X_{r(min)}$ are the minimum and maximum value in current region, and the value of B is a parameter that denotes the number of blocks; $h(i,s)$ denotes the statistical values in current region r. The total number of ranges of each region is counted by histogram.

Probability distribution entropy is an effective method to measure the distribution information of feature maps. Herein, we utilize the distributional probability to compute probability distribution entropy for all these regions. The distributional probability of each region can be calculated by following formula:

$$P_i = h(i,s) / \sum_{i=1}^{B} h(i,s) \tag{7}$$

where $h(i,s)$ denotes the statistical values i-th block. P_i is the distributional probability of i-th block in in current region r.

After that, according to the distributional probability of each region that computes the statistical values of distribution histogram h, the probability distribution entropy of each region is calculated using the following formulation:

$$H_r = -\sum_{i=1}^{B} P_i \log P_i \tag{8}$$

where P_i denotes the distributional probability of block i.

The probability distribution entropy of a certain region computed by our proposed scheme reflects the distribution of the pixel values of the feature regions. The FDE measures the distribution of feature values in the feature maps. The more concentrated the feature values of the feature maps in a certain region, the smaller the entropy value, and vice versa, as shown in Figure 1. This can reflect the distribution of information in different regions and make descriptors more distinguished. It could be significant to focus on areas that are more useful and eliminate the influence of useless background noises.

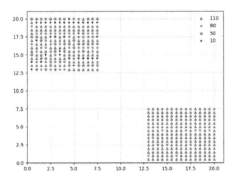

Figure 1. The two regions with different distribution in a feature map. The x-axis and y-axis denote the position of the feature values. The dots with different colors are the feature values. To be specific, the blue, red, green, and black dots are the values of 110, 80, 50, and 10, respectively.

As is shown in Figure 1, we assume the large square is a certain feature map generated by the convolutional layer and there are four different feature values within the first region and two different feature values within the second region. The dots of different colors represent feature values in different ranges. The largest value is represented as a blue dot. It is obvious that the distribution in the two regions is inconsistent, but when using the R-MAC algorithm, the MAC feature of the two regions are both 110. R-MAC does not highlight the difference in information distribution between the two regions. We utilize our proposed scheme 1 to calculate the distribution entropy, which could be applied to reflect the difference of the two regions as the values of FDE are different. Then, computed FDE is combined with R-MAC features to improve the effectiveness of image retrieval.

Herein, we discuss the impact of different noises to our proposed FDE scheme. We conduct a set of comparative experiments on computing the values of FDE for a set of 50 images distilled randomly from Oxford5k dataset and after applying four kinds of noises on these images. We give the results in Figure 2. We could make a conclusion as follows. The values of FDE calculated by our proposed method are hardly sensitive to poisson, salt, and speckle noises. Gaussian noise might increase the value of FDE slimly in a very small number of images. We could conclude that our proposed FDE is robust to the most types of noise.

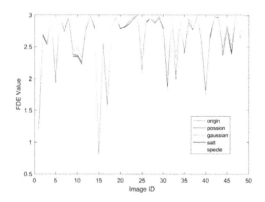

Figure 2. The feature distribution entropy (FDE) curve of 50 images applied four kinds of noise. The x-axis represents the image ID and y-axis denotes the values of FDE. The red curve is the FDE values of original images. The other four curves denote the FDE values of the same images with Poisson, Gaussian, salt, and speckle noise, respectively.

3.3. Fusing R-MAC Features with FDE

We aim to use FDE to reflect distribution information of feature regions to overcome the shortcomings of R-MAC. However, the format of FDE is quite different from R-MAC features. It is cautious for us to combine the FDE with R-MAC features. We give more detailed description of the combination schemes, as follows.

The simplest fusing method is to directly concatenate entropy with R-MAC features. There are two ways to concatenate them together. One is to concatenate the regional FDE with R-MAC features directly, which is represented as strategy 1. This strategy will increase the dimension of regional features from $R \times C$ to $2R \times C$. There is another way to fuse them, which would be called strategy 2. The main idea of strategy 2 is to sum the FDE for the whole feature map and then concatenate FDE vector with R-MAC features. The dimension would be increased to $(R + 1) \times C$. The two strategies will both increase the dimension of feature vector, which could further increase computational overhead in the process of image retrieval. We design strategy 3, which adopts the weighted summation to fuse the R-MAC features with the FDE without increasing the feature dimension. In strategy 3, the final feature vectors could be obtained by adjusting the weight parameter α. The experiment's results indicate that strategy 3 will induce better performance of image retrieval. We describe strategy 3 in detail, as follows.

In order to reduce the difference of R-MAC features and FDE, we perform an operation of L2 and power normalization on them separately. We define p_1 and p_2 as the power normalization parameter for R-MAC and FDE, respectively. After the normalization operation, the two parts are weighted summed together. The regional feature, which is produced by MAC pooling, is fused with the FDE for a certain feature map by weighted summation as follows:

$$f_c = [f_{c1}, \ldots, f_{cr}, \ldots, f_{cR}]^T, f_{cr} = f_{cr}^m + \alpha H_{cr} \tag{9}$$

where f_c denotes the regional feature vector of with FDE of the c-th channel and f_{cr} is the element in the regional feature vector of f_c, which is weighted summed with an element of R-MAC regional feature f_{cr}^m and an element of regional FDE H_{cr} of one feature map. α is an adjustable parameter. We set $\alpha = 0.5$ according to analysis in Section 3.4.

Then, we concatenate all these regional features to generate a feature vector. The concatenation of the element f_{cr} in the vector f_c yields the feature descriptors f_{cE} of each channel, and is calculated by the following formula:

$$f_E = [f_{1E}, \ldots, f_{cE}, \ldots, f_{CE}]^T, f_{cE} = \sum_{r=1}^{R} f_{cr} \tag{10}$$

where f_{cE} is computed by adding all these regional features

Lastly, we perform L2, power normalization, and PCA to generate more effective feature representations, which is significant to improve the performance of image retrieval. We define p_3 as the power normalization parameter for the fused features. The retrieval process with our fused feature descriptor is illustrated in Figure 3.

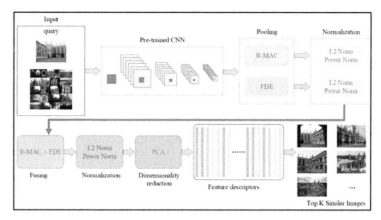

Figure 3. The process of image retrieval with our proposed fused features. The pre-trained convolutional neural network (CNN) is used to produce feature maps on the input query image and a retrieval set. The region maximum activations of convolutions (R-MAC) pooling and the proposed FDE are applied on feature maps and the normalization is applied on them separately. Then, weighted fusion strategy is utilized to produce raw features. We finally perform normalization and principal component analysis (PCA) on the fused features to generate the final feature descriptors, which are used in the following retrieval stage.

As is shown in Figure 3, we follow the deep framework to perform image retrieval. Firstly, we utilize the pre-trained network to extract feature maps. Then, we utilize the R-MAC pooling and FDE scheme mentioned in Section 3.2 to produce deep features for all regions. Then, we fuse them after the process of L2 and power normalization and add up all these regional fused vectors in one feature map to get compact vectors by using the strategy 3 mentioned in Section 3.3. Then, we obtain the final feature descriptors after the process of L2 and power normalization, and we apply the dimensionality reduction of PCA on the final compact feature descriptors, which are then used to perform image retrieval. The experiment results described in Section 4 demonstrate the effectiveness of our proposed method.

3.4. Parameter Analysis

In this section, we analyze the main parameters of our algorithm on Paris6k dataset [42] and measure the retrieval accuracy by mean average precision (mAP) [27] as the same to R-MAC. mAP is a commonly evaluation method to measure the performance of image retrieval. It is defined as follows: During the testing phase, we rank all these samples in the testing database by computing their Euclidean distance with the query. Then, we utilize the ranked list to get the average precision (AP) for each query image. Then, we use the AP to compute mAP as follows.

$$mAP = \frac{1}{|Q|} \sum_{i=1}^{|Q|} \frac{1}{n_i} \sum_{j=1}^{n_i} pr(i,j), pr(i,j) = \frac{k}{n} \tag{11}$$

where $|Q|$ is the amount of the query images in the dataset, n_i is the volume of images in testing dataset which is relevant to i-th query image, $Pr(i,j)$ is the precision of j-th retrieved image to i-th query image, k is the result images returned relevant to the query image, and n is the volume of returned images during retrieving.

In FDE computation, the number of blocks B in the histogram is an adjustable parameter. When fusing the R-MAC features with FDE, the weight α will affect the descriptive ability of the finally generated feature descriptors. In order to fuse the R-MAC features with the FDE, L2 and power normalization processing is required to make them range on the same level. After fusing, we perform a power normalization and L2 normalization again to facilitate subsequent training of PCA. As mentioned in Section 3.1 during the stage of testing, we use multiscale strategy to extract features. We conduct the experiment with different scale size L. And we would give the analysis as follows.

As mentioned in Section 3.2, the number of blocks B plays an important role in computing the FDE. We conduct the experiments on different B with range from 2 to 275 on AlexNet and we set α to 0.5 and p_1, p_2, p_3 to 1.0, 1.1, and 1.1. The results are presented in Table 1.

Table 1. The influence of B on retrieval.

.B	2	15	50	100	175	275
mAP(%)	75.01	74.44	74.21	74.12	74.06	74.03

From Table 1 we can learn that when the value of B becomes larger, the result gradually decreases. The mAP reaches the maximum with 75.01% when B is equal to 2, and we set B to 2 in the later experiments.

Factor α is a parameter used to fuse the computed FDE and R-MAC features. We perform the experiments with α from 0.2 to 1.2 on AlexNet and set B to 2, p_1, p_2, p_3 to 1.0, 1.1, and 1.1, respectively. The results are shown in Table 2.

Table 2. The influence of α on retrieval.

α	0.2	0.4	0.5	0.6	0.8	1.0	1.2
mAP(%)	73.69	73.95	74.05	73.95	73.80	73.58	73.31

From Table 2, we can know that the mAP (%) obtains the maximum value when $\alpha = 0.5$, and the maximum mAP value is bolded. According to the data in the table, $\alpha = 0.5$ is finally selected for the following experiments.

As described in Section 2.3, we utilize L2 and power normalization on R-MAC features and FDE. We conduct the following experiment to find the best values for power normalization. We conduct the experiments with different values from 0.2 to 2 on p_1, p_2, and p_3 separately with $\alpha = 0.5$ and $B = 2$. Then, we show the results in Figure 4.

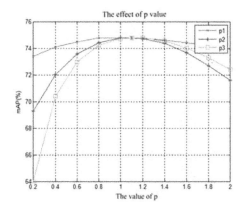

Figure 4. The changes of mean average precision (mAP) with p_1, p_2, p_3. The x-axis is the value of p, which could be used to denote the values of p_1, p_2, p_3. The y-axis is the value of mAP (%).

We could conclude that the three curves are convex function, which increase and then decrease monotonously. Figure 4 shows that p_3 is most sensitive to the values of mAP. The maximum values are obtained at 1.0, 1.1, and 1.1, respectively. We set the p_1, p_2, p_3 equal to 1.0, 1.1, and 1.1, respectively, in the rest of the experiments.

We conducted an analysis about different scale size to choose the most appropriate value for L. We performed the following experiment with different scale size for $L = 1, 2, 3, 4$ on AlexNet and VGG16 separately. E indicates feature distribution entropy

We can conclude from Table 3 that when L is smaller than 3, the mAP (%) will increase with the increase of L. However, when L is larger than 3, the mAP (%) will decrease with the increase of L. The best result is obtained when $L = 3$. We set the scale size L to 3 for the following experiments.

Table 3. The results with different scale size.

Methods.	AlexNet				VGG16			
L	1	2	3	4	1	2	3	4
R-MAC	47.9	54.6	56.1	55.6	57.3	64.5	66.9	67.44
R-MAC+E	54.00	57.03	**57.15**	56.38	64.04	67.97	**69.64**	69.09
MAC		44.83				55.01		

4. Experiments and Evaluation

In this chapter, we discuss the implementation details and evaluate our algorithm on different datasets. More details are shown as follows.

4.1. The Details of Implementation

All our experiments are implemented on Ubuntu 16.04 with a GPU NVIDIA TITAN X and memory of GPU is 64 GB. We use the deep learning tool MatConvNet [43] to realize the convolutional neural network. The experiments use AlexNet [44] and VGG16 [11] pre-trained on ImageNet [45]. We also finetune the network to achieve better performance. We use stochastic gradient descent (SGD) to train Alexnet and Adam to train VGG16. We initialize the learning rate with $l_0 = 10^{-3}$ for SGD and $l_0 = 10^{-6}$ for Adam. The channel of convolutional activation is 256 on AlexNet and 512 on VGG16. We set the input image of resolution to 1024×768. The cosine similarity is used to measure the similarity between the features. The experiments are performed on the Oxford5k [46] and Paris6k [42], Holidays [46], Oxford105k [27], and Paris106k [42]. We utilize mAP to measure the performance of image retrieval.

We conduct our experiments on the following benchmark datasets frequently used for image retrieval. Herein, we give the details of these datasets as follows.

Oxford5k [27] is a dataset provided by Flickr with a total of 5062 images. It contains 11 different landmarks of Oxford. Oxford5k [27] owns five query areas for each landmark building. Each image is labeled as one of four tags: Good, okay, junk, bad. The first two match the current query area; not good means that the error is matched.

Paris6k [42] is usually used in conjunction with Oxford5k [27] and it is also provided by Flickr with 11 classes. Each class has five query areas with a total of 6412 images about Paris buildings. It is also labeled with a four-category label, which is similar to Oxford.

Flickr100k [27] is made up of 1,000,071 high-resolution images from 145 of the most popular tags on Flickr, and is late added to the Oxford5k and Paris6k to become Oxford105k [27] and Paris106k [42] for large-scale image retrieval.

Holidays [46] mainly contains a variety of landscape pictures. It consists of 1491 images with 500 groups of similar images; each group has a query image. Unlike Oxford5k [27] and Paris6k [42], the query image on Holidays is the entire image rather than the region of interest (ROI).

4.2. The Calculation and Fusing Schemes of Entropy

In this section, we analyze the effect of different fusion strategies. As mentioned in Section 3.3, we show three strategies to fuse feature distribution entropy with R-MAC features. Strategy 1 stitches the region entropy directly to R-MAC features of each region. Strategy 2 is to stitch the entropy of the entire feature map to the R-MAC feature. Strategy 3 uses weighted summation. We use the pre-trained AlexNet and VGG16 to perform a series of comparative experiment with the three strategies on Paris6k, respectively. The results are shown in Table 4.

Table 4. The strategies of fusing with R-MAC features.

Network	Fusing	mAP(%)
	1	71.15
	2	73.56
AlexNet	3	**75.01**
	R-MAC	72.95
	1	79.46
	2	83.02
VGG16	3	**83.50**
	R-MAC	83.02

From the results shown in Table 4, we can make the following conclusions. The result of strategy 3 achieves 75.01% and is better than other fusion schemes when the experiment is conducted on AlexNet. Then, we perform the experiment of the three schemes on VGG16. We find that scheme 3 still gains the best result and obtains 83.50%. It can be seen from Table 4 that both scheme 2 and scheme 3 have been improved, but scheme 3 is more effective, and scheme 3 is adopted for the following experiments.

4.3. Compact Representation Comparison

In this section, we conduct the experiments with our proposed method and verify the compatibility of our proposed algorithm. To gain more, we perform PCA whitening to reduce the influence of noise with no dimensionality reduction. We use MAC, R-MAC, and our proposed method for the comparative experiments; the calculation and fusion method of FDE is selected in Section 4.1.

PCA whitening is one of the most important post-processing methods, which can reduce noise influence and improve retrieval efficiency. In order to verify the impact of PCA whitening on the retrieval results, we conducted comparative experiments on AlexNet [10] and VGG16 [47] on Paris6k [42], Oxford5k [27], and Holidays [46]. The results are shown in Table 5.

Table 5. Performance (mAP) comparison with or without PCA whitening. P: Performing PCA whitening (+P), E: Fusing with feature distribution entropy (+E), or without entropy. The best result is highlighted in bold.

Pooling.	AlexNet			VGG16		
	Paris6k	Oxford5k	Holidays	Paris6k	Oxford5k	Holiday
MAC+P	54.42	44.83	68.75	74.73	55.01	75.23
R-MAC	66.82	50.99	75.75	75.31	56.23	81.26
R-MAC+P	72.95	56.06	80.99	83.02	66.71	84.04
R-MAC+E	66.35	49.71	76.90	74.36	57.58	81.66
R-MAC+E+P	**75.01**	**57.15**	**82.76**	**83.56**	**69.64**	**86.90**

Table 5 shows the results of different feature descriptors before and after PCA whitening. It should be noted that whether using PCA whitening or not, the feature dimension in AlexNet is 256, the same as VGG16 is 512. We train PCA on Oxford5k and then use it to test on Holidays or Paris6k and similarly we train PCA on Paris6k to test on Oxfor5k. From this table we can learn that when we test on AlexNet, R-MAC+E+P on Paris6k, Oxford6k, and Holidays achieve 75.01%, 57.15%, and 82.76%, respectively, and the best retrieval results are obtained. When we perform the experiments on VGG16, the best results are 83.56%, 57.58%, and 86.90% on Paris6k, Oxford5k, and Holidays, respectively. We can also see that whether it is MAC, R-MAC, or R-MAC+E, the mAP value of using PCA compared with the one without using PCA has been significantly improved in most cases, with only a slight drop in the MAC method using AlexNet on Holidays. It is fully proved that PCA whitening can effectively improve retrieval performance in most cases.

Fusion representations. In order to verify the compatibility of the algorithm, we designed four sets of comparative experiments, using pre-trained networks on VGG16 and AlexNet to perform experiments. Oxford5k and Paris6k use the query area specified in the 55 query images given by the datasets. We compare the MAC and R-MAC features with features after fusing entropy by mAP (%). The results are shown in Table 6.

Table 6. Performance (mAP (%)) comparison between fusing with FDE or without. E: Fusing with FDE (+E), or without entropy. The best result is highlighted in bold.

Network.	Pooling	Oxford5k	Oxford105k	Paris6k	Paris106k	Holidays
AlexNet	MAC	44.83	34.84	54.42	37.09	68.75
	MAC+E	51.73	45.54	64.78	50.34	78.01
	R-MAC	56.06	46.85	72.95	60.07	80.99
	R-MAC+E	**57.15**	**50.19**	**75.01**	**63.29**	**82.76**
VGG16	MAC	55.01	48.50	74.73	62.46	75.23
	MAC+E	62.57	58.75	79.51	72.57	82.07
	R-MAC	66.71	62.35	83.02	76.28	84.04
	R-MAC+E	**69.64**	**64.91**	**83.56**	**77.89**	**86.90**

The conclusions drawn from Table 6 are as follows. The results in the table indicate that when we experiment on AlexNet, R-MAC+E obtains the best results on all these datasets with 57.15%, 50.19%, 75.01%, 63.29%, and 82.76%, which has been bolded. The same conclusion is obtained when we perform the experiments on VGG16. We get the results of 69.64%, 64.91%, 83.56%, 77.89%, and 86.90%, which are the maximum values on the different datasets. We have bolded these maximum values in Table 6. We can know that the fused feature representations generated by using our proposed method could be more effective and gain a better performance.

Re-ranking and QE. As mentioned in Section 2.4, re-ranking and QE can further improve the performance of image retrieval. In this section, we use the pooling methods above to calculate feature descriptors. Then, we examine the advantage of re-ranking and QE on Oxford5k, Paris6k, Oxford105k, and Paris106k. We show the results in Table 7.

Table 7. Performance (mAP(%)) comparison using re-ranking and query expansion or without, R: Using re-ranking (+R), QE: Using query expansion (+QE). E: Fusing with feature distribution entropy (+E), or without entropy. The best result is highlighted in bold.

Network	Pooling	Oxford5k	Oxford105k	Paris6k	Paris106k
AlexNet	MAC+R	59.01	46.26	65.67	45.61
	MAC+R+QE	63.92	50.21	69.13	49.08
	MAC+E+R	64.02	54.03	73.95	58.51
	MAC+E+R+QE	70.46	59.67	76.60	61.16
	R-MAC+R	61.13	55.16	77.52	65.63
	R-MAC+R+QE	66.85	60.68	80.42	69.02
	R-MAC+E+R	62.37	56.46	78.35	67.49
	R-MAC+E+R+QE	**67.83**	**61.01**	**81.04**	**70.75**
	MAC+R	70.57	60.39	81.16	64.29
	MAC+R+QE	74.21	63.65	82.84	69.01
	MAC+E+R	76.09	67.97	84.36	76.37
VGG16	MAC+E+R+QE	78.95	71.47	85.22	76.99
	R-MAC+R	74.54	70.89	85.16	79.29
	R-MAC+R+QE	77.33	74.69	86.45	80.73
	R-MAC+E+R	75.96	72.97	85.33	79.97
	R-MAC+E+R+QE	**78.48**	**75.79**	**86.53**	**81.22**

From Table 7, we could know that the best results are achieved when we apply re-ranking and QE on our proposed method with the mAP (%) being 67.83%, 61.01%, 81.04%, and 70.75% on the four different datasets on AlexNet. We get the same conclusion on VGG16 with the maximum values of mAP (%) being 78.48%, 75.79%, 86.53%, and 81.22%. We could conclude from Table 7 that the results of the features would likely increase when we apply the operation of re-ranking and QE on four different datasets on AlexNet and VGG16.

Comparison with state-of-the-art algorithms. In order to demonstrate the effectiveness and superiority of our algorithm, our experimental results are compared with other state-of-the-art algorithms. We conduct experiments not only with raw image representations, but also use representations performed with re-ranking and query expansion. The results can be seen in Table 8.

Table 8. Performance (mAP (%)) comparison with the state-of-the-art algorithms. Dim: Dimensionality of final compact image feature descriptors, Not Applicable (N/A) for the bag-of-visual-words (BoW)-CNN due to its sparse representations. R: Using re-ranking (+R), QE: Using QE (+QE). E: Fusing with feature distribution entropy (+E), or without entropy. The best result is highlighted in bold.

Network	Pooling	Dim	Oxford5k	Oxford105k	Paris6k	Paris106k	Holidays
			Original retrieval results				
	MAC [12]	256	44.24	34.84	54.42	37.09	68.75
AlexNet	R-MAC [24]	256	56.06	46.85	72.95	60.07	80.99
	R-MAC+E	256	57.15	50.19	75.01	63.29	82.76
VGG16	SPOC [15]	256	53.1	50.1	-	-	80.2
	uCrow [23]	256	66.7	61.2	73.9	65.8	81.5
	MFC [32]	256	68.4	62.9	83.4	-	-
	MAC [17]	512	55.01		74.73		75.23
	SPOC [15]	512	56.4	47.8	72.3	58.0	79.0
	uCrow [23]	512	69.7	64.1	78.6	71.0	83.9
	BoW-CNN [26]	N/A	73.9	59.3	82.0	64.8	-
	NetVLAD [16]	4096	55.5	-	67.7	-	82.1
	MFC [32]	512	70.6	65.3	83.3	-	-
	R-MAC [24]	512	66.71	62.35	83.02	76.28	84.04
	R-MAC+E	512	69.64	64.91	83.56	77.89	85.90
			After re-ranking (R) and query expansion (QE)				
AlexNet	MAC+R+QE [12]	256	63.92	50.21	69.13	49.08	-
	R-MAC+R+QE [24]	256	66.85	60.68	80.42	69.02	-
	R-MAC+E+R+QE	256	**67.83**	**61.01**	**81.04**	**70.75**	-
VGG16	Crow+QE [23]	512	74.9	70.6	84.8	79.4	-
	MAC+R+QE [12]	512	74.21	63.65	82.84	69.01	-
	R-MAC+R+QE [24]	512	77.33	74.69	86.45	80.73	-
	BoW-CNN+R+QE [26]	512	**78.8**	65.1	84.8	64.1	-
	R-MAC+E+R+QE	512	78.48	**75.79**	**86.53**	**81.22**	-

We can see from Table 8 that our proposed method obtains the best results in most cases. After performing re-ranking and query expansion, the performance has been significantly improved. According to the experimental data, our algorithm after re-ranking and query expansion achieves the best results in almost all categories, which fully demonstrates the effectiveness and superiority of the improved algorithm.

Test on fine-tuned network. The network proposed by Filip et al. [25] uses the contrastive loss function to train the parameters in the network. In order to adapt the network parameters to our algorithm and achieve better performance, four sets of comparative experiments were performed on Oxford5k, Oxford105k, Paris6k, and Paris106k and Holidays. The results are shown in Table 9.

Table 9. The results of image retrieval on five different datasets with fine-tuned network on AlexNet and VGG16. R: Using re-ranking (+R), QE: Using query expansion (+QE). E: Fusing with feature distribution entropy (+E), or without entropy.

Network	Pooling	Oxford5k	Oxford105k	Paris6k	Paris106k	Holidays
AlexNet	MAC	61.20	50.29	68.25	53.40	73.42
	MAC+E	**66.57**	**58.14**	74.13	61.61	78.95
	R-MAC	63.68	53.12	73.35	60.02	78.96
	R-MAC+E	64.96	54.96	**75.79**	**63.01**	**79.72**
VGG16	MAC	81.34	75.29	83.90	75.22	80.11
	MAC+E	**84.23**	**78.65**	**86.80**	**80.36**	82.25
	R-MAC	80.73	72.67	85.08	77.64	82.43
	R-MAC+E	81.91	73.82	85.91	79.06	**83.39**

We can make the following conclusion from the results achieved on fine-tuned network. When we experiment on AlexNet, MAC+E gains the best results on Oxford5k and Oxford105k. R-MAC+E achieve the best results on Paris6k, Paris106k, and Holidays. When we test on the fine-tuning network initialized with VGG16, the best results were obtained on MAC+E on Oxford5k, Oxford105k, Paris6k, and Paris106k, and on R-MAC+E on Holidays. The experimental results indicate that the feature distribution entropy can also be used in the fine-tuned network to promote the performance of image retrieval.

Experiment on medical dataset. To further demonstrate our proposed method, we conduct a set of experiments on medical dataset with the methods of MAC, MAC+E, R-MAC, and R-MAC+E on the AlexNet, which is pre-trained on ImageNet. The dataset for performing our experiments is composed of two public medical datasets of Brain_Tumor_Dataset [48] and Origa [49]. We present the results in Table 10.

Table 10. Performance (mAP (%)) comparison between fusing with FDE or without. E on medical dataset: Fusing with FDE (+E), or without entropy. The best result is highlighted in bold.

Pooling	mAP(%)
MAC	86.64
MAC+E	91.11
R-MAC	92.89
R-MAC+E	**92.93**

We can learn from Table 10 that our proposed method of fusing our FDE with R-MAC features obtains the best result with mAP is 92.93%, which is higher than R-MAC by 0.04%. The mAP of fusing MAC with our FDE would be increased by nearly 4.5% compared to MAC. The results in Table 10 shows that our proposed FDE is effective in improving the performance of image retrieval. Furthermore, the results demonstrate that our proposed method of fusing FDE with R-MAC outperforms the existing methods.

4.4. Discussion

We would like to give some discussion for our proposed method. Taking the distribution information in different regions into consideration and combining the distribution information with R-MAC features could obtain remarkable performance in image retrieval. We design an effective scheme to calculate FDE, which is significant for promoting the performance of image retrieval. Here, we propose a superior strategy of weighted summation to fuse our proposed FDE with R-MAC feature descriptors to generate more informative feature representations. Furthermore, the post-processing of re-ranking and QE would be helpful to promote the effectiveness of our proposed method. When we test the five public datasets on AlexNet, our method can achieve state-of-the-art performance in image retrieval. When we test on VGG 16, we obtain the best results for most datasets and acceptable results on Oxford5k lower than BoW-CNN.

5. Conclusions

In this paper, we proposed to make full use of the regional distribution information to generate more informative feature representations to promote the performance of image retrieval. We proposed to utilize FDE to reflect the difference of distribution information in different regions. We designed an effective scheme to calculate our proposed FDE, and the experimental results show that our FDE is effective to improve the performance of image retrieval. Then, we proposed a superior strategy to fuse the proposed FDE with R-MAC features to generate more effective deep representations, which could achieve prominent performance in the task of image retrieval. In order to demonstrate the compatibility of our proposed method, we also conducted the experiments with the fused features on different datasets on the pre-trained network. The results show that the performance with our proposed method outperforms that of the existing state-of-the-art methods. Furthermore, we used the post-process methods of re-ranking and QE to further improve the performance. Finally, we used the fine-tuned network and medical dataset to verify the effectiveness of our proposed method. We obtained state-of-the-art results with our proposed method on five different datasets.

Our method mainly focuses on how to make full use of the feature maps output from the pre-trained network. We would like to pay attention to train more suitable network for the task of image retrieval, and we would concentrate on improving the effectiveness and robust of our network by designing more effective loss function and network architecture in our following works.

Author Contributions: P.L. conceived the research subject of this paper, revised the paper, and directed this study. H.G. carried out the calculation of the FDE. G.G. drafted the paper and approved the final version to be published. D.Z., Q.Z., and H.Z. validated the results.

Funding: This work was supported by the Nature Science Foundation of China, under Grants 61841602, General Financial Grant from China Postdoctoral Science Foundation, under Grants 2015M571363 and 2015M570272,the Provincial Science and Technology Innovation Special Fund Project of Jilin Province, under Grant 20190302026GX, the Jilin Province Development and Reform Commission Industrial Technology Research and Development Project, under Grant 2019C054-4, and the State Key Laboratory of Applied Optics Open Fund Project, under Grant20173660.

Conflicts of Interest: The authors declare no conflict of interest.

References

1. Singhai, N.; Shandilya, S.K. A Survey On: Content Based Image Retrieval Systems. *Int. J. Comput. Appl.* **2010**, *4*, 22–26. [CrossRef]
2. Smeulders, A.; Worring, M.; Santini, S.; Gupta, A.; Jain, R. Content-based image retrieval at the end of the early. *IEEE Trans. Pattern Anal. Mach. Intell.* **2000**, *22*, 1349–1380. [CrossRef]
3. Zheng, L.; Yang, Y.; Tian, Q. SIFT Meets CNN: A Decade Survey of Instance Retrieval. *IEEE Trans. Pattern Anal. Mach. Intell.* **2018**, *40*, 1224–1244. [CrossRef] [PubMed]
4. Duanmu, X. Image Retrieval Using Color Moment Invariant. In Proceedings of the International Conference on Information Technology: New Generations, Las Vegas, NV, USA, 12–14 April 2010; pp. 200–203.

5. Lowe, D.G. Distinctive Image Features from Scale-Invariant Keypoints. *Int. J. Comput. Vis.* **2004**, *60*, 91–110. [CrossRef]
6. Sivic, J.; Zisserman, A. Video Google: a text retrieval approach to object matching in videos. In Proceedings of the Ninth IEEE International Conference on Computer Vision 2003, Nice, France, 13–16 October 2003; pp. 1470–1477.
7. Jegou, H.; Douze, M.; Schmid, C.; Perez, P. Aggregating local descriptors into a compact image representation. In Proceedings of the 2010 IEEE Conference on Computer Vision and Pattern Recognition (CVPR), San Francisco, CA, USA, 13–18 June 2010; pp. 3304–3311. [CrossRef]
8. Jegou, H.; Perronnin, F.; Douze, M.; Sanchez, J.; Perez, P.; Schmid, C. Aggregating Local Image Descriptors into Compact Codes. *Proc. IEEE Trans. Pattern Anal. Mach. Intell.* **2012**, *34*, 1704–1716. [CrossRef]
9. He, K.; Zhang, X.; Ren, S.; Sun, J. Spatial Pyramid Pooling in Deep Convolutional Networks for Visual Recognition. *IEEE Trans. Pattern Anal. Mach. Intell.* **2015**, *37*, 1904–1916. [CrossRef]
10. Krizhevsky, A.; Sutskever, I.; Hinton, G.E. ImageNet Classification with Deep Convolutional Neural Networks. *Neural Inf. Process. Syst.* **2012**, *141*, 1097–1105. [CrossRef]
11. Ng, J.Y.-H.; Yang, F.; Davis, L.S. Exploiting local features from deep networks for image retrieval. In Proceedings of the 2015 IEEE Conference on Computer Vision and Pattern Recognition Workshops (CVPRW), Boston, MA, USA, 7–12 June 2015; pp. 53–61.
12. Razavian, A.S.; Josephine, S.; Stefan, C.; Atsuto, M. Visual Instance Retrieval with Deep Convolutional Networks. *ITE Trans. Media Technol. Appl.* **2016**, *4*, 251–258. [CrossRef]
13. Szegedy, C.; Liu, W.; Jia, Y.Q.; Sermanet, P.; Reed, S.; Anguelov, D.; Erhan, D.; Vanhoucke, V.; Rabinovich, A. Going Deeper with Convolutions. In Proceedings of the 2015 IEEE Conference on Computer Vision and Pattern Recognition (CVPR), Boston, MA, USA, 7–12 June 2015; pp. 1–9.
14. Alzu'Bi, A.; Amira, A.; Ramzan, N. Content-based image retrieval with compact deep convolutional features. *Neurocomputing* **2017**, *249*, 95–105. [CrossRef]
15. Yandex, A.B.; Lempitsky, V. Aggregating Local Deep Features for Image Retrieval. In Proceedings of the IEEE International Conference on Computer Vision ICCV, Santiago, Chile, 7–13 December 2015.
16. Arandjelovic, R.; Gronat, P.; Torii, A.; Pajdla, T.; Sivic, J. NetVLAD: CNN architecture for weakly supervised place recognition. In Proceedings of the 2016 IEEE Conference on Computer Vision and Pattern Recognition (CVPR), Las Vegas, NV, USA, 27–30 June 2016. [CrossRef]
17. Razavian, A.S.; Azizpour, H.; Sullivan, J.; Carlsson, S. CNN Features off-the-shelf: An Astounding Baseline for Recognition. In Proceedings of the 2014 IEEE Conference on Computer Vision and Pattern Recognition Workshops (CVPRW), Columbus, OH, USA, 23–28 June 2014. [CrossRef]
18. Ren, S.; He, K.; Girshick, R.; Sun, J. Faster R-CNN: Towards Real-Time Object Detection with Region Proposal Networks. In Proceedings of the Neural Information Processing Systems Conference, Montreal, QC, Canada, 7–12 December 2015.
19. Girshick, R. Fast R-CNN. *arXiv* **2015**, arXiv:1504.08083.
20. Gong, Y.C.; Wang, L.W.; Guo, R.Q.; Lazebnik, S. Multi-scale Orderless Pooling of Deep Convolutional Activation Features. *Comput. Vis. ECCV* **2014**, *8695*, 392–407.
21. Mousavian, A.; Kosecka, J. Deep Convolutional Features for Image Based Retrieval and Scene Categorization. *arXiv* **2015**, arXiv:1509.06033.
22. Babenko, A.; Lempitsky, V. Aggregating Deep Convolutional Features for Image Retrieval. *arXiv* **2015**, arXiv:1510.07493.
23. Kalantidis, Y.; Mellina, C.; Osindero, S. Cross-Dimensional Weighting for Aggregated Deep Convolutional Features. In Proceedings of the European Conference on Computer Vision 2016, Amsterdam, The Netherlands, 8–16 October 2016; pp. 685–701.
24. Tolias, G.; Sicre, R.; Jégou, H. Particular object retrieval with integral max-pooling of CNN activations. *arXiv* **2015**, arXiv:1511.05879.
25. Radenović, F.; Tolias, G.; Chum, O. Fine-tuning CNN Image Retrieval with No Human Annotation. *IEEE Trans. Pattern Anal. Mach. Intell.* **2017**, *41*, 1655–1668. [CrossRef]
26. Mohedano, E.; Salvador, A.; Mcguinness, K.; Marques, F.; O'Connor, N.E.; Giro-I-Nieto, X. Bags of Local Convolutional Features for Scalable Instance Search. In Proceedings of the ACM on International Conference on Multimedia Retrieval, New York, NY, USA, 6–9 June 2016.

27. Philbin, J.; Chum, O.; Isard, M.; Sivic, J.; Zisserman, A. Object retrieval with large vocabularies and fast spatial matching. In Proceedings of the IEEE Conference on Computer Vision & Pattern Recognition, Minneapolis, MN, USA, 17–22 June 2007.

28. Chum, O.; Philbin, J.; Sivic, J.; Isard, M.; Zisserman, A. Total Recall: Automatic Query Expansion with a Generative Feature Model for Object Retrieval. In Proceedings of the IEEE International Conference on Computer Vision, Rio de Janeiro, Brazil, 14–21 October 2007.

29. Chum, O.; Mikulik, A.; Perdoch, M.; Matas, J. Total recall II: Query expansion revisited. In Proceedings of the Computer Vision and Pattern Recognition, Providence, RI, USA, 20–25 June 2011; pp. 889–896.

30. Chen, L.-C.; Zhu, Y.; Papandreou, G.; Schroff, F.; Adam, H. *Encoder-Decoder with Atrous Separable Convolution for Semantic Image Segmentation*; Springer: Berlin/Heidelberg, Germany, 2018; pp. 833–851.

31. Long, J.; Shelhamer, E.; Darrell, T. Fully convolutional networks for semantic segmentation. In Proceedings of the Computer Vision and Pattern Recognition, Providence, RI, USA, 20–25 June 2011; pp. 3431–3440.

32. Hao, J.; Wei, W.; Jing, D.; Tan, T. MFC: A multi-scale fully convolutional approach for visual instance retrieval. In Proceedings of the IEEE International Conference on Multimedia & Expo Workshops, Hong Kong, China, 10–14 July 2017.

33. Objects, O.; Illumination, V.I. A Novel Image Retrieval Based on a Combination of Local and Global Histograms of Visual Words. *Math. Probl. Eng.* **2016**, *2016*, 1–12.

34. Krapac, J.; Verbeek, J.; Jurie, F. Modeling spatial layout with fisher vectors for image categorization. In Proceedings of the International Conference on Computer Vision, Barcelona, Spain, 6–13 November 2011; pp. 1487–1494.

35. Koniusz, P.; Mikolajczyk, K. Spatial Coordinate Coding to reduce histogram representations, Dominant Angle and Colour Pyramid Match. In Proceedings of the IEEE International Conference on Image Processing, Brussels, Belgium, 11–14 September 2011.

36. Sánchez, J.; Perronnin, F.; Campos, T.D. Modeling the spatial layout of images beyond spatial pyramids. *Pattern Recogn. Lett.* **2012**, *33*, 2216–2223. [CrossRef]

37. Liu, P.; Zhuang, M.; Guo, H.; Wang, Y.; Ni, A. Adding spatial distribution clue to aggregated vector in image retrieval. *EURASIP J. Image Video Process.* **2018**, *2018*, 9. [CrossRef]

38. Shannon, C.E. A mathematical theory of communication. *Bell Syst. Tech. J.* **1948**, *27*, 379–423. [CrossRef]

39. Jegou, H.; Chum, O. Negative evidences and co-occurences in image retrieval: The benefit of PCA and whitening. In Proceedings of the European Conference on Computer Vision, Firenze, Italy, 7–13 October 2012; pp. 774–787.

40. Gordo, A.; Larlus, D. Beyond instance-level image retrieval: Leveraging captions to learn a global visual representation for semantic retrieval. In Proceedings of the 30th IEEE Conference on Computer Vision and Pattern Recognition (CVPR 2017), Honolulu, HI, USA, 21–26 July 2017. [CrossRef]

41. Mikolajczyk, K.; Matas, J. Improving Descriptors for Fast Tree Matching by Optimal Linear Projection. In Proceedings of the International Conference on Computer Vision, Rio De Janeiro, Brazil, 14–21 October 2007; pp. 1–8.

42. Philbin, J.; Chum, O.; Isard, M.; Sivic, J.; Zisserman, A. Lost in Quantization: Improving Particular Object Retrieval in Large Scale Image Databases. In Proceedings of the 2008 IEEE Conference on Computer Vision and Pattern Recognition, Anchorage, AK, USA, 23–28 June 2008; pp. 1–8.

43. Vedaldi, A.; Lenc, K. MatConvNet—Convolutional Neural Networks for MATLAB. *arXiv* **2014**, arXiv:1412.4564.

44. Krizhevsky, A.; Sutskever, I.; Hinton, G.E. ImageNet Classification with Deep Convolutional Neural Networks. *Commun. ACM* **2017**, *60*, 84–90. [CrossRef]

45. Deng, J.; Dong, W.; Socher, R.; Li, L.J.; Li, K.; Li, F.F. ImageNet: A Large-Scale Hierarchical Image Database. In Proceedings of the IEEE Conference on Computer Vision & Pattern Recognition, Miami, FL, USA, 20–25 June 2009.

46. Jegou, H.; Douze, M.; Schmid, C. Hamming Embedding and Weak Geometric Consistency for Large Scale Image Search. In Proceedings of the European Conference on Computer Vision, Marseille, France, 12–18 October 2008; pp. 304–317.

47. Simonyan, K.; Zisserman, A. Very Deep Convolutional Networks for Large-Scale Image Recognition. *arXiv* **2014**, arXiv:1409.1556.

48. Jun, C. Brain Tumor Dataset. *Figshare* **2017**. [CrossRef]
49. Zhang, Z.; Yin, F.S.; Liu, J.; Wong, W.K.D.; Tan, N.M.; Lee, B.H.; Cheng, J.; Wong, T.Y. ORIGA-light: An online retinal fundus image database for glaucoma analysis and research. In Proceedings of the International Conference of the IEEE Engineering in Medicine and Biology Society, Buenos Aires, Argentina, 31 August–4 September 2010; pp. 3065–3068.

 © 2019 by the authors. Licensee MDPI, Basel, Switzerland. This article is an open access article distributed under the terms and conditions of the Creative Commons Attribution (CC BY) license (http://creativecommons.org/licenses/by/4.0/).

Article

Improvement of Image Binarization Methods Using Image Preprocessing with Local Entropy Filtering for Alphanumerical Character Recognition Purposes

Hubert Michalak and Krzysztof Okarma *

Faculty of Electrical Engineering, West Pomeranian University of Technology, Szczecin, 70-313 Szczecin, Poland; michalak.hubert@zut.edu.pl
* Correspondence: okarma@zut.edu.pl

Received: 26 May 2019; Accepted: 2 June 2019; Published: 4 June 2019

Abstract: Automatic text recognition from the natural images acquired in uncontrolled lighting conditions is a challenging task due to the presence of shadows hindering the shape analysis and classification of individual characters. Since the optical character recognition methods require prior image binarization, the application of classical global thresholding methods in such case makes it impossible to preserve the visibility of all characters. Nevertheless, the use of adaptive binarization does not always lead to satisfactory results for heavily unevenly illuminated document images. In this paper, the image preprocessing methodology with the use of local image entropy filtering is proposed, allowing for the improvement of various commonly used image thresholding methods, which can be useful also for text recognition purposes. The proposed approach was verified using a dataset of 140 differently illuminated document images subjected to further text recognition. Experimental results, expressed as Levenshtein distances and F-Measure values for obtained text strings, are promising and confirm the usefulness of the proposed approach.

Keywords: image binarization; optical character recognition; local entropy filter; thresholding; image preprocessing; image entropy

1. Introduction

Image binarization is one of the most relevant preprocessing steps leading to significant decrease in the amount of information subjected to further analysis and allowing for an increase of its speed. Such an operation is typically applied in many systems which utilize mainly shape recognition methods and do not require the colour or texture analysis. Some good examples might be some robotic applications, including line followers and visual navigation in corridors and labyrinths, advanced driver-assistance systems (ADAS) and autonomous vehicles with lane tracking, as well as widely used optical character recognition (OCR) methods. Binary image analysis may also be applied successfully in embedded systems with limited amount of memory and low computational power.

Nevertheless, the appropriate results of binary image analysis, in particular text recognition, depend on the correct prior binarization. In some applications, where the uniform illumination of the scene can be ensured, e.g., popular flatbed scanners or some non-destructive automated book scanners, even with additional infrared cameras allowing for software straightening the scanned book pages [1], the simplest global thresholding may be sufficient. However, in many other situations the illumination may be non-uniform, especially in natural images captured by cameras, and therefore more sophisticated adaptive methods should be applied.

One of the most challenging problems related to the influence of image thresholding on further analysis is document image binarization and therefore newly developed algorithms are typically validated by using intentionally prepared document images containing various distortions. For this reason well-known document image binarization competitions (DIBCO) datasets are typically used to verify the usefulness and validate the advantages of binarization methods. These databases are prepared for yearly document image binarization competitions organized during two leading conferences in this field—the International Conference on Document Analysis and the Recognition (ICDAR) [2] and International Conference on Frontiers in Handwriting Recognition (ICFHR) [3], where the H-DIBCO datasets are used, containing only handwritten document images without machine printed samples. All DIBCO datasets contain not only the distorted document images but also "ground truth" binary images and therefore the binarization results can be compared with them at the pixel level analysing the numbers of correctly and improperly classified pixels [4,5].

Despite the fact that image binarization is not a new topic, some enhancements of algorithms are still proposed, particularly for historical document image binarization, as well as unevenly illuminated natural images. A proposal of such an improvement based on the image entropy filter, possible to apply in many commonly known binarization methods, is presented in this paper.

The rest of the paper consists of the short overview of the most widely used image binarization methods, description of the proposed approach based on the use of local entropy filter, presentation and discussion of results and final conclusions.

2. Brief Overview of Image Binarization Algorithms

Probably the most popular image thresholding method was proposed in 1979 by Nobuyuki Otsu [6], who delivered the idea of minimizing the sum of intra-class variances of two groups of pixels classified as foreground and background, assuming the bi-modal histogram of the image pixels' intensity. Hence, this approach leads to maximization of inter-class variance and therefore a good separation of two classes of pixels, represented finally as black and white, is achieved. Due to the operations on the histograms, this method is fast, although it works properly only for uniformly illuminated images with bi-modal histograms.

A similar approach, utilizing the entropy of the histogram instead of variances was proposed by Kapur et al. [7], whereas the idea of combining the global and local Otsu and Kapur methods was presented in the paper [8]. An extended adaptive version of Otsu method, known as AdOtsu, proposed by Moghaddam and Cheriet [9], assumed some additional operations such as multi-scale background estimation and calculation of average stroke widths and line heights. Since some images with unimodal histograms cannot be properly binarized using the above mentioned histogram-based methods another interesting idea was presented by Paul Rosin [10], who proposed to determine the threshold as the corner of the histogram curve.

Since the images containing some shadows being the result of non-uniform illumination should not be binarized using a single global threshold, some adaptive algorithms, which require the analysis of each pixels' neighbourhood, were proposed as well. The most popular approach developed by Wayne Niblack [11] assumed the determination of the local threshold as the average local intensity lowered by the local standard deviation scaled by the constant parameter k. A further modification of this approach, utilizing the additional normalization of the local standard deviation by its division by its maximum value in the image, is known as Sauvola method [12]. Its multi-scale version was further developed by Lazzara and Géraud [13].

A simple choice of the local threshold as the average of the minimum and the maximum intensity within the local window (so called midgray value) was proposed by John Bernsen [14], whereas Bradley and Roth [15] developed the method using the integral image for the calculation of the local mean intensity of the neighbourhood. The implementation of this method, also in the modified versions utilising the local median and Gaussian weighted mean, is available as MATLAB *adaptthresh* function.

Some other adaptive binarization methods were proposed by Wolf and Jolion [16], who used a relatively simple contrast maximization approach as a modification of Niblack's method, as well as Feng and Tan [17], where a similar idea based on the maximization of local contrast was used, however significantly slower due to the application of additional median filtering and bilinear interpolation. Another method proposed by Gatos et al. [18] utilizes a low-pass Wiener filtering and background estimation, followed by the use of Sauvola's thresholding with additional interpolation and post-processing using so called shrink and swell filters to remove noise and fill some foreground gaps and holes.

More recent document image binarization methods include the idea of region-based thresholding using Otsu's method with additional use of support vector machines (SVM) presented by Chou et al. [19] as well as faster region-based approaches [20,21]. Another method utilising the SVM-based approach with local features was presented recently by Xiong et al. [22].

The algorithm proposed by Howe [23] utilizes a Laplacian operator, Canny edge detection and graph cut method to find the threshold minimizing the energy. Erol et al. [24] proposed a more general approach related to the localization of text on a document captured by mobile phone camera using morphological operations for background estimation. Another background suppression method, although working properly mainly for evenly illuminated document images, was proposed by Lu et al. [25], whereas another attempt to the application of morphological operations was presented by Okamoto et al. [26].

Lelore and Bouchara [27] proposed the extended fast algorithm for document image restoration (FAIR) algorithm based on rough text localization and likelihood estimation followed by simple thresholding of the obtained super-resolution likelihood image. A multi-scale adaptive–interpolative method was proposed by Bag and Bhowmick [28], useful for faint characters. A method proposed by Su et al. [29] exploited adaptive image contrast map combined with results of Canny edge detection, whereas an attempt to use multiple thresholding methods was presented by Yoon et al. [30].

Some faster ideas of image thresholding based on the Monte Carlo method were proposed as well [31–33], where the simplified histogram of the image was approximated using the limited number of randomly chosen pixels. On the other hand, Khitas et al. [34] developed recently an algorithm based on median filtering used for estimation of the background information. An application of local features with Gaussian mixtures was examined in the paper [35], whereas Chen and Wang [36] used extended non-local means method followed by adaptive thresholding with additional postprocessing.

Bataineh et al. [37] developed an algorithm inspired by Niblack's and Sauvola's methods with additional application of dynamic windows. Further modifications of Niblack's method were proposed by Khurshid et al. [38], Kulyukin et al. [39] and recently by Samorodova and Samorodov [40]. A direct binarization scheme of colour document images based on multi-scale mean-shift algorithm with the use of modified Niblack's method was recently proposed by Mysoret al. [41]. A review of many modifications of Niblack inspired algorithms can be found in Saxena's paper [42], whereas many other approaches are discussed in some other survey papers [43–45]. Some earlier methods can also be found in *BinarizationShop* software developed by Deng et al. [46].

Some recent trends in image binarization are related to the use of variational models [47] and deep learning methods [48]. Recently, Vo et al. [49] proposed another supervised approach based on hierarchical deep neural networks. A comprehensive overview of many document image binarization algorithms can be found in the survey paper written by Sulaiman et al. [50].

An interesting method of binarization of non-uniformly illuminated images based on Curvelet transform followed by Otsu's thresholding was proposed by Wen et al. [51]. However, the application of this algorithms requires the additional nonlinear enhancement functions and time-consuming multi-scale processing.

Some of the binarization methods utilize the calculation of histogram entropy as well as image entropy. The most widely known approach proposed by Kapur et al. [7] may be considered as the modification of the classical Otsu's thresholding, which is based on earlier ideas presented by Thierry Pun [52,53]. Fan et al. [54] proposed a method maximizing the 2D temporal entropy, whereas Abutaleb [55] developed a method which uses pixel's grey level as well the average of its neighbourhood for minimization of two-dimensional entropy. Brink and Pendock [56] used the cross-entropy instead of distance or similarity between the original image and the result of binarization to optimize the threshold. Some similar multilevel methods have been further developed as well for image segmentation [57], also with the use of genetic methods [58]. A ternary entropy-based method [59], based on the classification of pixels into text, near-text, and non-text regions was proposed as well, which utilized Shannon entropy, whereas Tsallis entropy was used by Tian and Hou [60]. Nevertheless, entropy-based methods are generally less popular than simple histogram-based thresholding or some adaptive binarization methods. Apart from the typical image binarization, one can find some other applications of entropy related to classification of signals or images obtained as the results of measurements or some other experiments, e.g., in a gearbox testing system presented by Jiang et al. [61], where Shannon entropy of the vibration signal is used to detect worn and cracked gears.

Development of any new image processing algorithms usually requires their reliable validation based on the comparison of the obtained results with the other methods. Stathis et al. [62] proposed a method of evaluation of binarization algorithms based on comparison of individual pixels, using the pixel error rate (PERR), peak signal to noise ratio (PSNR) and similar metrics, whereas some other approaches were presented in the survey paper by Sezgin and Sankur [63]. A much more popular approach is the use of typical classification metrics based on precision, recall, sensitivity, specificity or F-Measure [4,5], as well as the application of misclassification penalty metric (MPM) [64] or distance reciprocal distortion (DRD) [65]. Another binarization assessment method was presented by Lins et al. [66], which utilizes a dataset of synthetic images for comparison of various thresholding algorithms. Nevertheless, considering the final results of the document image recognition as the recognized text strings, a more useful approach would be the application of metrics calculated for characters instead of individual pixels. Apart from F-Measure, some metrics dedicated for text strings, such as Levenshtein distance, defined as the number of character operations necessary to convert one string into another, may be applied as well.

3. Proposed Method and Its Experimental Verification

3.1. Description of the Method

Analysing the unevenly illuminated document images, important information can be achieved with the use of the local image entropy, which may be calculated using the MATLAB entropyfilt function. Using its default parameters the local measure of randomness of the grey levels of the neighbourhood defined by the 9×9 pixels mask was achieved and stored as the result for the central pixel. Such an approach may be useful for image forgery detection, switching purposes in adaptive median filtering as well as for image preprocessing followed by comparison of properties of image regions. Hence, the local entropy filter was considered in the proposed method as one of the preprocessing steps for adaptive image binarization of unevenly illuminated document images subjected to further optical text recognition.

It is worth noting that most of the OCR engines used some "built-in" thresholding procedures and therefore their results are dependent also on the quality of the input data. For example, widely used freeware Tesseract OCR developed by Google utilized global Otsu's thresholding, whereas the commercial ABBYY FineReader software employed the adaptive Bradley's method. Therefore, the application of some other image binarization methods may improve or decrease the recognition accuracy, since the OCR "internal" thresholding does not change the input binary image. Hence, prior image thresholding may be considered as a replacement of the default methods used in the OCR engines.

The proposed method caused the equalization of illumination of an image, increasing also its contrast, making it easier to conduct the proper binarization and further recognition of alphanumerical characters. It is based on the analysis of the local entropy, assuming its noticeably higher values in the neighbourhood of the characters. Hence, only the relatively high entropy regions should be further analyzed as potentially containing some characters, whereas low entropy regions may be considered as the background. The proposed algorithm consists of the following steps:

- entropy filter—calculation of the local entropy using the predefined mask (in our experiments the most appropriate size is 19×19 pixels) leading to the local entropy map;
- negative—simple negation leads to more readable dark characters on a bright background; assuming the maximum entropy value equal to eight (considering eight bits necessary to store 256 grey levels), the additional normalization can be applied with the formula $Y = 1 - \frac{X}{8}$, where X is the local entropy map and the final range of the output image Y is $\langle 0; 1 \rangle$;
- thresholding—one of the global binarization methods may be used for this purpose, in our experiments the classical Otsu's thresholding was used, leading to the image M with segmented regions containing text and representing the background;
- masking—the obtained binary image M was used as the mask for the original input image, leading to the background image B with removed text regions;
- morphological dilation—the purpose of this operation was to fill the gaps containing the characters making it possible to obtain a full estimate of the background; a critical element of this step is an appropriate choice of the size of the structuring element (in our experiments the square 20×20 pixels one was sufficient and larger structuring elements caused an increase of the computation time);
- background subtraction—the expected result of the subtraction of the background estimate from the original input image should contain a bright text and the dark background with equalized illumination;
- negation with increase of contrast—a simple operation leading to the dark text and the bright background with improved readability;
- final binarization—the last step conducted after pre-processing, which can utilize any of commonly used binarization methods (in our experiments good results were obtained using adaptive Bradley's and Niblack's thresholding).

The simplified flowchart of the method is shown in Figure 1, whereas the illustration of results obtained after consecutive steps of the algorithm is presented in Figure 2.

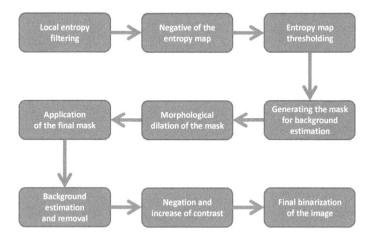

Figure 1. The simplified flowchart of the proposed method.

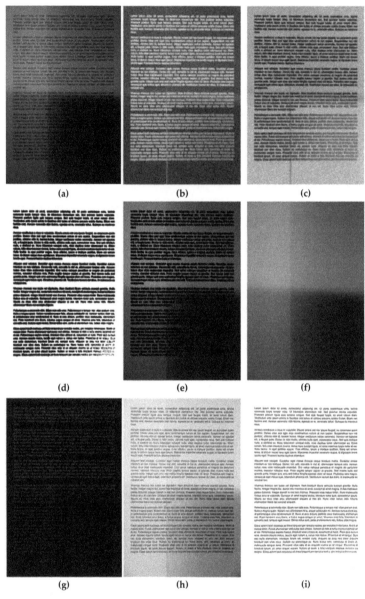

Figure 2. Results of the consecutive steps of the proposed algorithm obtained for an exemplary document image: (**a**) original input image, (**b**) local entropy map, (**c**) normalized negative entropy image, (**d**) binarized entropy image, (**e**) result of masking, (**f**) dilated masked image being the full background estimate, (**g**) result of background subtraction, (**h**) negative with eliminated background, and (**i**) final result of adaptive Niblack's thresholding after preprocessing.

3.2. Practical Verification

The verification of the proposed method was conducted using the database of document images, prepared applying various illuminations (uniform lighting and six types of non-uniform or directional shadows). The well-known quasi-Latin text *Lorem ipsum*, used as the basis for the generated sample pages containing 536 words, was printed using five various font shapes (Arial, Times New Roman, Calibri, Verdana and Courier) and their style modifications (normal, bold, italics and bold+italics). Such printed 20 sheets of paper were photographed applying 7 types of illuminations mentioned above (six unevenly illuminated examples are shown in Figure 3). These 140 captured images were binarized in two scenarios: with and without the proposed preprocessing. In both cases several binarization algorithms were applied to verify the proposed approach in practice. All the obtained binary images were used as the input data for the Google Tesseract OCR engine. For each of the images, the number of correctly and incorrectly recognized characters were determined, allowing for the calculation of some typical classification metrics, such as F-Measure defined as:

$$FM = 2 \cdot \frac{PR \cdot RC}{PR + RC}, \tag{1}$$

where PR and RC stand for the precision (true positives to sum of all positives ratio) and recall (ratio of true positives to sum of true positives and false negatives). Hence, they can be expressed as:

$$PR = \frac{TP}{TP + FP} \quad \text{and} \quad RC = \frac{TP}{TP + FN}, \tag{2}$$

where TP are true positives and FN false negatives, respectively. All positive and negative values are considered as the numbers of correctly and incorrectly recognized characters.

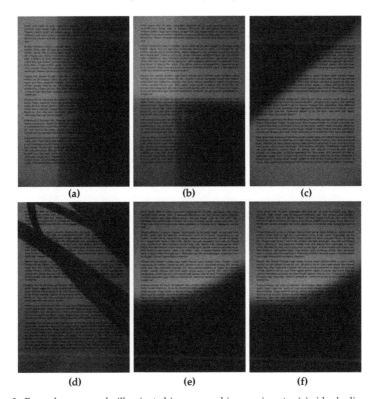

Figure 3. Exemplary unevenly illuminated images used in experiments: (**a**) side shading—series #2, (**b**) shading from the bottom—series #3, (**c**) diagonal shading—series #4, (**d**) irregular sharp shadow edges—series #5, (**e**) arc type shadows—series #6, (**f**) overexposure in the central part with underexposed boundaries—series #7.

The additional metric, which may be applied for the evaluation of text similarity, is known as Levenshtein distance, representing the minimum number of text changes (insertions, deletions or substitutions of individual characters) necessary to change the analyzed text into another. This metric was also applied for evaluation purposes, assuming the knowledge of the original text string (Lorem ipsum-based in these experiments).

4. Results and Discussion

The development of the final preprocessing algorithm allowing for the increase of the final OCR accuracy required an appropriate choice of some parameters mentioned earlier. The first of them is the size of the block used for the entropy filter which influences significantly the obtained results. Too small size of the filter would not be efficient due to its sensitivity to small details and noise whereas too big windows would be vulnerable to averaging effects. Since the default size of the filter in MATLAB entropyfilt function is 9×9 pixels, the first experiments were conducted using various windows to verify the influence of their size on the OCR results. The obtained results are presented in Figure 4, where the best values can be observed for 19×19 pixels filter. Therefore, the application of the default values would be inappropriate, particularly for the series #5 containing the non-uniformly illuminated images with sharp shadow edges as shown in Figure 3d.

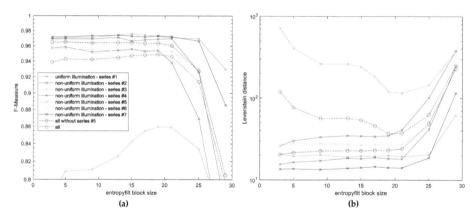

Figure 4. Experimental OCR results obtained for various size of blocks applied in the entropy filter: (**a**) F-Measure values, (**b**) Levenshtein distance.

A similar difference may be observed during the choice of the most appropriate size of the structuring element applied during the morphological dilation, since the results obtained for the series #5 differ significantly from the others. Nevertheless, in all cases the choice of a similar size of the structuring element to the size of the block in the entropy filter leads to the best results as illustrated in Figure 5 (in our experiments 20×20 pixels structuring element was chosen).

The additional reason of the choice of such structuring element was the processing time, which increased noticeably for bigger structuring elements as shown in Figure 6, where its values normalized according to the computation time obtained using the selected 20×20 pixels structuring element are presented. Unfortunately, relatively shorter processing did not guarantee good enough OCR accuracy, whereas increase of the structuring element's size and computation time did not enhance the obtained results significantly. Since the experiments were conducted using a personal computer, some processes running in background (including the Tesseract OCR engine) might have influenced the obtained results. Nonetheless, the relation between the size of structuring element and the processing time can be considered as nearly linear. Hence, the most reasonable choice was the smallest possible structuring element not affecting the acceptable OCR accuracy level.

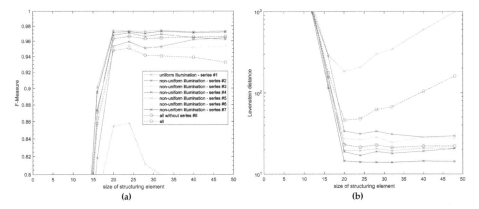

Figure 5. Experimental optical character recognition (OCR) results obtained for various size of structuring element applied for morphological dilation: (**a**) F-Measure values, (**b**) Levenshtein distance.

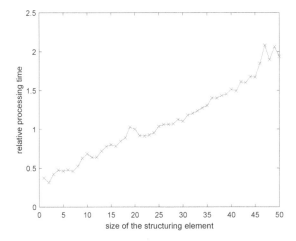

Figure 6. Normalized processing time for various size of structuring elements used in morphological dilation relatively to the time obtained applying the 20 × 20 pixels structuring element.

Having chosen the most appropriate parameters of the proposed preprocessing method, the obtained F-Measure values and Levenshtein distances for the whole dataset and each of the illumination types, as well as individual font faces and style modifications, were compared with some other methods applied without the proposed preprocessing. The comparison of the influence of the proposed preprocessing method on the F-Measure values is presented in Table 1, whereas respective Levenshtein distances are shown in Table 2. Analysing the results, a significant decrease of the Levenshtein distance, as well as the increase of the F-Measure values, may be observed for all methods, proving the usefulness of the proposed approach. The best results were achieved for Niblack, Sauvola and Wolf thresholding, as well as the simple Meanthresh method, which was significantly improved by the use of the entropy filtering-based preprocessing.

Table 1. Comparison of F-Measure values obtained for various binarization methods with and without the proposed preprocessing.

Binarization Method	Series							
	#1	#2	#3	#4	#5	#6	#7	All
None	0.9638	0.6201	0.8139	0.6650	0.6693	0.7460	0.6260	0.7291
+ preprocessing	0.9728	0.6475	0.8729	0.7272	0.8167	0.8027	0.9584	0.8283
Otsu (global) [6]	0.9614	0.6281	0.7908	0.6662	0.6841	0.7598	0.6583	0.7355
+ preprocessing	0.9737	0.6400	0.8573	0.7312	0.8049	0.7947	0.9561	0.8226
Region-based [20]	0.9616	0.7579	0.8661	0.8407	0.7737	0.8318	0.9528	0.8550
+ preprocessing	0.9525	0.8377	0.8861	0.8254	0.7438	0.8468	0.9104	0.8575
Niblack [11]	0.9614	0.7920	0.8668	0.8444	0.8510	0.8567	0.9589	0.8759
+ preprocessing	0.9596	0.9439	0.9451	0.9516	0.8878	0.9436	0.9674	0.9427
Sauvola [12]	0.9709	0.9581	0.9646	0.9722	0.7660	0.9655	0.9721	0.9385
+ preprocessing	0.9674	0.9635	0.9665	0.9668	0.8401	0.9671	0.9694	0.9487
Wolf [16]	0.9661	0.9482	0.9513	0.9514	0.7614	0.9594	0.9703	0.9297
+ preprocessing	0.9691	0.9661	0.9643	0.9662	0.8561	0.9621	0.9657	0.9499
Bradley (mean) [15]	0.9665	0.9191	0.9093	0.8484	0.7369	0.8976	0.9699	0.8925
+ preprocessing	0.9666	0.8896	0.9169	0.9262	0.8040	0.9103	0.9642	0.9111
Bradley (Gaussian) [15]	0.9663	0.8521	0.8295	0.7528	0.7267	0.7907	0.9489	0.8381
+ preprocessing	0.9678	0.8863	0.8991	0.8741	0.7521	0.8786	0.9124	0.8815
Feng [17]	0.9110	0.3782	0.7924	0.6312	0.7292	0.7938	0.8461	0.7285
+ preprocessing	0.9261	0.4418	0.7990	0.6489	0.7103	0.8076	0.8688	0.7432
Bernsen [14]	0.6948	0.6414	0.6844	0.6467	0.6286	0.7122	0.7245	0.6764
+ preprocessing	0.6971	0.6688	0.6938	0.6752	0.6312	0.7047	0.7141	0.6836
Meanthresh	0.9597	0.7348	0.8314	0.7921	0.8317	0.7947	0.9308	0.8393
+ preprocessing	0.9651	0.9570	0.9596	0.9602	0.8970	0.9606	0.9684	0.9525

Table 2. Comparison of Levenshein distances obtained for various binarization methods with and without the proposed preprocessing.

Binarization Method	Series							
	#1	#2	#3	#4	#5	#6	#7	All
None	56.40	1897.20	1031.80	1362.40	1548.30	1387.90	1815.50	1299.93
+ preprocessing	10.90	1665.10	718.40	1045.20	512.55	1063.85	68.15	726.31
Otsu (global) [6]	62.75	1878.20	1039.80	1393.40	1514.55	1358.55	1715.80	1280.44
+ preprocessing	12.60	1671.85	720.05	1047.05	514.20	1066.75	76.10	729.80
Region-based [20]	27.30	537.40	388.35	217.50	294.35	423.60	44.75	276.18
+ preprocessing	27.40	133.55	78.90	141.60	378.55	166.30	48.15	139.21
Niblack [11]	30.50	560.55	359.95	388.00	222.10	398.05	31.15	284.33
+ preprocessing	26.00	42.90	35.40	25.55	79.45	32.20	16.55	36.86
Sauvola [12]	20.30	22.85	17.35	14.80	651.60	17.75	12.40	108.15
+ preprocessing	22.40	30.25	23.05	17.35	197.55	19.75	15.95	46.61
Wolf [16]	21.35	54.90	69.90	74.05	923.65	58.55	17.60	174.29
+ preprocessing	21.45	27.75	19.75	23.50	202.75	17.80	16.65	47.10
Bradley (mean) [15]	26.45	63.15	157.15	389.45	1231.95	188.25	17.35	296.25
+ preprocessing	26.30	75.60	52.05	44.10	312.15	54.05	19.05	83.33
Bradley (Gaussian) [15]	27.10	355.80	731.00	950.75	1282.40	1136.70	32.25	645.14
+ preprocessing	25.75	91.05	193.00	219.95	700.15	149.50	19.95	199.91
Feng [17]	66.20	2518.00	1069.50	1507.50	1030.50	1037.10	174.20	1057.57
+ preprocessing	59.15	2385.25	1015.75	1435.10	887.75	945.70	142.30	981.57
Bernsen [14]	467.75	1471.25	1071.00	1273.65	1634.15	1167.40	623.10	1101.19
+ preprocessing	490.40	1178.10	1046.75	1011.85	1402.25	1093.35	687.35	987.15
Meanthresh	20.85	776.30	529.10	519.85	250.70	763.35	72.40	418.94
+ preprocessing	21.95	26.10	21.55	17.25	81.65	20.00	14.20	28.96

Entropy **2019**, *21*, 562

Some exemplary results obtained using the proposed preprocessing as well as its application for Bradley binarization with Gaussian kernel are illustrated in Figure 7. The additional illustration of its advantages for three exemplary images with the use of Niblack and Sauvola methods is shown in Figure 8, whereas another such comparison for Bernsen and Meanthresh methods is presented in Figure 9.

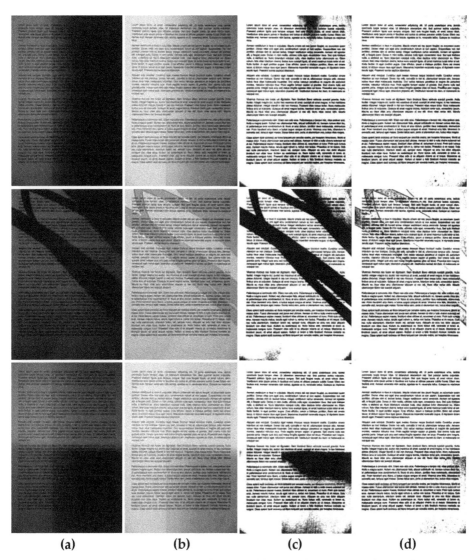

| **(a)** | **(b)** | **(c)** | **(d)** |

Figure 7. Comparison of binarization results obtained for exemplary unevenly illuminated images before the binarization: (**a**) without preprocessing, (**b**) with the proposed preprocessing, as well as using the Bradley method with a Gaussian kernel: (**c**) without preprocessing, (**d**) with the proposed preprocessing.

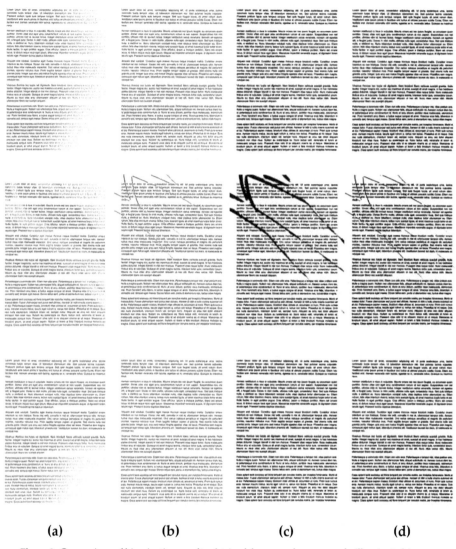

(a)   (b)   (c)   (d)

Figure 8. Comparison of binarization results obtained for exemplary unevenly illuminated images using the Niblack method: (**a**) without preprocessing, (**b**) with the proposed preprocessing, as well as Sauvola thresholding: (**c**) without preprocessing, (**d**) with the proposed preprocessing.

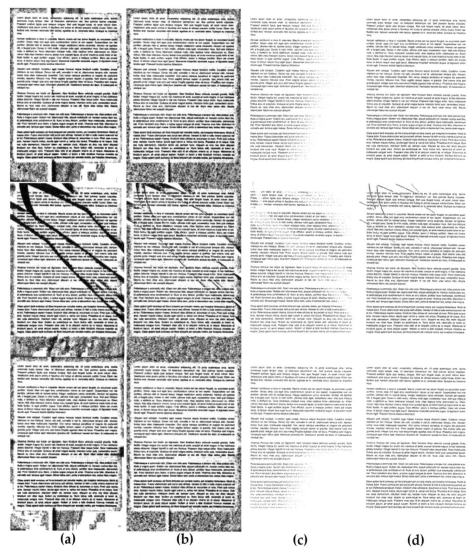

| (a) | (b) | (c) | (d) |

Figure 9. Comparison of binarization results obtained for exemplary unevenly illuminated images using the Bernsen method: (**a**) without preprocessing, (**b**) with the proposed preprocessing, as well as using the Meanthresh: (**c**) without preprocessing, (**d**) with the proposed preprocessing.

Since the properties of the proposed method may differ for various font shapes and styles, particularly for some of the thresholding algorithms, more detailed results are presented for them in Tables 3 and 4, where F-Measure values can be compared for the same methods with and without the proposed entropy-based preprocessing method.

Table 3. Comparison of F-Measure values obtained for various binarization methods with and without the proposed preprocessing for various font faces.

Binarization Method	Font Face				
	Arial	Times New Roman	Calibri	Courier	Verdana
None	0.7556	0.7432	0.7374	0.6483	0.7612
+ preprocessing	0.8541	0.8277	0.7886	0.8173	0.8539
Otsu (global) [6]	0.7489	0.7528	0.7525	0.6598	0.7637
+ preprocessing	0.8506	0.8214	0.7802	0.8058	0.8548
Region-based [20]	0.8726	0.8799	0.8738	0.7970	0.8514
+ preprocessing	0.8513	0.8689	0.8590	0.8425	0.8659
Niblack [11]	0.8776	0.9012	0.8729	0.8499	0.8777
+ preprocessing	0.9463	0.9550	0.9475	0.9195	0.9452
Sauvola [12]	0.9395	0.9476	0.9412	0.9239	0.9402
+ preprocessing	0.9555	0.9540	0.9450	0.9395	0.9495
Wolf [16]	0.9399	0.9507	0.9355	0.8826	0.9400
+ preprocessing	0.9567	0.9558	0.9551	0.9310	0.9511
Bradley (mean) [15]	0.9036	0.9004	0.8946	0.8676	0.8963
+ preprocessing	0.9158	0.9186	0.9158	0.8906	0.9147
Bradley (Gaussian) [15]	0.8475	0.8448	0.8434	0.8087	0.8463
+ preprocessing	0.9004	0.8992	0.8873	0.8609	0.8599
Feng [17]	0.7137	0.7430	0.7113	0.7528	0.7210
+ preprocessing	0.7368	0.7462	0.7304	0.7540	0.7487
Bernsen [14]	0.6735	0.6970	0.6938	0.6213	0.6971
+ preprocessing	0.7062	0.6917	0.6956	0.6041	0.7202
Meanthresh	0.8251	0.8698	0.8483	0.8197	0.8337
+ preprocessing	0.9511	0.9623	0.9516	0.9429	0.9548

Table 4. Comparison of F-Measure values obtained for various binarization methods with and without the proposed preprocessing for various font styles.

Binarization Method	Font Style			
	Normal	Bold	Italic	Bold + Italic
None	0.6945	0.7497	0.7221	0.7291
+ preprocessing	0.8049	0.8455	0.8076	0.8283
Otsu (global) [6]	0.7095	0.7544	0.7272	0.7355
+ preprocessing	0.7980	0.8426	0.8038	0.8226
Region-based [20]	0.8631	0.8444	0.8700	0.8550
+ preprocessing	0.8621	0.8590	0.8593	0.8575
Niblack [11]	0.8781	0.8898	0.8669	0.8759
+ preprocessing	0.9396	0.9444	0.9424	0.9427
Sauvola [12]	0.9366	0.9377	0.9340	0.9385
+ preprocessing	0.9464	0.9545	0.9463	0.9487
Wolf [16]	0.9165	0.9430	0.9223	0.9297
+ preprocessing	0.9428	0.9567	0.9467	0.9499
Bradley (mean) [15]	0.8888	0.8942	0.8916	0.8925
+ preprocessing	0.9031	0.9230	0.9099	0.9111
Bradley (Gaussian) [15]	0.8342	0.8418	0.8370	0.8381
+ preprocessing	0.8801	0.8738	0.8754	0.8815
Feng [17]	0.7333	0.7342	0.7391	0.7285
+ preprocessing	0.7368	0.7458	0.7518	0.7432
Bernsen [14]	0.6722	0.6786	0.6718	0.6764
+ preprocessing	0.6656	0.7060	0.6573	0.6836
Meanthresh	0.8379	0.8454	0.8381	0.8393
+ preprocessing	0.9547	0.9519	0.9541	0.9525

Comparing the influence of the proposed approach on the obtained OCR accuracy expressed as the F-Measure values calculated for individual text characters, relatively smaller enhancement may be observed for adaptive binarization methods, which achieve good results even without the proposed preprocessing method, such as Niblack or Sauvola. Nevertheless, in all cases the improvements may be noticed, also for the binarization method proposed by Wolf, which achieved much worse results for Courier fonts without the presented preprocessing method. A great improvement may also be observed for the simple mean thresholding as well as the direct usage of OCR engine's built-in binarization, whereas the proposed method caused a small decrease of recognition accuracy after Bernsen thresholding for some font shapes (Courier and Times New Roman). It is worth to note that the proposed entropy-based preprocessing method always leads to better text recognition of bold fonts.

5. Conclusions

Binarization of unevenly illuminated and degraded document images is still an open and challenging field of research. Considering the necessity of fast image processing, many sophisticated methods, which cannot be effectively applied in many applications, may be replaced by simpler thresholding supported by less complicated preprocessing methods without the necessity of shape analysis or training procedures.

The approach proposed in the paper may be efficiently applied as the preprocessing step for many binarization methods in the presence of non-uniform illumination of document images, increasing significantly the accuracy of further text recognition, as shown in experimental results. Since its potential applicability is not limited to binarization of document images for OCR purposes, our further research may concentrate on the development of similar approaches for some other applications related to binarization of natural images and machine vision in robotics, particularly in unknown lighting conditions.

Author Contributions: H.M. worked under the supervision of K.O., H.M. prepared the data and sample document images, H.M. and K.O. designed the concept and methodology and proposed the algorithm, H.M. implemented the algorithm, performed the calculations and made the data visualization, K.O. validated the results and wrote the final version of the paper.

Funding: This research received no external funding.

Conflicts of Interest: The authors declare no conflict of interest.

Abbreviations

The following abbreviations are used in this manuscript:

ADAS	advanced driver-assistance system
DIBCO	document image binarization competition
DRD	distance reciprocal distortion
FAIR	fast algorithm for document image restoration
ICDAR	International Conference on Document Analysis and Recognition
ICFHR	International Conference on Frontiers in Handwriting Recognition
MPM	misclassification penalty metric
OCR	optical character recognition
PERR	pixel error rate
PSNR	peak signal to noise ratio
SVM	support vector machines

References

1. Guizzo, E. Superfast Scanner Lets You Digitize Book by Flipping Pages. Available online: https://spectrum. ieee.org/automaton/robotics/robotics-software/book-flipping-scanning (accessed on 3 June 2019).
2. Pratikakis, I.; Zagoris, K.; Barlas, G.; Gatos, B. ICDAR2017 Competition on Document Image Binarization (DIBCO 2017). In Proceedings of the 2017 14th IAPR International Conference on Document Analysis and Recognition (ICDAR), Kyoto, Japan, 9–15 November 2017; Volume 1, pp. 1395–1403. [CrossRef]
3. Pratikakis, I.; Zagori, K.; Kaddas, P.; Gatos, B. ICFHR 2018 Competition on Handwritten Document Image Binarization (H-DIBCO 2018). In Proceedings of the 2018 16th International Conference on Frontiers in Handwriting Recognition (ICFHR), Niagara Falls, NY, USA, 5–8 August 2018; pp. 489–493. [CrossRef]
4. Ntirogiannis, K.; Gatos, B.; Pratikakis, I. Performance evaluation methodology for historical document image binarization. *IEEE Trans. Image Process.* **2013**, *22*, 595–609. [CrossRef] [PubMed]
5. Sokolova, M.; Lapalme, G. A systematic analysis of performance measures for classification tasks. *Inf. Process. Manag.* **2009**, *45*, 427–437. [CrossRef]
6. Otsu, N. A threshold selection method from gray-level histograms. *IEEE Trans. Syst. Man Cybern.* **1979**, *9*, 62–66. [CrossRef]
7. Kapur, J.; Sahoo, P.; Wong, A. A new method for gray-level picture thresholding using the entropy of the histogram. *Comput. Vis. Graph. Image Process.* **1985**, *29*, 273–285. [CrossRef]
8. Lech, P.; Okarma, K.; Wojnar, D. Binarization of document images using the modified local-global Otsu and Kapur algorithms. *Przegląd Elektrotechniczny* **2015**, *91*, 71–74. [CrossRef]
9. Moghaddam, R.F.; Cheriet, M. AdOtsu: An adaptive and parameterless generalization of Otsu's method for document image binarization. *Pattern Recognit.* **2012**, *45*, 2419–2431. [CrossRef]
10. Rosin, P.L. Unimodal thresholding. *Pattern Recognit.* **2001**, *34*, 2083–2096. [CrossRef]
11. Niblack, W. *An Introduction to Digital Image Processing*; Prentice Hall: Englewood Cliffs, NJ, USA, 1986.
12. Sauvola, J.; Pietikäinen, M. Adaptive document image binarization. *Pattern Recognit.* **2000**, *33*, 225–236. [CrossRef]
13. Lazzara, G.; Géraud, T. Efficient multiscale Sauvola's binarization. *Int. J. Doc. Anal. Recognit.* **2014**, *17*, 105–123. [CrossRef]
14. Bernsen, J. Dynamic thresholding of grey-level images. In Proceedings of the 8th International Conference on Pattern Recognition (ICPR), Paris, France, 27–31 October 1986; pp. 1251–1255.
15. Bradley, D.; Roth, G. Adaptive thresholding using the integral image. *J. Graph. Tools* **2007**, *12*, 13–21. [CrossRef]
16. Wolf, C.; Jolion, J.M. Extraction and recognition of artificial text in multimedia documents. *Form. Pattern Anal. Appl.* **2004**, *6*, 309–326. [CrossRef]
17. Feng, M.L.; Tan, Y.P. Adaptive binarization method for document image analysis. In Proceedings of the 2004 IEEE International Conference on Multimedia and Expo (ICME), Taipei, Taiwan, 27–30 June 2004; Volume 1, pp. 339–342. [CrossRef]
18. Gatos, B.; Pratikakis, I.; Perantonis, S. Adaptive degraded document image binarization. *Pattern Recognit.* **2006**, *39*, 317–327. [CrossRef]
19. Chou, C.H.; Lin, W.H.; Chang, F. A binarization method with learning-built rules for document images produced by cameras. *Pattern Recognit.* **2010**, *43*, 1518–1530. [CrossRef]
20. Michalak, H.; Okarma, K. Region based adaptive binarization for optical character recognition purposes. In Proceedings of the International Interdisciplinary PhD Workshop (IIPhDW), Swinoujscie, Poland, 9–12 May 2018; pp. 361–366. [CrossRef]
21. Michalak, H.; Okarma, K. Fast adaptive image binarization using the region based approach. In *Artificial Intelligence and Algorithms in Intelligent Systems*; Silhavy, R., Ed.; Springer International Publishing: Cham, Switzerland, 2019; Volume 764, AISC, pp. 79–90. [CrossRef]
22. Xiong, W.; Xu, J.; Xiong, Z.; Wang, J.; Liu, M. Degraded historical document image binarization using local features and support vector machine (SVM). *Optik* **2018**, *164*, 218–223. [CrossRef]
23. Howe, N.R. A Laplacian energy for document binarization. In Proceedings of the 2011 International Conference on Document Analysis and Recognition (ICDAR), Beijing, China, 18–21 September 2011; pp. 6–10. [CrossRef]
24. Erol, B.; Antúnez, E.R.; Hull, J.J. HOTPAPER: multimedia interaction with paper using mobile phones. In Proceedings of the 16th International Conference on Multimedia 2008, Vancouver, BC, Canada, 26–31 October 2008; pp. 399–408. [CrossRef]

25. Lu, S.; Su, B.; Tan, C.L. Document image binarization using background estimation and stroke edges. *Int. J. Doc. Anal. Recognit. (IJDAR)* **2010**, *13*, 303–314. [CrossRef]

26. Okamoto, A.; Yoshida, H.; Tanaka, N. A binarization method for degraded document images with morphological operations. In Proceedings of the 2013 IAPR International Conference on Machine Vision Applications (MVA 2013), Kyoto, Japan, 20–23 May 2013; pp. 294–297.

27. Lelore, T.; Bouchara, F. Super-resolved binarization of text based on the FAIR algorithm. In Proceedings of the 2011 International Conference on Document Analysis and Recognition (ICDAR), Beijing, China, 18–21 September 2011; pp. 839–843. [CrossRef]

28. Bag, S.; Bhowmick, P. Adaptive-interpolative binarization with stroke preservation for restoration of faint characters in degraded documents. *J. Vis. Commun. Image Represent.* **2015**, *31*, 266–281. [CrossRef]

29. Su, B.; Lu, S.; Tan, C.L. Robust document image binarization technique for degraded document images. *IEEE Trans. Image Process.* **2013**, *22*, 1408–1417. [CrossRef] [PubMed]

30. Yoon, Y.; Ban, K.D.; Yoon, H.; Lee, J.; Kim, J. Best combination of binarization methods for license plate character segmentation. *ETRI J.* **2013**, *35*, 491–500. [CrossRef]

31. Lech, P.; Okarma, K. Optimization of the fast image binarization method based on the Monte Carlo approach. *Elektronika Ir Elektrotechnika* **2014**, *20*, 63–66. [CrossRef]

32. Lech, P.; Okarma, K. Fast histogram based image binarization using the Monte Carlo threshold estimation. In *Computer Vision and Graphics*; Chmielewski, L.J., Kozera, R., Shin, B.S., Wojciechowski, K., Eds.; Springer International Publishing: Cham, Switzerland, 2014; Volume 8671, pp. 382–390. [CrossRef]

33. Lech, P.; Okarma, K. Prediction of the optical character recognition accuracy based on the combined assessment of image binarization results. *Elektronika Ir Elektrotechnika* **2015**, *21*, 62–65. [CrossRef]

34. Khitas, M.; Ziet, L.; Bouguezel, S. Improved degraded document image binarization using median filter for background estimation. *Elektronika ir Elektrotechnika* **2018**, *24*, 82–87. [CrossRef]

35. Mitianoudis, N.; Papamarkos, N. Document image binarization using local features and Gaussian mixture modeling. *Image Vis. Comput.* **2015**, *38*, 33–51. [CrossRef]

36. Chen, Y.; Wang, L. Broken and degraded document images binarization. *Neurocomputing* **2017**, *237*, 272–280. [CrossRef]

37. Bataineh, B.; Abdullah, S.N.H.S.; Omar, K. An adaptive local binarization method for document images based on a novel thresholding method and dynamic windows. *Pattern Recognit. Lett.* **2011**, *32*, 1805–1813. [CrossRef]

38. Khurshid, K.; Siddiqi, I.; Faure, C.; Vincent, N. Comparison of Niblack inspired binarization methods for ancient documents. In Proceedings of the Document Recognition and Retrieval XVI, San Jose, CA, USA, 18–22 January 2009; Volume 7247, pp. 7247:1–7247:9. [CrossRef]

39. Kulyukin, V.; Kutiyanawala, A.; Zaman, T. Eyes-free barcode detection on smartphones with Niblack's binarization and Support Vector Machines. In Proceedings of the 16th International Conference on Image Processing, Computer Vision, and Pattern Recognition (IPCV'2012), Las Vegas, NV, USA, 16–19 July 2012; Volume 1, pp. 284–290.

40. Samorodova, O.A.; Samorodov, A.V. Fast implementation of the Niblack binarization algorithm for microscope image segmentation. *Pattern Recognit. Image Anal.* **2016**, *26*, 548–551. [CrossRef]

41. Mysore, S.; Gupta, M.K.; Belhe, S. Complex and degraded color document image binarization. In Proceedings of the 2016 3rd International Conference on Signal Processing and Integrated Networks (SPIN), Noida, India, 11–12 February 2016; pp. 157–162. [CrossRef]

42. Saxena, L.P. Niblack's binarization method and its modifications to real-time applications: A review. *Artif. Intell. Rev.* **2017**, 1–33. [CrossRef]

43. Leedham, G.; Yan, C.; Takru, K.; Tan, J.H.N.; Mian, L. Comparison of some thresholding algorithms for text/background segmentation in difficult document images. In Proceedings of the 7th International Conference on Document Analysis and Recognition (ICDAR 2003), Edinburgh, UK, 3–6 August 2003; pp. 859–864. [CrossRef]

44. Shrivastava, A.; Srivastava, D.K. A review on pixel-based binarization of gray images. In *ICICT 2015*; Springer: Singapore, 2016; Volume 439, pp. 357–364. [CrossRef]

45. Mustafa, W.A.; Kader, M.M.M.A. Binarization of document images: A comprehensive review. *J. Phys. Conf. Ser.* **2018**, *1019*, 012023. [CrossRef]

46. Deng, F.; Wu, Z.; Lu, Z.; Brown, M.S. Binarizationshop: A user assisted software suite for converting old documents to black-and-white. In Proceedings of the Annual Joint Conference on Digital Libraries, Gold Coast, Queensland, Australia, 21–25 June 2010; pp. 255–258. [CrossRef]

47. Feng, S. A novel variational model for noise robust document image binarization. *Neurocomputing* **2019**, *325*, 288–302. [CrossRef]

48. Tensmeyer, C.; Martinez, T. Document image binarization with fully convolutional neural networks. In Proceedings of the 14th IAPR International Conference on Document Analysis and Recognition (ICDAR 2017), Kyoto, Japan, 9–15 November 2017; pp. 99–104. [CrossRef]

49. Vo, Q.N.; Kim, S.H.; Yang, H.J.; Lee, G. Binarization of degraded document images based on hierarchical deep supervised network. *Pattern Recognit.* **2018**, *74*, 568–586. [CrossRef]

50. Sulaiman, A.; Omar, K.; Nasrudin, M.F. Degraded historical document binarization: A review on issues, challenges, techniques, and future directions. *J. Imaging* **2019**, *5*, 48. [CrossRef]

51. Wen, J.; Li, S.; Sun, J. A new binarization method for non-uniform illuminated document images. *Pattern Recognit.* **2013**, *46*, 1670–1690. [CrossRef]

52. Pun, T. A new method for grey-level picture thresholding using the entropy of the histogram. *Signal Process.* **1980**, *2*, 223–237. [CrossRef]

53. Pun, T. Entropic thresholding, a new approach. *Comput. Graph. Image Process.* **1981**, *16*, 210–239. [CrossRef]

54. Fan, J.; Wang, R.; Zhang, L.; Xing, D.; Gan, F. Image sequence segmentation based on 2D temporal entropic thresholding. *Pattern Recognit. Lett.* **1996**, *17*, 1101–1107. [CrossRef]

55. Abutaleb, A.S. Automatic thresholding of gray-level pictures using two-dimensional entropy. *Comput. Vis. Graph. Image Process.* **1989**, *47*, 22–32. [CrossRef]

56. Brink, A.; Pendock, N. Minimum cross-entropy threshold selection. *Pattern Recognit.* **1996**, *29*, 179–188. [CrossRef]

57. Li, J.; Tang, W.; Wang, J.; Zhang, X. A multilevel color image thresholding scheme based on minimum cross entropy and alternating direction method of multipliers. *Optik* **2019**, *183*, 30–37. [CrossRef]

58. Tang, K.; Yuan, X.; Sun, T.; Yang, J.; Gao, S. An improved scheme for minimum cross entropy threshold selection based on genetic algorithm. *Knowl.-Based Syst.* **2011**, *24*, 1131–1138. [CrossRef]

59. Le, T.H.N.; Bui, T.D.; Suen, C.Y. Ternary entropy-based binarization of degraded document images using morphological operators. In Proceedings of the 2011 International Conference on Document Analysis and Recognition, Beijing, China, 18–21 September 2011; pp. 114–118. [CrossRef]

60. Tian, X.; Hou, X. A Tsallis-entropy image thresholding method based on two-dimensional histogram obique segmentation. In Proceedings of the 2009 WASE International Conference on Information Engineering, Taiyuan, China, 10–11 July 2009; Volume 1, pp. 164–168. [CrossRef]

61. Jiang, Y.; Zhu, H.; Malekian, R.; Ding, C. An improved quantitative recurrence analysis using artificial intelligence based image processing applied to sensor measurements. *Concurr. Comput. Pract. Exp.* **2019**, *31*, e4858. [CrossRef]

62. Stathis, P.; Kavallieratou, E.; Papamarkos, N. An evaluation technique for binarization algorithms. *J. UCS* **2008**, *14*, 3011–3030. [CrossRef]

63. Sezgin, M.; Sankur, B. Survey over image thresholding techniques and quantitative performance evaluation. *J. Electron. Imaging* **2004**, *13*, 146–165. [CrossRef]

64. Young, D.P.; Ferryman, J.M. PETS metrics: On-line performance evaluation service. In Proceedings of the 2005 IEEE International Workshop on Visual Surveillance and Performance Evaluation of Tracking and Surveillance, Beijing, China, 15–16 October 2005; pp. 317–324. [CrossRef]

65. Lu, H.; Kot, A.; Shi, Y. Distance-reciprocal distortion measure for binary document images. *IEEE Signal Process. Lett.* **2004**, *11*, 228–231. [CrossRef]

66. Lins, R.D.; de Almeida, M.M.; Bernardino, R.B.; Jesus, D.; Oliveira, J.M. Assessing binarization techniques for document images. In Proceedings of the 2017 ACM Symposium on Document Engineering (DocEng), Valletta, Malta, 4–7 September 2017; pp. 183–192. [CrossRef]

© 2019 by the authors. Licensee MDPI, Basel, Switzerland. This article is an open access article distributed under the terms and conditions of the Creative Commons Attribution (CC BY) license (http://creativecommons.org/licenses/by/4.0/).

Article

PGNet: Pipeline Guidance for Human Key-Point Detection

Feng Hong [1,2], Changhua Lu [1], Chun Liu [1,*], Ruru Liu [2,*], Weiwei Jiang [1], Wei Ju [1] and Tao Wang [1]

1   College of computer and Information, Hefei University of Technology, Hefei 230009, China; hfeng255@sina.cn (F.H.); jsdzlch@hfut.edu.cn (C.L.); cttjww@126.com (W.J.); juwei@mail.hfut.edu.cn (W.J.); wtustc@mail.ustc.edu.cn (T.W.)
2   College of Electrical and Mechanical Engineering, Chizhou University, Chizhou 247000, China
*   Correspondence: dqlch03@hfut.edu.cn (C.L.); lruru@sina.cn (R.L.)

Received: 16 February 2020; Accepted: 20 March 2020; Published: 24 March 2020

Abstract: Human key-point detection is a challenging research field in computer vision. Convolutional neural models limit the number of parameters and mine the local structure, and have made great progress in significant target detection and key-point detection. However, the features extracted by shallow layers mainly contain a lack of semantic information, while the features extracted by deep layers contain rich semantic information but a lack of spatial information that results in information imbalance and feature extraction imbalance. With the complexity of the network structure and the increasing amount of computation, the balance between the time of communication and the time of calculation highlights the importance. Based on the improvement of hardware equipment, network operation time is greatly improved by optimizing the network structure and data operation methods. However, as the network structure becomes deeper and deeper, the communication consumption between networks also increases, and network computing capacity is optimized. In addition, communication overhead is also the focus of recent attention. We propose a novel network structure PGNet, which contains three parts: pipeline guidance strategy (PGS); Cross-Distance-IoU Loss (CIoU); and Cascaded Fusion Feature Model (CFFM).

Keywords: object detection; key-point detection; IoU; feature fusion

1. Introduction

Deep-learning methods have been successfully applied to many fields, such as image recognition and analysis, speech recognition, and natural language processing, due to their automatic learning and continuous learning capabilities. Detection of human key points is a fundamental step in expounding human behavior, such as action analysis, action prediction, and behavior judgment. In addition, behavior prediction needs to capture the fine details of an object, such as video tracking and behavior prediction. A fast and effective key-point detection is of great practical value in predicting and tracking people's behavior under special scenarios.

Human key-point detection is a considerable undertaking in computer vision. Before 2014, researchers mainly solved the task by using SIFT, HOG, and other feature operators to extract features, and combined them with graph structure models to detect joint point positions. With the combination of deep learning and many tasks of computer vision achieving remarkable results, researchers have begun to try to combine it with human key-point detection tasks.

The main application of human body key-point detection is human body pose estimation. These methods involve detecting the location of human body key points and distinguishing artificially set key-point locations on the human body, separating human body key points from a given image. In [1], a novel method for the maintenance of temporal consistency is proposed, and maintained the temporal consistency of the video by the structured space learning and halfway temporal evaluation methods. Wang et al. [2] proposed a method for estimating 3D human poses from single images or video sequences. [3] explored the human action analysis in a specified situation, based on the human posture extraction by pose-estimation algorithm. Deep neural network (DNN) methods were used, composed of residual learning blocks for feature extraction and recurrent neural network for time-series data learning. However, although this method performs predictive analysis on the behavior of people in the video, using deep convolutional networks, the trade-offs in computational consumption and real-time performance are not fully considered, meanwhile showing that human pose estimation is an important research field of computer vision, and that human key-point detection is a front-end research of human pose estimation. In [4] it was illustrated that human body pose recognition is performed by comparing the shadow of the projection with the shadow of the human body under special circumstances, and proposed a normalization technique to bridge the gap and help the classifier better generalize with real data. Zhang et al. [5] proposed three effective training strategies, and exploited four useful postprocessing techniques and proposed a cascaded context mixer (CCM). [6] proposed an end-to-end architecture for joint 2D and 3D human pose estimation in natural images. However, the above uses deep convolutional networks for training and positioning. However, the down-sampling makes for a lack of spatial information at the deep level and a lack of semantic information at the shallow level. At the same time, the trade-off between calculation volume and efficiency also makes it difficult to consider performance of the network in terms of practicality. There are deficiencies in real-time and computational burden. Figure 1 below shows the detection results of the method proposed in this paper.

Figure 1. The proposed network to find key points of the human body.

Substantial research has been done before in human key-point detection. The purpose of human key-point detection is to estimate the key points of a human body from pictures or videos; it is also an important link in some downstream applications prior to preprocessing, e.g., [4,7–11]. At present, convolutional neural networks show strong advantages in feature extraction. Various models have been proposed for features, as well as various evolutionary networks, some for extracting high-semantic information, and more attention to shallow spatial information. The structure of the model is also the focus of many scholars; coding-decoder, fusion mechanism, and feedback mechanism are responsible for the optimization and supplement of the network structure. [5] depicted a key-point graph network designed to extract object detection and object segmentation of key points. There was excellent performance, but easy overlap of key points when separating small objects. [6,12] proposed improved network mainly using anchor center points to detect small objects, but the efficiency of the whole network was reduced. In the feature extraction process, there are two main methods of feature extraction. One is box-of-free feature extraction [13–15], in which target detection is accomplished by embedding a cosine function or embedding a class of clusters in pixels. The other is based on frame-based feature extraction, but this method of embedding clusters has two major disadvantages in the extraction process [16]. One is that the global information of the picture cannot be fully considered, and the other is that the embedded information is mainly a cosine function, so there are many restrictions before embedding, and this method must be limited in the use process. Another feature extraction and positioning method is based on bounding box object detection. [13,14,17–20]. [13] addressed two limitations brought up by conventional anchor-based detection: (1) heuristic-guided feature selection; and (2) overlap-based anchor sampling. Specifically, an anchor-free branch is attached to each level of the feature pyramid, allowing box encoding and decoding in the anchor-free manner at an arbitrary level [14]. Han et al. proposed an efficient framework for real-time object tracking which is an end-to-end trained offline Fully Conventional Anchor-Free Siamese network; the network consists of correlation section, implemented by depth-wise cross correlation, and supervised section which has two branches, one for classification and the other for regression. [17] presented a monocular 3D object detection method with feature enhancement networks; 3D geometric features of RoI point clouds are further enhanced by the proposed point feature enhancement (PointFE) network, which can be served as an auxiliary module for an autonomous driving system.

The current popular framed object detection method is based on anchored framed feature extraction. This method maps the density of the anchored frame onto the feature heat map and further improves the border of the anchored image by predicting the offset. An important metric for framed object detection is intersection over union (IoU). [18] used the IoU of the union of the bounding boxes for multiple objects predicted by images taken at different times, termed mIOU, and the corresponding estimated number of vehicles to estimate the multi-level traffic status. [19] generated a tight oriented bounding box for elongated object detection which achieves a large margin of improvement for both detection and localization of elongated objects in images. [20] used multi-label classification as an auxiliary task to improve object detection, and the box-level features and the image-level features of multi-label are fused to improve accuracy. [21–23] demonstrated that the main problems of current IoU loss are the speed of convergence and the inaccuracy of iterative regression. Zhao et al. [24] proposed that Distance-IoU mainly predicts the target frame based on normalized data, which makes convergence speed of the network itself and the accuracy of feature extraction better, compared with the previous methods IoU and Genaralized-IoU.

In this paper, we propose a novel network of human key-point detection. The main backbone of the network is Resnet50, in the way our model can accurately locate the key points of the human body; the model adopts the pipeline structure, which effectively optimizes communication and network computing before contradiction. By using the form of bus pipeline, the features extracted at each stage are recombined, so that efficiency and speed are greatly improved. With the optimized network, the features of each stage can be shared to a greater extent, and the contradiction between the semantic information of the shallow features and the spatial information of the deep features is solved.

The improved PGNet network has excellent performance on the COCO datasets. We use the image-guided method to accurately extract the key points of the human body to complete the positioning, and consider the combination of the network structure features extracted by the shallow network and the semantic features extracted by the deep network. A good feature extraction actuator should contain two common features; one is spatial information and edge information with sufficient shallow features, and the extraction of such information is mainly done through multiple convolution and iterative convolution operations; the other is with abundant semantic information for more accurate localization to complete the classification. In addition, we use the cross-loss function in the design of the loss function, which performs well on the COCO dataset, and our main contributions are as follows:

1. We introduce a kind of pipeline guiding strategy (PGS) to share the extracted features to all layers (shallow layers and deep layers) in the form of a pipeline. This allows each layer to better separate the background noise, and at the same time share the weight of the opposite transfer between each other.

2. We propose a cross-fusion feature extraction mode. Combining this model with PGS enables shallow spatial information and deep semantic information to be combined through a pipelined

bus strategy, so that the computational efficiency and the network's separation of foreground and background can effectively remove the foreground noise and the background effective information at the edges is fully considered.

3. We developed a crossed Distance-IoU loss function. To obtain the region of interest, we calculated the convergence and speed of the border regression. The Cross-Distance-IoU loss function used is based on the distance between the center points and the overlap area, and shows excellent results in the rectangular anchor border regression. The pipeline is used to guide the network to use the Distance-IoU loss function and the backbone network uses the GIoU loss function.

Table 1 shows that the backbone structure of the [12] network is the same, but due to the different processing methods of subsequent decoding fusion, the performance on the COCO dataset is different. In the case of the same encoding method, this paper uses deep convolution and 1*1 convolution, so that a trade-off between calculation volume and speed is satisfied in feature extraction. The proposed method improves the accuracy of the COCO dataset by 0.2% over the previous method. Table 1 demonstrates that our algorithm makes full use of the pipeline guidance method, and the accuracy of the COCO dataset exceeds the previous advanced algorithms.

Table 1. Performance comparison of various network structures.

Method	Backbone	Decoder	Postprocessing	Performance
Mask-R-CNN [25]	ResNet-50-FPN	conv+deconv	offset regression	63.1AP@COCO
DHN [26]	ResNet-152	deconv	Flip/sub-pixel shift	73.7 AP@COCO
CNN [27]	VGG-19	conv	Flip/sub-pixel shift	61.8 AP@COCO
PGNN [28]	ResNet-50	GlobalNet	Flip/sub-pixel shift	68.7 AP@COCO
DetNet [29]	ResNet-50	deconv	Flip/sub-pixel shift	69.7 AP@COCO
DENSENETS [30]	ResNet-50	deconv	-	61.8AP@COCO
LCR-Net++ [12]	ResNet-50	deconv	Flip/sub-pixel shift	73.2AP@COCO
HRNet [6]	HRNet-152	1×1conv	Flip/sub-pixel shift	77.0AP@COCO
[5]	ResNet-101	deconv	Flip/sub-pixel shift	69.9 AP@COCO
PFAN [24]	VGG-19	multi-stage CNN	Flip/sub-pixel shift	70.2 AP@COCO
Proposed method	ResNet-50-Pipeline	Deconv+1×1conv	offset regression	77.2AP@COCO

2. Materials and Methods

In this section, we mainly introduce some studies related to this article, including the recent key-point detection method, the characteristics of the pipeline structure and the working principle and the loss function of border regress-IoU.

2.1. Key-Point Detection Method

Previous methods mainly optimized network structure improvement and used deeper network structures. However, these methods achieved satisfactory results in key-point detection. [6,12,24–30].

2.2. Pipeline Guidance Strategy

As the layers of the network become deeper and deeper, the joint parallel computing of multiple GPUs provides the possibility of speeding up the network. Multi-GPU is divided into multiple stages of the network, and the convolution operations and rectangular transformations of the network are performed in parallel, and communication between various operations is performed. The guidance mechanism is to use its own characteristics to supervise and complete the further optimization of its own information feature extraction. For example, shallow rich-edge information is used to guide the deep layer to better extract deep semantic information, feedback the deep semantic information to supervise the extraction of shallow edge information, and ultimately complete the performance optimization of the network structure, so that the receptive fields of different layers can play to their own advantages [31–34].

2.3. IoU Loss

In deep convolutional neural networks, the mainstream method in the process of feature extraction is the frame regression method. An important index for measuring this method is the loss function. The function of the loss function is to predict the distance between the target frame and the prediction frame. Current popular networks such as YOLOv3, SSD, and Faster R-CNN use GIoU, CIoU, or some improved loss functions combined with them [22,23]. Figure 2 shows that the number of positive bounding boxes after the NMS, grouped by their IoU with the matched ground truth.

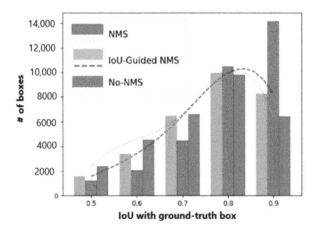

Figure 2. The number of positive bounding boxes after the NMS, grouped by their IoU with the matched ground truth. In traditional NMS (blue bar), a significant portion of accurately localized bounding boxes get mistakenly suppressed due to the misalignment of classification confidence and localization accuracy, while IoU-guided NMS (yellow bar) preserves more accurately localized bounding boxes [35].

IoU is an important indicator for neural networks to measure between ground truths and predicted images. In object detection of a bounding box, the object being detected is the minimum value of the rectangular border through multiple iterations.

$$IoU = \frac{B \cap B^{gt}}{B \cup B^{gt}} \tag{1}$$

B^{gt}, B respectively ground truth and predicted images.

Table 2 shows IoU operation logic, realized target point detection, and key fixed positioning through the same iterative operation multiple times. Despite the detection network frameworks being different, the regression calculation logic for predicting the borders to locate the borders in the target object is the same. Shengkai et al. [36] propose IoU-balanced loss functions that consist of IoU-balanced classification loss and IoU-balanced localization loss to solve poor localization accuracy, and this is harmful for accurate localization. [37] proposed visible IoU to explicitly incorporate the visible ratio in selecting samples, which included a box regressor to separately predict the moving direction of training samples. [23] Yan et al. proposed a novel IoU-Adaptive Deformable R-CNN framework for multi-class object detection, i.e., IoU-guided detection framework to reduce the loss of small-object information during training. Zheng et al. [38] proposed a Distance-IoU (DIoU) loss by incorporating the normalized distance between the predicted box and the target box, which converges much faster in training than IoU and GIoU losses.

Table 2. Logic operation based on bounding box regression.

Alogrithm1 IoU for two axis-Aligned BBox.
Require: -Corners of the bounding boxes: $A_1(x_1,y_1),B_1(x_2,\ y_1),C_1(x_2,\ y_2),D_1(x_1,\ y_2),$ $A_2\ (x_1',y_1'),B_2(x_2',y_1'),\ C_2(x_2',y_2'),\ D_2(x_1',y_2'),$ Where $x_1 \le x_2, y_2 \le y_1,$ and $x_1' \le x_2',\ y_2' \le y_1'$ Ensure: - IoU value;
1:▲The area of B_g : $\textbf{Area}_g = (x_1 - x_2) \times (y_1 - y_2);$ 2:▲The area of B_d : $\textbf{Area}_d = \left(x_2' - x_1'\right) \times \left(y_1' - y_2'\right);$ 3:▲The area of overlap: $Area_{overlap} = (\max\left(x_2, x_2'\right) - \min(x_1, x_1')) \times \left(\max\left(y_1, y_1'\right) - \min\left(y_2, y_2'\right)\right);$ 4:▲ $IOU = \dfrac{Area_{ocerlap}}{Area_g + Area_d - Area_{overlap}};$

3. Results

Based on the above, our model solves the accuracy and efficiency of key points in positioning, and optimizes the communication consumption due to many iterative operations and convolution operations. In the process of extracting features, the feature fusion mechanism is used to combine high-latitude semantic information with low-latitude spatial information, which makes for great efficiency in the process of locating key points of the human body. Figure 1 shown our proposed framework, which consists of three parts, in three branches—ResNet-51 is selected as the backbone network for picture feature extraction; there is adaptive strategy using pipeline guidance; and a cascaded feature fusion model. The framework of the network is shown in Figure 3.

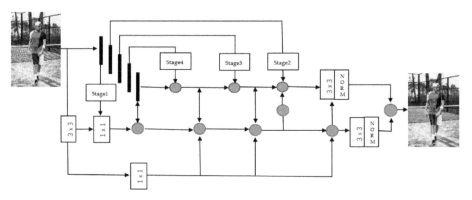

Figure 3. An of overview of proposed PGNet.ResNet-50 is used as the backbone. Using the cascaded fusion feature model (CFFM), the backbone network is divided into 5 stages, and the feature-guided network after the image is convolved is used to extract key-point features.

3.1. Cascaded Fusion Feature Model

The main task of the Cascaded Fusion Feature Model (CFFM) is to extract the multi-layer features of the input picture and generate regions where key points are located. The traditional method is to directly use the multi-layer features to generate the prediction anchor frame and compare the ground truth picture to generate the key-point coordinates.

We propose the cascade fusion feature model to extract high-level features and low-level features; the high layers are rich in semantic correlation information and lack low-level spatial information. In contrast, the low layers are rich in edge and spatial features and lack semantic information. In particular, we build CFFM on ResNet-50, which will extract its features using conv1–5 layers.

Considering the shallow layers simultaneously use a lot of computing resources, there is no significant improvement in performance, lack of edge, and spatial information during deep feature processes. We use the middle three layers to avoid the consumption of a large amount of spatial information during convolution calculations.

3.2. Pipeline Guidance Strategies

The process of locating key points of the human body is mainly done by analyzing human characteristics, locate key parts, and prepare for downstream video surveillance. We proposed combined traditional data parallelism with model parallelism enhanced with pipelining [39]. Through the structure of the pipelining, separate processing is performed on feature extraction and feature guidance, which effectively saves the resource consumption of the network structure in the communication process.

Pipeline-parallel training partitions the layers of the object being trained into multiple stages. Each layer contains a continuous set of structures in the model, as shown in Figure 3. The pipeline-type structure is used to guide the feature extraction at each stage. After the feature extraction, a convolution operation is used to fuse the features of the two branches on the pipeline and after the feature extraction to complete the key points. Figure 2 shows a network structure based on pipeline guidance, and Figure 3 is a diagram of key points of the human body using the PGNet network. [39]. Figure 4 shows that an example pipeline-parallel assignment with four machines and an example timeline at one of machines.

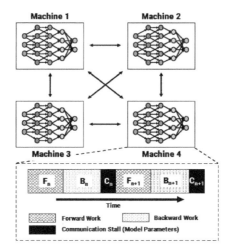

Figure 4. An example pipeline-parallel assignment with four machines and an example timeline at one of machines, highlighting the temporal overlap of computation and activation/gradient communication.

Because IoU loss can only be effective when the bounding box is reattached during the training process, there are steps where gradient optimization cannot be performed without any coincidence [36]. To overcome the disadvantage that the border boxes must have coincidence to take advantage of IoU losses, GIoU was proposed. Both these losses can make key-point detectors more powerful for accurate localization. According to Equation (1), IoU loss can be defined such that

$$\mathcal{L}_{IoU} = 1 - \frac{B \cap B^{gt}}{B \cup B^{gt}} \tag{2}$$

According to Equation (2), it can be known that the calculation of $\mathcal{L}_{IoU}$ must be performed in an iterative manner only if there is intersection between the predicted target and the ground truth. GIoU was proposed to improve the gradient descent prediction operation of two bounding boxes without intersection. GIoU defines a distance, between which two bounding boxes can exist without crossing. GIoU is defined as such that

$$GIoU = IoU - \frac{C - B \cup B^{gt}}{|C|} \tag{3}$$

where C is the smallest box covering B and B^{gt}. Due to the introduction of the penalty term, the predicted box will move towards the target box in non-overlapping cases [40]. From Equation (3), we get $\mathcal{L}_{GIoU}$ Loss such that

$$\mathcal{L}_{GIoU} = 1 - IoU + \frac{\left|C - B \cup B^{gt}\right|}{|C|} \tag{4}$$

$\mathcal{L}_{GIoU}$ loss aims to reduce the distance between the center point of the predicted box and the real box.

The cross-distance loss functions we propose inherit some of their inherent properties and are defined as

$$\mathcal{L}_{CDIoU} = 1 - IoU + \frac{\rho^2\left(\delta, \delta^{gt}\right)}{D^2} \tag{5}$$

Where, in Equation (5), δ, δ^{gt} denote the central points of B and B^{gt}, ρ is the Euclidean distance, and D is the distance of the B and B_g. Figure 5 shows that $\mathcal{L}_{CDIoU}$ Distribution of bounding boxes for iterative training.

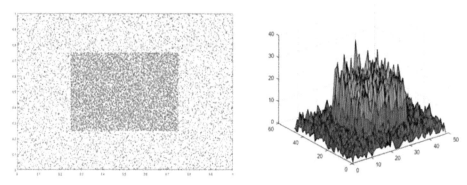

Figure 5. $\mathcal{L}_{CDIoU}$ Distribution of bounding boxes for iterative training.

4. Discussion

4.1. Datasets and Evaluation Metrics

Our method of evaluating our designed network is on the COCO-2017 database, which is a large image dataset designed for object detection, segmentation, human key-point detection, thing segmentation, and subtitle generation.

In addition, the average precision (AP) metric is used to measure and evaluate the performance of PGNet. To illustrate the performance between the key-point location of the detection object and the key-point of the ground truth object, the results show that the method performs well.

4.2. Ablation Studies

The ablation experiment uses different backbone networks to regularize the method separately and unreasonably, and experiments on the network structure of this problem are based on six indicators. The benchmark database of the experiment is COCO val-2017. The experimental results shown below are obtained. The experimental results show that the network structure proposed in this paper is superior to other network structures in performance, as shown in Table 3.

Table 3. Parameter comparison of various network structures after different regularization processing.

Backbone	Norm	AP^{bbox}	AP_{50}^{bbox}	AP_{75}^{bbox}	AP_S^{bbox}	AP_M^{bbox}	AP_L^{bbox}
	GN	37.8	59.0	40.8	22.3	41.2	48.4
ResNet50+FPN	syncGN	37.7	58.5	41.1	22.3	40.2	48.9
	CBN	37.8	59.8	40.3	22.5	40.5	49.1
	GN	39.3	60.6	42.7	22.5	42.5	48.8
ResNet101+FPN	syncGN	39.3	59.8	43.0	22.3	42.9	51.6
	CBN	39.2	60.0	42.2	22.3	42.6	51.8
	GN	39.3	60.7	42.6	22.5	43.2	48.1
ResNet50+proposed	syncGN	39.3	59.8	43.5	23.4	43.7	51.9
	CBN	39.4	59.8	43.2	23.1	42.9	52.6

Another part of the ablation experiment is to compare the results of the Eproch training using a pipelined structure. On object detection and image classification with small mini-batch sizes, CBN is found to outperform the original batch normalization and a direct calculation of statistics over previous iterations without the proposed compensation technique [41] in COCO val-2017. Figure 6 shows the training and test results.

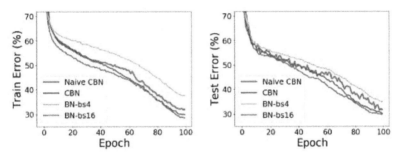

Figure 6. Comparison of epoch trained by this method and epoch of other training methods.

5. Conclusions

In this paper, we propose an up-to-date type of human body key-point positioning network structure Piple-Guidance NeT. Considering that different layers contain incomprehensible features, the use of pipelined guidance in the structure allows the network to achieve a balance between the convolution calculations and the communication time between the layers, which improves the training speed of the network. In addition, the Cross-Distance-IoU mode is used in the training process, and the results are pleasing in different network backbones. Finally, regarding the COCO2017 dataset, the effectiveness of the algorithm is measured by the six parameters of the AP, and the effects demonstrate that the algorithm performs well. Compared with the current most advanced algorithms, the method improves the accuracy by 0.2%.

Author Contributions: Conceptualization, F.H. and C.L. (Changhua Lu); methodology, F.H.; software, R.L.; validation, C.L. (Chun Liu); formal analysis, W.J. (Wei Ju); investigation, F.H.; resources, R.L.; data curation, F.F.; writing—original draft preparation, F.H.; writing—review and editing, W.J. (Weiwei Jiang) and T.W.; visualization, F.H.; supervision, F.H.; project administration, F.H.; funding acquisition, C.L. (Changhua Lu). All authors have read and agreed to the published version of the manuscript.

Acknowledgments: Major National Science and Technology Projects (NO.JZ2015KJZZ0254); National High Technology Research Development Plan (863) (NO.2014AA06A503); Major National Science and Technology Projects (NO.JZ2015KJZZ0254); National High Technology Research Development Plan (863) (NO.2014AA06A503); Anhui Province of Outstanding Young Talent Project (NO.gxyq2018110, gxyq2019111), ChiZhou University Natural Key Project (NO.cz2019zrz07) fund.

Conflicts of Interest: The authors declare no conflict of interest.

References

1. Liu, S.; Li, Y.; Hua, G. Human Pose Estimation in Video via Structured Space Learning and Halfway Temporal Evaluation. *IEEE Trans. Circuits Syst. Video Technol.* **2019**, *29*, 2029–2038. [CrossRef]
2. Wang, C.; Wang, Y.; Lin, Z.; Yuille, A.L. Robust 3D Human Pose Estimation from Single Images or Video Sequences. *IEEE Trans. Pattern Anal. Mach. Intell.* **2019**, *41*, 1227–1241. [CrossRef]
3. Yunqing, Z.; Fok, W.W.T.; Chan, C.W. Video-Based Violence Detection by Human Action Analysis with Neural Network. *Proc. SPIE* **2019**, *11321*, 113218. [CrossRef]
4. Gouiaa, R.; Meunier, J. Learning Cast Shadow Appearance for Human Posture Recognition. *Pattern Recognit. Lett.* **2017**, *97*, 54–60. [CrossRef]
5. Zhang, J.; Chen, Z.; Tao, D. Towards High Performance Human Keypoint Detection. *arXiv* **2020**, arXiv:2002.00537.
6. Ke, S.; Bin, X.; Dong, L.; Jingdong, W. Deep High-Resolution Representation Learning for Human Pose Estimation. *arXiv* **2019**, arXiv:1902.09212.
7. He, S. A Sitting Posture Surveillance System Based on Kinect. In Proceedings of the 2018 International Conference on Electronics, Communications and Control Engineering, Avid College, Maldives, 6–8 March 2018; Kettani, H., Sivakumar, R., Song, I., Eds.; Volume 1026.
8. Liang, G.; Lan, X.; Chen, X.; Zheng, K.; Wang, S.; Zheng, N. Cross-View Person Identification Based on Confidence-Weighted Human Pose Matching. *IEEE Trans. Image Process.* **2019**, *28*, 3821–3835. [CrossRef]
9. Patel, P.; Bhatt, B.; Patel, B. Human Body Posture Recognition A Survey. In Proceedings of the 2017 International Conference on Innovative Mechanisms for Industry Applications (ICIMIA), Bangalore, India, 21–23 February 2017; pp. 473–477.
10. Raja, M.; Hughes, A.; Xu, Y.; Zarei, P.; Michelson, D.G.; Sigg, S. Wireless Multifrequency Feature Set to Simplify Human 3-D Pose Estimation. *IEEE Antennas Wirel. Propag. Lett.* **2019**, *18*, 876–880. [CrossRef]
11. Rasouli, M.S.D.; Payandeh, S. A Novel Depth Image Analysis for Sleep Posture Estimation. *J. Ambient Intell. Humaniz. Comput.* **2019**, *10*, 1999–2014. [CrossRef]
12. Rogez, G.; Weinzaepfel, P.; Schmid, C. LCR-Net++: Multi-Person 2D and 3D Pose Detection in Natural Images. *IEEE Trans. Pattern Anal. Mach. Intell.* **2019**. [CrossRef]
13. Chenchen, Z.; Yihui, H.; Savvides, M. Feature Selective Anchor-Free Module for Single-Shot Object Detection. *arXiv* **2019**, 10arXiv:1903.00621.
14. Han, G.; Du, H.; Liu, J.; Sun, N.; Li, X. Fully Conventional Anchor-Free Siamese Networks for Object Tracking. *IEEE Access* **2019**, *7*, 123934–123943. [CrossRef]
15. Zhen, H.; Jian, L.; Daxue, L.; Hangen, H.; Barber, D. Tracking by Animation: Unsupervised Learning of Multi-Object Attentive Trackers. *arXiv* **2019**, arXiv:1809.03137.
16. Wang, J.; Ding, J.; Guo, H.; Cheng, W.; Pan, T.; Yang, W. Mask OBB: A Semantic Attention-Based Mask Oriented Bounding Box Representation for Multi-Category Object Detection in Aerial Images. *Remote Sens.* **2019**, *11*. [CrossRef]
17. Bao, W.; Xu, B.; Chen, Z. MonoFENet: Monocular 3D Object Detection with Feature Enhancement Networks. *IEEE Trans. Image Process. Publ. IEEE Signal Process. Soc.* **2019**. [CrossRef]
18. Chan-Tong, L.; Ng, B.; Chi-Wang, C. Real-Time Traffic Status Detection from on-line Images Using Generic Object Detection System with Deep Learning. In Proceedings of the 2019 IEEE 19th International Conference on Communication Technology (ICCT), Xi'an, China, 16–19 October 2019; 2019; pp. 1506–1510. [CrossRef]
19. Fang, F.; Li, L.; Zhu, H.; Lim, J.-H. Combining Faster R-CNN and Model-Driven Clustering for Elongated Object Detection. *IEEE Trans. Image Process.* **2020**, *29*, 2052–2065. [CrossRef]
20. Gong, T.; Liu, B.; Chu, Q.; Yu, N. Using Multi-Label Classification to Improve Object Detection. *Neurocomputing* **2019**, *370*, 174–185. [CrossRef]
21. Dingfu, Z.; Jin, F.; Xibin, S.; Chenye, G.; Junbo, Y.; Yuchao, D.; Ruigang, Y. IoU loss for 2D/3D Object Detection. *arXiv* **2019**, arXiv:1908.03851.
22. Fagui, L.; Dian, G.; Cheng, C. IoU-Related Arbitrary Shape Text Scoring Detector. *IEEE Access* **2019**, *7*, 180428–180437. [CrossRef]
23. Yan, J.; Wang, H.; Yan, M.; Diao, W.; Sun, X.; Li, H. IoU-Adaptive Deformable R-CNN: Make Full Use of IoU for Multi-Class Object Detection in Remote Sensing Imagery. *Remote Sens.* **2019**, *11*, 286. [CrossRef]
24. Zhao, T.; Wu, X. Pyramid Feature Attention Network for Saliency detection. *arXiv* **2019**, arXiv:1903.00179.

25. He, K.; Gkioxari, G.; Dollar, P.; Girshick, R. Mask R-CNN. *IEEE Int. Conf. Comput. Vis.* **2017**, 2980–2988. [CrossRef]

26. Bin, X.; Haiping, W.; Yichen, W. Simple Baselines for Human Pose Estimation and Tracking. *arXiv* **2018**, arXiv:1804.06208.

27. Zhe, C.; Simon, T.; Shih-En, W.; Sheikh, Y. Realtime Multi-Person 2D Pose Estimation using Part Affinity Fields. *arXiv* **2016**, arXiv:1611.08050.

28. Zhang, H.; Ouyang, H.; Liu, S.; Qi, X.; Shen, X.; Yang, R.; Jia, J. Human Pose Estimation with Spatial Contextual Information. *arXiv* **2019**, arXiv:1901.01760.

29. Zeming, L.; Chao, P.; Gang, Y.; Xiangyu, Z.; Yangdong, D.; Jian, S. DetNet: Design Backbone for Object Detection. *arXiv* **2018**, arXiv:1804.06215.

30. Huang, G.; Liu, Z.; Pleiss, G.; van der Maaten, L.; Weinberger, K.Q. Convolutional Networks with Dense Connectivity. *arXiv* **2020**, arXiv:2001.02394. [CrossRef]

31. Alyafeai, Z.; Ghouti, L. A Fully-Automated Deep Learning Pipeline for Cervical Cancer Classification. *Expert Syst. Appl.* **2020**, *141*. [CrossRef]

32. Hsu, Y.S. Finite Element Approach of the Buried Pipeline on Tensionless Foundation under Random Ground Excitation. *Math. Comput. Simul.* **2020**, *169*, 149–165. [CrossRef]

33. Qiu, R.; Zhang, H.; Zhou, X.; Guo, Z.; Wang, G.; Yin, L.; Liang, Y. A multi-objective and multi-scenario optimization model for operation control of CO2-flooding pipeline network system. *J. Clean. Prod.* **2020**, *247*. [CrossRef]

34. Zhang, Y.; Lobo-Mueller, E.M.; Karanicolas, P.; Gallinger, S.; Haider, M.A.; Khalvati, F. CNN-based survival model for pancreatic ductal adenocarcinoma in medical imaging. *BMC Med. Imaging* **2020**, *20*, 11. [CrossRef] [PubMed]

35. Shengkai, W.; Xiaoping, L. IoU-balanced Loss Functions for Single-stage Object Detection. *arXiv* **2019**, arXiv:1908.05641.

36. Ruiqi, L.; Huimin, M. Occluded Pedestrian Detection with Visible IOU and Box Sign Predictor. In Proceedings of the 2019 IEEE International Conference on Image Processing (ICIP), Taipei, Taiwan, 22–25 September 2019; pp. 1640–1644. [CrossRef]

37. Zheng, Z.; Wang, P.; Liu, W.; Li, J.; Ye, R.; Ren, D. Distance-IoU Loss: Faster and Better Learning for Bounding Box Regression. *arXiv* **2019**, arXiv:1911.08287.

38. Jiang, B.; Luo, R.; Mao, J.; Xiao, T.; Jiang, Y. Acquisition of Localization Confidence for Accurate Object Detection. *Springer Int. Publ.* **2018**, 816–832. [CrossRef]

39. Harlap, A.; Narayanan, D.; Phanishayee, A.; Seshadri, V.; Devanur, N.; Ganger, G.; Gibbons, P. PipeDream: Fast and Efficient Pipeline Parallel DNN Training. *arXiv* **2018**, arXiv:1806.03377.

40. Rezatofighi, H.; Tsoi, N.; JunYoung, G.; Sadeghian, A.; Reid, I.; Savarese, S. Generalized Intersection Over Union: A Metric and a Loss for Bounding Box Regression. *arXiv* **2019**, arXiv:1902.09630.

41. Yao, Z.; Cao, Y.; Zheng, S.; Huang, G.; Lin, S. Cross-Iteration Batch Normalization. *arXiv* **2020**, arXiv:2002.05712.

 © 2020 by the authors. Licensee MDPI, Basel, Switzerland. This article is an open access article distributed under the terms and conditions of the Creative Commons Attribution (CC BY) license (http://creativecommons.org/licenses/by/4.0/).

Article

Application of Continuous Wavelet Transform and Convolutional Neural Network in Decoding Motor Imagery Brain-Computer Interface

Hyeon Kyu Lee and Young-Seok Choi *

Department of Electronics and Communications Engineering, Kwangwoon University, Seoul 01897, Korea;
skgusrb12@kw.ac.kr
* Correspondence: yschoi@kw.ac.kr; Tel.: +82-2-940-5186

Received: 8 November 2019; Accepted: 3 December 2019; Published: 5 December 2019

Abstract: The motor imagery-based brain-computer interface (BCI) using electroencephalography (EEG) has been receiving attention from neural engineering researchers and is being applied to various rehabilitation applications. However, the performance degradation caused by motor imagery EEG with very low single-to-noise ratio faces several application issues with the use of a BCI system. In this paper, we propose a novel motor imagery classification scheme based on the continuous wavelet transform and the convolutional neural network. Continuous wavelet transform with three mother wavelets is used to capture a highly informative EEG image by combining time-frequency and electrode location. A convolutional neural network is then designed to both classify motor imagery tasks and reduce computation complexity. The proposed method was validated using two public BCI datasets, BCI competition IV dataset 2b and BCI competition II dataset III. The proposed methods were found to achieve improved classification performance compared with the existing methods, thus showcasing the feasibility of motor imagery BCI.

Keywords: brain-computer interface (BCI); electroencephalography (EEG); motor imagery (MI); continuous wavelet transform (CWT); convolutional neural network (CNN)

---

1. Introduction

Brain-Computer Interface (BCI) translates brain signals into an interpretable output without the direct use of peripheral nerves and muscles. The primary purpose of BCI is to create a communication system through brain signals without physical movement for people with severe motor disabilities [1]. Non-invasive BCI model consists of a variety of paradigms on the basis of experimental processes and types of electroencephalography (EEG) recordings. As representative models, event-related potential (e.g., P300), steady-state visual evoked potential (SSVEP), and motor imagery (MI) have attracted attention in the BCI research community. Among them, the MI BCI model has been widely used since it can be easily applied to control external devices [2]. The MI BCI approach is increasingly being applied in various fields, including games [3] and assistive technology [4]. However, despite the growing interest, MI BCI has limitations in real-life applications due to the following issues: First, about 20% of BCI users have difficulties controlling the system compared to others, which we term "BCI illiteracy" [5]. Second, even for the remaining BCI users, particularly those with motor disabilities, MI BCI might not offer the best mental option for BCI control [1,6]. Thus, effective improvement of MI BCI performance remains a challenging issue [7].

In recent years, various MI studies using EEG recordings have been conducted [8,9]. The conventional method of MI BCI with EEG signals consists of the extraction of hidden features and subsequent classification based on various machine learning methods. The common spatial pattern (CSP) algorithm is one of the most popular feature extraction methods [10,11]. CSP is commonly used to analyze spatial patterns of multichannel MI EEG signals. However, CSP is considerably dependent on the frequency band of EEG signals. To deal with this issue, variant algorithms, which are extended versions of the CSP, such as filter bank CSP (FBCSP) [12] or filter bank regularized CSP (FBRCSP) [13], are presented. In addition, feature extraction methods that reduce the dimension of EEG signals are becoming popular due to their advanced classification performance. Typical examples include principal component analysis (PCA) [14] and independent component analysis (ICA) [15]. Other widely-known methods, such as Short Time Fourier Transform (STFT) and Wavelet Transform (WT) [16], which possess outstanding time-frequency localization characteristics and multi-resolution properties, have been extensively applied to the feature extraction process. These provide the ability to capture dynamic time-frequency properties of MI EEG signals.

To classify MI EEG signals, a variety of machine learning algorithms have appeared in the literature [17]; they include support vector machine (SVM) [18], linear discriminant analysis (LDA) [19, 20], and restricted Boltzmann machines (RBM) [21]. The deep neural network (DNN) approach has recently shown excellent classification performance in this field. The convolutional neural network (CNN) is one of the most famous methods in the DNN model and has various applications. It has successfully achieved object detection tasks [22] and text recognition [23,24]. A recent study by Tabar and Halici [25] presented a classification model based on STFT and 1D CNN, which made use of the 1D CNN with a one-dimensional kernel for images generated by using STFT for MI EEG signals. However, STFT has difficulty in generating images with high quality information about signals due to the trade-off between time and frequency resolution. Thus, this might degrade the classification performance of MI BCI.

To address this issue, we propose an advanced MI EEG-decoding method using continuous wavelet transform (CWT) and the subsequent 1D CNN with low computational complexity. In order to alleviate the limitation of STFT, a new MI EEG image is formed by CWT with three features (time, frequency, and electrode), which contain a highly informative spectrum without loss of time and frequency features in EEG signals. The input image is composed of the frequency domain with distinct electrodes in the horizontal axis and the time domain in the vertical axis. It is widely known that the characteristics of MI EEG signals are mainly reflected in two frequency bands of EEG signals, i.e., mu (μ)-band (8–13 Hz) and beta (β)-band (13–30 Hz). By utilizing CWT, images using the power spectrum of EEG signals—time, frequency, and electrode information—are obtained. Thus, it is more capable of detecting specific MI patterns in EEG signals compared to Fourier transform. In addition, the use of 1D CNN leads to efficient discrimination of the MI-related patterns that are shown as temporal variations in mu and beta bands. The performance of the proposed method was validated using a public BCI competition dataset [26].

The rest of this paper is organized as follows: Section 2 introduces the experimental datasets used in this study and methodology of the proposed method. The results are presented in Section 3. Section 4 concludes this work.

2. Method

2.1. Motor Imagery EEG Datasets

Among the datasets provided by BCI competitions, we used the BCI competition IV dataset 2b [27,28] and the BCI competition II dataset III [29] for validation. Both datasets consist of left and right hand MI, and EEG signals were recorded at C3, Cz, and C4 channels.

As shown in Leeb et al. [27], the first MI EEG dataset, i.e., BCI competition IV dataset 2b, is composed of EEG signals from nine healthy subjects with a sampling frequency of 250 Hz. Each subject consists of two sessions without feedback, three sessions with online smiley feedback, and a total of five sessions. In this study, we utilized previously studied three sessions in whole sessions. In the first two sessions, after 2 s from the beginning on the fixation cross in the 3 s interval, short acoustic stimulus indicates the start of the trial. A visual cue, which is an arrow pointing to the left and right hand, appears for

1.25 s. The subject then performs the MI task on the hand corresponding to the visual cue for 4 s. In the last session, the scheme of the experiment is similar to the preceding sessions. However, the difference is that the visual cue and MI task parts in the preceding sessions were replaced by cue and smiley feedback. The smiley feedback changed to green when user feedback moved in the correct direction; else, it turned to red. The number of trials in each session is 120, 120, and 160, respectively.

The second MI EEG dataset, i.e., BCI competition II dataset III, consists of one subject and 7 runs with 40 trials for each run, which was sampled at 128 Hz. During acquisition of the EEG signals, the subject starts the MI experiment and rests for 2 s. After that, the acoustic stimulus is activated to indicate the beginning of the trials. Then, a cross '+' was displayed for 1 s, and the hand MI task, depending on where the arrow is pointed, is carried out from 3 s to 9 s. The details of the datasets are summarized in Table 1.

Table 1. Details of the datasets.

Dataset	Subjects	Channels	Trials	Sampling Frequency (Hz)	MI Class
BCI competition IV dataset 2b	9	C3, Cz, C4	400	250	2 (left/right hands)
BCI competition II dataset III	1	C3, Cz, C4	280	128	

2.2. Motor Imagery EEG Image Form Using Continuous Wavelet Transform

We developed a new two-dimensional image by extracting MI features that appear in a specific frequency band during the MI task. The resulting image was obtained through the time-frequency representation of the MI EEG signals. STFT, which is widely used in the time-frequency representation, is ineffective in interpreting MI EEG signals because of a trade-off in resolution between time and frequency. When the size of the window in STFT is short, it results in good time and poor frequency. A wide window offers the opposite results. To resolve this problem, we make use of continuous wavelet transform (CWT) [30] to develop an image of the EEG signal. CWT and Fourier Transform (FT) have similar methods. FT yields correlation coefficients between the original signal and a sinusoidal signal. Similarly, CWT obtains correlation coefficients between the original signal and a mother wavelet. However, unlike the FT, where the signal is decomposed into a frequency domain, CWT assigns the signal to a time-frequency domain by controlling the shape of the mother wavelet. Here, the shape of the mother wavelet is controlled by scaling and shifting parameters. The mathematical formula of CWT is given in Equation (1):

$$\text{CWT}(\omega, s) = \frac{1}{\sqrt{|s|}} \int x(t) \psi\left(\frac{t-\omega}{s}\right) dt \tag{1}$$

where $x(t)$ is MI EEG signal in this paper, ψ is the mother wavelet, ω denotes a time shifting parameter or translation, and s denotes a scaling parameter. $\text{CWT}(\omega, s)$ represents the correlation coefficients of CWT. The MI EEG signal $x(t)$ can be recovered by an inverse CWT, as following:

$$x(t) = \frac{1}{C} \int \int \text{CWT}(\omega, s) \frac{\psi_{\omega,s}(t)}{|s|^{3/2}} ds d\omega \tag{2}$$

where C indicates the normalization constant, which depends on the choice of wavelet.

Unlike STFT with a constant window function, CWT, with a smooth analytical mother wavelet, is capable of identifying the dynamic frequency properties over MI EEG signals at different scales. The CWT coefficients in Equation (1), by applying various scales and translations to the mother wavelet, reflect the similarity of the signal to the wavelet at the current scale. We use three types of mother wavelet, i.e., Morlet, Mexican hat, and Bump wavelets, provided in MATLAB.

The mathematical expressions of Morlet, Mexican hat, and Bump mother wavelets are given by Equations (3)–(5), respectively [31–33]. First, as shown in Equations (3) and (4), Morlet wavelet originated from a Gaussian function, while the Mexican hat wavelet, which is also called the Ricker wavelet, is a special case of the second derivative of a Gaussian function. Next, as another type of mother wavelet different from the Gaussian function-based ones, Bump wavelet with scale s and window w is defined in the Fourier domain as Equation (5):

$$\psi_{\text{Morl}}(t) = e^{2\pi it}e^{-t^2/2\sigma^2} = (\cos 2\pi t + i\sin 2\pi t)e^{-t^2/2\sigma^2} \tag{3}$$

$$\psi_{\text{Mexh}}(t) = \left(1 - \frac{t^2}{\sigma^2}\right)e^{-t^2/2\sigma^2} \tag{4}$$

$$\psi_{Bump}(sw) = e^{\left(1 - \frac{1}{1-(sw-\mu)^2/\sigma^2}\right)}\chi[\mu - \sigma, \mu + \sigma] \tag{5}$$

where ψ_{Morl}, ψ_{Mexh}, and ψ_{Bump} denote Morlet, Mexican hat, and Bump mother wavelets, respectively. The parameter σ plays a role in transshaping the mother wavelet. μ in Equation (5) admits the peak frequency defined by $sw_\psi := \underset{sw}{\text{argmax}}\left|\psi_{Bump}(sw)\right|$ and χ denotes the indicator function.

CWTs with three mother wavelets are employed to a duration of 2 s of the MI EEG signals. We set the frequency range of CWT—from a minimum frequency of 0.1 Hz to a maximum frequency of 50 Hz. Then, we extract the time-frequency image of the mu and beta bands from the overall frequency range. The image of the MI EEG signal is obtained by the following two methods.

First, an input image was obtained from the CWT results of the mu and beta bands. The sizes of the image extracted were 26×500 and 37×500, respectively. In order to prevent the extracted features from being biased by one dominant frequency band, both bands were resized to have similar size using the cubic spline interpolation method. The deformation of the input image is conducted not only on the frequency axis, but also on the time axis. In the labeled 2 s MI task, the smallest part of the output spectrum of the 0.5 s interval was extracted from the mu-band image. Then, the obtained samples corresponding to 0.5 s were resized to 32 samples by using the cubic spline interpolation method. Subsequently, we obtained a MI EEG image with $N_f = 31$ and $N_t = 32$ for one electrode by combining samples generated from the mu and beta bands. The same procedure was repeated for three electrodes, i.e., C3, Cz, and C4, and the resultant three MI EEG are stacked as one MI image. As a result, an MI EEG image has a size of $N_v \times N_t$, where $N_v = N_f \times 3$. This overall process is carried out for the three mother wavelets.

Second, we make use of a time-frequency image of only the mu-band in the frequency axis. Similar to the previous method, the modification process of the MI EEG image on the time axis is carried out. However, in this method, the frequency axis is not resized to avoid loss of frequency information. The MI EEG image constructed by the proposed method is used as an input to the proposed CNN architecture.

2.3. Convolutional Neural Networks Architecture

We conduct MI task recognition based on a variant of CNN. The conventional CNN has shown considerable performance for 2D image classification. CNN consists of input, output, and several hidden layers, which contain several pairs of convolutional-pooling layers and a fully connected layer. The standard CNN extracts the features of the image through the 2D kernel in the convolutional layer and subsampled them to a smaller size in the pooling layer. The reduced image is then classified in the fully connected layer.

It is to be noted that our input MI EEG image is different from the existing input data format for CNN. Since the input image contains three specific details (time, frequency, and electrode locations), our aim is to classify the hand MI tasks through features in the vertical axis corresponding to the electrode and frequency, whereas the time axis (horizontal axis) is not of critical interest. Therefore, we propose a new MI classification method based on CNN with 1D kernels rather than standard 2D kernels to capture the features of frequency and electrode location on the same time axis.

The proposed CNN architecture with four layers is shown in Figure 1. The input layer is the 2D input MI EEG image with a size of 93 × 32. The input layer is followed by the second and third hidden layers, which consist of one convolutional layer and one max-pooling layer. The following fourth layer is a fully connected layer that differentiates between left or right hand MI tasks. In the convolution layer, the input image is convolved using $N_F = 30$ kernels with a size of 93 × 3 for the same time axis. The output of convolution between an input image and a kernel is given by Equation (6):

$$y_i^k = f(a) = f\left(\left(W^k * x\right)_i + b_k\right) \tag{6}$$

where x is an input image, W^k is a convolution kernel, b_k is a bias for $k = 1, 2, \ldots, N_F$, and $i = 1, 2, \ldots,$ $N_t - 2$. As a result, the output applied with stride 1 in the convolutional layer yields 30 feature maps with a size of $(N_t - 2) \times 1$. $f(\cdot)$ denotes an activation function; here, the rectified linear unit (ReLU) function is used. The ReLU function is carried out between the convolutional layer and max-pooling layer by the following Equation (7):

$$f(a) = \text{ReLU}(a) = \max(a, 0) = \begin{cases} a, & \text{if } a > 0 \\ 0, & \text{otherwise} \end{cases} \tag{7}$$

where a is defined in Equation (6).

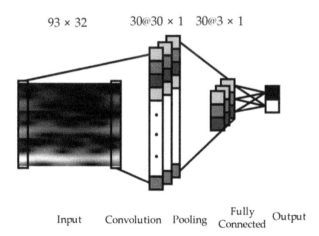

93 × 32   30@30 × 1   30@3 × 1

Input   Convolution   Pooling   Fully Connected   Output

Figure 1. Convolutional Neural Network architecture consists of a convolutional layer with 93 × 3 kernels, a max-pooling layer with sampling factor of 10, and a fully connected layer. The input image with 93 × 32 is passed through the proposed neural network. In the convolution layer, 30 kernels are convolved with the input image. Through a pooling procedure, each feature map is shrunk to 3 × 1. Refer to Section 2.3 for more details.

The output of the convolutional layer with size of $(N_t - 2) \times 1$ is applied to the input of the max-pooling layer. The max-pooling is carried out with a sampling factor of 10 and zero padding is not used. Therefore, the size of the output of the convolutional layer is subsampled to 3 × 1 dimension for 30 kernels by the max-pooling layer. Finally, the fully connected layer is performed to classify two classes, i.e., left and right hand MI tasks. In this approach, the labeled training data is used for training of the proposed CNN model and the classification error is calculated as the difference between the CNN output and target data. The weights of the neural network during training are updated by the back-propagation algorithm, which is conducted using gradient descent to minimize errors. Using the trained weights of the neural network, MI classification performance with the test MI EEG signals is computed.

In the proposed method, we make use of the continuous wavelet transform to produce the transient EEG image with improved time and frequency resolution. The use of a one-dimensional convolution neural network results in the improved discrimination capability of temporal variations of motor imagery patterns of an EEG image. In addition, since the neural network consists of not only a one-dimensional kernel to extract MI patterns of the input image but also shallow layers compared to conventional models, it has the advantage of low computation complexity in training. Since the proposed method utilizes an input image with time, frequency, and electrode information, the training of the neural network is robust to variations or abnormal patterns of MI EEG signals.

3. Results

3.1. Quantification of the Event-Related Desynchronization/Event-Related Synchronization Pattern

In the literature, it has been widely known that MI features are reflected in the mu-band (8–13 Hz) and beta-band (13–30 Hz) of EEG signals [34,35]. In the case of imagination of left and right motor movements, the power decrease of mu and beta bands of EEG signals, named event-related desynchronization (ERD), is observed in the contralateral brain region. In addition, the phenomenon in which the power of both frequency bands of the EEG signals is restored after the MI tasks is called event-related synchronization (ERS). To reflect ERD and ERS, we utilize a method to quantify the ERD/ERS patterns of the MI tasks done in this work [34].

The ERD/ERS patterns are reflected as a variation of power in the MI EEG signal, compared to a reference interval, prior to the start of motor movement imagery. Firstly, each MI EEG channel is averaged over all subjects and trials in the dataset. The ERD/ERS patterns are then calculated as the rate of the change of power with respect to the reference signals, which are given in Equations (8)–(10) [36]:

$$\text{EEG}_{avg(j)} = \frac{1}{N} \sum_{i=1}^{N} s_{ij}^2 \tag{8}$$

$$\text{EEG}_{ref} = \frac{1}{k} \sum_{j=t}^{t+k} \text{EEG}_{avg(j)} \tag{9}$$

$$\text{ERD/ERS} \ (\%) = \left(\frac{\text{EEG}_{avg(j)} - \text{EEG}_{ref}}{\text{EEG}_{ref}} \right) \times 100 \ (\%) \tag{10}$$

where N is the total number of trials and s_{ij} is the jth sample of the ith trial of the bandpass filtered MI EEG signals. $\text{EEG}_{avg(j)}$ is the average power of MI EEG signals for all trials. EEG_{ref} is the average power of MI EEG signals measured on the reference interval.

To extract a reliable MI task interval, we detect the ERD/ERS patterns in typical motor movement related frequency bands, e.g., mu-band (8–13 Hz), beta-band (13–30 Hz), and combined mu and beta band (8–30 Hz). In general, ERD/ERS patterns are distinctly observed in the motor cortex region. Figure 2 shows the ERD/ERS patterns in mu-band, which are recorded from C3, Cz, and C4 electrodes, respectively. The relative amplitude indicates the values calculated by Equation (10) across all subjects in the first dataset. Details on the dataset are described in Section 2.1. As shown in Figure 2, the ERD patterns occur bilaterally during the MI tasks (second 2–5), which are lateralized to the contralateral hemisphere. However, ERS patterns on the motor cortex are not observed clearly in mu-band. The ERD and ERS patterns in the Cz channel are not clearly distinguished, compared to other electrodes. Therefore, in order to extract a common MI task interval from datasets used in this paper, we chose a 2 s MI task interval (0.5 s~2.5 s after the visual cue is displayed), where the ERD patterns actively appear.

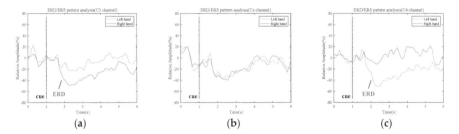

Figure 2. Event-related desynchronization/Event-related synchronization patterns over each channel (C3, Cz, and C4) during the hand Motor Imagery tasks in mu-band: (**a**) C3 electrode; (**b**) Cz electrode; (**c**) C4 electrode.

3.2. Classification Results

We validated the classification performance of the proposed CNN for two input image types and three mother wavelets. Here, we show MI EEG images for left hand MI tasks by utilizing CWTs to extract mu and beta bands or only mu-band of EEG signals. Figures 3–5 show the MI EEG images generated by the first method, i.e., using the mu and beta bands, for a left hand MI task using three distinct mother wavelets, i.e., Morlet, Mexican, and Bump wavelets, respectively. In each figure, the left figure denoted by (a) shows the resized image on the frequency axis, and the right figure denoted by (b) denotes the resized image both on the frequency and the time axes. Compared to FT and STFT, the use of CWT helps reveal ERD patterns of the mu-band from an EEG input image more clearly without loss of information in terms of time and electrode-frequency due to its superior time-frequency resolution. Figure 6a–c show the MI EEG images generated by using only the mu-band and three mother wavelets, respectively. As stated before, ERD patterns are shown in the MI EEG recorded from C4 electrode contralaterally in case of left hand MI tasks. Thus, in the figures, the ERD patterns are represented as a decrease of mu-band power in the C4 electrode, compared to other electrodes, i.e., Cz and C3. As a result, as shown in Figures 3–6, the generated EEG images depict the ERD patterns of mu-band EEG signals; the mu-band power of the C4 electrode is lower than those of other electrodes, regardless of mother wavelets.

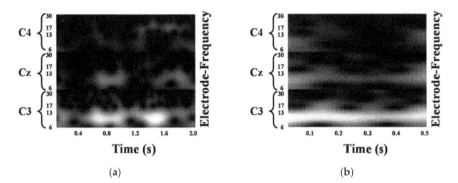

Figure 3. Motor Imagery EEG image for left hand Motor Imagery task using Morlet wavelet: (**a**) size of 93×500; (**b**) size of 93×32.

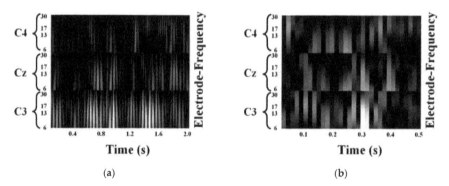

Figure 4. Motor Imagery EEG image for left hand Motor Imagery task using Mexican hat wavelet:
(a) size of 93 × 500; (b) size of 93 × 32.

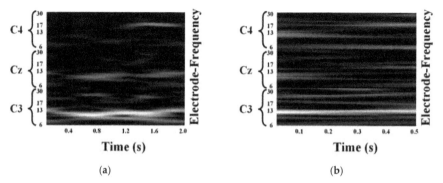

Figure 5. Motor Imagery EEG image for left hand Motor Imagery task using Bump wavelet: (a) size of
93 × 500; (b) size of 93 × 32.

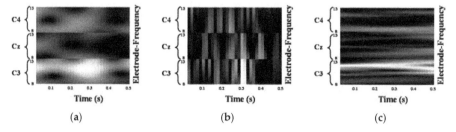

Figure 6. Motor Imagery EEG image with mu-band for left hand Motor Imagery task: (a) Morlet
wavelet; (b) Mexican hat wavelet; (c) Bump wavelet.

Tables 2 and 3 indicate the average accuracy and standard deviation across all subjects in the two
datasets used in this paper. We compared classification performance with the previous CNN-based MI
classification method, which used STFT and CNN [25]. The first dataset, BCI competition IV dataset 2b,
is comprised solely of the labeled training set. To evaluate the accuracy of the dataset, we divided it into
training and test sets for each subject using 10-fold cross-validation. Thus, 90% of the total 400 trials per
subject was randomly selected as the training set and the rest as the test set. This process was repeated
10 times and a total of 100 test sets were produced to reduce the effect of within-subject variation.

Table 2. Average classification accuracy and standard deviation (%) for BCI competition IV dataset 2b.

| Subjects | STFT [25] | CWT | | | | | |
| | | Morlet | | Mexican Hat | | Bump | |
	mu + beta	mu + beta	mu	mu + beta	mu	mu + beta	mu
1	74.5 ± 4.6	85.6 ± 1.3	84.7 ± 1.6	81.8 ± 1.3	81.7 ± 1.6	83.2 ± 1.4	82.4 ± 1.1
2	64.3 ± 2.0	72.8 ± 1.4	72.7 ± 2.0	70.6 ± 2.1	71.9 ± 2.0	73.8 ± 2.1	72.5 ± 2.0
3	71.8 ± 1.6	78.0 ± 1.9	79.5 ± 2.1	76.4 ± 1.8	74.7 ± 2.1	71.5 ± 2.1	73.6 ± 1.8
4	94.5 ± 0.2	95.4 ± 1.0	96.4 ± 0.5	96.0 ± 0.4	95.0 ± 0.9	96.2 ± 0.8	97.4 ± 0.5
5	79.5 ± 2.5	82.6 ± 1.7	79.6 ± 2.1	78.7 ± 1.9	75.6 ± 2.0	81.0 ± 1.0	73.1 ± 1.7
6	75.0 ± 2.4	79.8 ± 2.1	77.9 ± 1.6	75.5 ± 2.2	76.9 ± 1.5	80.6 ± 1.8	81.0 ± 1.3
7	70.5 ± 2.3	82.9 ± 1.2	81.0 ± 1.6	82.1 ± 1.2	81.4 ± 1.8	78.9 ± 2.0	81.7 ± 1.9
8	71.8 ± 4.1	85.0 ± 1.9	85.7 ± 1.7	84.7 ± 1.4	83.5 ± 1.4	83.5 ± 1.5	83.1 ± 1.6
9	71.0 ± 1.1	85.3 ± 1.9	84.9 ± 1.4	84.6 ± 1.2	85.1 ± 1.7	86.6 ± 1.4	84.0 ± 2.2
Mean	74.8 ± 2.3	83.0 ± 1.6	82.5 ± 1.6	81.2 ± 1.5	80.6 ± 1.7	81.7 ± 1.6	81.0 ± 1.6

Table 3. Average classification accuracy (%) for BCI competition II dataset III. N/A denotes 'not available'.

| Frequency Band | Accuracy (%) | | | |
	STFT [25]	Morlet	Mexican Hat	Bump
Mu + beta	89.3	89.3	90.0	92.9
mu	N/A	91.4	89.2	91.4

The test sets were evaluated by the proposed CNN, which is described in Section 2.3. The CNN structure consists of one convolutional layer, one max-pooling layer, and one fully connected layer. CNN was trained by using a batch training method with a batch size of 50 for 300 epochs. Data normalization was applied to the input MI EEG image in an interval between 0 and 1.

Table 2 shows the classification performance depending on input image types using the BCI competition IV dataset 2b. As can be seen, for the first input MI EEG image type with the mu and beta band, the proposed method yields average accuracy of 83.0%, 81.2%, and 81.7% for Morlet, Mexican hat, and Bump mother wavelet, respectively. Since the previous study using STFT has 74.8% accuracy, the proposed method achieves an improvement for classification of MI tasks. In addition, the proposed method with an input image using mu-band outperformed the STFT-based method, while it is comparable with the use of an input image using the mu and beta bands. For distinct mother wavelets, using the Molet wavelet for combined mu and beta bands results in the best classification accuracy.

Table 3 indicates the classification performance using the second dataset, i.e., BCI competition II datasets III. This dataset consists of a labeled training set and an unlabeled test set. The number of trials for each set is 140. This dataset with sampling rate of 128 Hz was likewise selected for the samples for 2 s after 0.5 s from the start of the cue appearance, as described in Section 2.2. The parameters used in the CNN model are the same as those applied to the first dataset. The results in Table 3 show that the proposed method is superior to the STFT-based method for both input image types. Note that STFT for an input image using mu-band is not available to compute accuracy, whereas the proposed method achieves comparable performance with the use of both mu and beta bands.

4. Conclusions

In this study, we propose a new continuous wavelet transform and convolutional neural network based decoding scheme to classify motor imagery tasks. The proposed method is comprised of two stages: image generation using continuous wavelet transform for motor imagery EEG signals and motor imagery tasks classification using the proposed one-dimensional convolutional neural network. By employing continuous wavelet transform, highly informative input motor imagery EEG image with

Entropy **2019**, *21*, 1199

time, frequency, and electrode location is generated. We confirm that the resultant motor imagery EEG image contains event-related desynchronization patterns, along with frequency and electrode location. Next, using the motor imagery EEG image as input, a one-dimensional convolutional neural network with four layers is developed to decode the two distinct motor imagery tasks. The network aims to capture the one-dimensional dynamics of event-related desynchronization patterns of the input motor imagery EEG image.

In the proposed method, the use of continuous wavelet transform yields a more detailed representation of motor imagery-related EEG patterns, compared to Fourier transform-based methods. In addition, by utilizing a one-dimensional convolutional neural network, it is capable of discriminating between temporal variations of EEG signals of motor imagery tasks, in terms of specific frequency and electrode. The combinational use of wavelet transform and neural network may lead to the development of advanced signal processing and machine learning tools to analyze EEG signals in motor imagery BCI research.

Author Contributions: H.K.L. and Y.-S.C. conceived and designed the methodology, and were responsible for analyzing and writing the paper. Both the authors have read and approved the final manuscript.

Funding: This work was supported by a research grant from Kwangwoon University in 2018 and the Institute for Information & communications Technology Promotion (IITP) grant funded by the Korean Government (MISP) (No. 2018-0-00735, Media Interaction Technology based on Human Reaction and Intention to Content in UHD Broadcasting Environment).

Conflicts of Interest: The authors declare no conflict of interest.

References

1. Gao, S.; Wang, Y.; Gao, X.; Hong, B. Visual and Auditory Brain–Computer Interfaces. *IEEE Trans. Biomed. Eng.* **2014**, *61*, 1436–1447. [PubMed]

2. Pfurtscheller, G.; Neuper, C.; Flotzinger, D.; Pregenzer, M. EEG-based discrimination between imagination of right and left hand movement. *Electroencephalogr. Clin. Neurophysiol.* **1997**, *103*, 642–651. [CrossRef]

3. Bonnet, L.; Lotte, F.; Lécuyer, A. Two Brains, One Game: Design and Evaluation of a Multiuser BCI Video Game Based on Motor Imagery. *IEEE Trans. Comput. Intell. AI Games* **2013**, *5*, 185–198. [CrossRef]

4. Yu, Y.; Zhou, Z.; Liu, Y.; Jiang, J.; Yin, E.; Zhang, N.; Wang, Z.; Liu, Y.; Wu, X.; Hu, D. Self-Paced Operation of a Wheelchair Based on a Hybrid Brain-Computer Interface Combining Motor Imagery and P300 Potential. *IEEE Trans. Neural Syst. Rehabil. Eng.* **2017**, *25*, 2516–2526. [CrossRef] [PubMed]

5. Blankertz, B.; Sannelli, C.; Halder, S.; Hammer, E.M.; Kübler, A.; Müller, K.-R.; Curio, G.; Dickhaus, T. Neurophysiological predictor of SMR-based BCI performance. *Neuroimage* **2010**, *51*, 1303–1309. [CrossRef]

6. Kosmyna, N.; Lindgren, J.T.; Lécuyer, A. Attending to Visual Stimuli versus Performing Visual Imagery as a Control Strategy for EEG-based Brain-Computer Interfaces. *Sci. Rep.* **2018**, *8*, 13222. [CrossRef]

7. Wolpaw, J.R.; Birbaumer, N.; Heetderks, W.J.; McFarland, D.J.; Peckham, P.H.; Schalk, G.; Donchin, E.; Quatrano, L.A.; Robinson, C.J.; Vaughan, T.M. Brain-computer interface technology: A review of the first international meeting. *IEEE Trans. Rehabil. Eng.* **2000**, *8*, 164–173. [CrossRef]

8. Nicolas-Alonso, L.F.; Gomez-Gil, J. Brain Computer Interfaces, a Review. *Sensors* **2012**, *12*, 1211–1279. [CrossRef]

9. Dai, M.; Zheng, D.; Na, R.; Wang, S.; Zhang, S. EEG Classification of Motor Imagery Using a Novel Deep Learning Framework. *Sensors* **2019**, *19*, 551. [CrossRef]

10. Ramoser, H.; Muller-Gerking, J.; Pfurtscheller, G. Optimal spatial filtering of single trial EEG during imagined hand movement. *IEEE Trans. Rehabil. Eng.* **2000**, *8*, 441–446. [CrossRef]

11. Martín-Clemente, R.; Olias, J.; Thiyam, D.B.; Cichocki, A.; Cruces, S. Information Theoretic Approaches for Motor-Imagery BCI Systems: Review and Experimental Comparison. *Entropy* **2018**, *20*, 7. [CrossRef]

12. Ang, K.K.; Chin, Z.Y.; Zhang, H.; Guan, C. Filter Bank Common Spatial Pattern (FBCSP). In Proceedings of the International Joint Conference on Neural Networks (IJCNN), Hong Kong, China, 1–8 June 2008; pp. 2390–2397.

13. Park, S.; Lee, D.; Lee, S. Filter Bank Regularized Common Spatial Pattern Ensemble for Small Sample Motor Imagery Classification. *IEEE Trans. Neural Syst. Rehabil. Eng.* **2018**, *26*, 498–505. [CrossRef] [PubMed]

14. Jolliffe, I. Principal component analysis. In *International Encyclopedia of Statistical Science*; Springer: Berlin, Germany, 2011; pp. 1094–1096.

15. Comon, P. Independent component analysis, A new concept? *Signal Process.* **1994**, *36*, 287–314. [CrossRef]

16. Hsu, W.-Y.; Sun, Y.-N. EEG-based motor imagery analysis using weighted wavelet transform features. *J. Neurosci. Methods* **2009**, *176*, 310–318. [CrossRef] [PubMed]

17. Lotte, F.; Bougrain, L.; Cichocki, A.; Clerc, M.; Congedo, M.; Rakotomamonjy, A.; Yger, F. A review of classification algorithms for EEG-based brain–computer interfaces: A 10 year update. *J. Neural Eng.* **2018**, *15*, 031005. [CrossRef]

18. Kang, H.; Nam, Y.; Choi, S. Composite common spatial pattern for subject-to-subject transfer. *IEEE Signal Process. Lett.* **2009**, *16*, 683–686. [CrossRef]

19. Fazli, S.; Popescu, F.; Danóczy, M.; Blankertz, B.; Müller, K.-R.; Grozea, C. Subject-independent mental state classification in single trials. *Neural Netw.* **2009**, *22*, 1305–1312. [CrossRef]

20. Cho, H.; Ahn, M.; Kim, K.; Jun, S.C. Increasing session-to-session transfer in a brain–computer interface with on-site background noise acquisition. *J. Neural Eng.* **2015**, *12*, 066009. [CrossRef]

21. Lu, N.; Li, T.; Ren, X.; Miao, H. A Deep Learning Scheme for Motor Imagery Classification based on Restricted Boltzmann Machines. *IEEE Trans. Neural Syst. Rehabil. Eng.* **2017**, *25*, 566–576. [CrossRef]

22. Ren, S.; He, K.; Girshick, R.; Sun, J. Faster R-CNN: Towards Real-Time Object Detection with Region Proposal Networks. *IEEE Trans. Pattern Anal. Mach. Intell.* **2017**, *39*, 1137–1149. [CrossRef]

23. Simard, P.; Steinkraus, D.; Platt, J.C. Best Practices for Convolutional Neural Networks Applied to Visual Document Analysis. In *Seventh International Conference on Document Analysis and Recognition*; IEEE: Piscataway, NJ, USA, 2003; pp. 958–963.

24. Bengio, Y.; LeCun, Y. Scaling Learning Algorithms towards AI. In *Large-Scale Kernel Machines*; MIT Press: Cambridge, MA, USA, 2007; pp. 1–41. ISBN 1002620262.

25. Tabar, Y.R.; Halici, U. A novel deep learning approach for classification of EEG motor imagery signals. *J. Neural Eng.* **2017**, *14*, 016003. [CrossRef] [PubMed]

26. BCI Competitions. Available online: http://www.bbci.de/competition/ (accessed on 25 January 2019).

27. Leeb, R.; Lee, F.; Keinrath, C.; Scherer, R.; Bischof, H.; Pfurtscheller, G. Brain–Computer Communication: Motivation, Aim, and Impact of Exploring a Virtual Apartment. *IEEE Trans. Neural Syst. Rehabil. Eng.* **2007**, *15*, 473–482. [CrossRef] [PubMed]

28. Leeb, R.; Brunner, C.; Mueller-Put, G.; Schloegl, A.; Pfurtscheller, G. *BCI Competition 2008-Graz Data Set b*; Graz University of Technology: Graz, Austria, 2008.

29. Bashar, S.K.; Bhuiyan, M.I.H. Classification of motor imagery movements using multivariate empirical mode decomposition and short time Fourier transform based hybrid method. *Eng. Sci. Technol. Int. J.* **2016**, *19*, 1457–1464. [CrossRef]

30. Gómez, M.J.; Castejón, C.; García-Prada, J.C. Review of Recent Advances in the Application of the Wavelet Transform to Diagnose Cracked Rotors. *Algorithms* **2016**, *9*, 19. [CrossRef]

31. Auger, F.; Patrick, F.; Paulo, G.; Olivier, L. *Time-Frequency Toolbox*; CNRS France-Rice University: Paris, France, 1996.

32. Meignen, S.; Oberlin, T.; McLaughlin, S. A New Algorithm for Multicomponent Signals Analysis Based on SynchroSqueezing: With an Application to Signal Sampling and Denoising. *IEEE Trans. Signal Process.* **2012**, *60*, 5787–5798. [CrossRef]

33. Landau, R.H.; Paez, J.; Bordeianu, C.C. *A Survey of Computational Physics: Introductory Computational Science*; Princeton University Press: Princeton, NJ, USA, 2008.

34. Pfurtscheller, G.; Lopes da Silva, F.H. Event-related EEG/MEG synchronization and desynchronization: Basic principles. *Clin. Neurophysiol.* **1999**, *110*, 1842–1857. [CrossRef]

35. Tang, Z.; Sun, S.; Zhang, S.; Chen, Y.; Li, C.; Chen, S. A Brain-Machine Interface Based on ERD/ERS for an Upper-Limb Exoskeleton Control. *Sensors* **2016**, *16*, 2050. [CrossRef]

36. Jeon, Y.; Nam, C.S.; Kim, Y.-J.; Whang, M.C. Event-related (De)synchronization (ERD/ERS) during motor imagery tasks: Implications for brain–computer interfaces. *Int. J. Ind. Ergon.* **2011**, *41*, 428–436. [CrossRef]

© 2019 by the authors. Licensee MDPI, Basel, Switzerland. This article is an open access article distributed under the terms and conditions of the Creative Commons Attribution (CC BY) license (http://creativecommons.org/licenses/by/4.0/).

Article

Evolution of Neuroaesthetic Variables in Portrait Paintings throughout the Renaissance

Ivan Correa-Herran [1,2], Hassan Aleem [3] and Norberto M. Grzywacz [1,3,4,5,]*

[1] Department of Neuroscience, Georgetown University, Washington, DC 20057, USA; iac16@georgetown.edu
[2] Facultad de Artes, Universidad Nacional de Colombia, Bogotá 110111, Colombia
[3] Interdisciplinary Program in Neuroscience, Georgetown University, Washington, DC 20057, USA; ha438@georgetown.edu
[4] Department of Physics, Georgetown University, Washington, DC 20057, USA
[5] Graduate School of Arts and Sciences, Georgetown University, Washington, DC 20057, USA
* Correspondence: norberto@georgetown.edu

Received: 2 December 2019; Accepted: 16 January 2020; Published: 26 January 2020

Abstract: To compose art, artists rely on a set of sensory evaluations performed fluently by the brain. The outcome of these evaluations, which we call neuroaesthetic variables, helps to compose art with high aesthetic value. In this study, we probed whether these variables varied across art periods despite relatively unvaried neural function. We measured several neuroaesthetic variables in portrait paintings from the Early and High Renaissance, and from Mannerism. The variables included symmetry, balance, and contrast (chiaroscuro), as well as intensity and spatial complexities measured by two forms of normalized entropy. The results showed that the degree of symmetry remained relatively constant during the Renaissance. However, the balance of portraits decayed abruptly at the end of the Early Renaissance, that is, at the closing of the 15th century. Intensity and spatial complexities, and thus entropies, of portraits also fell in such manner around the same time. Our data also showed that the decline of complexity and entropy could be attributed to the rise of chiaroscuro. With few exceptions, the values of aesthetic variables from the top of artists of the Renaissance resembled those of their peers. We conclude that neuroaesthetic variables have flexibility to change in brains of artists (and observers).

Keywords: neuroaesthetics; symmetry; balance; complexity; chiaroscuro; normalized entropy; renaissance; portrait paintings; art history; art statistics

---

1. Introduction

Aesthetic emotions are not arbitrary. For instance, people exhibit aesthetic preference for visual art with high degrees of symmetry across many cultures [1,2]. Other such visual properties with universal impact on aesthetic values are balance, contrast, and complexity (measured as normalized entropy—[3,4]. Why do these properties have universal aesthetic impact? The Processing Fluency Theory provides a simple answer by postulating that sensory variables processed by the brain with ease facilitate positive aesthetic emotions [5,6]. Hence, if the brain has specialized mechanisms to deal with a sensory variable, it will tend to be aesthetically valuable. This is the case for symmetry, balance, contrast, and complexity variables, which have dedicated neural circuitries, because of their evolutionary importance. We call such properties like symmetry, balance, contrast, and complexity neuroaesthetic (or fluency) variables, since their importance emerges directly from neural constraints [7].

Because specialized brain mechanisms constrain neuroaesthetic variables, one may expect that they remain relatively constant over time, especially across art periods [8]. However, a recent study found that artists exhibited an appropriate bias towards these variables, but did not optimize them. Moreover, artists also exhibited individuality with respect to these variables [4]. Because of this individuality, a certain degree of flexibility appears to exist with respect to neuroaesthetic variables. Therefore, they could potentially evolve across different periods of art. The possibilities that neuroaesthetic variables could either remain constant or evolve across art periods raised a series of questions in our minds: Are changes in art periods occurring in the absence of evolution of neuroaesthetic variables? Conversely, if these variables evolve over time, in what directions are the changes? For example, would the degrees of symmetry, balance, contrast, and complexity necessarily increase over time following the improvement of artistic techniques? And would the evolution in neuroaesthetic variables at the boundaries of different art periods (as determined by art historians) be abrupt? So far, there has been limited research aimed at answering such questions. One notable study looked at changes in fractal dimension and Shannon entropy in western paintings ranging from years 1285 to 2008 [9]. The research found that both measures remained relatively stable over time, except for an abrupt change around the late nineteenth century. The author speculates that this change may indicate the transition from pre-Modern to Modern Art.

In the study reported here, we probed what happened to the neuroaesthetic variables across the three periods of the Renaissance. These periods were the Early and High Renaissance, and the Mannerism (Late Renaissance—[8]). Art historians have characterized the differences between these art periods in terms of key artistic concepts. New concepts were continuously discovered or rediscovered during the Renaissance, and introduced in the work of artists. For example, Alberti's books on painting [10] and architecture [11] introduced new concepts that influenced the theory of the arts during the Renaissance itself. These concepts included ideas that evolved throughout the Renaissance, such as harmony, golden ratio, naturalism, anatomical studies, linear perspective, aerial perspective, and *chiaroscuro* [8]. These ideas are related to the neuroaesthetic variables mentioned above. Harmony and golden sections have to do with balance and symmetry. In turn, naturalism, anatomical studies, and the two forms of perspective produce realism and thus, complexity. Finally, *chiaroscuro* (translates to 'bright and dark' in Italian) is related to contrast. In *chiaroscuro*, strong tonal contrast between light and dark in different regions of a painting helps highlight its important parts, often with a dramatic effect. Furthermore, *chiaroscuro* helps to model three-dimensional forms through shades. Consequently, *chiaroscuro* along with the other concepts were elements of a new theory that transformed art from the practices of the Middle Ages.

Finally, our study focused on portrait paintings during these three periods. Our rationale for focusing on portraits was their relative simplicity, as they centered solely on the depiction of the human subject as opposed to other forms of art. In addition, portraits tended to have a vertical composition, simplifying the measurement of symmetries in the canvas. Therefore, focusing on portraits helped us constrain our study in a simpler set of measurements and statistics. Another reason to focus on portraits was that they encouraged interesting evolutionary tendencies across time during the Renaissance. Such evolution happened because portraits set up a competition among artists and their workshops to gain the favor of patrons and get the commissions [12]. This competition led the painters to explore new artistic forms to represent the character of the individual subject. Thus, portrait paintings evolved and improved over time.

In this article, we begin with a series of statistical measurements on symmetry, balance, and complexity. We chose these variables because they have a direct relationship to processing fluency [13,14]. In this study, we expand the measurements to all three periods of the Renaissance and compare Italy with the rest of Europe. We also attempt to compare the most famous painters of those times (as judged today) with other Renaissance painters. This comparison allows studying whether these two cohorts of painters differ significantly in terms of neuroaesthetic variables. Finally, some of the findings related to the evolution of balance and complexity were surprising. We attempted to explain them statistically through both the emergence of perspective and new measurements of the degree of *chiaroscuro*.

2. Materials and Methods

Methods for image selection, and measurements of symmetry, balance, and complexity appear in detail elsewhere [4]. In this section, we mainly focus on methods that are unique to this article. New qualitative methods include the choice of portrait paintings (Section 2.1), selection of top portrait painters of the Renaissance (Section 2.2), and classification of paintings into stylistic characteristics (Section 2.3). New quantitative methods are the statistical analyses (Section 2.4) and the development of indices of *chiaroscuro* (Section 2.5). In Section 2.6, we rewrite the equations of symmetry, balance, and complexity in [4] using the notation of Section 2.5.

2.1. Portrait Paintings

We studied 456 portrait paintings from 53 painters. We only included a painting in the study if it displayed one main individual as the subject. The portraits were painted in oil, tempera, frescos, or a mixture of these materials. We obtained all paintings from the "Artstor Digital Library" (library.artstor.org). If the painting had a frame, we removed it before performing our analyses, except if the painter had painted it. While all images were originally in color, we converted them into an 8-bit grayscale by rounding the average of the red, green, and blue values.

A complete list of the painters, paintings (by Artstor file name), and their classification into periods and stylistic characteristics appears in the Supplementary Materials. To each painting, we assigned a date of completion, giving the median values in case art historians are uncertain about the exact times. We excluded paintings with more than 50 years of uncertainty from the study. We decided to allow such an uncertainty in the determination of the time of completion, because such uncertainty happened rarely and we used robust statistics for all our conclusions (Section 2.4). The classification into Early Renaissance, High Renaissance, and Mannerism used the date of birth of the painters. Thus, we took painters of the Early Renaissance as those born between 1370 and 1450. In turn, painters of the High Renaissance were born between 1452 and 1489. Finally, Mannerists were those born from 1494 until 1571. This manner of classification into periods can be debated, especially at their borders. However, as shown in Supplementary Materials, the classification yields results accepted by art historians [8]. Furthermore, border errors were removed by the robust statistics in our articles (Section 2.4).

2.2. Selection of Top Portrait Painters

We wished to compare the most renowned painters of the Renaissance (as judged today) with other good painters from the same period. We know of no objective way to make a list of the most famous artists and different people may disagree on it. However, some rankings do exist and we decided to use one of them. We used the ranking developed by Ranker, a digital media company that produces polls on entertainment, brands, sports, and culture (Ranker.com). The list that Ranker has produced on Renaissance artists appears in http://bit.ly/RankerRenaissance (accessed 11 April 2018). Table 1 shows the top-ten vote getters in this list on 11 April 2018:

Table 1. Top Ten Renaissance Artists According to Ranker.com.

	Name	# of Votes	Period	In this Study
1	da Vinci	1322	High Renaissance	Yes
2	Michelangelo	1071	High Renaissance	No
3	Raphael	713	High Renaissance	Yes
4	Donatello	599	High Renaissance	No
5	Titian	413	High Renaissance	Yes
6	Botticelli	407	Early Renaissance	Yes
7	Caravaggio	296	Mannerism	Yes
8	van Eyck	275	Early Renaissance	Yes
9	Brunelleschi	258	Early Renaissance	No
10	Dürer	257	High Renaissance	Yes

As the table indicates, we used seven of these ten artists in our study. The other three were renowned for other forms of art not portraits. Further support that these seven portrait artists are among the leading ones from the Renaissance comes from [15].

2.3. Classification of Paintings into Stylistic Characteristics

To try understanding some surprising findings related to the evolution of balance and complexity, we divided the portrait paintings into four categories, namely, *chiaroscuro*, linear perspective, aerial perspective, and none of those. This classification was performed by eye by one of us, ICH. He used the following definitions for the classification:

- Linear Perspective: perspective in which the relative size, shape, and position of architectural objects are determined by imagined lines converging at a point on the horizon.
- Aerial Perspective: the technique of representing distant objects as fainter and bluer.
- *Chiaroscuro*: an effect of contrasted light and shadow created by light falling unevenly on something.

Fortunately, no portrait painting seemed to belong to more than one of these categories.

2.4. Statistical Analyses

All analyses comparing neuroaesthetic variables across locations and art periods used two-way ANOVA followed by post-hoc two-sided *t*-tests. In turn, we compared the probability of artistic styles (Section 2.3) across art periods with the Fisher's exact test. Finally, the comparison of neuroaesthetic variables for different artistic styles employed one-way ANOVA followed by post-hoc one-sided *t*-tests.

We performed tests of temporal trends of neuroaesthetic variables with the robust Kendall's τ correlation coefficient. The probability that this coefficient is different from zero and estimates of error are as developed by [16]. We used the Kendall's τ correlation coefficient, since the data exhibited outliers and trends often did not seem linear. To quantify the trends, we attempted to obtain robust fits with each of the following functions:

$$\varnothing_C(t : \alpha_1) = \alpha_1, \tag{1}$$

$$\varnothing_L(t : \alpha_1, \alpha_2) = \alpha_1 + \alpha_2(t - t_0), \tag{2}$$

$$\varnothing_{exp}(t : \alpha_1, \alpha_2, \alpha_3) = \alpha_2 + (\alpha_1 - \alpha_1)e^{-(t-t_0)/\alpha_3}, \tag{3}$$

$$\varnothing_{erf}(t : \alpha_1, \alpha_2, \alpha_3, \alpha_4) = \alpha_1 + \alpha_2 \ \text{erf}\big(((t - t_0) - \alpha_3)/\big(\sqrt{2}\alpha_4\big)\big), \tag{4}$$

where t was time, t_0 was the year of the first painting in our dataset, $\alpha_1, \alpha_2, \alpha_3$ and α_4 were the parameters of the functions, and *erf* was the error function [17]. Equations (1)–(4) represent constant, linear, exponential, and error-function trends respectively. The parameters of these functions have familiar interpretations. For the linear trend, α_1 is the estimated value of the neuroaesthetic variable at t_0 and α_2 is the slope of the change. Similarly, for the exponential trend, α_1 is the estimated value of the neuroaesthetic variable at t_0. However, α_2 is the value at long t's, and α_3 is the rate constant of the change. Finally, for the *erf* trend, $\varnothing_{erf}$ is a sigmoidal function [18]. Its point of fastest rate of change is α_3, with α_1 being the value of the function at that point. In turn, $2\alpha_2$ would be the output range of $\varnothing_{erf}$ if t were to vary from $-\infty$ to $+\infty$, with α_4 setting the rate of the transition.

To obtain robust fits of these functions to the values of neuroaesthetic variables, we proceeded as follows: Let v_i be the values of paintings completed at times t_i where $1 \leq i \leq N$ and N is the number of paintings in the dataset. To reduce the effect of outliers, we computed the M medians $\bar{v}_i$ of non-overlapping subsets of the data comprising temporally consecutive paintings. The number of paintings included in the medians ranged from 25 to 35 depending on the noise in the data. We indicate this number for each case when describing the results. We also computed the median time, $\bar{t}_i$ for each of the $\bar{v}_i$. For these pairs, we then computed:

$$\vec{\alpha} = \text{argmin}_{\vec{\alpha}^*} \sum_{i=1}^{M} \big(\bar{u}_i - \varnothing_x\big(\bar{t}_i : \vec{\alpha}^*\big)\big)^2, \tag{5}$$

where $\vec{\alpha}$ was the optimal set of parameters for each fit (e.g., $\vec{\alpha} = (\alpha_1, \alpha_2, \alpha_3, \alpha_4)$ for Equation (4)), and the subscript x captured one of the functions in Equations (1)–(4) (e.g., $x = erf$ for Equation (4)). To perform this computation, we used a trust-region-reflective algorithm [19,20]. Each fit computation used five initial conditions to minimize local-minimum trapping. The initial conditions were random and chosen from the ranges of either v_i or t_i (the latter for variables with time dimensions). Means and standard errors of fit parameters were obtained for those initial conditions yielding errors (Equation (5)) of not more than 10% of the minimum. Although this fit was a least-squares procedure, the estimates were robust because of the median steps to obtain $\bar{t}_i$, and $\bar{v}_i$.

To probe the quality of the fits provided by Equations (1)–(4) and to compare them, we used a regression test for arbitrary fits [21,22]. In this test, a linear regression was performed in the data-versus-fitted-model scatter plot. If the model was good, this regression should be close to a straight line, with slope = 1, and intercept = 0. Consequently, the regression test probed whether the correlation coefficient, slope, and intercept were statistically significantly larger than zero, not different from 1, and not different from 0, respectively. We used the *p*-values of the tests to compare the fits (*t*-tests). The calculations of the *p*-values considered the number of parameters of the equations through the degrees of freedom (d.f.). Hence, for example, if the linear and exponential fits gave similar results, the linear one was better, because it had fewer parameters.

Finally, we tested whether particular painters behaved differently from the population for each neuroaesthetic variable. For this purpose, we used a two-sided *t*-test of whether the neuroaesthetic variable was statistically significantly above or below the optimal trend line.

For each neuroaesthetic variable, we removed outliers with a median-absolute-deviation (MAD) method [23] (MAD > 3.5) before beginning the statistical analyses.

All statistical tests, and the computations described in Sections 2.5 and 2.6 were performed with MATLAB R2015a (MathWorks, Natick, MA, USA), using code specially developed for this project.

2.5. Chiaroscurro Indices

We developed two computational indices of *chiaroscuro*. We begin the description of each with a paragraph providing the physical intuition of the proposed calculations. We hope that these paragraphs will allow the reader to understand the rationale even by skipping the equations, which in turn, appear after the introductory paragraphs.

The *chiaroscuro* technique tries to use intensities close to the extremes, i.e., some regions bright and others dark. In its extreme form, *chiaroscuro* would make all the points of the painting either black or white. To measure the Index of *Chiaroscuro* Extremes, we thus calculate the distance from the distribution of intensities in the canvas to distributions in which the intensities are at either minimal or maximal possible values. (In our study, these intensities are 0 and 255 respectively.) To measure this distance, we begin by obtaining the midway point between the minimal or maximal possible intensities. We then perform a sum with two components: 1. sum of the subtractions of the intensities below the midway point and the minimal possible intensity; 2. sum of the subtractions of the maximal intensity and the intensities above the midway point. This sum should be zero only if the canvas has the desired property. In contrast, this sum reaches a maximal value when the image is homogeneous at the mid gray. We then construct the Index of *Chiaroscuro* Extremes, which is linearly related with this sum, being 0 and 1 when the sum 1 and 0 respectively.

Let $I\left(p_{k,j}\right)$ be the intensity of the pixel $p_{k,j}$ in Row k and Column j of the image. Let the number of rows and columns be N_r and N_c respectively. (We ensure that N_c is even). Consequently, the total number of pixels is $N = N_r N_c$. Finally, let I^* be the maximal possible intensity (the minimum being 0).

Denote the set of all pixels in the image as $S = \left\{p_{k,j}\right\}$. Define two subsets of S, namely S_- and S_+, with the following properties:

$$S_- \cup S_+ = S, S_- \cap S_+ = \varnothing,$$
$$p_{k,j} \in S_- \implies I\left(p_{k,j}\right) < \tfrac{I^*}{2}, p_{k,j} \in S_+ \implies I\left(p_{k,j}\right) > \tfrac{I^*}{2} \tag{6}$$

Hence, S_- and S_+ contain all pixels with intensities below and above $I^*/2$ respectively. (Because intensities are integers and $I^*/2 = 127.5$, S_- and S_+ contain all the pixels in S.) We use Equation (6) to define the Index of *Chiaroscuro* Extremes through the sum:

$$C_E = \sum_{p \in S_-} I(p) + \sum_{p \in S_+} (I^* - I(p)) \qquad (7)$$

This sum would have an upper bound at $NI^*/2$, if the image were homogeneous with intensity $I^*/2$, the middle gray. We thus define the Index of *Chiaroscuro* Extremes as:

$$i_{c_E} = 1 - \frac{2C_E}{NI^*} \qquad (8)$$

Therefore, $0 \le i_{c_E} \le 1$. The value of i_{c_E} would be 0 if the image were homogeneous with intensity $I^*/2$, and would be 1 if all the pixels of the image were black or white.

A possible limitation of the Index of *Chiaroscuro* Extremes was that some images could be entirely very dark or entirely very bright, and the index would still suggest the presence of *chiaroscuro*. If for example, an image was homogeneously back, the Index of *Chiaroscuro* Extremes would be 1. Although this limitation was unimportant for images of Renaissance portrait paintings, we decided to devise an alternate index of *chiaroscuro*. The new index measured how different the highest and lowest intensities were. To make this measurement, we again divided the set of points in the image into two sets. One set had all the points with intensities above a given percentile, while the other had the points with intensities below this percentile. We then subtracted the median intensity in the dark set from the median intensity in the bright set. As the result of this subtraction increases, we have more evidence of *chiaroscuro*. Thus, we use the result of this subtraction to construct an Index of *Chiaroscuro* Intensities. This index is linearly related to the result of the subtraction. The index is 0 when the subtraction yields 0, and 1 when the subtraction is equal to the difference between the maximal and minimal possible intensities. The higher this index is, therefore, the higher is the *chiaroscuro* intensity difference.

We again split S into two subsets, but this time based on a given percentile. We define $S_{-,f}$ and $S_{+,f}$ with the following properties:

$$\begin{aligned} S_{-,f} \cup S_{+,f} = S, S_{-,f} \cap S_{+,f} = \varnothing, \\ |S_{-,f}| = fN, |S_{+,f}| = (1-f)N, \\ p_{k,j} \in S_{-,f}, p_{l,m} \in S_{+,f} \implies I(p_{k,j}) \le I(p_{l,m}), \end{aligned} \qquad (9)$$

where $0 \le f \le 1$, i.e., a fraction of 1. Hence, $S_{-,f}$ contains fN elements below the fth percentile of S and $S_{+,f}$ contains all the $(1-f)N$ elements above the fth percentile of S. We use Equation (9) to define the Index of *Chiaroscuro* Intensities through the subtraction:

$$C_\Delta = \tilde{S}_{+,f} - \tilde{S}_{-,f}, \qquad (10)$$

where the wiggles denote medians. The result of this subtraction reaches its maximum, I^*, when the medians of $\tilde{S}_{+,f}$ and $\tilde{S}_{-,f}$ are the extremes, i.e., I^* and 0 respectively. We thus define the Index of *Chiaroscuro* Intensities as:

$$i_{c_\Delta} = \frac{C_\Delta}{I^*} \qquad (11)$$

Consequently, $0 \le i_{c_\Delta} \le 1$. The value of i_{c_Δ} would be 0 if $\tilde{S}_{+,f} = \tilde{S}_{-,f}$, i.e., the intensities at the top and the bottom were similar, and would be 1 if $\tilde{S}_{+,f} = I^*$ and $\tilde{S}_{-,f} = 0$, i.e., the intensities at the top and the bottom are maximally different.

2.6. Brief Descriptions of Symmetry, Balance, & Complexity

Elsewhere, we developed indices to quantify symmetry, balance, and complexity with methods and arguments like those in Section 2.5 [4]. Here, we describe these indices briefly to give the reader an intuitive understanding.

2.6.1. Symmetry

Our measure of symmetry is bilateral. This measure is taken as a comparison of intensities of pixels equidistant from the vertical midline of the whole image. The difference between a pair of pixels can range from 0 (perfectly symmetric) to 255 (highly asymmetric). To compute our final measure, we take the root mean square of all of the pixel-wise computations and normalize by the maximum intensity. The result is a measure of asymmetry ranging from 0 to 1, which is what we use in all of our analyses.

2.6.2. Balance

Like symmetry, we calculate our measure of balance across the vertical midline of the whole image. However, unlike symmetry, the computation for balance involves the integrals of the two sides rather than being pixel by pixel. The left and right integrals are then subtracted, and the absolute value of the result normalized by their sum. This measure gives an index of imbalance also ranging from 0 (full balance) to 1 (full imbalance).

2.6.3. Thickness of Balance Line

We further extend the balance measure above to catch finer details of balance composition. Artists often compose an image by parts as well as a whole. For example, while an image may be perfectly balanced at the bottom of the canvas, it may not be at the top. The overall balance measure in Section 2.6.2 does not capture this difference. To do so, we take the same computation of balance but in a row-by-row manner. This gives us a row-wise vertical balance line. We measure the "thickness" of this line as the relative median absolute deviation of all the points on the line divided by the horizontal size of the canvas. Thus, this measure expresses the thickness of the balance line as fraction of the size of the canvas, and thus being similar for small and large paintings. The measure is such that the thicker the line is, the greater is the amount of changes in balance across the image.

2.6.4. Complexity of Order 1

This form of complexity is the relative intensity entropy of an image. Thus, Complexity of Order 1 is the entropy of the distribution of intensities normalized by the maximal possible intensity entropy for an image of the same dimensions. An image with a wider distribution of intensities leads to greater Complexity of Order 1. This measure gives an index of complexity ranging from 0 (no complexity—only one intensity) to 1 (maximal complexity—all possible intensities equally likely).

2.6.5. Complexity of Order 2

This form of complexity captures Complexity of Order 1 minus the loss of complexity due to the spatial organization of the image. While the distribution of pixel intensities may be the same in two images, the relative spatial organization can be different. For example, an image with relatively large regions of isometric intensities (for example, large objects, shadows, or walls) will be less complex due to the spatial grouping. In contrast, images with finer details (for example, embroidered clothing or smaller objects) will have greater complexity. In the latter example, Complexity of Order 2 will be larger than in the former. We measure the spatial organization underlying Complexity of Order 2 through the ability to explain the image with isometric transformations. (They are translations, rotations, reflections, and their compositions). Consequently, the resulting entropy comes from the two-dimensional distribution of intensities obtained by juxtaposing an image with all possible isometric transformations of it. We then divide the outcome by the maximal possible Entropy of Order 2 for an image of the same dimensions. Therefore, Complexity of Order 2 ranges from 0 to 1. An index of 0 for Complexity of Order 2 occurs in single-intensity images, while 1 happens for spatially random images with all possible intensities equally likely.

2.6.6. Spatial Simplicity

As explained in Section 2.6.5, Complexity of Order 2 is Complexity of Order 1 minus the loss of complexity due to the spatial organization of the image. Hence, Complexity of Order 2 depends and is never greater than Complexity of Order 1. To obtain an index that captures spatial organization independently of Complexity of Order 1, we subtract from it Complexity of Order 2. We call this index Spatial Simplicity. To understand why, consider that the more spatially organized is the image (large homogeneous regions) the larger is Spatial Simplicity. From the definitions of Complexities of Order 1 and 2, Spatial Simplicity ranges from 0 to 1.

3. Results

3.1. Evolution of Neuroaesthetic Variables Throughout the Renaissance

In this study, we were interested in the drift of values of neuroaesthetic variables in relation to the passing of time and the evolution of art. Because specialized brain mechanisms constrain these variables, one may expect that they remain relatively constant over time. In contrast, a recent study demonstrated that a certain degree of flexibility appears to exist with respect to neuroaesthetic variables [4]. Therefore, they could potentially evolve across different periods of art. We thus asked whether changes across art periods can occur in the absence of evolution of neuroaesthetic variables. To answer this question, we first performed computational measurements of asymmetry, imbalance, and complexity (normalized entropy—[3]). In particular, we probed the changes that happened to these variables in Italy and in the rest of Europe. Our study involved a time span of close to 200 years of Renaissance. In Figure 1, we see the results divided to the periods of Early Renaissance, High Renaissance, and Mannerism.

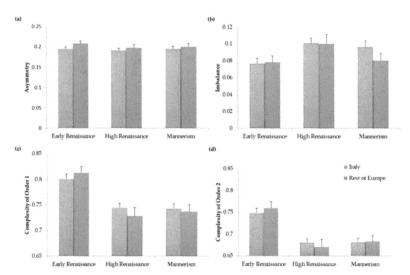

Figure 1. Spatiotemporal Evolution of Neuroaesthetic Variables throughout the Renaissance. (**a**) Index of Asymmetry; (**b**) Index of Imbalance; (**c**) Index of Complexity of Order 1; (**d**) Index of Complexity of Order 2. Error bars are standard errors. Whereas asymmetry remained statistically constant throughout the Renaissance, imbalance rose and complexities fell, especially in the Early Renaissance.

Our analysis revealed no significant changes in symmetry across the Early Renaissance, High Renaissance, and Mannerism (Figure 1a). In contrast, imbalance rose significantly between the Early and High Renaissance (Figure 1b, two-way ANOVA and post-hoc two-sided *t*-test, 298 d.f., t = 3.19, $p < 0.002$). The degree of imbalance grew by almost 30% in the span of 80 years. Complexity (i.e., normalized entropy) also evolved over time. We found falls in Complexities of Order 1 and Order 2 between the Early and High Renaissance (Figure 1c,d). These falls were significant for both Order 1 (432 d.f., t = 6.90, $p < 2 \times 10^{-11}$) and Order 2 (449 d.f., t = 7.26, $p < 2 \times 10^{-12}$). The falls reduced complexities of both orders by about 10%. Interestingly, however, no changes occurred in imbalance or complexity from High Renaissance to Mannerism. Hence, all changes in neuroaesthetic variables took place during the Early Renaissance. Finally, although we detected temporal changes in these variables throughout the Renaissance, we found no significant differences between Italy and the rest of Europe.

In conclusion, although symmetry was constant throughout the Renaissance, balance and complexities fell during the Early Renaissance.

3.2. Abrupt Transitions at the End of the 15th Century

To quantify these results further and to compare top artists with the other painters in our dataset, we produced scatter plots of the data. An example of the analysis of these plots appears in Figure 2 for Complexity of Order 2 (normalized spatial entropy).

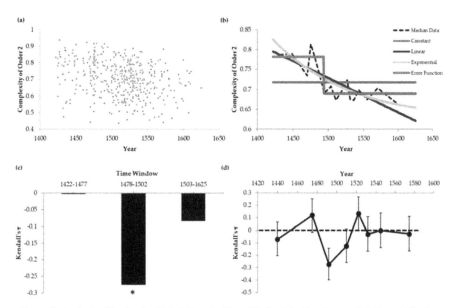

Figure 2. Analysis of the Scatter Plot of Complexities of Order 2 for Renaissance Paintings (**a**) Scatter Plot for All paintings (**b**) Best Fits of Four Models to Medians of the Data (25 paintings per median) (**c**) Kendall's τ in Three Non-overlapping Time Windows During the Renaissance. The star indicates a Kendall's τ statistically significantly different from zero (**d**) Kendall's τ and their standard errors for non-overlapping, consecutive periods with 55 paintings each. The Complexity of Order 2 appeared to fall abruptly in the last decade of the 15th century.

The basic scatter plot for Complexity of Order 2 appears in Figure 2a, in which each point represents an individual painting. The abscissas correspond to the years of painting completion and the ordinates are the measured complexities. The data show great variability in each moment of portrait evolution. To quantify the variability, we calculated the ratio between the standard deviation and the mean of the Complexity of Order 2. Overall, this variability ratio was 15%. Despite the variability, the results in Figure 2a confirmed and extended the temporal trends in Figure 1. One observes that the Complexity of Order 2 falls during the Renaissance. The Kendall's τ correlation coefficient was statistically significantly negative for Complexity of Order 2 (Kendall's $\tau = -0.211$, $p < 3 \times 10^{-11}$).

We attempted to characterize the fall of Complexity of Order 2 throughout the Renaissance by fitting four models (Equations (1)–(4); Figure 2b). Figure 1d had suggested that this fall was nonlinear and thus, we attempted exponential and error-function fits (Equations (3)–(5)). The latter seemed especially relevant, because no fall was apparent from the High Renaissance to the Mannerism. For completeness, we also attempted constant and linear fits. All the fits were statistically robust, by first extracting median complexities in small sections of the data (black dashed line in Figure 2b—obtained from the scatter plot with medians from 25 paintings).

The median Complexity-of-order-2 curve appeared to exhibit an abrupt fall around 1490. Not surprisingly, therefore, the Error-function model (Equation (4)) provided the best fit to the data. For example, the sums of squared errors for the optimal fits were 0.036, 0.014, 0.013, and 0.0068 for the Constant, Linear, Exponential, and Error-function models respectively. However, that the fit was better for the Error-function model was perhaps not surprising, because it had more parameters and could subsume some of the other models. Hence, we tested the quality of the fits with a regression test for arbitrary fits (Section 2.4; [21,22]). This test considered the number of parameters of the models. The test first plotted model predictions against the data and then analyzed the statistics of the resulting linear regression. The predictions were of positive correlation, with an intercept of 0 and a slope of 1. For all models, except the Constant one, we could not reject the null hypothesis that the correlation was positive. But the correlation was highest for the Error-function model ($R^2 = 0$, 0.61, 0.65, and 0.81 for the Constant, Linear, Exponential, and Error-function models respectively). Furthermore, we could reject that the intercept was zero for the Constant, Linear, and Exponential models ($p < 0.0001$, $p < 0.005$, and $p < 0.008$ respectively). In contrast, we could not reject this null hypothesis for the Error-function fit. Similarly, although we could not reject that the slope was 1 for the Error-function model, we could reject this null hypothesis for the Constant, Linear, and Exponential models ($p < 0.0001$, $p < 0.005$, and $p < 0.008$ respectively). Consequently, the Error-function model provided a superior fit than did the others. This superiority was true for all the fits in this article for data exhibiting trends. Moreover, we could not reject the Error-function model for any of these data.

The excellent error-function fit reinforced the conclusion of an abrupt fall of Complexity of Order 2 around 1490. The optimal transient year parameter ($t_0 + \alpha_3$ in Equation (4)) was 1493. In addition, the optimal transition was indeed abrupt as shown by the red line Figure 2b. However, the transition was not as abrupt as suggested by the red line. This line was obtained by fitting the curve of medians from 25 paintings, corresponding to a span of 12 years around 1493, namely [1488–1500]. Therefore, all that we could say was that 12 years was the upper bound for the duration of the transition. We call this time window ([1488–1500] for this example) the upper-bound transition interval.

Such an abrupt transition was surprising, because it was not immediately apparent in the scatter plot (Figure 2a). We thus wished to obtain model-independent evidence for such a transition. This evidence is what Figure 2c,d show. In Figure 2c, we show the Kendall's τ correlation coefficients for three non-overlapping section of the data. The middle section has 50 paintings around 1493. The other sections have all the paintings before and after the middle section. As the figure shows, although the middle section is far smaller than are the others, only it has a statistically significantly negative Kendall's τ. The Kendall's τ for the middle section is -0.275, with $p < 0.004$ that the correlation coefficient is 0. Comparison of this Kendall's τ with that obtained for the entire data (-0.211) suggests that most of the fall of Complexity of Order 2 happens during the period encompassed by the middle section [1478–1502]. This result is compatible with the upper-bound transition interval estimated above.

Further confirmation of the conclusion of abrupt transition appears in Figure 2d. This figure plots the Kendall's τ's and their standard errors for non-overlapping, consecutive sections of the data comprising 55 paintings. Only one point in the plot is statistically significantly negative, namely, the one centered on 1492.

In conclusion, our data indicate an abrupt fall in Complexity of Order 2 in the last decade of the 15th century. Similar analyses have shown abrupt transitions in most other variables studied in this paper, except as indicated.

3.3. Dynamics of Neuroaesthetic Variables and the Top Painters

We extended the scatter-plot analysis of Section 3.2 to Asymmetry, Imbalance, and Complexity of Order 1. In particular, we superimposed on the scatter plots temporal trend lines to help quantify the time courses of the drift of values of neuroaesthetic variables (Equations (4) and (5)). Finally, we colored the points of the top artists (Section 2.2) to compare them with peers from their periods. The results appear in Figure 3.

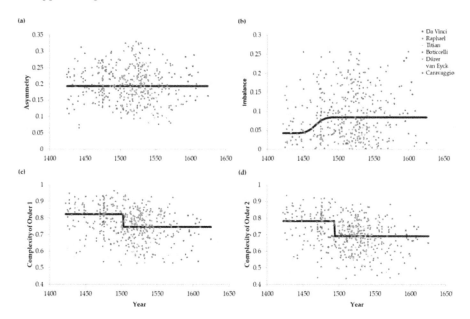

Figure 3. Scatter Plots of the Evolution of Neuroaesthetic Variables throughout the Renaissance with Error-function Fit (**a**) Index of Asymmetry (**b**) Index of Imbalance (**c**) Index of Complexity of Order 1 (**d**) Index of Complexity of Order 2. Data points from portraits of the top painters of the Renaissance are colored for ease of identification. Except for asymmetry, all aesthetic variables considered here undergo an abrupt transition at the end of the 15th century. For the most part, the statistics for the top painters are like those of their peers.

As for Complexity of Order 2 (Figure 2), Figure 3 shows great variability in each moment of portrait evolution. This variability holds for the results of both the whole group of painters and the work of the leading masters. The ratio between the standard deviation and the mean is 28%, 73%, and 13% for asymmetry, imbalance, and Complexity of Order 1 respectively. Consequently, the variability is lowest for complexities (see also Section 3.2) and highest for balance.

Also, like Figure 2, despite the variability, the results in Figure 3 confirmed and extended the temporal trends in Figure 1. The Kendall's τ correlation coefficient was not significantly different from zero for asymmetry (see its robust constant regression in Figure 3a). However, the coefficient was statistically significantly positive for Imbalance (Kendall's $\tau = 0.0738$, $p < 0.03$). In contrast, the coefficient was statistically significantly negative for Complexity of Order 1 (Kendall's $\tau = -0.204$, $p < 2 \times 10^{-10}$), as it was for Order 2 (Section 3.2). The Error-function regression lines (Equation (4)) attempted to capture these positive and negative tendencies in the evolution of the neuroaesthetic variables. The line rose for imbalance (Figure 3b; obtained from the scatter plot with medians from 30 paintings). But the line fell for complexities (Figure 3c,d; obtained from the scatter plot with medians from 25 paintings). The best-fit transition times for Imbalance and Complexities of Order 1 were 1467 and 1502 respectively. The corresponding upper-bound transition intervals (see discussion of Figure 2b) were [1451–1473] for imbalance and [1499–1505] for Complexity of Order 1. Therefore, the transition was about 25 years earlier for imbalance than Complexity of Order 2 (see also Section 3.2). The transition was also slower for imbalance. Moreover, the transition may have been about 10 years later for Complexity of Order 1 than of Order 2.

The top painters did not generally appear to behave differently from their peers in terms of neuroaesthetic variables. Statistical comparisons of the values of their neuroaesthetic variables show relatively equal distribution above and below the trend lines. Titian was the only exception for Complexity of Order 1 ($t = 3.48$, 17 d.f., and $p < 0.003$). In turn, two painters were exceptions for Complexity of Order 2. They were van Eyck ($t = 3.42$, 5 d.f., and $p < 0.02$) and Titian ($t = 3.11$, 17 d.f., and $p < 0.007$). These painters produced portraits that were less complex than were those of the peers, as evaluated by the Error-function model.

In sum, balance and complexity declined abruptly towards the end of the 15th century, but this fall was not generally due to the top painters of those times. Together with the results of Sections 3.1 and 3.2, we thus established that neuroaesthetic variables were not constant throughout the Renaissance.

3.4. Evolution of the Thickness of Balance Lines

To understand the decline of balance over time, we must start from the definition of imbalance. We defined it relative to the midline of the canvas. Elsewhere, we also considered the position of balance, i.e., the place for which the integrals of intensities to the right and left of it were equal [4]. Thus, the decline of balance in the Renaissance meant that the distance between the midline and the position of balance tended to increase over time (see examples in Figure 4a,b). However, this decline did not imply a rise in the sloppiness of balancing different parts of the portrait. Painters could continue to balance portraits delicately but simply do it in a position of balance away from the midline. In an earlier publication, we reported that painters in the Early Renaissance not only balanced their portraits, but also did so at every row of the canvas [4]. We thus decided to test if this delicate form of balance was also diminished as the Renaissance progressed. To do so, we measured the positions of balance at every height of the painting. All these points together formed the Balance Line (Figure 4c,d). We then measured the thickness of this line as a fraction of the horizontal size of the canvas. The results appear in Figure 4e,f.

The results show that the thickness of the balance line also rises in the Renaissance. We can appreciate an example by comparing a portrait by Domenico Veneziano in the Early Renaissance with one by Giovanni Battista Moroni during Mannerism (Figure 4c,d, respectively). In Veneziano's portrait, the balance line shows that the distributions of intensities on the two sides of the midline of the canvas are similar. The balance line is close to midline at every height analyzed. In contrast, in Moroni's portrait, the balance line has more variation across vertical positions. Hence, the balance line in Moroni's portrait has more thickness (0.165) than in Veneziano's (0.018). Thus, Moroni was "sloppier" in balancing different parts of his portrait than was Veneziano.

This difference held when we analyzed the thicknesses of balance lines throughout the Renaissance. In the Early Renaissance, the thickness of the balance line was significantly lower in Italy than in the rest of Europe (two-way ANOVA and post-hoc two-sided post-hoc t-test, 149 d.f., $t = 4.05$, $p < 9 \times 10^{-5}$—Figure 4e).

Afterwards the thickness of the balance line grew in Italy from the Early to High Renaissance (211 d.f., $t = 7.00$, $p < 4 \times 10^{-11}$—Figure 4e), catching up with the values in the rest of Europe. In contrast, the thickness of the balance line was statistically constant in the rest of Europe throughout the Renaissance. Therefore, portraits in the rest of Europe were more prescient of future trends of balance than were Italian ones.

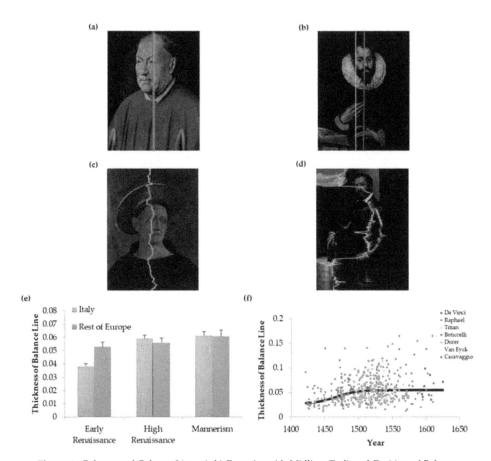

Figure 4. Balance and Balance Line. (**a,b**) Portraits with Midline (Red) and Position of Balance (Green) Marked. (**a**) Jan van Eyck, Portrait of Cardinal Niccolo Albergati Papal Envoy in the Spanish Netherlands, 1431–1432, Kunsthistorisches Museum, Viena, Austria. Photo Credit: Erich Lessing/Art Resource, N.Y. (**b**) El Greco, Portrait of a Man (possibly Alonso de Herrera) 1600, Musée de Picardie, Amiens and Picardy, France. (**c,d**) Portraits with Balance Lines Marked. (**c**) Domenico Veneziano, Head of a Tonsured, Beardless Saint 1440-4, The National Gallery, London, Great Britain. Photograph: ©The National Gallery, London National Gallery Picture Library, The National Gallery Company. (**d**) Giovanni Battista Moroni; Portrait of Mario Benvenuti 1560, The John and Mable Ringling Museum of Art, the State Art Museum of Florida, a division of Florida State University. (**e**) Spatiotemporal Evolution of the Thicknesses of Balance Lines throughout the Renaissance. (**f**) Scatter Plot of the Evolution of the Thicknesses of Balance Line throughout the Renaissance. Conventions for Panels E and F are as in Figures 1 and 3. The thickness of balance line was low in Italy during the Early Renaissance, but rose before the High Renaissance.

The scatter plots showed that the thicknesses of the balance lines (Figure 4f) followed a trend like that of imbalance (Figure 3d). The data in Figure 4f show great variability of thicknesses in each moment of portrait evolution. The ratio between the standard deviation and the mean of the thicknesses of balance lines was 50%. Despite the variability of thicknesses, the results in Figure 4f confirmed and extended the temporal trends in Figure 4e. Portraits tended to be carefully balanced in the Early Renaissance, but exhibit sloppier balances in the High Renaissance and Mannerism. Accordingly, the Kendall's τ correlation coefficient was statistically significantly positive for the thicknesses of balance lines (Kendall's $\tau = 0.190$, $p < 1.21 \times 10^{-9}$). The best-fit transition time was 1467, confirming that most change happened in the Early Renaissance. However, the change was much slower for the thickness of balance line than for other aesthetic variables. Consequently, its change was not abrupt. Finally, as for imbalance (Figure 3b), top painters did not generally produce portraits with thicker balance lines than those of peers (Figure 4f). In conclusion, as the Early Renaissance progressed, painters tended to become "sloppier" in balancing different parts of the portrait.

3.5. Evolution of Spatial Complexity

How are we to understand the decline of complexity over time? The Complexities of Order 1 and 2 in Figures 1 and 3 have different types of interpretation [4]. Complexity of Order 1 measures the normalized entropy in the distribution of intensities in the image. In turn, Complexity of Order 2 begins from Complexity of Order 1 and then discounts the reduction of entropy due to spatial organization. Consequently, if we want to isolate the loss of complexity due to spatial organization alone, we must calculate Complexity of Order 1 minus Complexity of Order 2. We call this quantity the Spatial Simplicity [4]. The temporal drift of the Spatial Simplicity throughout the Renaissance appears in Figure 5.

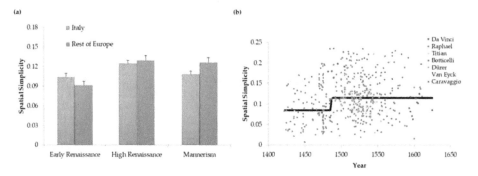

Figure 5. Evolution of Spatial Simplicity throughout the Renaissance. (**a**) Spatiotemporal Evolution. (**a**) Scatter Plot of the Evolution. Conventions for Panels (**a**) and (**b**) are as in Figures 1 and 3. Spatial simplicity was lower in the Early Renaissance, but rose abruptly at the end of the 15th century.

Portrait paintings tended to become spatially simpler as the Renaissance progressed. Thus, in High Renaissance and Mannerist periods, spatial complexity was lower than in the Early Renaissance (Figure 5a; $t = 4.38$, 447 d.f., $p < 0.00002$). However, as for Figure 1, although we detected temporal changes in Spatial Simplicity throughout the Renaissance, we found no significant differences between Italy and the rest of Europe. The scatter plots confirmed the rise of spatial simplicity (Figure 5b). Accordingly, the Kendall's τ correlation coefficient was statistically significantly positive for spatial simplicity ($\tau = 0.0994$, $p < 0.002$). The Error-function regression lines (Equation (4)) rose abruptly for Spatial Simplicity (Figure 5b; obtained from the scatter plot with medians from 35 paintings). The best-fit transition time was 1486, with the upper-bound transition interval lasting 17 years, namely, [1477–1494]. Finally, four of the seven top painters produced portraits with different spatial-simplicity distributions than those of their peers (Figure 5b). Botticelli ($t = 2.55$, 8 d.f., $p < 0.04$), van Eyck ($t = 3.19$, 4 d.f., $p < 0.04$), and Raphael ($t = 3.76$, 18 d.f., $p < 0.002$) exhibited more spatial simplicity than did their peers. In contrast, Caravaggio exhibited less ($t = 3.60$, 7 d.f., $p < 0.009$).

Hence, spatial complexity (the component of entropy due to spatial organization) fell abruptly towards the end of the Early Renaissance. This fall mimicked the decline of the complexity due to the distribution of intensities (Complexity of Order 1—Figure 3c).

3.6. Chiaroscuro and the Fall of Complexity

The decline of complexity over time (Figure 1c,d, Figure 2, Figure 3c,d and Figure 5) was surprising to us. We had expected complexity to increase as paintings became more realistic in the Renaissance. Ideas that evolved throughout the Renaissance, such as naturalism, anatomical studies, linear perspective, and aerial perspective should perhaps have made portraits more complex. Therefore, we wondered why complexity fell. We hypothesized that portrait paintings got simpler with the invention of *chiaroscuro*. It introduced large dominant regions with fewer colors and homogeneous intensities. We can appreciate an example of such regions by comparing a portrait by Andrea Mantegna in the Early Renaissance with one by Caravaggio during Mannerism (Figure 6a,b, respectively). Mantegna's portrait has no dominant regions in terms of blacks and white, and thus has no or very little *chiaroscuro*. In contrast, Caravaggio's portrait is a good example of *chiaroscuro*, with some bright regions contrasting against a large, dark background. Hence, the indices of *chiaroscuro* extremes (Equation (8)) and *chiaroscuro* intensities (Equation (11)) in Mantegna's portrait (0.33 and 0.094 respectively) were lower than those in Caravaggio's (0.75 and 0.83 respectively). Thus, our measurements confirm the common knowledge that Caravaggio used *chiaroscuro* more than did Mantegna (and most other painters—[8]). A quantitative study of these *chiaroscuro* indices across the Renaissance appears in Figure 6c–f.

Figure 6c revealed that the index of *chiaroscuro* extremes (Equation (8)) rose between the Early Renaissance and the High Renaissance periods (two-way ANOVA and post-hoc two-sided *t*-test, 303 d.f., $t = 8.28$, $p < 4 \times 10^{-15}$). However, no such rise occurred from the High Renaissance to Mannerism. Figure 6d also showed that this index grew abruptly towards the end of the Early Renaissance (Kendall's $\tau = 0.231$, $p < 3 \times 10^{-13}$; best-fit transition time = 1487, upper-bound transition interval lasting 15 years, namely, [1479, 1494]). These findings were replicated for the index of *chiaroscuro* intensities (Equation (11)) in Figure 6e,f (313 d.f., $t = 5.92$, $p < 9 \times 10^{-9}$; Kendall's $\tau = 0.194$, $p < 8 \times 10^{-10}$; best-fit transition time = 1481, transition interval lasting 14 years, namely, [1476, 1490]). Another finding in Early Renaissance was the statistical similarity of Italy with the rest of Europe in terms of *chiaroscuro* tendencies. Finally, we again detected little difference from top painters and their peers (Figure 6d,f). The only exceptions were van Eyck for the index of *chiaroscuro* extremes ($t = 3.75$, 5 d.f., $p < 0.02$), and Caravaggio for the index of *chiaroscuro* intensities ($t = 6.67$, 7 d.f., $p < 3 \times 10^{-4}$). Both van Eyck and Caravaggio exhibited more chiaroscuro than did their peers.

That the degree of *chiaroscuro* usage went up in the Renaissance was consistent with our hypothesis for the decline of complexity. However, we still had to demonstrate that more *chiaroscuro* in a portrait tended to lead to less complexity. In Figure 7, we classified portrait paintings in three compositional concepts that could affect complexity: linear perspective, aerial perspective, and *chiaroscuro* (Section 2.3). We also included a class for those portraits that do not belong to any of these three categories. Finally, we quantified Complexity of Order 1, spatial simplicity, and the index of *chiaroscuro* extremes in these four categories.

In Figure 7a, we observe that the prevalence of *chiaroscuro* portraits increases as time progresses in the Renaissance. This increase occurs specially from the Early to the High Renaissance (Fisher's exact test, odds ratio = 0.30, $p < 5 \times 10^{-4}$). Although we categorized these portraits by hand (Section 2.3), Figure 7b supported the idea that our *chiaroscuro* category was correct. The index of *chiaroscuro* extremes was higher for this category than was for the others (one-way ANOVA followed by post-hoc one-sided *t*-tests; linear perspective, 114 d.f., $t = 6.37$, $p < 3 \times 10^{-9}$; aerial perspective, 151 d.f., $t = 6.52$, $p < 6 \times 10^{-10}$; None, 381 d.f., $t = 8.28$, $p < 2 \times 10^{-15}$). Consequently, the use of *chiaroscuro* techniques increased over time. In contrast, the prevalence of portraits with the other tested compositional categories, namely, linear and aerial perspective, was statistically constant throughout the Renaissance. Hence, of the compositional elements studied, *chiaroscuro* is the only candidate available to explain the fall of complexity over time. Is *chiaroscuro* contributing to the simplification of portraits? The answer to this question appears in Figure 7c,d. The former shows that Complexity of Order 1 is significantly lower in portrait paintings with *chiaroscuro* than in portraits in the other categories (linear perspective, 106 d.f.,

$t = 3.83, p < 2 \times 10^{-4}$; aerial perspective, 143 d.f., $t = 6.09, p < 5 \times 10^{-9}$; None, 378 d.f., $t = 5.40, p < 6 \times 10^{-8}$). Consequently, *chiaroscuro* reduces the complexity of the distribution of intensities. Furthermore, in Figure 7d, we see that spatial simplicity is significantly lower in portraits with *chiaroscuro* than is in portraits of the other categories (linear perspective, 106 d.f., $t = 3.30, p < 2 \times 10^{-3}$; aerial perspective, 142 d.f., $t = 2.57, p < 6 \times 10^{-3}$; None, 372 d.f., $t = 3.37, p < 5 \times 10^{-4}$). Therefore, *chiaroscuro* tends to reduce the spatial complexity of portraits. We conclude that the reduction in complexity in the Renaissance may be due to the rise of *chiaroscuro*.

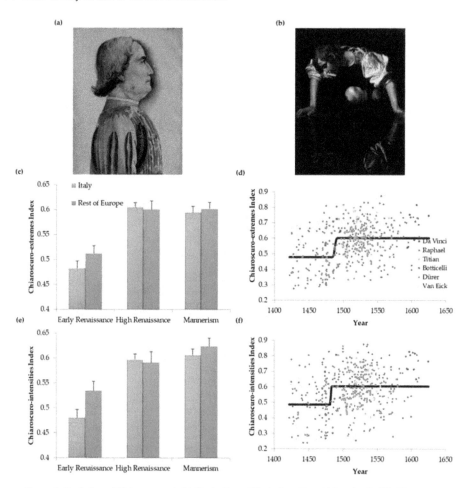

Figure 6. Evolution of Chiaroscuro. (**a,b**) Illustration of Portraits without (**a**) and with (**b**) chiaroscuro. (**a**) Andrea Mantegna, Portrait of Jacopo Antonio Marcello, 1453, Bibliothèque de l'Arsenal, Paris, France. Photo Credit: Erich Lessing/Art Resource, N.Y. (**b**) Michelangelo Merisi da Caravaggio, Narcissus, 1597, Galleria Nazionale D'arte Antica nel Palazzo Corsini, Rome, Italy © 2006, Scala, Florence/Art Resource, N.Y. C. Spatiotemporal Evolution of the Index of Chiaroscuro Extremes. (**d**) Scatter Plot of the Evolution of the Index of Chiaroscuro Extremes. (**e**) Spatiotemporal Evolution of the Index of Chiaroscuro Intensities. (**f**) Scatter Plot of the Evolution of the Index of Chiaroscuro Intensities. Conventions for Panels A and B are as in Figures 1 and 3. The degree of chiaroscuro rose abruptly in the Early Renaissance.

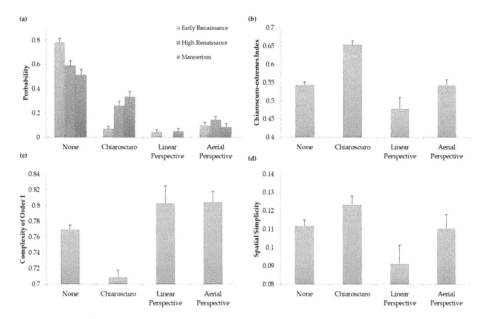

Figure 7. Link between Chiaroscuro and Complexity. (**a**) Evolution of Various Compositional Elements throughout the Renaissance. (**b**) Link between these Compositional Elements and the Index of Chiaroscuro Extremes. (**c**) Link between these Compositional Elements and Complexity of Order 1. (**d**) Link between these Compositional Elements and Spatial Simplicity. Error bars are standard errors. The emergence and rise of chiaroscuro accounted for the fall of complexity in the Renaissance.

4. Discussion

4.1. Neuroaesthetic Variables Evolved Throughout the Renaissance

In the Introduction, we raised the hypothesis that specialized brain mechanisms might constrain neuroaesthetic variables to remain relatively constant across art periods. However, our analysis of the Renaissance ruled out this hypothesis, showing that neuroaesthetic variables evolved.

This evolution tended to be abrupt towards the end of the Early Renaissance (Figures 1–3, Figures 5 and 6). We observed abrupt transitions centered around 1465 for balance-related variables, around 1490 for spatial complexity and *chiaroscuro*, and around 1500 for intensity complexity. These different transition centers were significant, because their corresponding upper-bound transition intervals did not overlap. The transition was significantly slower for balance-related variables than for the others (see for example, Figure 4f). As for the other variables, the transitions were fast. They lasted less than about 15 years for spatial complexity and chiaroscuro, and no more than about 6 years for intensity complexity.

What are possible explanations for these abrupt transitions? The median data on Figure 2b and the model fits in Figures 3, 5 and 6 suggest that these abrupt changes correspond to phase transitions. Phase transitions are common in natural sciences (for example, the abrupt transition from ice to liquid water as a function of temperature), but may also occur in the social sciences [24]. In our data, the phases are demonstrated by the relatively constant median values of the aesthetic variables before and after the transitions. However, although the transition between the phases is fast, it is not discontinuous. Hence, the abrupt changes of aesthetic values are second-order phase transitions (like the magnetization in iron) and not first-order (like the ice-to-water transformation). The essential components of such transitions are nonlinear interactions between the basic components of a system, which is under the influence of changing external conditions. We propose that the basic elements are the values associated with different instantiations of aesthetic variables by individual artists. In turn, the nonlinear interactions are due to mutual influence between artists learning from each other [7]. Finally, the changing external environment could exert social pressure to innovate and lead to cultural trends ([25]—for example, the desire for increasing realism during the Renaissance—[8]). With such external pressure, the phase transitions of aesthetic variables could happen as follow: Under pressure to innovate, an artist invents a technique, whose aesthetic value is superior to extant pieces of art (for instance, *chiaroscuro*). A small number of other artists are exposed to the new technique, using and perfecting it. As the number of artists using the technique increase, the probability that others learn increases exponentially. Therefore, the use of technique accelerates rapidly, causing the phase transition.

4.2. What Explains the Unexpected Declines of Balance and Complexity

Why did balance decrease throughout the Renaissance? Because balance is highly salient and preferred [26–28], we might have expected balance to increase over time. An explanation for the decay of balance and the thickening of balance lines might be the rise of dynamism in paintings, leading to the Baroque period [8]. Representation of motion brings with itself new neuroaesthetic variables that might compete with balance. Such competition exists between complexity and balance [29]. Computer simulations with a new theory for the learning of aesthetic values suggest that learning under the influence of motivated behaviors may have a role in generating these competitions [7].

Why did complexity fall throughout the Renaissance? Our expectation before starting this study was that the addition of details in paintings should perhaps have increased their entropy and thus, complexity. After all, Renaissance painters were striving to make their portraits more realistic, by studying nature and adding features such as perspective [30]. Perhaps with the realism, complexity would rise. We considered different factors that could fight complexity. One factor would be the invention of aerial perspective, with its tendency to smear details from distant objects [8]. However, our results argued against the aerial-perspective factor (Figure 7). An alternate factor for the fall of complexity would be the gradual emergence of *chiaroscuro*, with its large dominant regions with relatively homogeneous colors or intensities. Our data confirm the relevance of this factor, showing a strong link between the fall of complexity and the rise of *chiaroscuro* (Figure 7). This link makes an interesting point: Fall in complexity does not imply fall in realism. Part of the reason to use *chiaroscuro* is to increase the sense of realistic three-dimensional space through shadow effects [31].

4.3. Temporal Constancy of Symmetry

Different from balance and complexity, we did not detect temporal trends in the degree of symmetry in portrait paintings across the Renaissance (Figures 1a and 3a). However, this lack of trend did not mean that symmetry was frozen. We used the Kendall's τ statistic to test for such trends. This statistic is used to measure the ordinal association between two quantities [16]. However, different aspects of the distribution of symmetries could have changed over time without affecting this association. For example, inspection of Figure 3a suggests a temporal change of the variance in the distribution of the Index of Asymmetry.

Why did the central tendency of symmetry not exhibit a temporal trend? The emphasis on symmetry appeared as a rebirth of the classical ideas of composition from antiquity [31]. Hence, perhaps the cultural force of the classic ideals kept symmetry strong throughout the Renaissance. An alternative is that symmetry is an important variable in the brain, thus constraining what painters do. We believe that this is not the case, because from the Early Renaissance, portrait painting is not mostly frontal, therefore with an automatic break of symmetry [32]. Instead, we propose that our mathematics of symmetry is delicate, because of the point-by-point comparisons. Consequently, the variability of the data could have destroyed temporal trends of symmetry. One component of the variability could have arisen by our choice of using the Artstor Digital Library. Therefore, the digital images were acquired by different institutions without common standards for color balance or lighting conditions. Future studies of temporal trends of symmetry should try to standardize the methods of image acquisition. Other factors of variability beyond image acquisition are individuality of artists, image degradation over time, and image restoration. Those factors are harder to control and may cause the permanent loss of any possible small temporal trends of symmetry. Fortunately, all these factors of variability do not invalidate statistically significant temporal trends such as those observed for balance, complexities, and chiaroscuro. This is especially fortunate for the latter, because degradation and restoration could have had a specially devastating effect on chiaroscuro trends.

In contrast to symmetry, the mathematics of balance and complexity involve integrations that make the measurement of these variables more robust. Consequently, variations across paintings could impair detection of any changes of symmetry that may be occurring during the Renaissance. Future efforts could try to bypass this potential limitation by using either multiple measures of symmetry (for example, radial—[33]) or object-wise, local symmetry [34,35] instead of global measures obtained from the whole picture.

4.4. Italy Versus the Rest of Europe

A surprising result in our study was obtained when comparing the evolution of balance in Italy versus the rest of Europe. When measuring the progression of the thickness of balance lines, the rest of Europe seemed to be more prescient about their future trends than Italy was. For example, the thicknesses of balance lines rose over time during the Renaissance (Figure 4e,f). These thicknesses were already larger in the rest of Europe than in Italy in the Early Renaissance (Figure 4e). Similar results (although without strong statistical significance) held for Spatial Simplicity and both indices of *chiaroscuro* (Figures 5a and 6c,e). How could we understand these results given that for most scholars, Italy was the most influential site of the Renaissance [36]. In truth, the Northern Renaissance remained relatively independent of Italy until the end of the 15th century [8]. In addition, some scholars even suggest a North-to-South direction of influence [37,38]. Venetians had much in common with the Flemish in their oil technique and representation of light. Only after 1500, the Italian Renaissance began influencing the rest of Europe. Therefore, the rest of Europe could develop a style with both *chiaroscuro* and relatively low spatial complexity before did the Italian Renaissance.

4.5. Top Master Painters Versus Peers

We were curious whether the top master painters of the Renaissance distinguished themselves by having different values of neuroaesthetic variables. When we probed this issue, we found that for the most part, the top artists had similar statistics as the rest of their contemporaries. (The standardization of the methods of image acquisition proposed in Section 4.3 could help reveal more delicate differences between artists.) However, there were some interesting and important exceptions. One of the most important examples was van Eyck, who was ahead of his time in terms of various artistic trends of the Renaissance. He led specially in trends related to *chiaroscuro* and spatial complexity of the portraits (Figures 3d, 5b and 6d). These van Eyck results are compatible with those discussed in Section 4.4. We pointed out in that section that Flanders was ahead of Italy in terms of neuroaesthetic variables in the Early Renaissance. Another worthwhile painter to mention in terms of uniqueness of neuroaesthetic variables was Caravaggio, whose portraits showed high spatial complexity (Figure 5b). Thus, although he painted in *chiaroscuro* [8], the bright portions of his paintings were highly complex.

A possibly surprising negative result was that da Vinci's portraits did not yield *chiaroscuro* indices statistically significantly higher than did those of his peers (Figure 6d,f). This is surprising, because many consider that in European painting, he was the one who brought the technique to its full potential [8,39]. He painted some clearly *chiaroscuro* pieces (for example, The Adoration of the Magi, 1481). However, he also had non-*chiaroscuro* portraits (for example, the Mona Lisa, 1517). Figure 6f illustrates this variety of da Vinci styles. This figure shows that his portraits yield indices of *chiaroscuro* intensities in roughly equal amounts above and below the fit line. This distribution contrasts sharply with that for Caravaggio, for whom the fit line is entirely below the corresponding indices.

Apropos Leonardo da Vinci's paintings, a modification of our techniques may be able to reveal some special statistics as compared to his contemporaries. da Vinci is famous for his sfumato technique [10,27]. If we increased the spatial resolution of our analysis, we could perhaps gauge sfumato through Complexity of Order 2 in small translations of the image. However, we would have to perform this analysis near the edges of image objects. To achieve this goal, we would have to add shape analysis, such as edge detection to our study [40]. Such shape analysis could enhance our studies in the future in other ways. In our paper, the analysis of aesthetic variables was performed with global tools, accounting for pixel statistics obtained from the entire image. An interesting question concerns the analysis of these variables in shapes in the centerpiece. These shapes would be faces in the case of portraits. Using edge-detection and machine-learning algorithms [41]), the outlines of faces in paintings could be selected. Then, the same measures used in this paper could in principle be applied to these selections. Based on our previous study [4], we believe that these focused measures might reveal interesting results. In that study, we did a manual pose classification of the subjects of portrait paintings and found that artists seldom painted their subjects in frontal poses, instead opting for a side-on or $\frac{3}{4}$ pose. Often, the pose was such that the head would be turned in relation to the torso to create greater variation. Our analysis showed that those paintings with complex poses had greater Complexity of Order 2, and lesser balance and symmetry, which is what we would expect here as well.

4.6. Implications of Evolution across Art Periods

Why does the distribution of the values of neuroaesthetic variables drift over time if the brain constrains them? This is only possible if these variables have flexibility to change in the brains of artists (and other people). Elsewhere, we propose that the range of values of these variables form a space, which we termed neuroaesthetic space [4]. The aesthetic choices of each artist would reside in a sub-region of this space. The locations of this sub-region depend on both the life experience of each artist and the materials and techniques available him or her. Consequently, artists learn from their social and cultural background, and especially, from other artists of the cultural moment. Thus, we propose two principles as guides for how neuroaesthetic variables evolved in portrait painting throughout the Renaissance: (1) New materials, techniques, or ideas by artists, propelled other artists to change through learning from the cultural environment. In this paper, the best example was the evolution of chiaroscuro. Our data set contained a portrait from as early as 1438 that belonged to the chiaroscuro category (Portrait of Giovanni Arnolfini, Jan van Eyck). Other painters liking the result were compelled to produce more and better chiaroscuro pieces (Figure 7a). (2) The necessity of artistic innovation was accelerated by different workshops competing for the favor of patrons [12]. For example, many artists included in this study, such as Pollaiuolo, da Vinci, Botticelli, and Ghirlandaio competed for Lorenzo de' Medici's. Thus, such artists reciprocally affected each other's artistic evolution, pushing aesthetic variables to evolve rapidly. Hence, the artistic influence of aesthetic variables may have evolved in a manner related to biological co-evolution [42].

Supplementary Materials: The supplementary materials for this article are available online at https://github.com/ha554n/Neuroaesthetic-Variables-Measures.

Author Contributions: Conceptualization, I.C.-H. and N.M.G.; Methodology, N.M.G.; Software, H.A. and N.M.G.; formal analysis, I.C.-H. and N.M.G.; data curation, I.C.-H. and N.M.G.; writing—original draft preparation, N.M.G.; writing—review and editing, I.C.-H., H.A. and N.M.G.; visualization, H.A. and N.M.G.; supervision, N.M.G. All authors have read and agreed to the published version of the manuscript.

Funding: This research received no external funding.

Acknowledgments: We would like to thank Helen Ryan, Joyce Gray, and Richard Pike for administrative support during the performance of this project. We would also like to thank the Provost's Office of Georgetown University for partial funding of this project.

Conflicts of Interest: The authors declare no conflict of interest.

References

1. Bode, C.; Helmy, M.; Bertamini, M. A cross-cultural comparison for preference for symmetry: Comparing British and Egyptians non-experts. *Psihologija* **2017**, *50*, 383–402. [CrossRef]
2. Rhodes, G.; Yoshikawa, S.; Clark, A.; Lee, K.; Mckay, R.; Akamatsu, S. Attractiveness of facial averageness and symmetry in non-Western cultures: In search of biologically based standards of beauty. *Perception* **2001**, *30*, 611–625. [CrossRef] [PubMed]
3. Gray, R.M. *Entropy and Information Theory*; Springer Science & Business Media: Berlin/Heidelberg, Germany, 2011.
4. Aleem, H.; Correa-Herran, I.; Grzywacz, N.M. Inferring master painters' esthetic biases from the statistics of portraits. *Front. Hum. Neurosci.* **2017**, *11*, 94. [CrossRef] [PubMed]
5. Reber, R.; Schwarz, N.; Winkielman, P. Processing fluency and aesthetic pleasure: Is beauty in the perceiver's processing experience? *Personal. Soc. Psychol. Rev.* **2004**, *8*, 364–382. [CrossRef] [PubMed]
6. Winkielman, P.; Chwarz, N.; Fazendeiro, T.; Reber, R. The hedonic marking of processing fluency: Implications for evaluative judgment. In *The Psychology of Evaluation: Affective Processes in Cognition and Emotion*; Lawrence Erlbaum Associates Publishers: Mahwah, NJ, USA, 2003; Volume 189, p. 217.
7. Aleem, H.; Pombo, M.; Correa-Herran, I.; Grzywacz, N.M. Is Beauty in the Eye of the Beholder or an Objective Truth? A Neuroscientific Answer. In *Mobile Brain–Body Imaging and the Neuroscience of Art, Innovation and Creativity*; Contreras-Vidal, J.L., Robleto, D., Cruz-Garza, J.G., Azorín, J.M., Nam, C., Eds.; Springer: Basel, Switzerland, 2019.
8. Janson, H.W.; Janson, A.F.; Marmor, M. *History of Art*; Thames and Hudson London: London, UK, 1997.
9. Mather, G. Visual image statistics in the history of Western art. *Art Percept.* **2018**, *6*, 97–115. [CrossRef]
10. Alberti, L.B. *On Painting*; Kemp, M., Ed.; Penguin: London, UK, 1991.
11. Alberti, L.B. *On the Art of Building in Ten Books*; MIT Press: Cambridge, MA, USA, 1988.
12. Chambers, D. *Patrons and Artists in the Italian Renaissance*; Springer: Berlin/Heidelberg, Germany, 1970.
13. Mayer, S.; Landwehr, J.R. Quantifying visual aesthetics based on processing fluency theory: Four algorithmic measures for antecedents of aesthetic preferences. *Psychol. Aesthet. Creat. Arts* **2018**, *12*, 399. [CrossRef]
14. Graf, L.K.; Mayer, S.; Landwehr, J.R. Measuring processing fluency: One versus five items. *J. Consum. Psychol.* **2018**, *28*, 393–411. [CrossRef]
15. Vasari, G. *The Lives of the Most Excellent Painters, Sculptors, and Architects*; Modern Library: New York, NY, USA, 2007.
16. Bonett, D.G.; Wright, T.A. Sample size requirements for estimating Pearson, Kendall and Spearman correlations. *Psychometrika* **2000**, *65*, 23–28. [CrossRef]
17. Andrews, L.C.; Andrews, L.C. *Special Functions of Mathematics for Engineers*; McGraw-Hill: New York, NY, USA, 1992.
18. Marreiros, A.C.; Daunizeau, J.; Kiebel, S.J.; Friston, K.J. Population dynamics: Variance and the sigmoid activation function. *Neuroimage* **2008**, *42*, 147–157. [CrossRef]
19. Coleman, T.F.; Li, Y. On the convergence of interior-reflective Newton methods for nonlinear minimization subject to bounds. *Math. Program.* **1994**, *67*, 189–224. [CrossRef]
20. Coleman, T.F.; Li, Y. An interior trust region approach for nonlinear minimization subject to bounds. *SIAM J. Optim.* **1996**, *6*, 418–445. [CrossRef]
21. Dunn, O.J.; Clark, V.A. *Applied Statistics: Analysis of Variance and Regression*; Wiley New York: New York, NY, USA, 1987.
22. Amthor, F.R.; Grzywacz, N.M. Nonlinearity of the inhibition underlying retinal directional selectivity. *Vis. Neurosci.* **1991**, *6*, 197–206. [CrossRef] [PubMed]
23. Sprent, P. *Data Driven Statistical Methods*; Chapman & Hall/ CRC Press: Boca Raton, FL, USA, 2019.
24. Solé, R.; Princeton, U. *Phase Transitions*; Princeton U. Press: Princeton, NJ, USA, 2011.
25. Barnett, H.G. *Innovation: The Basis of Cultural Change*; McGraw-Hill Book Company: New York, NY, USA, 1953.
26. Arnheim, R. *Art and Visual Perception: A Psychology of the Creative Eye*; Univ of California Press: Berkeley, CA, USA, 1965.
27. Itti, L.; Koch, C.; Niebur, E. A model of saliency-based visual attention for rapid scene analysis. *IEEE Trans. Pattern Anal. Mach. Intell.* **1998**, *20*, 1254–1259. [CrossRef]

28. Locher, P.; Gray, S.; Nodine, C. The structural framework of pictorial balance. *Perception* **1996**, *25*, 1419–1436. [CrossRef]

29. Donderi, D.C. Visual complexity: A review. *Psychol. Bull.* **2006**, *132*, 73. [CrossRef]

30. Kubovy, M. *The Psychology of Perspective and Renaissance Art*; CUP Archive: Cambridge, UK, 1988.

31. McManus, I.C. Symmetry and asymmetry in aesthetics and the arts. *Eur. Rev.* **2005**, *13*, 157–180. [CrossRef]

32. Pope-Hennessy, J. *The Portrait in the Renaissance*; John Pope-Hennessy: Florence, Italy, 1968.

33. Lockwood, E.H.; Macmillan, R.H. *Geometric Symmetry*; CUP Archive: Cambridge, UK, 1978.

34. Tyler, C.W. The symmetry magnification function varies with detection task. *J. Vis.* **2001**, *1*, 7. [CrossRef]

35. Tyler, C.W. *Human Symmetry Perception and Its Computational Analysis*; Psychology Press: Hove, UK, 2003.

36. Baxandall, M. *Painting and Experience in Fifteenth Century Italy: A Primer in the Social History of Pictorial Style*; Oxford University Press: New York, NY, USA, 1988.

37. Friedländer, M.J. *Early Netherlandish Painting*; Albertus Willem Sijthoff: Leiden, The Netherlands, 1975; Volume 12.

38. Panofsky, E. *Early Netherlandish Painting*; Routledge: New York, NY, USA, 2019; Volume 1.

39. Shearman, J. Leonardo's colour and chiaroscuro. *Z. Für Kunstgesch.* **1962**, *25*, 13–47. [CrossRef]

40. Perona, P.; Malik, J. Scale-space and edge detection using anisotropic diffusion. *IEEE Trans. Pattern Anal. Mach. Intell.* **1990**, *12*, 629–639. [CrossRef]

41. Yang, M.-H.; Kriegman, D.J.; Ahuja, N. Detecting faces in images: A survey. *IEEE Trans. Pattern Anal. Mach. Intell.* **2002**, *24*, 34–58. [CrossRef]

42. Wong, J.T.-F. A co-evolution theory of the genetic code. *Proc. Natl. Acad. Sci. USA* **1975**, *72*, 1909. [CrossRef] [PubMed]

 © 2020 by the authors. Licensee MDPI, Basel, Switzerland. This article is an open access article distributed under the terms and conditions of the Creative Commons Attribution (CC BY) license (http://creativecommons.org/licenses/by/4.0/).

Article

Invariant Image-Based Currency Denomination Recognition Using Local Entropy and Range Filters

Hafeez Anwar [1,2], Farman Ullah [2], Asif Iqbal [3], Anees Ul Hasnain [4], Ata Ur Rehman [2], Peter Bell [1,*] and Daehan Kwak [5,*]

[1] Interdisciplinary Center for Digital Humanities and Social Sciences, Friedrich-Alexander-Universität Erlangen-Nürnberg, 91052 Erlangen, Germany; hafeez.anwar@fau.de
[2] Department of Electrical & Computer Engineering, COMSATS University Islamabad-Attock Campus, Attock 43600, Pakistan; farmankttk@cuiatk.edu.pk (F.U.); dr.ataurrehman@cuiatk.edu.pk (A.U.R.)
[3] Department of Information and Communication Engineering, Inha University, Incheon 22212, Korea; asifsoul@inha.ac.kr
[4] EPAS Engineering, Topi 23460, Pakistan; anees@epas.com.pk
[5] Department of Computer Science, Kean University, Union, NJ 07083, USA
[*] Correspondence: peter.bell@fau.de (P.B.); dkwak@kean.edu (D.K.)

Received: 20 September 2019; Accepted: 2 November 2019; Published: 6 November 2019

Abstract: We perform image-based denomination recognition of the Pakistani currency notes. There are a total of seven different denominations in the current series of Pakistani notes. Apart from color and texture, these notes differ from one another mainly due to their aspect ratios. Our aim is to exploit this single feature to attain an image-based recognition that is invariant to the most common image variations found in currency notes images. Among others, the most notable image variations are caused by the difference in positions and in-plane orientations of the currency notes in images. While most of the proposed methods for currency denomination recognition only focus on attaining higher recognition rates, our aim is more complex, i.e., attaining a high recognition rate in the presence of image variations. Since, the aspect ratio of a currency note is invariant to such differences, an image-based recognition of currency notes based on aspect ratio is more likely to be translation- and rotation-invariant. Therefore, we adapt a two step procedure that first extracts a currency note from the homogeneous image background via local entropy and range filters. Then, the aspect ratio of the extracted currency note is calculated to determine its denomination. To validate our proposed method, we gathered a new dataset with the largest and most diverse collection of Pakistani currency notes, where each image contains either a single or multiple notes at arbitrary positions and orientations. We attain an overall average recognition rate of 99% which is very encouraging for our method, which relies on a single feature and is suited for real-time applications. Consequently, the method may be extended to other international and historical currencies, which makes it suitable for business and digital humanities applications.

Keywords: image entropy; image processing; image segmentation

1. Introduction

In this paper, we propose an image-based framework for the denomination recognition of Pakistani currency notes of the 2005 series. Such systems are important as they are used at several situations such as automatic vending machines and supporting visually impaired people [1] to identify the denomination of a given currency note. Consequently, such image-based denomination recognition of currency notes has become an active area of research where the proposed solutions can be divided into two main groups. The first group of methods [2] deals with currency images that are acquired with scanners. Such methods are more suitable for applications like automatic teller machines (ATM).

However, the most common variations that are found in scanned images are the ones induced by the variations in currency notes orientation, scale and position. The second group of methods deals with currency note images that are taken with a camera in a cluttered environment. Such a system installed on a smart phone can be helpful for visually impaired people in their daily life. However, these methods face challenges due to the image variations caused by non-uniform illumination, background clutter, and partial occlusions.

The proposed solutions are broadly divided into three different groups, where the first group uses the so-called local features matching such as scale-invariant feature transform (SIFT) [3]. The second group makes use of supervised machine learning algorithms such as artificial neural networks (ANNs) [4]. Lastly, the third group of methods uses pure image processing techniques such as template matching [5].

We take a different approach to the problem of image-based denomination recognition of scanned currency notes. We are interested in utilizing simple image filtering, rather than using a complex handcrafted local image descriptor or a complicated machine learning algorithm such as a convolutional neural network (CNN). In addition to simplicity, such filtering is also invariant to the most commonly found variations in the currency notes images. These image variations are not explicitly identified and dealt with in any of the previously proposed methods. Figure 1 depicts the variations that are found in the currency note images. The orientation differences found among the currency notes of any given denomination can cause variations in their images. This makes images of the same currency note look different from one another. This can clearly be observed in the third column of Figure 1. Similarly, the variations caused by position and scale differences of the currency notes induce variations in their images which are depicted in the first and second columns of Figure 1. Lastly, the image variations can also be caused by multiple currency notes in a single image as shown in fourth column of Figure 1. Apart from image variations, the recognition rate is also likely to be affected by the variations caused by the condition of currency notes themselves. The standard aspect ratios of currency notes are based on the freshly printed and unused notes as shown in Table 1. However, the excessive and rough use of the currency notes causes wear and tear, due to which, their boundaries become irregular. Since, we propose to use the aspect ratio as a single feature for recognizing these notes, such irregularities in the boundaries induce variations in the aspect ratios as well. This is demonstrated in Figure 2 where currency notes of multiple frequently used denominations are shown. It can be observed that the aspect ratios of currency notes belonging to a single denomination vary from one another. However, it should be noted that the measurements are based on our proposed image-based method. To summarize, in order to achieve an accurate denomination recognition of currency notes, it is important to address these variations. This has already been proved in other object categories such as ancient coins [6], and butterflies and fish [7], where image variations caused by changes in object orientation, scale, and position are very common.

Table 1. The Pakistani currency notes of the 2005 Series along with their width, height, and aspect ratios. The aspect ratio is calculated as (height/width) × 10, 000.

Denomination	10	20	50	100	500	1000	5000
Height	65	65	65	65	65	65	65
Width	115	123	131	139	147	155	163
Aspect ratio	5652	5285	4962	4676	4422	4194	3988

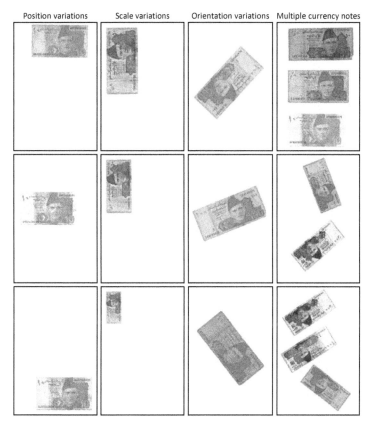

Figure 1. Common image variations found in currency note images.

Figure 2. Variations in aspect ratios due to irregular currency note boundaries. The aspect ratio of each currency note is shown below it.

1.1. Related Work

Based on their approaches, the proposed solutions for currency denomination recognition can broadly be divided in the following three classes.

Machine learning-based methods:

The first class of methods uses machine learning techniques such as artificial neural networks (ANN) and support vector machines (SVMs). For instance Takeda and Omatu [4] used a mask of slabs that is convoluted with a local pixel neighborhood such that the coefficients of the masks are randomly chosen to be either 1 or 0. The aggregated results of several such convolutions are then provided as input to a three layered neural network (NN). Similarly, in Frosini et al. [8], the light refracted from the bank notes was captured with arrays of opto-electronic sensors and then provided to a multi-layered perceptron. Takeda and Nishikagi [9] extracted the information of the currency notes via the so-called axis symmetric masks and then provided it to an ANN. However, more recently, a state-of-the-art performing convolutional neural network (CNN) was used by Pham et al. [10] where they reported a result of 100% on an image dataset of 64,000 images belonging to 64 classes. In addition to the neural networks and its variants, other machine learning algorithms such SVM were also used along with its various types of kernels. In Chang et al. [11], the features constructed from sensors were used to represent the key regions of a currency note to predict whether it is real or fake. The task of classification is performed by a support vector machine (SVM) where different kernels were evaluated for classification accuracy. Similarly, He et al. [12] used principle component analysis (PCA) features to represent an edge image of a single banknote. A genetic algorithm (GA) was further used for feature selection. Such image representation was then used to train an SVM model, and later for testing it. Other machine learning algorithms used for currency recognition include the hidden Markov model (HMM) [13], k nearest neighbors (kNN) [14] and Gaussian mixture model (GMM) [15].

Local feature matching-based methods:

A local feature descriptor is a compact representation of the pixel intensity distribution within a local image patch. As a common practice, the feature descriptor is an N dimensional vector. For object recognition via feature matching, the local features such as scale-invariant feature transform (SIFT) [3] are extracted from the representative images of a given object and stored in a database. Then, from a given image, local features are extracted and matched with the ones using a distance measure such as the Euclidean distance. The local features of the stored image that are the nearest to the local features of the test image are declared to belong to the same object class. The local feature matching is also used for currency recognition. SIFT and its color variant also known as color SIFT are used for currency recognition by [16]. Similarly, Hasanuzzaman et al. [1] used speedup robust features (SURF) for currency recognition. More recently, Yousry et al. [17] used binary local feature descriptor oriented FAST and rotated BRIEF (ORB) to represent the input image and then compared with the representative images in the database using the Hamming distance. Other local features used for currency recognition include local binary patterns (LBP) [18] and a gray-level co-occurrence matrix (GLCM) [19].

Image processing-based methods:

The image processing based methods are purely based on pixel level manipulation techniques such as extracting the region of interest (ROI) of a banknote and then applying various measures such as the correlation [20]. Such methods are more relevant for real-time applications where relatively less processing resources are available. Similarly, Youn et al. [5] used the banknote size information along with multiple-template matching for a multiple currency recognition system.

The proposed paper currency recognition method is based on a two step process. The first step deals with the extraction of paper currency note from the image via local entropy and range filters. The extracted note region is then normalized and rotated to achieve scale- , translation-, and rotation-invariance. The second step is recognition of the extracted and processed region. This is done by computing the aspect ratio of the extracted region as currency notes differ from one another based on this feature. Since, the aspect ratio of a rectangular object is not affected by the scale, position, and in-plane orientation of the object, it is also invariant to these transformations. Hence, our proposed paper currency recognition method is invariant to changes in scale, position, and in-plane orientations.

2. Methodology

The proposed method is inspired by an automatic ancient coin segmentation [21]. Following their method, we also segment a currency note that is imaged on a homogeneous background by the following assumptions.

1. The image region depicting a currency note contains the highest information content.
2. The most rectangular object found in the image is a currency note that has a predefined but slightly varying aspect ratio depending on its condition.

Therefore, the segmentation becomes a two step process. The first step deals with the extraction of image regions with the highest information contents. The second step consists of finding rectangular regions among the extracted ones and calculating their aspect ratios. In the following, we elaborate on both of these steps.

2.1. Informative Region Extraction

Since, the currency note is imaged on a homogeneous background, the image region depicting the note is likely to have more variations in terms of pixel values than the background. We employ a local image neighborhood processing strategy to extract informative regions. The local image neighborhood is simply a subset of image pixels arranged in rows and columns. The image filters that process a local image neighborhood are called the local filters. In our proposed method, we use the following two local filters.

Local entropy filter:

Entropy gives the measurement of information content in an event, signal, or in our case, an image. Concretely, entropy is inversely proportional to the probability of a random variable where it is maximum if the value of probability is close to zero and vice versa. In the context of image, the information content is represented by the pixel intensity values that range from 0 to $L - 1$ where $L = 256$. The histogram of pixel intensity values [22] represents the number of pixels per intensity value as shown in Equation (1).

$$h(i) = n_i \quad i \in 0, 1, 2, ..., 255. \tag{1}$$

Each bin in this histogram gives the count of that particular intensity value as $n_0, n_1, ..., n_{(L-1)}$. In other words, the total number of pixels in the local neighborhood having intensity value i, is n_i.

In order to convert this count into their respective probabilities, the histogram of intensity values is normalized by dividing the value of each bin over the total number of pixels in the local neighborhood. For instance, if the total number of rows in a local neighborhood is M and the total number of columns is N, then, the normalized histogram is given as,

$$p(i) = h(i)/(M * N) = n_i/(M * N), \quad i \in 0, 1, 2, ..., 255 \tag{2}$$

where, $p(0), p(1), ..., p(255)$ give the probabilities of each intensity value in the local neighborhood.

The entropy of a local neighborhood Ω in an image is then found via this normalized histogram using Equation (3),

$$H(\Omega) = -\sum_{i=0}^{255} p(i) . \log_2(p(i)). \tag{3}$$

Local range filter:

The range of a local neighborhood is simply calculated by finding the difference of maximum intensity value in that particular neighborhood and the minimum.

To summarize, the response of local entropy filter for homogeneous image regions will be minimum whereas it will be maximum for those regions having higher variations of pixel intensity values. A similar result will be achieved for the local range filter. Figure 3 shows the same effect where we depict the responses of both the filters on a one dimensional signal that is a single row of a currency note image. The flat part of the intensity signal corresponds to the homogeneous background while its fluctuating part shows the variations in pixel intensity values of the currency note. Consequently, both the local filters achieve results that are adequate for both the homogeneous and fluctuating part. It can be observed that they do not respond to the homogeneous part while their responses are more pronounced to the part of signal representing the pixel intensity variations.

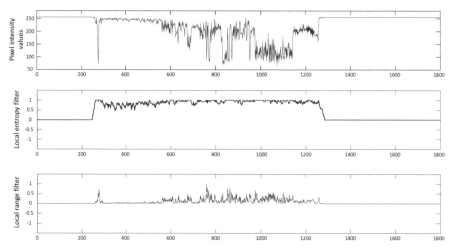

Figure 3. Responses of local entropy and range filters on a single row of currency note images (where the values at *x*-axis show the number of rows).

Sum of both filters:

Both these local filters are applied to a given note image. For each filter, the size of a local circular neighborhood is empirically selected as 3. The resultant response of each filter is scaled to the range between 0 and 1. Finally, both the responses are summed to get the combined response of both the filters. The individual and cumulative responses of both filters are shown in Figure 4. The region of images where both the filters give high response shows more informativeness than the background. However, the border region of the currency note in the combined response is more pronounced, which proves supportive at the stage of segmentation.

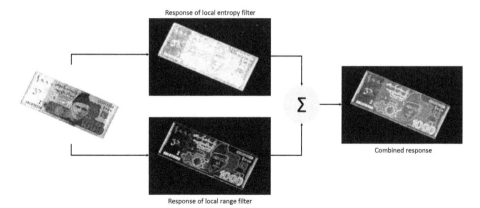

Figure 4. Responses of local entropy and range filters on a currency note image.

2.2. Currency Note Segmentation and Recognition via Aspect Ratio

For the current imaging conditions, a set of empirically defined thresholds (0.3, 0.4, 0.5, and 0.6) is applied to the combined response image for segmenting the currency note region. All the binarized images produced by each threshold are summed to get the final segmentation mask for the currency note. This whole procedure is shown in Figure 5. The resulting binarized masks generated by applying each threshold are very similar except for the last one. This effect is more or less observed on all the images of the dataset. As a last step, the aspect ratio of the generated mask is calculated to determine the denomination of currency note. To this end, we evaluated two kinds of aspect ratios for each denomination that are further explained in Section 4.

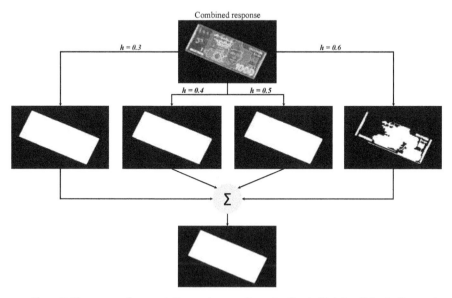

Figure 5. The process of segmentation mask generation using threshold-defined binarization masks.

3. A New Pakistani Currency Notes Dataset (PCND):

We collect an image dataset of Pakistani notes with images having high variations with respect to in-plane orientation and position. These images are taken with a desktop flatbed scanner where the background is completely homogeneous. All the seven denominations of the Pakistani Currency of the 2005 series are represented in the dataset. We divide our dataset into two disjoint subsets. The first subset that we call the "training set" is used to calibrate the aspect ratio for each note. This is further elaborated in Section 4. We call the second dataset the "test dataset", where currency notes of various denominations are scanned at arbitrary positions and orientations in the following three different settings.

1. This is the simplest setting with only one currency note per image.
2. The images in this setting contain two currency notes of different denominations. However, the notes are separated from each other to such an extent that they are not overlapped. Overlapping will cause their boundaries to merge thus resulting in a wrong segmentation. For this setting, the notes of consecutive denominations are chosen as they are more likely to get confused.
3. Finally, three notes per image of consecutive denominations are imaged together.

It should be noted that the images with multiple notes are simply made by combining the single note images such that they are scaled, rotated and positioned arbitrarily. Table 2 shows the total number of images per denomination in both subsets of the image dataset.

Table 2. Details of image dataset.

Number of Images per Denomination			
Training Set		Test Set	
Denomination	No. of Images	Denomination	No. of Images
10	105	10	100
20	101	20	97
50	101	50	96
100	94	100	97
500	101	500	77
1000	101	1000	90
5000	9	5000	41

4. Results and Discussion

Since the proposed currency note recognition method is based on the aspect ratios of the banknotes, as a first step, the values of aspect ratios for each denomination have to be established. We simply divide the shorter side (width) of the note over its longer side (length) and then scale the number by multiplying it with 10, 000. For instance the standard width of the currency note of 10 rupees is 65 mm and its standard length is 115 mm. Therefore, its aspect ratio is $10,000 \times (65/115)$, which is 5652. We adapt the following two different methods to obtain the aspect ratio for each denomination.

1. Standard aspect ratio: This is the aspect ratio of an original and new currency note.
2. Calibrated aspect ratio: The majority of the notes used in the market undergo wear and tear due to their age and usage. These currency notes are not in their original shape and thus their aspect ratios are more likely to differ from the new ones. We use currency note images in the training set to "calibrate" the aspect ratio for each denomination. To this end, we find the aspect ratios of currency notes for each denomination in the training set that include both new and used notes. The mean of all these aspect ratios is then considered as the aspect ratio of the respective denomination.

Both the standard and the calibrated aspect ratios for each denomination are shown in Table 3. The difference between standard and calibrated aspect ratios for each denomination can be observed. This is due to the fact that training set consists of used currency notes whose aspect ratios vary from one another due to the deterioration in their shapes. Once, the aspect ratios are established for each note type, the next step is to use them to label a given test note image. From a given image, the note image is extracted and then its aspect ratio is calculated using the proposed method. This aspect ratio is then compared with the aspect ratios of all the notes and the one nearest to it is assigned the label.

Table 3. Values of standard and calibrated aspect ratios for different denominations.

	10	20	50	100	500	1000	5000
Standard aspect ratio	5652	5285	4962	4676	4422	4194	3988
Calibrated aspect ratio	5750	5356	5055	4762	4526	4295	4083
Difference	98	71	93	86	104	101	95

Table 4. Denomination recognition rates achieved on standard and calibrated aspect ratios where the recognition rates of calibrated aspect ratios (in bold) are better than those of the standard aspect ratios

	10	20	50	100	500	1000	5000	Overall
Standard aspect ratio	100%	97%	96%	97%	77%	90%	41%	89%
Calibrated aspect ratio	100%	**99%**	**98%**	**99%**	**100%**	**99%**	**100%**	**99%**

The images per note type vary from 60 to 100, for a total of 598 images. In each test image, there is a single instance of a note that is displayed at an arbitrary scale, position, and orientation. The results for both the methods are shown in Table 4 while the confusion matrices for each note type are shown in Figure 6. The aspect ratios based on calibration clearly outperform the standard aspect ratios. This is quite realistic as the notes in the test image data contain both new and old notes. Due to this reason, the calibration-based aspect ratios give flexible values for both new and old notes to be recognized. The denomination recognition results for currency notes images with different variation are shown in Figure 7. Our method successfully recognizes denominations of each currency note under the translation, scale, and rotation variations. It also very accurately recognizes the denomination of multiple currency notes that are imaged together. However, a failure on the currency note of 500 can be observed where it is wrongly recognized as a note with the denomination of 100.

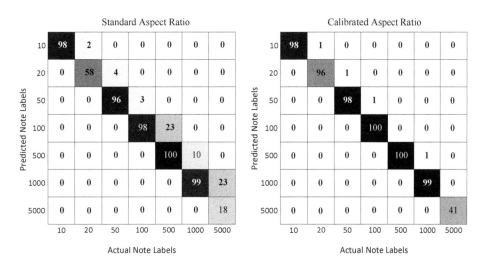

Figure 6. Confusion matrices for both the standard and calibrated aspect ratios.

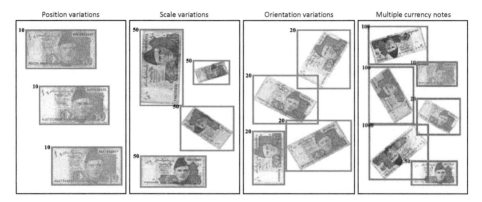

Figure 7. Visualization of denomination recognition of currency notes imaged at various positions, scales and orientations.

5. Conclusions and Future Work

We performed denomination recognition of Pakistani currency notes from their images. Such recognition was achieved despite the image variations that are commonly found in currency images. These image variations are mainly caused by changes in currency note position, scale, and orientation. To achieve a denomination recognition that is invariant to these image variations, a two step process was proposed where, in the first step, informative regions from the images are extracted via local entropy and range filters. As a second step, these regions are extracted via binarization and their aspect ratios are calculated for final denomination recognition. The aspect ratio for each denomination is established by using a training set where, for each denomination, notes are used with various degrees of deterioration based on their usage. The proposed method is evaluated on a novel dataset of Pakistani currency notes of the 2005 series, where it achieves a recognition rate of 99% on a total of 598 images. The usage of a single feature, i.e. aspect ratio, for recognition makes our method more feasible for real-time application which we plan to test in the future by implementing this method on a Raspberry Pi.

Author Contributions: Methodology, H.A.; Validation, F.U.; Conceptualization, A.I.; Data curation, A.U.H.; Conceptualization, A.U.R.; Editing and funding acquisition, P.B.; Methodology, writing–review and editing, D.K.

Funding: We acknowledge support by Deutsche Forschungsgemeinschaft and Friedrich-Alexander-Universität Erlangen-Nürnberg (FAU) within the funding program Open Access Publishing. The research was also partially funded by the Untenured Faculty Research Initiative (UFRI), Kean University.

Conflicts of Interest: The authors declare no conflict of interest.

References

1. Hasanuzzaman, F.M.; Yang, X.; Tian, Y Robust and Effective Component-Based Banknote Recognition for the Blind. *IEEE Trans. Syst. Man Cybern. Part C (Appl. Rev.)* **2012**, *42*, 1021–1030. doi:10.1109/TSMCC.2011.2178120. [CrossRef] [PubMed]
2. Sargano, A.B.; Sarfraz, M.; Haq, N. Robust features and paper currency recognition system. In Proceedings of the 6th International Conference on Information Technology (ICIT 2013), Amman, Jordan, 8–10 May 2013; pp. 8–10.
3. Lowe, D.G. Distinctive Image Features from Scale-Invariant Keypoints. *Int. J. Comput. Vision* **2004**, *60*, 91–110. doi:10.1023/B:VISI.0000029664.99615.94. [CrossRef]
4. Takeda, F.; Omatu, S. High Speed Paper Currency Recognition by Neural Networks. *Trans. Neur. Netw.* **1995**, *6*, 73–77. doi:10.1109/72.363448. [CrossRef] [PubMed]
5. Youn, S.; Choi, E.; Baek, Y.; Lee, C. Efficient multi-currency classification of CIS banknotes. *Neurocomputing* **2015**, *156*. doi:10.1016/j.neucom.2015.01.014. [CrossRef]

Entropy **2019**, *21*, 1085

6. Anwar, H.; Zambanini, S.; Kampel, M. A Bag of Visual Words Approach for Symbols-Based Coarse-Grained Ancient Coin Classification. *arXiv* **2013**, arXiv:1304.6192. Available online: https://arxiv.org/abs/1304.6192 (accessed on 30 October 2019).

7. Anwar, H.; Zambanini, S.; Kampel, M. Encoding Spatial Arrangements of Visual Words for Rotation-Invariant Image Classification. In *Pattern Recognition*; Jiang, X., Hornegger, J., Koch, R., Eds.; Springer International Publishing: Cham, Germany, 2014; pp. 443–452.

8. Frosini, A.; Gori, M.; Priami, P. A neural network-based model for paper currency recognition and verification. *IEEE Trans. Neural Networks* **1996**, *7*, 1482–1490. doi:10.1109/72.548175. [CrossRef] [PubMed]

9. Takeda, F.; Nishikage, T. Multiple Kinds of Paper Currency Recognition Using Neural Network and Application for Euro Currency. In Proceedings of the IEEE-INNS-ENNS International Joint Conference on Neural Networks. IJCNN 2000. Neural Computing: New Challenges and Perspectives for the New Millennium, Como, Italy, 27 July 2000; IEEE Computer Society: Washington, DC, USA, 2000. IJCNN '00. pp. 143–147.

10. Pham, T.; Eun Lee, D.; Ryoung Park, K. Multi-National Banknote Classification Based on Visible-light Line Sensor and Convolutional Neural Network. *Sensors* **2017**, *17*, 1595. doi:10.3390/s17071595. [CrossRef] [PubMed]

11. Chen Chang, C.; Xing Yu, T.; Yen Yen, H. Paper Currency Verification with Support Vector Machines. In Proceedings of the 2007 Third International IEEE Conference on Signal-Image Technologies and Internet-Based System, Shanghai, China, 16–18 December 2008; pp. 860–865. doi:10.1109/SITIS.2007.146. [CrossRef]

12. He, J.B.; Zhang, H.M.; Liang, J.; Jin, O.; Li, X. Paper Currency Denomination Recognition Based on GA and SVM. Paper Currency Denomination Recognition Based on GA and SVM. In Proceedings of the Chinese Conference on Image and Graphics Technologies, Beijing, China, 19–20 June 2015; pp. 366–374.

13. Hassanpour, H.; Farahabadi, P.M. Using Hidden Markov Models for Paper Currency Recognition. *Expert Syst. Appl.* **2009**, *36*, 10105–10111. doi:10.1016/j.eswa.2009.01.057. [CrossRef]

14. HLAING, K.N.N. First order statistics and GLCM based feature extraction for recognition of Myanmar paper currency. In Proceedings of the IIER International Conference, Bangkok Thailand, 17 June 2015; pp. 1–6.

15. Jin, Y.; Song, L.; Tang, X.; Du, M. A Hierarchical Approach for Banknote Image Processing Using Homogeneity and FFD Model. *IEEE Signal Process. Lett.* **2008**, *15*, 425–428. [CrossRef]

16. Doush, I.A.; Sahar, A.B. Currency recognition using a smartphone: Comparison between color SIFT and gray scale SIFT algorithms. *J. King Saud Univ. Comput. Inf. Sci.* **2017**, *29*, 484–492.

17. Yousry, A.; Taha, M.; Selim, M. Currency Recognition System for Blind people using ORB Algorithm. *Int. Arab J. Inf. Technol.* **2018**, *5*, 34–40.

18. Sharma, B.; Kaur, A. Recognition of Indian paper currency based on LBP. *Int. J. Comput. Appl.* **2012**, *59*, 24–27. doi:10.5120/9514-3913. [CrossRef]

19. Yan, W.Q.; Chambers, J.; Garhwal, A. An empirical approach for currency identification. *Multimedia Tools Appl.* **2015**, *74*, 4723–4733. [CrossRef]

20. Semary, N.; Fadl, S.; Eissa, M.; Gad, A. Currency Recognition System for Visually Impaired: Egyptian Banknote as a Study Case. In Proceedings of the 5th International Conference on Information & Communication Technology and Accessibility, Marrakesh, Morocco, 21–23 December 2015; pp. 1–6.

21. Zambanini, S.; Kampel, M. Robust Automatic Segmentation of Ancient Coins. In Proceedings of the VISAPP, Liboa, Portugal, 5–8 February 2009; pp. 273–276.

22. Ullah, F.; Anwar, H.; Shahzadi, I.; Ur Rehman, A.; Mehmood, S.; Niaz, S.; Mahmood Awan, K.; Khan, A.; Kwak, D. Barrier Access Control Using Sensors Platform and Vehicle License Plate Characters Recognition. *Sensors* **2019**, *19*, 3015. [CrossRef] [PubMed]

© 2019 by the authors. Licensee MDPI, Basel, Switzerland. This article is an open access article distributed under the terms and conditions of the Creative Commons Attribution (CC BY) license (http://creativecommons.org/licenses/by/4.0/).

Article

Using Entropy for Welds Segmentation and Evaluation

Oto Haffner *, Erik Kučera, Peter Drahoš and Ján Cigánek

Faculty of Electrical Engineering and Information Technology, Slovak University of Technology in Bratislava, 841 04 Bratislava, Slovakia; erik.kucera@stuba.sk (E.K.); peter.drahos@stuba.sk (P.D.); jan.ciganek@stuba.sk (J.C.)
* Correspondence: oto.haffner@stuba.sk

Received: 24 October 2019; Accepted: 26 November 2019; Published: 28 November 2019

Abstract: In this paper, a methodology based on weld segmentation using entropy and evaluation by conventional and convolution neural networks to evaluate quality of welds is developed. Compared to conventional neural networks, there is no use of image preprocessing (weld segmentation based on entropy) or data representation for the convolution neural networks in our experiments. The experiments are performed on 6422 weld image samples and the performance results of both types of neural network are compared to the conventional methods. In all experiments, neural networks implemented and trained using the proposed approach delivered excellent results with a success rate of nearly 100%. The best results were achieved using convolution neural networks which provided excellent results and with almost no pre-processing of image data required.

Keywords: weld segmentation; local entropy filter; weld evaluation; convolution neural network; image entropy; Python; Keras; RSNNS; MXNet

---

1. Introduction

The Fourth Industrial Revolution (Industry 4.0) has opened space for research and development of new manufacturing methods, systems and equipment based on innovations such as computing intelligence, autonomous robots, big data, augmented reality, process simulation, quality management systems, etc. [1].

Weld evaluation is very important quality control process in many manufacturing processes. Without this technological process, it would be almost impossible to produce welded constructions with current efficiency—whether we are talking about time, price, or material consumption. It is therefore necessary to welds be inspected to meet the specified quality level. In order to detect the possible presence of different weld defects, proper sensing, monitoring and inspection methods are necessary for quality control. Very effective and non-destructive method for weld evaluation is visual inspection. Inspection process using this method can be in certain level automated and done by computer systems [2,3].

Visual inspection of a weld is an important non-destructive method for weld quality diagnostics that enables to check welded joint and its various parameters. This examination is carried out as a first examination and able to detect various defects [4].

In this paper, we focus on indirect visual evaluation due to which the evaluation process can be automated. Indirect inspection can be applied also in places that are not directly accessible, for example the inner surface of a pipeline, the interior of pressure vessels, car body cavities etc. It also eliminates errors of human judgment and removes errors caused by workers for such reasons as e.g., fatigue, inattention or lack of experience.

The improved beamlet transformation for weld toe detection described in [5,6] considers images which are corrupted by noise. The authors aim at detecting edge borders of welds. The dynamic thresholding is performed in one of the beamlet algorithm steps. The algorithm predicts the directional characteristics of the weld allows to filtrate unsuitable edges. Using this method, it is possible to directly extract weld seam edges from highly noisy welding images without any pre-processing or post-processing steps.

In [7], the authors work with pipeline weld images with a very low contrast and corrupted by noise; this causes problems to conventional edge detectors. At first, the image is noise-filtered using a morphological operation of opening and closing. Next, the improved algorithm of fuzzy edge detection is applied. Multi-level fuzzy image improvement is based on interactive searching of optimal threshold level and multi-directional edge detector which convolution kernel is 5×5 with 8 directions based on gradient searching. The result of the algorithm is compared with detectors as Sobel, canny FED and fast FED.

Edge detection and histogram projection are used in [8], where histogram projections of tested welds are compared with a specified similarity threshold used to evaluate quality of the tested welds. The loaded image pattern has the same specifications (width and position) as the tested image. Always one vertical line from the pattern and the tested images is compared. Line histograms of pattern and tested images are computed, the correlation degree of two histograms is computed using the Tukey HSD difference. A lower correlation degree than the specified correlation threshold indicates edge defects in this part of the examined image. The procedure is repeated over the entire width of image.

Evaluation of metal cans welds is dealt with in [9]. Can's weld defects may not be directly related to welding (they can be brought about by rest of glue, dust, etc.). Therefore, authors use probability evaluation of two evaluation methods; the Column Gray-Level Accumulation Inspection represents histogram projection in general. The histogram projections of the pattern and the tested weld are compared. The comparison of first derivation for making better results is also performed. This method can detect defects of wider surface. The overall evaluation is done using Dampster-Shafer theory of evidence.

In another work [10], the above authors deal with edge detection based on pixel intensity difference of the foreground and the background. The background pixels' intensity occurs with a maximum probability and the distribution of the background pixels fits the Gauss distribution.

The weld visual inspection process performed through image processing on the image sequence to improve data accuracy is presented in [11]. The Convolution Neural Network (CNN) as an image processing technique can determine the feature automatically to classify the variation of each weld defect pattern. A classification using CNN consists of two stages: image extraction using image convolution, and image classification using neural network. The proposed evaluation system has obtained classification for four different types of weld defects with validation accuracy of 95.83%.

A technique for automatic endpoint detection of weld seam removal in a robotic abrasive belt grinding process using a vision system based on deep learning is demonstrated in [12]. The paper presents results of the first investigative stage of semantic segmentation of weld seam removal states using encoder-decoder convolutional neural networks (EDCNN). The prediction system based on semantic segmentation is able to monitor weld profile geometry evolution taking into account the varying belt grinding parameters during machining which allows further process optimization.

Utilizing computing intelligent using support vector machine (SVM) is presented in [13,14]. Authors developed real-time monitoring system to automatically evaluate the welding quality during high-power disk laser welding. Fifteen features were extracted from images of laser-induced metal vapor during welding. To detect the optimal feature subset for SVM, a feature selection method based on the SFFS algorithm was applied. An accuracy of 98.11% by 10-fold cross validation was achieved for the SVM classifier generated by the ten selected features. The authors declare the method has the potential to be applied in the real-time monitoring of high-power laser welding.

The authors of [15–18] deal with the development of a system for automatic weld evaluation using new information technologies based on cloud computing and single-board computer in the context of Industry 4.0. The proposed approach is based on using a visual system for weld recognition, and a

Entropy **2019**, *21*, 1168

neural network cloud computing for real-time weld evaluation, both implemented on a single-board low-cost computer. The proposed evaluation system was successfully verified on welding samples corresponding to a real welding process. The system considerably contributes to the weld diagnostics in industrial processes of small- and medium-sized enterprises. In [18], the same authors use a single-board computer able to communicate with an Android smartphone which is a very good interface for a worker or his shift manager. The basic result of this paper is a proposal of a weld quality evaluation system that consists of a single-board computer in combination with Android smartphone.

This paper deals with development of a software system for visual weld quality evaluation based on weld segmentation using entropy and evaluation by conventional and convolution neural networks. The evaluation of the performance results is compared to the conventional methods (weld segmentation based on entropy and evaluation using conventional neural networks with and without weld segmentation). Most experiments of proposed method apply on weld metal, however, one experiment with convolution neural networks applies also on weld adjected zones. 6422 real and adjusted laboratory samples of welds are used for experiments. The paper is organized in five sections: Section 2 deals with preparation of input data for the neural network. Section 3 describes configuration of used neural networks and their training process. In Section 4 the results of experiments are presented. In Section 5 we discuss the results.

2. Preparation of Input Data for the Neural Network

The input data for the proposed diagnostic system were represented in the form of grayscale laboratory samples of metal sheet welds in JPEG format. The samples were pre-classified as OK (correct) and NOK (incorrect) (Figures 1 and 2). Defective weld samples (NOK) include samples of various surface defects such as irregular weld bead, excess weld metal, craters, undercut, etc. Welds images are captured under the same illumination and have the same resolution 263×300 pixels. The total number of evaluated sample images was 6422.

However, for several reasons the image resolution 263×300 pixels is not suitable for a conventional neural network due to the necessity of large amount of allocated memory (about gigabytes for thousands of frames even in a relatively low resolution) and time-consuming network training time.

Figure 1. Laboratory sample of an OK weld.

Entropy **2019**, *21*, 1168

Figure 2. Laboratory sample of NOK weld.

Several suitable options for data processing that eliminate the above problems are presented next. At first, the background weld segmentation is described. Segmentation provides two outputs - the weld mask and the segmented weld itself. Three transformations of the weld mask into a one-dimensional feature vector are described further. Feature vectors are useful as inputs for the multilayer perceptron (MLP)/radial basis function (RBF) neural networks. Finally, the size of the segmented/unsegmented weld image is reduced when applied in the conventional neural network (if CNN is applied, no size reduction is needed).

2.1. Weld Segmentation

The sample images depict the weld itself and the background—metal sheet. The background does not affect the evaluation of the weld and is masked from the images by the proposed algorithm. The simplified flowchart of the algorithm is shown in Figure 3.

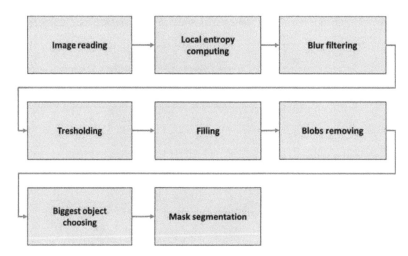

Figure 3. The simplified flowchart of the segmentation algorithm.

After reading the images, local entropy of each pixel is computed according to [19]:

$$\sum_{i=1}^{K} \sum_{j=1}^{K} p_{ij} \log_2 p_{ij}, \tag{1}$$

where p_{ij} represents the probability function for the pixel $[i, j]$.

This value contains information about the complexity/unevenness around the pixel. The neighbourhood radius was set to 8 pixels. To compute the entropy, the filters.rank.entropy function from the Python library scikit-image was used. The resulting local entropy matrix effectively finds the edges and texture complexity in the image. The results of filtering can be seen in Figure 4.

As the entropy resolution values were too detailed for our application, the blur filtering was applied. The anisotropic blur filter from the imager library was implemented, which removes noise/unimportant details while preserving edges better than other types of blur filters. The blur filter with an amplitude of 250 was applied (Figure 5).

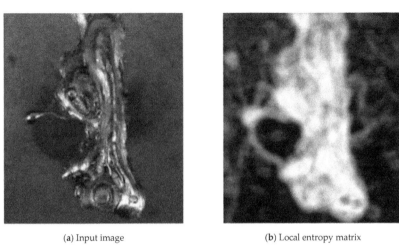

(a) Input image   (b) Local entropy matrix

Figure 4. Step 1—local entropy computing.

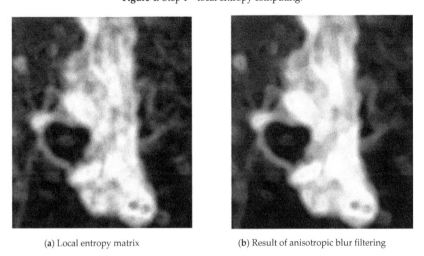

(a) Local entropy matrix   (b) Result of anisotropic blur filtering

Figure 5. Step 2—blur filtering.

The next step is thresholding. In the image matrix, the value 1 (white) represents weld pixels, the value 0 (black) represents background. Thresholding was implemented using the function threshold from the imager library. The optimal threshold value was computed automatically using the kmeans method (Figure 6).

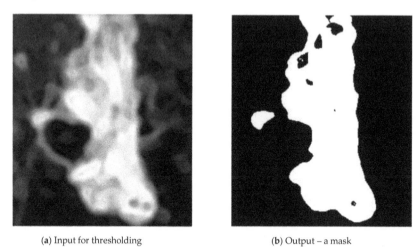

(a) Input for thresholding   (b) Output – a mask

Figure 6. Step 3—thresholding.

The thresholding result may have some imperfections—small blobs and unfilled areas. Unfilled areas are removed using the inverted output of the function bucketfill (imager library). It is applied on the background of the weld and it finds all pixels of the background. The remaining the pixels are filled with value 1 (white) (Figure 7a).

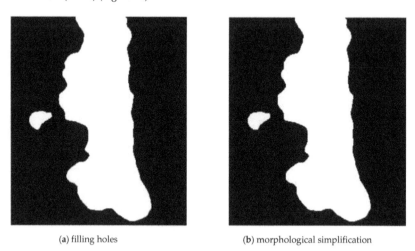

(a) filling holes   (b) morphological simplification

Figure 7. Step 4—filling holes (a) and morphological simplification (b).

Very small blobs were removed using the function clean (imager library). This function reduces objects size using morphological erosion, and then increases it. This causes, that very small objects are removed and the shape of larger object is simplified (Figure 7b).

However, larger blobs were not removed in the previous step. To find the largest object in the image, the function split_connected (imager library) was used (Figure 8).

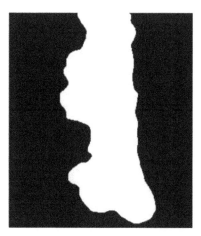

Figure 8. Step 5—Finding the largest object.

The segmentation result—the mask and the masked weld can be seen in Figure 9.

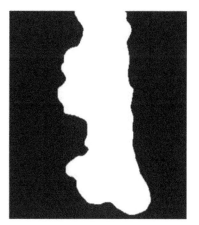

(**a**) The resulting weld mask

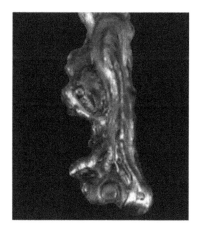

(**b**) The resulting segmented weld

Figure 9. Results of segmentation.

2.2. Vector of Sums of Subfields in the Mask

The first representation of the mask is a vector which entries are sums of subfields. For input images of resolution 263 × 300 pixels, was selected a subfield of 50 × 50 pixels, which corresponds to 36 values. The function for vector calculation is shown in the Algorithm 1.

The function ceiling rounds a number to the next higher integer. Using division of the index (i, j) by the size of the subfield, and subsequently the function ceiling, we obtained *indI*/*indJ* for the selected index i/j. The function as.vector retypes the resulting two-dimensional array into a vector by writing the matrix elements column-wise into a vector. Example of retyping can be understood from Figures 10 and 11.

Graphs for OK and NOK welds (Figure 12) can be compared in Figure 13: the OK mask graph has every third value (representing the subfields in the image center) maximal. Values of the NOK weld graph are distributed into more columns and the values do not achieve maximum values. The main drawback of this representation is that it can be used only for images with the same size. The benefit is a multiple reduction of input data (number of mask pixels in our case has been reduced 50^2-times).

Algorithm 1. Computing of subfields sums of the mask

```
procedure MaskToSums(img, size)
    xLen ←length(img[ ,1])
    yLen ←length(img[1, ])
    nRows ← ceiling(xLen/size)
    nCols ← ceiling(yLen/size)
    res ← matrix(0, nRows, nCols)
    for i in 1:xLen do
        for j in 1:yLen do
            if img[i,j] == TRUE then
                indI ← ceiling(i/size)
                indJ ← ceiling(j/size)
                res[indI, indJ] ++
            end if
        end for
    end for
    return as.vector(res)
end procedure
```

```
        [,1]  [,2]  [,3]  [,4]  [,5]  [,6]
[1,]       0     0     0     0     0     0
[2,]       6  1755   185     0     0     0
[3,]     689  1699  1696   216     0     0
[4,]    1583     0   621   452     0     0
[5,]       0     0     0     0     0     0
[6,]       0     0     0     0     0     0
```

Figure 10. Two-dimensional array of sums.

```
[1]    0    6  689 1583    0    0    0 1755 1699    0    0    0
[13]   0  185 1696  621    0    0    0    0  216  452    0    0
[25]   0    0    0    0    0    0    0    0    0    0    0    0
```

Figure 11. Resulting vector of sums.

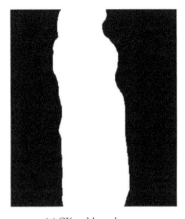

(a) OK weld mask

(b) NOK weld mask

Figure 12. Weld masks.

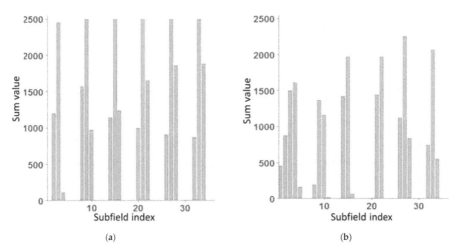

Figure 13. Graphs of vector of sums of subfields in the mask for OK (**a**) and NOK (**b**) weld.

2.3. Histogram Projection of the Mask

A histogram projection is a vector containing sums of columns and rows of the input image matrix (Figure 14). In the case of an image mask, these are amounts representing numbers of white pixels. Thus, the length of the vector corresponds to the vector of the height and width of the image.

In the graphs (Figures 15 and 16) showing the histogram projection of the mask, the difference between correct and wrong welds is visible. The projection of the correct weld mask is more even, the sums by columns have an even increase and slope, and the sums per line have small variations. On the other hand, the histogram projection of the wrong weld mask has a lot of irregularities. The disadvantage of this representation consists in that it cannot be used for input images of different resolutions. The resulting projection vector is much larger than other representations. The advantage is easy implementation and calculation.

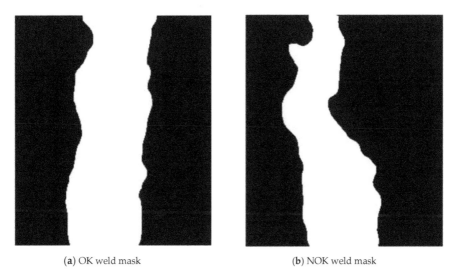

(**a**) OK weld mask   (**b**) NOK weld mask

Figure 14. Weld masks for histogram projection.

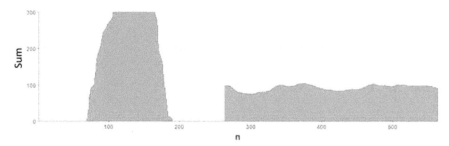

Figure 15. Graph of histogram projection of an OK weld.

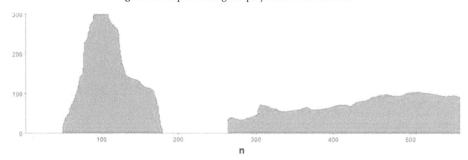

Figure 16. Graph of histogram projection of a NOK weld.

2.4. Vector of Polar Coordinates of the Mask Boundary

A next representation of a weld mask in this paper is the vector of polar coordinates of the mask boundary. To transform weld masks, an algorithm has been proposed and implemented. Its main steps are described below.

The first step is to find the x, y coordinates of the mask boundary using the function boundary (imager library). Then, coordinates of the center of the object $[cx, cy]$ are calculated according to:

$$c_x = \frac{max(x) - min(x)}{2} + min(x), \tag{2}$$

$$c_y = \frac{max(y) - min(y)}{2} + min(y), \tag{3}$$

In the next step, the position of the object is normalized (the center is moved to the position $[0, 0]$) according to the found coordinates. Then, for each boundary point, the coordinates are converted from Cartesian to polar $[r, \alpha]$ (i.e., distance from center, angle). According to the Pythagorean theorem, the distance is calculated as follows:

$$r = \sqrt{x^2 + x^2}, \tag{4}$$

Calculation of the angle is realized by Algorithm 2:

Algorithm 2. Calculation of angle from Cartesian coordinates

---

procedure Angle(x, y)
    z ← x + 1i * y
    a ← 90 - arg(z) / π * 180
    return round(a mod 360)
end procedure

If the resulting number of coordinates is less than 360, the missing angle values are completed and the corresponding distances are calculated from the surrounding values by linear interpolation using the na_approx function (zoo library). The result is a vector with 360 elements, which indices correspond to the angle values in degrees, and the value is the distance r. The resulting graphs of OK and NOK weld masks (Figure 17) are in Figures 18 and 19.

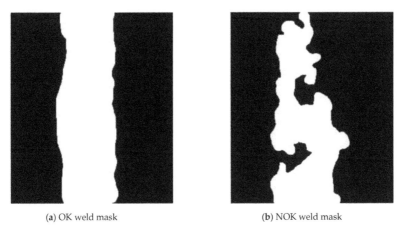

(a) OK weld mask   (b) NOK weld mask

Figure 17. Mask of OK and NOK weld.

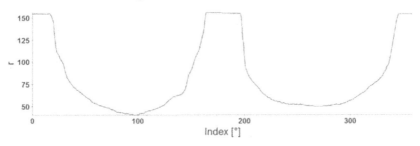

Figure 18. Graph of polar coordinates vector of an OK weld mask.

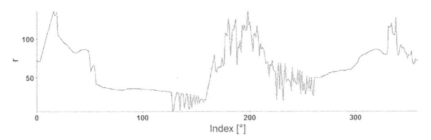

Figure 19. Graph of polar coordinates vector of a NOK weld mask.

The representation in the form of polar coordinates for the OK weld visibly differs from the NOK one. The big jumps and variations on the graph are caused by large irregularities in the weld shape. The advantage of such representation is that it can be used for any input mask resolution. The disadvantage is a complicated calculation. Generally, mask representations contain information only about the shape of the weld, which can be considered as a disadvantage because texture information is important input data for the neural network.

2.5. Data Preparation for Neural Network

Weld images and feature vectors were stored in two data structures of type list. The first list represented welds classified as NOK (incorrect); the second list welds classified as OK (correct). For neural networks, it was necessary to combine data, i.e., to transform and randomly mix them. For MLP and RBF networks, each input vector has to have assigned a classification value 0 (incorrect) or 1 (correct). Then, the vectors were merged together and with randomly mixed elements. Next, the L2-normalization was applied to the data. Finally, 85% of training and 15% of test samples were selected randomly. For convolution neural networks, the images were 5-times reduced, then the data type was converted to a three-dimensional array data structure. In the arrays, the dimensions were transposed to represent to correspond to the following structure: $[number\ of\ images * length * height]$. The vector of zeros with the same length as the first dimension corresponded to the first array (array of NOK welds). The vector of ones corresponded to the second array (array of OK welds). The arrays and vectors were merged into a common list and their elements were mixed randomly. Then, 77% of training samples, 15% of test samples and 8% of validation samples were selected.

3. Configuration and Training of Neural Networks

Several neural network architectures were configured for comparison and testing. Their parameters were changed during the experiments and the experiment results were compared and evaluated. Both RBF and MLP networks were configured in The Stuttgart Neural Network Simulator for R language - RSNNS library, the MLP networks were configured in the Keras library, and the convolution networks were configured in the Keras and the MXNet libraries.

3.1. RBF Network

To implement the RBF network, the RSNNS library was chosen (just in this one the RBF network template is available). Three RBF networks were configured using the function rbf (RSNN library). The set parameters were the number of units in the hidden layer and the number of epochs, the initial parameters had default values. The best configurations were chosen experimentally. Configuration details are in Figures 20–22.

```
Class: rbf->rsnns
Number of inputs: 36
Number of outputs: 2
Maximal iterations: 50
Initialization function: RBF_Weights
Initialization function parameters: 0 1 0 0.02 0.04
Learning function: RadialBasisLearning
Learning function parameters: 1e-05 0 1e-05 0.1 0.8
Update function:Topological_Order
Update function parameters: 0
Patterns are shuffled internally: TRUE
Compute error in every iteration: TRUE
Architecture Parameters:
$size
[1] 72
```

Figure 20. Settings for RBF network—for the vector of sums of subfields in the mask.

```
Class: rbf->rsnns
Number of inputs: 563
Number of outputs: 2
Maximal iterations: 50
Initialization function: RBF_Weights
Initialization function parameters: 0 1 0 0.02 0.04
Learning function: RadialBasisLearning
Learning function parameters: 1e-05 0 1e-05 0.1 0.8
Update function:Topological_Order
Update function parameters: 0
Patterns are shuffled internally: TRUE
Compute error in every iteration: TRUE
Architecture Parameters:
$size
[1] 50
```

Figure 21. Settings for RBF network—for the histogram projection vector.

```
Class: rbf->rsnns
Number of inputs: 360
Number of outputs: 2
Maximal iterations: 50
Initialization function: RBF_weights
Initialization function parameters: 0 1 0 0.02 0.04
Learning function: RadialBasisLearning
Learning function parameters: 1e-05 0 1e-05 0.1 0.8
Update function:Topological_Order
Update function parameters: 0
Patterns are shuffled internally: TRUE
Compute error in every iteration: TRUE
Architecture Parameters:
$size
[1] 60
```

Figure 22. Settings for RBF network—for the polar coordinates vector.

3.2. MLP Network

Experiments with training and testing of MLP networks showed, that a one-layer architecture is sufficient for our data representation. The performance of the network was very good and the difference from multiple hidden layers was negligible. To keep the objectivity, MLP networks had the same configuration in both libraries. The sigmoid activation function and the randomize weights initialization functions were used. For the NN training, the error backpropagation algorithm with learning parameter 0,1 was used.

The implementation in the RSNNS library uses the mlp function for configuration and training. Configuration details are in Figures 23–25.

```
Class: mlp->rsnns
Number of inputs: 36
Number of outputs: 2
Maximal iterations: 50
Initialization function: Randomize_weights
Initialization function parameters: -0.3 0.3
Learning function: Std_Backpropagation
Learning function parameters: 0.1
Update function:Topological_Order
Update function parameters: 0
Patterns are shuffled internally: TRUE
Compute error in every iteration: TRUE
Architecture Parameters:
$size
[1] 6
```

Figure 23. Settings for MLP network—for vector of sums of subfields in the mask.

```
Class: mlp->rsnns
Number of inputs: 563
Number of outputs: 2
Maximal iterations: 40
Initialization function: Randomize_weights
Initialization function parameters: -0.3 0.3
Learning function: Std_Backpropagation
Learning function parameters: 0.1
Update function:Topological_Order
Update function parameters: 0
Patterns are shuffled internally: TRUE
Compute error in every iteration: TRUE
Architecture Parameters:
$size
[1] 12
```

Figure 24. Settings for MLP network—for histogram projection vector.

```
Class: mlp->rsnns
Number of inputs: 360
Number of outputs: 2
Maximal iterations: 40
Initialization function: Randomize_weights
Initialization function parameters: -0.3 0.3
Learning function: Std_Backpropagation
Learning function parameters: 0.1
Update function:Topological_Order
Update function parameters: 0
Patterns are shuffled internally: TRUE
Compute error in every iteration: TRUE
Architecture Parameters:
$size
[1] 36
```

Figure 25. Settings for MLP network - for the polar coordinates vector.

The implementation of the MLP network in the Keras library required a detailed list of layers in the code. Two layer_dense layers were used; the first one defines the hidden layer with the ReLU activation function, and the second one defines the output layer with the size 2 (two output categories) using the softmax activation function (Figure 26).

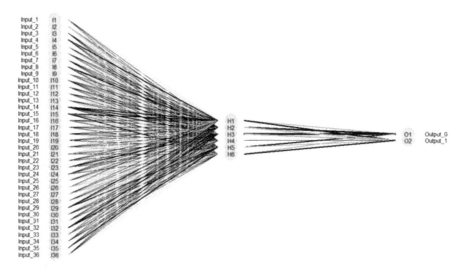

Figure 26. MLP network architecture—for vector of sums of subfields in the mask.

3.3. Convolution Neural Network

For an objective comparison of the Keras and MXNet libraries, the same convolution network architecture in both libraries was used at first, however in the MXNet library, training such a neural network was too slow. Thus, we designed our own architecture with a better learning time performance. The discussion about the results is provided in the next Section 4.

The architecture of the convolution network 1 is shown in Figure 27 and visualized in Figure 28. The architecture includes a list of all layers and the size of output structures for both NN. Two pairs of convolution and pooling layers were used, the convolution being applied twice before the first pooling layer. The input image size was 56 × 60. The number of convolution filters was 32 at the beginning, in further convolution filters it rose to 64. A dropout was used between some layers to prevent overtraining of the neural network by deactivating a certain percentage of randomly selected neurons. At the end, the flatten layer was used to convert the resulting structure into a one-dimensional vector used as an input for a simple MLP network with one hidden layer containing 256 neurons.

```
Model
_____
Layer (type)                        Output Shape
=================================================================
conv2d_1 (Conv2D)                   (None, 27, 30, 32)
_____
activation_1 (Activation)           (None, 27, 30, 32)
_____
conv2d_2 (Conv2D)                   (None, 14, 15, 64)
_____
activation_2 (Activation)           (None, 14, 15, 64)
_____
max_pooling2d_1 (MaxPooling2D)      (None, 7, 7, 64)
_____
dropout_1 (Dropout)                 (None, 7, 7, 64)
_____
conv2d_3 (Conv2D)                   (None, 7, 7, 64)
_____
activation_3 (Activation)           (None, 7, 7, 64)
_____
max_pooling2d_2 (MaxPooling2D)      (None, 3, 3, 64)
_____
dropout_2 (Dropout)                 (None, 3, 3, 64)
_____
flatten_1 (Flatten)                 (None, 576)
_____
dense_1 (Dense)                     (None, 256)
_____
activation_4 (Activation)           (None, 256)
_____
dropout_3 (Dropout)                 (None, 256)
_____
dense_2 (Dense)                     (None, 2)
_____
activation_5 (Activation)           (None, 2)
=================================================================
```

Figure 27. Architecture of the convolution neural network 1.

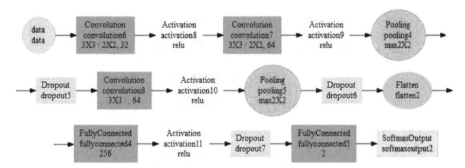

Figure 28. Architecture visualization of the convolution neural network 1.

Parameters of individual layers are shown in the diagram in Figure 28. For example, the convolution layer (red) contains a list of 3 × 3 - filter size, 3 × 3 - stride, 32 - number of filters.

The architecture of the convolution network 2 is visualized in Figure 29. Two pairs of convolution and pooling layers were used, however in this case a double convolution occurs only in the second layer. There is also a difference in the design of the convolution, where the parameter stride (step of the filter) is 3,3. Dropout was used only in two places.

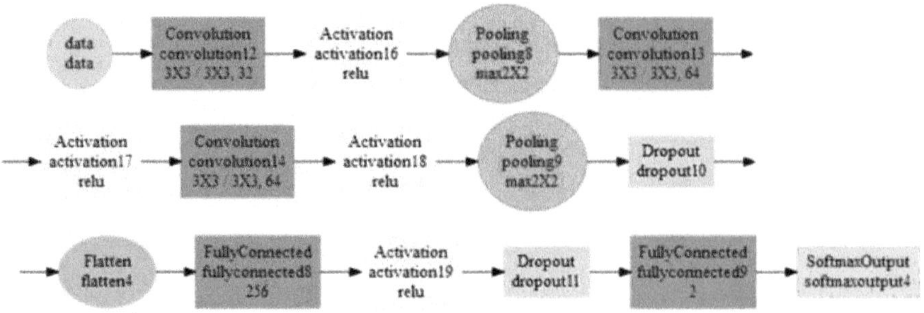
data
data
Convolution
convolution12
3X3 / 3X3, 32
Activation
activation16
relu
Pooling
pooling8
max2X2
Convolution
convolution13
3X3 / 3X3, 64
Activation
activation17
relu
Convolution
convolution14
3X3 / 3X3, 64
Activation
activation18
relu
Pooling
pooling9
max2X2
Dropout
dropout10
Flatten
flatten4
FullyConnected
fullyconnected8
256
Activation
activation19
relu
Dropout
dropout11
FullyConnected
fullyconnected9
2
SoftmaxOutput
softmaxoutput4

Figure 29. Architecture visualization of the Convolution neural network 2.

4. Results

This chapter presents results of code profiling, weld segmentation and evaluation of neural networks.

4.1. Code Profiling

Profiling was done using the profvis library at the level of the code line. The output is an interactive visualization using memory listing in MB and computing time in ms for each code line. The example can be seen in Figure 30.

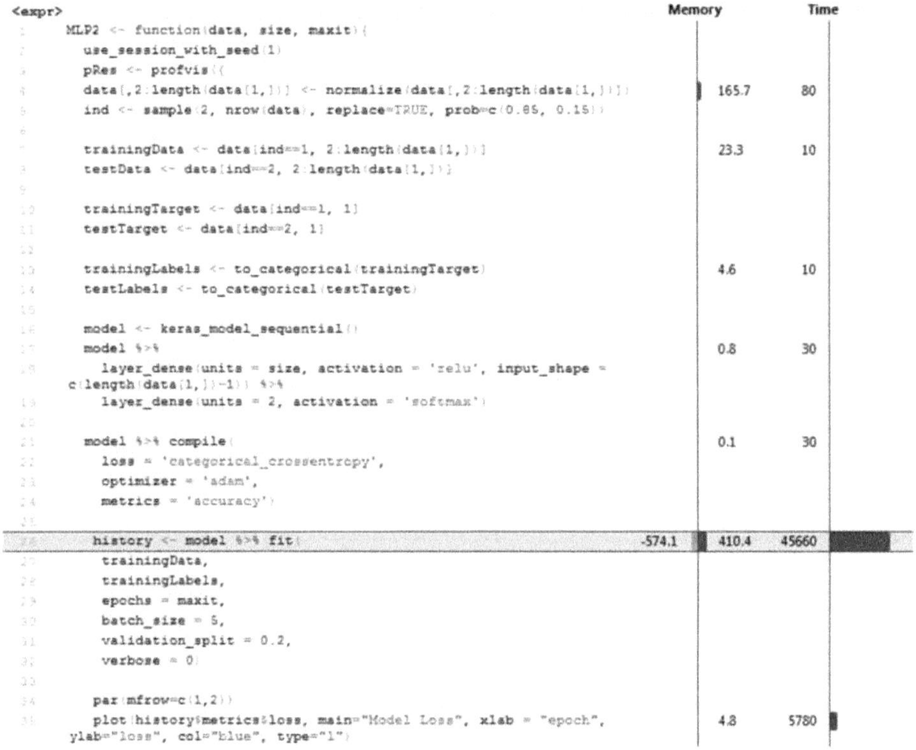
<expr>
Memory
Time
MLP2 <- function(data, size, maxit){
use_session_with_seed(1)
pRes <- profvis({
data[,2:length(data[1,])] <- normalize(data[,2:length(data[1,])])
165.7
80
ind <- sample(2, nrow(data), replace=TRUE, prob=c(0.85, 0.15))
trainingData <- data[ind==1, 2:length(data[1,])]
23.3
10
testData <- data[ind==2, 2:length(data[1,])]
trainingTarget <- data[ind==1, 1]
testTarget <- data[ind==2, 1]
trainingLabels <- to_categorical(trainingTarget)
4.6
10
testLabels <- to_categorical(testTarget)
model <- keras_model_sequential()
model %>%
0.8
30
layer_dense(units = size, activation = 'relu', input_shape = c(length(data[1,])-1)) %>%
layer_dense(units = 2, activation = 'softmax')
model %>% compile(
0.1
30
loss = 'categorical_crossentropy',
optimizer = 'adam',
metrics = 'accuracy')
history <- model %>% fit(
-574.1
410.4
45660
trainingData,
trainingLabels,
epochs = maxit,
batch_size = 5,
validation_split = 0.2,
verbose = 0)
par(mfrow=c(1,2))
plot(history$metrics$loss, main="Model Loss", xlab = "epoch", ylab="loss", col="blue", type="l")
4.8
5780

Figure 30. Example of profiling output using profvis.

Profiling was performed on a desktop computer with parameters listed in Table 1 (the graphic card was not used).

Table 1. Technical specifications of PC.

Operating System	Windows 7 Professional 64-bit
Processor	Intel Core i7-2600 CPU @ 3,40 GHz
Memory	16 GB DDR3
Disc	Samsung SSD 850 EVO 500 GB

4.2. Results of Data Preparation and Segmentation

Segmentation was successful for all tested weld samples. For some NOK defective welds which consisted of several parts or contained droplets, only the largest continuous weld surface was segmented, which was considered to be a correct segmentation for proposed methodology. Segmentation examples are shown in Figure 31.

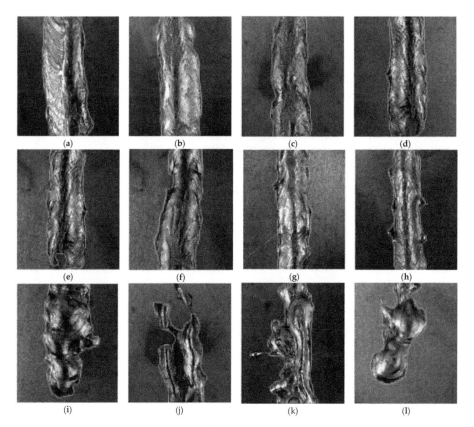

Figure 31. *Cont.*

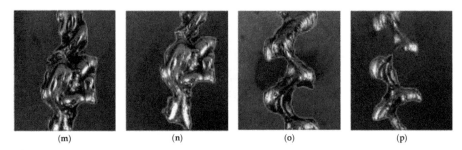

(m) (n) (o) (p)

Figure 31. Examples of weld segmentation results (**a–p**).

The segmentation time is an important indicator in comparison of results. Results of profiling different parts of the segmentation process can be seen in Figure 32. Code profiling was carried out using a computer with the technical specification shown in Table 1.

`<expr>`	Memory	Time
1 `profvis({`		
2 `    testIm01 <- load.image('./../zvary/zvary s doplnenym`		
`    pozadim/zvarNOKpozadie/zvarNOKpozadie_00133.jpg')`		
3 `    testIm01 <- grayscale(testIm01)`		
4 `    testIm01mask <- entropyFilter1(testIm01)`	43.6	140
5 `    testIm01mask <- createMask1(testIm01)`	25.8	20
6 `    testIm01segm <- segmentWeld(testIm01,testIm01mask)`		
7		
8 `    testIm02 <- load.image('./../zvary/zvary s doplnenym`	14.1	10
`    pozadim/zvarNOKpozadie/zvarNOKpozadie_00623.jpg')`		
9 `    testIm02 <- grayscale(testIm02)`		
10 `    testIm02mask <- entropyFilter1(testIm02)`	37.9	140
11 `    testIm02mask <- createMask1(testIm02)`	27.0	20
12 `    testIm02segm <- segmentWeld(testIm02,testIm02mask)`		
13		
14 `    testIm03 <- load.image('./../zvary/zvary s doplnenym`	19.5	20
`    pozadim/zvarOKpozadie/zvarOK_00023.jpg')`		
15 `    testIm03 <- grayscale(testIm03)`		
16 `    testIm03mask <- entropyFilter1(testIm03)`	36.1	160
17 `    testIm03mask <- createMask1(testIm03)`	14.1	20
18 `    testIm03segm <- segmentWeld(testIm03,testIm03mask)`		
19		
20 `    testIm04 <- load.image('./../zvary/zvary s doplnenym`	19.2	30
`    pozadim/zvarOKpozadie/zvarOK_01555.jpg')`		
21 `    testIm04 <- grayscale(testIm04)`		
22 `    testIm04mask <- entropyFilter1(testIm04)`	36.1	210
23 `    testIm04mask <- createMask1(testIm04)`	21.0	20
24 `    testIm04segm <- segmentWeld(testIm04,testIm04mask)`		
25		
26 `    testIm05 <- load.image('./../zvary/zvary s doplnenym`	8.7	10
`    pozadim/zvarOKpozadie/zvarOK_01555.jpg')`		
27 `    testIm05 <- grayscale(testIm05)`	16.3	10
28 `    testIm05mask <- entropyFilter1(testIm05)`	27.1	140
29 `    testIm05mask <- createMask1(testIm05)`	27.3	20
30 `    testIm05segm <- segmentWeld(testIm05,testIm05mask)`		
31		
32 `})`		
33		

Figure 32. Results of segmentation process profiling.

Segmentation was performed by concatenating the outputs from functions load.image, grayscale, entropyFilter, createMask, and segmentWeld. Almost all functions in this section of the program were performed very quickly (within 30 ms) except for the entropyFilter function, which took an average of 158 ms to be completed. This function is the most important part of the segmentation algorithm; the time was acceptable. The average time to complete the whole segmentation was 194 ms. The average amount of memory allocated was 74.76 MB. For MLP and RBF networks, the next step was to transform masks into feature vectors. The profiling results of functions performing three types of transformations can be seen in Figure 33.

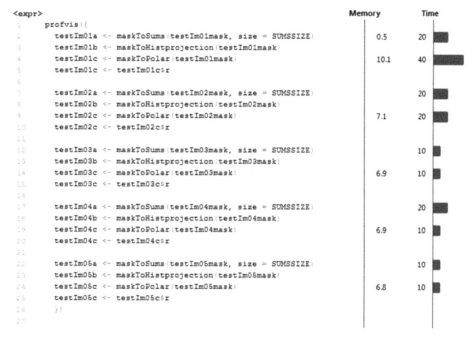

Figure 33. The profiling results of data transformation.

The results show that these functions are optimal, taking up minimal memory and time. The mean values for computing the vector of sums of subfields in the mask are 16 ms and 0.1 MB; for the histogram projection vector, it is less than 10 ms and less than 0.1 MB (estimation of profiling tool, real values are immeasurably small). Values for the polar coordinates vector are 18 ms and 7.56 MB. Presented results are also shown in Table 2.

Table 2. Algorithms results for transform masks into feature vectors.

Data Interpretation	Time [ms]	Memory [MB]
the vector of sums of subfields in the mask	16	0.1
histogram projection vector	10	0.1
polar coordinates vector	18	7.56

4.3. Criteria for Evaluation of Neural Network Results

As the main criterion for results evaluation the confusion matrix was chosen. The main diagonal of the confusion matrix contains the numbers of correctly classified samples, the antidiagonal contains the numbers of incorrectly classified samples; the smaller values in the antidiagonal, the more successful the prediction model. In a binary classification this matrix contains four values (Figure 34): TP—true positive; FP—false positive; FN—false negative; TN—true negative.

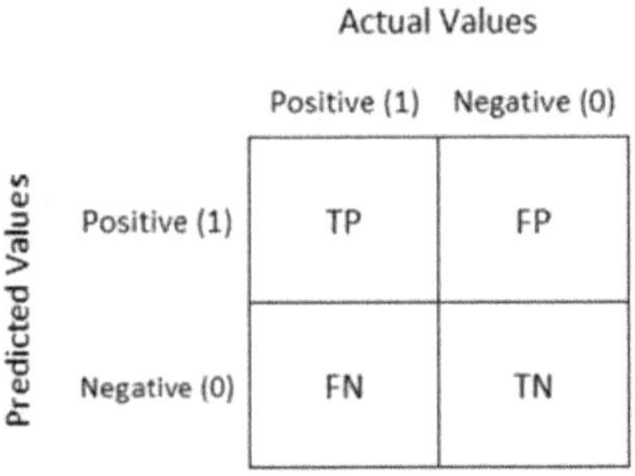

Actual Values

Positive (1)   Negative (0)

	Positive (1)	Negative (0)
Positive (1)	TP	FP
Negative (0)	FN	TN

(Predicted Values — left axis label)

Figure 34. Confusion matrix.

The accuracy was computed from the confusion matrix and is expressed as the ratio of correctly classified samples to all samples, see Equation (5) [20].

$$Accuracy = \frac{\sum TP + \sum TN}{\sum all\ samples},\tag{5}$$

Accuracy is an objective criterion only if the FN and FP values are similar.

A more objective criterion for comparing results is the F-score. The *F-score* is calculated as the harmonic average of the precision and the recall (sensitivity) values [20], the best score corresponds to F-score = 1:

$$Precision = \frac{\sum TP}{\sum TP + \sum FP},\tag{6}$$

$$Recall = \frac{\sum TP}{\sum TP + \sum FN},\tag{7}$$

$$F - score = \frac{2 * Recall * Precision}{\sum TP + \sum FN\ Recall + Precision},\tag{8}$$

To visualize the success of neural network classification, the ROC (Receiver operating characteristics) curve was chosen. It shows the recall (sensitivity) value depending on the value 1-specificity at the variable threshold [20] (Figure 35):

$$Specificity = \frac{\sum TN}{\sum TN + \sum FP},\tag{9}$$

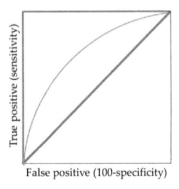

False positive (100-specificity)

(True positive (sensitivity) — left axis label)

Figure 35. ROC curves: excellent (blue); good (green); worthless (red).

The ROC curve for the best possible classifier is rectangular with the vertex [0,1].

4.4. Results of Neural Network Classificaton

We configured and tested neural networks for all data representations (in total 15 experiments). For a better clarity, the experiments results are labelled using labels from Table 3.

Table 3. Labels of neural network experiment.

Test Label	Network Type	Library	Data Format
rbf-rsn-sum01	RBF	RSNNS	Subfields sum
rbf-rsn-hpr02	RBF	RSNNS	Histogram projection
rbf-rsn-pol03	RBF	RSNNS	Polar coordinates
mlp-rsn-sum04	MLP	RSNNS	Subfields sum
mlp-rsn-hpr05	MLP	RSNNS	Histogram projection
mlp-rsn-pol06	MLP	RSNNS	Polar coordinates
mlp-ker-sum07	MLP	Keras	Subfields sum
mlp-ker-hpr08	MLP	Keras	Histogram projection
mlp-ker-pol09	MLP	Keras	Polar coordinates
cnn-ker-ori10	CNN 1	Keras	Original
cnn-ker-seg11	CNN 1	Keras	Segmented
cnn-mxn-ori12	CNN 1	MXNet	Original
cnn-mxn-seg13	CNN 1	MXNet	Segmented
cnn-mxn-ori14	CNN 2	MXNet	Original
cnn-mxn-seg15	CNN 2	MXNet	Segmented

The first tests were carried out for RBF and MLP networks with input data formats according to Table 3. Resulting confusion matrices for RBF networks are as follows:

$$rbf - rsn - sum01 = \begin{bmatrix} 502 & 14 \\ 15 & 433 \end{bmatrix},$$
$$rbf - rsn - hpr02 = \begin{bmatrix} 434 & 0 \\ 83 & 447 \end{bmatrix}, \tag{10}$$
$$rbf - rsn - pol03 = \begin{bmatrix} 435 & 0 \\ 82 & 447 \end{bmatrix},$$

From the matrices (10) it is evident that the RBF network performed bad when classifying NOK welds—they are often classified as OK. ROC curves of trained RBF networks are depicted in Figure 36.

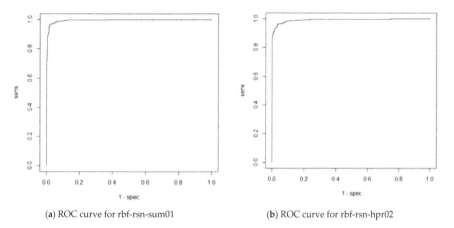

(a) ROC curve for rbf-rsn-sum01   (b) ROC curve for rbf-rsn-hpr02

Figure 36. *Cont.*

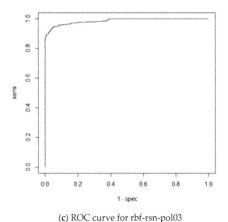

(c) ROC curve for rbf-rsn-pol03

Figure 36. ROC curves for experiments with RBF networks.

ROC curves for MLP networks are depicted Figure 37 and Resulting confusion matrices are as follows:

$$mlp - rsn - sum04 = \begin{bmatrix} 516 & 1 \\ 1 & 446 \end{bmatrix},$$
$$mlp - rsn - hpr05 = \begin{bmatrix} 5017 & 0 \\ 0 & 447 \end{bmatrix}, \qquad (11)$$
$$mlp - rsn - pol06 = \begin{bmatrix} 514 & 1 \\ 3 & 446 \end{bmatrix},$$

$$mlp - ker - sum07 = \begin{bmatrix} 517 & 15 \\ 17 & 446 \end{bmatrix},$$
$$mlp - ker - hpr08 = \begin{bmatrix} 511 & 2 \\ 23 & 459 \end{bmatrix}, \qquad (12)$$
$$mlp - ker - pol09 = \begin{bmatrix} 522 & 13 \\ 12 & 448 \end{bmatrix},$$

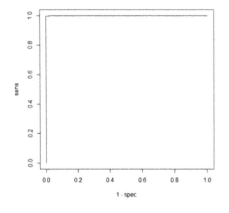

(a) ROC curve for mlp-rsn-sum04

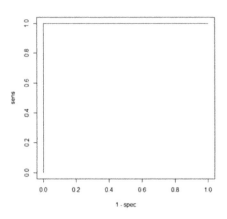

(b) ROC curve for mlp-rsn-hpr05

Figure 37. *Cont.*

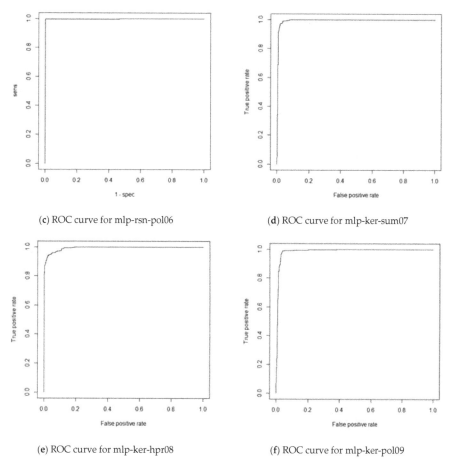

(**c**) ROC curve for mlp-rsn-pol06

(**d**) ROC curve for mlp-ker-sum07

(**e**) ROC curve for mlp-ker-hpr08

(**f**) ROC curve for mlp-ker-pol09

Figure 37. ROC curves for experiments with MLP networks.

The results show that the MLP implementation in the RSNNS library was more successful compared with the Keras library. The networks had no problem to classify correct (OK) or incorrect (NOK) welds. FP and FN values were approximately similar. The resulting calculated accuracy and F-scores shown in Table 4 describe the performance of the trained neural networks.

Table 4. Accuracy a F-score for RBF and MLP networks.

Test Label	Accuracy	F-Score
rbf-rsn-sum01	0.9699	0.9719
rbf-rsn-hpr02	0.9139	0.9127
rbf-rsn-pol03	0.9149	0.9139
mlp-rsn-sum04	0.9979	0.9981
mlp-rsn-hpr05	1.0000	1.0000
mlp-rsn-pol06	0.9959	0.9961
mlp-ker-sum07	0.9678	0.9700
mlp-ker-hpr08	0.9761	0.9761
mlp-ker-pol09	0.9766	0.9766

The results show that MLP networks are much more successful. Using default RBF initialization weights the RBF network less successful. From a practical point of view, MLP networks are more suitable for weld evaluation.

It was hard to compare the results for MLP networks, they provided similar results for all data representations. The RBF network achieved significantly better results in the vector of sums of subfields in the mask data representation.

It was found out, that using the same network configuration in the two libraries yields slightly different results. The implementation in the RSNNS library was almost 100% successful and therefore it was considered as the best candidate for practical use.

Training profiling for RSNN library was done next. Although training in the Keras library allocated less memory, the training time was several times longer than in case of the RSNNS library. Using vector of sums of subfields in the mask, the MLP network training time in RSNNS took less than one second, while using the Keras library was tens of seconds. The list of training profiling results is shown in Table 5.

Table 5. Profiling of RBF and MLP networks training.

Test Label.	Time [ms]	Memory [MB]
rbf-rsn-sum01	6660	687.6
rbf-rsn-hpr02	42,530	775.6
rbf-rsn-pol03	32,080	752.3
mlp-rsn-sum04	850	769.8
mlp-rsn-hpr05	9890	653.7
mlp-rsn-pol06	17,270	672.0
mlp-ker-sum07	52,830	485.2
mlp-ker-hpr08	45,660	410.4
mlp-ker-pol09	46,420	401.9

Comparison of convolution neural nets was again based on the confusion matrices, ROC curves, accuracy and F-scores. The input of the networks were just images of welds without any filtration and masked welds without background (black background). Confusion matrices are as follows:

$$cnn - ker - ori10 = \begin{bmatrix} 534 & 1 \\ 0 & 460 \end{bmatrix},$$
$$cnn - mxn - ori12 = \begin{bmatrix} 559 & 8 \\ 1 & 431 \end{bmatrix},$$
$$cnn - mxn - ori14 = \begin{bmatrix} 498 & 0 \\ 0 & 460 \end{bmatrix}, \tag{13}$$

$$cnn - ker - seg11 = \begin{bmatrix} 534 & 0 \\ 0 & 461 \end{bmatrix},$$
$$cnn - mxn - seg13 = \begin{bmatrix} 558 & 0 \\ 2 & 439 \end{bmatrix}, \tag{14}$$
$$cnn - mxn - seg15 = \begin{bmatrix} 498 & 0 \\ 0 & 460 \end{bmatrix},$$

Classification error in convolution neural networks was minimal, therefore the ROC curve was evaluated as ideal for all experiments with indistinguishable differences. For all neural nets, the ROC curve was the same (Figure 38).

The resulting accuracy and F-scores along with the number of epochs needed to train the networks are listed in Table 6.

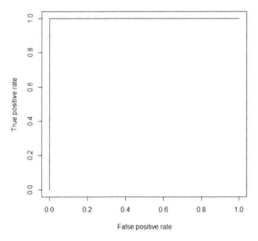

Figure 38. ROC curve for all convolution nets.

Table 6. Accuracy and F-scores for convolution neural network experiments.

Test Label	Epochs	Accuracy	F-Score
cnn-ker-ori10	5	0.9990	0.9991
cnn-ker-seg11	4	1.0000	1.0000
cnn-mxn-ori12	6	0.9910	0.9920
cnn-mxn-seg13	3	0.9980	0.9982
cnn-mxn-ori14	4	1.0000	1.0000
cnn-mxn-seg15	4	1.0000	1.0000

For convolution networks, changes of accuracy after each epoch for both training (blue line) and validation data (green line) are shown in Figure 39. The charts show that training with non-segmented weld images started at a lower accuracy and the learning was slower (Figure 40).

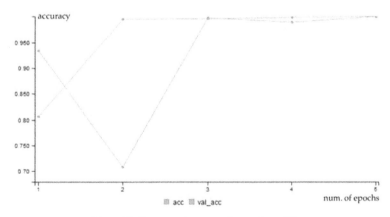

Figure 39. Progress of accuracy for cnn-ker-ori10.

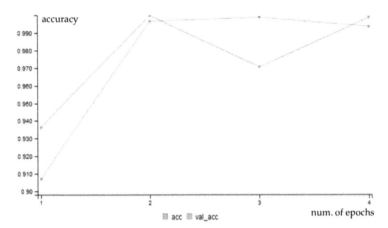

Figure 40. Progress of accuracy during epochs for cnn-ker-seg11.

The progress of training for the Keras library was more uniform, without steps. The graphs can be seen in Figures 41 and 42.

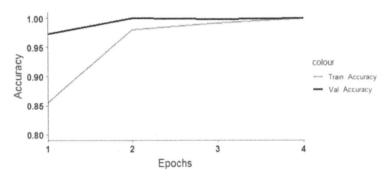

Figure 41. Progress of accuracy during epochs for cnn-mxn-ori14.

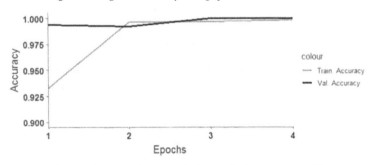

Figure 42. Progress of accuracy during epochs for cnn-mxn-seg15.

The success rate for all networks was higher than 99%. The decisive factor for comparison were the code profiling results shown in Table 7.

Table 7. Code profiling results for designed convolution neural networks.

Test Label	Epochs	Time [ms]	Memory [MB]
cnn-ker-ori10	5	38,610	186.9
cnn-ker-seg11	4	30,660	180.0
cnn-mxn-ori12	6	119,630	4.7
cnn-mxn-seg13	3	82,580	2.6
cnn-mxn-ori14	4	12,170	157.9
cnn-mxn-seg15	4	11,850	3.7

It can be concluded, that the network with the architecture shown in Figure 29 in Section 3.3 implemented using the MXNet library was the fastest. With a training time 12.170 ms and a 100% success also for non-segmented data it is considered the best choice for practical use.

Although the MLP network (mlp-rsn-sum04) was similarly successful and several times faster in training, the preparation of the representation in the form of the vector of sums of subfields in the mask took considerably more time. The number of training samples was approximately 5400, the average time to obtain a mask of one sample was 164 ms, and the vector calculation was 16 ms, in total 972 ms.

4.5. Profiling Single Weld Diagnostics

In practice, neural network training is not a frequent process. Usually, the network is trained once and then implemented for prediction. Therefore, at the end we decided to evaluate the prediction of one weld for the most successful models. The provided results represent the average of five independent tests. The list can be seen in Table 8 along with the average image preparation time and memory required to prepare the weld input image for the specific diagnostic model.

Table 8. Profiling results for single weld diagnostics.

Test Label	Image Time Preparation [ms]	Diagnostic Time [ms]	Memory [MB]
mlp-rsn-sum04	210	20	0.2
mlp-rsn-hpr05	194	240	3.0
mlp-rsn-pol06	198	105	1.8
cnn-mxn-ori14	14	14	0.5
cnn-mxn-seg15	194	4	0.5

The diagnostic profiling results confirmed that the best solution was the classification of the weld using the convolution net with the architecture shown in Figure 29 in Section 3.3. The average image loading time and its 5× reduction took only 14 ms on average, and evaluation time was 14 ms.

5. Discussion

The aim of this paper was to develop a neural network based methodology to evaluate quality of welds. Several types of neural networks implemented in several software libraries were compared with respect to performance. It was necessary to prepare the data (images of welds) into a format suitable for neural network processing. For some types of networks (convolution) the input data preparation was minimal (segmentation or no segmentation), while for other networks (MLP, RBF), a sophisticated data preprocessing was required (filtering, equalizing and segmenting the image based on entropy). Each library required its own input data format which also had to be taken into account during programming. The main result of the paper is confirmation, that the convolutional neural networks can be used for weld quality evaluation without using image preprocessing and in case of using no segmentation, they can be used for evaluation not only weld metal but also adjected zones.

Neural networks were configured experimentally to achieve the best performance and the obtained results were compared. In all cases, neural networks implemented and trained using the proposed approach delivered excellent results with a success rate of nearly 100%. Thus, we can recommend any of the tested libraries to solve the weld quality evaluation problem. The best results were achieved using convolution neural networks which provided excellent results and with almost no pre-processing of image data required. The longer training time of these networks is acceptable in practical usage.

In summary, based on achieved experimental results, convolution neural networks have shown to be a promising approach for weld evaluation and will be applied in the future research dealing with evaluation of images in the real welding processes. The convolutional neural networks can be used for weld quality evaluation without using image preprocessing.

Author Contributions: O.H. proposed the idea in this paper and prepared data; O.H., E.K., P.D. and J.C. designed the experiments, E.K. and P.D. performed the experiments; O.H., and E.K. analyzed the data; O.H. wrote the paper; E.K., P.D. and J.C. edited and reviewed the paper; All authors read and approved the final manuscript.

Funding: This research was supported by the Slovak research and Development Agency under the contract no. APVV-17-0190, by the Scientific Grant Agency of the Ministry of Education, Science, Research and Sport of the Slovak Republic under the grant VEGA 1/0819/17, and by the Cultural and Educational Grant Agency of the Ministry of Education, Science, Research and Sport of the Slovak Republic, KEGA 038STU-4/2018.

Acknowledgments: We would like to thank to Alena Kostuňová for helping with programming the implementation.

Conflicts of Interest: The authors declare no conflict of interest.

References

1. Akşit, M. The Role of Computer Science and Software Technology in Organizing Universities for Industry 4.0 and beyond. In Proceedings of the 2018 Federated Conference on Computer Science and Information Systems, FedCSIS 2018, Poznań, Poland, 9–12 September 2018.
2. Dahal, S.; Kim, T.; Ahn, K. Indirect prediction of welding fume diffusion inside a room using computational fluid dynamics. *Atmosphere* **2016**, *7*, 74. [CrossRef]
3. Huang, W.; Kovacevic, R. A laser-based vision system for weld quality inspection. *Sensors* **2011**, *11*, 506–521. [CrossRef] [PubMed]
4. Noruk, J. Visual weld inspection enters the new millennium. *Sens. Rev.* **2001**, *21*, 278–282. [CrossRef]
5. Deng, S.; Jiang, L.; Jiao, X.; Xue, L.; Deng, X. Image processing of weld seam based on beamlet transform. *Hanjie Xuebao/Trans. China Weld. Inst.* **2009**, *30*, 68–72.
6. Deng, S.; Jiang, L.; Jiao, X.; Xue, L.; Cao, Y. Weld seam edge extraction algorithm based on Beamlet Transform. In Proceedings of the 1st International Congress on Image and Signal Processing, CISP 2008, Hainan, China, 27–30 May 2008.
7. Zhang, X.; Yin, Z.; Xiong, Y. Edge detection of the low contrast welded joint image corrupted by noise. In Proceedings of the 8th International Conference on Electronic Measurement and Instruments, ICEMI 2007, Xi'an, China, 16–18 August 2007.
8. Hou, X.; Liu, H. Welding image edge detection and identification research based on canny operator. In Proceedings of the 2012 International Conference on Computer Science and Service Systems, CSSS 2012, Nanjing, China, 11–13 August 2012.
9. Shen, Z.; Sun, J. Welding seam defect detection for canisters based on computer vision. In Proceedings of the 6th International Congress on Image and Signal Processing, CISP 2013, Hangzhou, China, 16–18 December 2013.
10. Liao, Z.; Sun, J. Image segmentation in weld defect detection based on modified background subtraction. In Proceedings of the 6th International Congress on Image and Signal Processing, CISP 2013, Hangzhou, China, 16–18 December 2013.
11. Khumaidi, A.; Yuniarno, E.M.; Purnomo, M.H. Welding defect classification based on convolution neural network (CNN) and Gaussian Kernel. In Proceedings of the 2017 International Seminar on Intelligent Technology and Its Application: Strengthening the Link between University Research and Industry to Support ASEAN Energy Sector, ISITIA 2017, Surabaya, Indonesia, 28–29 August 2017.
12. Pandiyan, V.; Murugan, P.; Tjahjowidodo, T.; Caesarendra, W.; Manyar, O.M.; Then, D.J.H. In-process virtual verification of weld seam removal in robotic abrasive belt grinding process using deep learning. *Robot. Comput. Integr. Manuf.* **2019**, *57*, 477–487. [CrossRef]

13. Chen, J.; Wang, T.; Gao, X.; Wei, L. Real-time monitoring of high-power disk laser welding based on support vector machine. *Comput. Ind.* **2018**, *94*, 75–81. [CrossRef]
14. Wang, T.; Chen, J.; Gao, X.; Qin, Y. Real-time monitoring for disk laser welding based on feature selection and SVM. *Appl. Sci.* **2017**, *7*, 884. [CrossRef]
15. Haffner, O.; Kucera, E.; Kozak, S.; Stark, E. Proposal of system for automatic weld evaluation. In Proceedings of the 21st International Conference on Process Control, PC 2017, Štrbské Pleso, Slovakia, 6–9 June 2017.
16. Haffner, O.; Kučera, E.; Kozák, Š. Weld segmentation for diagnostic and evaluation method. In Proceedings of the 2016 Cybernetics and Informatics, K and I 2016—Proceedings of the the 28th International Conference, Levoca, Slovakia, 2–5 February 2016.
17. Haffner, O.; Kučera, E.; Kozák, Š.; Stark, E. Application of Pattern Recognition for a Welding Process. In Proceedings of the Communiation Papers of the 2017 Federated Conference on Computer Science and Information Systems, FedCSIS 2017, Prague, Czech Republic, 3–6 September 2017.
18. Haffner, O.; Kučera, E.; Bachurikova, M. Proposal of weld inspection system with single-board computer and Android smartphone. In Proceedings of the 2016 Cybernetics and Informatics, K and I 2016—Proceedings of the the 28th International Conference, Levoca, Slovakia, 2–5 February 2016.
19. Gajowniczek, K.; Ząbkowski, T.; Orłowski, A. Comparison of decision trees with Rényi and Tsallis entropy applied for imbalanced churn dataset. In Proceedings of the 2015 Federated Conference on Computer Science and Information Systems, FedCSIS 2015, Łódź, Poland, 13–16 September 2015.
20. Sokolova, M.; Lapalme, G. A systematic analysis of performance measures for classification tasks. *Inf. Process. Manag.* **2009**, *45*, 427–437. [CrossRef]

© 2019 by the authors. Licensee MDPI, Basel, Switzerland. This article is an open access article distributed under the terms and conditions of the Creative Commons Attribution (CC BY) license (http://creativecommons.org/licenses/by/4.0/).

Article

Investigating Detectability of Infrared Radiation Based on Image Evaluation for Engine Flame

Xia Li [1], Jun Wang [1,2], Meihui Li [2,3], Zhenming Peng [2,3,*] and Xingrun Liu [1]

[1] The Science and Technology on Optical Radiation Laboratory, Beijing 100854, China; lixia207@sina.com (X.L.); wangjunt@vip.sina.com (J.W.); liuxr207@126.com (X.L.)

[2] School of Information and Communication Engineering, University of Electronic Science and Technology of China, Chengdu 611731, China; meihuili@std.uestc.edu.cn

[3] Laboratory of Imaging Detection and Intelligent Perception, University of Electronic Science and Technology of China, Chengdu 610054, China

* Correspondence: zmpeng@uestc.edu.cn; Tel.: +86-1307-603-6761

Received: 04 September 2019; Accepted: 25 September 2019; Published: 27 September 2019

Abstract: Aiming at the application requirements of infrared detection, the influence of earth background interference on plume radiation detection is investigated and discussed in this article. The infrared image of the earth's atmospheric background radiation is simulated by the spectral correlation based on the conversion model of the surface radiation with different bands. The infrared radiation image of the jet flame and the background is generated by overlapping the infrared radiation of the engine flame and the background radiation according to the detection angle of view. Through the image quality evaluation model, the detectability of the flame is analyzed. The simulating results show that the comprehensive statistical features such as image information entropy, variance and signal-to-clutter ratio can be used to evaluate the detectability of the engine flame.

Keywords: atmosphere background; engine flame; infrared radiation; detectability; image quality evaluation

1. Introduction

The detectability investigation of infrared radiation gives very important guiding and reference for the performance evaluation and design of infrared detection system [1–3]. The detection of the rocket engine flame in flight state plays an important role in military affairs. Due to the complexity of the infrared imaging process, the rocket exhaust plume is presented as a small dim target in the image, which is hard to detect using existing techniques. Thus, many improvements have been made by researches in recent years, including shearlet features [4,5], high-order cumulant [6], local energy [7], non-convex optimization with regularization constraint [8–13].

The background of the flying rocket engine flame is mainly the earth's atmosphere background and deep space background. Deep space can be equivalent to 4K cold background and the atmosphere has strong selective absorption of infrared radiation from the plume. The background radiation of the earth also interferes with the detection of plume radiation. Therefore, it is of great significance to study the coupled radiative transfer characteristics between the plume and the earth's atmosphere. The jet flow field and its radiation characteristics are complex physical and chemical processes in many disciplines. The research involves the interaction and flow process between the jet and the accompanying flow, the secondary combustion of some components, the spectral characteristics of components and the radiation transfer process. Atmospheric radiation transfer involves scattering and absorption of molecules and aerosols, as well as radiation scattering of the sun and the moon. Since the last century, scholars at home and abroad have carried out relevant research and formed a series of special computing software. For example, Fluent, CFD and CFD++ are used to calculate the flow field of jet, LOWTRAN and MODTRAN are used to calculate the atmospheric radiation transmission. But usually the rocket engine works at a certain altitude. In order to achieve its detection and evaluation, it is necessary to consider the energy state after the coupling between the flame and the atmosphere and it is also related to the detection angle and altitude.

The research of flame radiation and flow field modeling has been published in many papers [1–3]. This paper focuses on analyzing the detectability of flame from the perspective of image by using the existing target and background data, and builds an image-based analysis model of detectability. First, we uses MODIS remote sensing data and MODTRAN calculation data to simulate the infrared image of the earth's atmospheric background radiation through the spectral correlation based conversion model of the surface radiation band. Then, the infrared radiation image of target and background is generated by overlapping the infrared radiation of engine exhaust flame and background radiation according to the detection angle of view. Finally, the detectability of plume radiation is studied and analyzed by some image quality evaluation model.

2. Infrared Radiation Calculation

2.1. Earth Background

In the process of sensor remote sensing imaging, the measured infrared image is the result of the interaction between surface, atmosphere and sensor. The imaging process is shown in Figure 1. The radiation received by the sensor is a comprehensive characterization of the solar radiation outside the atmosphere and the thermal radiation on the surface of the atmosphere. In this paper, the infrared radiation simulation method of the earth background in Reference [14], using MODIS remote sensing data and then, through adjacent channel band conversion [15], the infrared radiation image of the earth background is obtained.

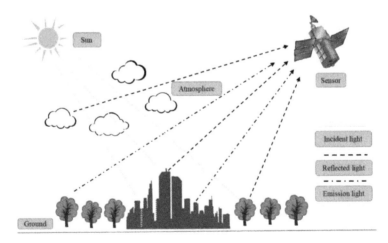

Figure 1. Diagram of the imaging process of a remote sensor.

2.2. Missile Flame

The line of sight (LOS) method combined with the single line group (SLG) model of single line group is used to solve the radiation transfer of jet flame. The transmission of L in radiation field is simplified to a problem of radiation transmission in multi-dimensional and multi-layered media. The flame region through which the line of sight passes is decomposed into N layers. The medium of each layer is considered homogeneous and isothermal. Considering the absorption and emission of each layer, the total infrared radiation intensity can be obtained by recursion step by step [16,17]. The formula is:

$$\bar{I}^i_{\Delta\eta} = \bar{I}^{i-1}_{\Delta\eta}\bar{\tau}^i_{\Delta\eta} + \bar{I}^i_{b,\Delta\eta}\left(1 - \bar{\tau}^i_{\Delta\eta}\right) \tag{1}$$

where $\bar{I}^i_{\Delta\eta}$ is average spectral radiation intensity, $\bar{I}^i_{b,\Delta\eta}$ is average spectral radiation intensity of blackbody, $\bar{\tau}^i_{\Delta\eta}$ is average transmittance.

3. Detectability Analysis of Flame Infrared Radiation

3.1. Generation of Infrared Radiation Data of Flame and Background

In order to analyze the detectability of engine exhaust plume, it is necessary to physically overlap the exhaust plume radiation with the background radiation. Firstly, the projection of the image plane is calculated according to the size of the jet and the spatial resolution of the sensor. Then the convolution calculation of the flame radiation spectrum observed by each pixel with the atmospheric transmittance spectrum and the sensor transmittance spectrum is carried out. Finally, the integration is carried out according to the band of the sensor. The energy distribution of the flame on the image plane of the sensor can be obtained. According to the maximum and minimum energy in the image, the gray level is linearly transformed and the energy infrared image is transformed into a gray level image. The determination of simulation band is based on the spectrum of flame and atmospheric transmission.

Figure 2 shows atmospheric transmittance spectra at different altitudes. We can see that 0.7–2.5 μm, 3–5 μm, 8–12 μm are three atmospheric windows, which are the range of electromagnetic wavelengths to which earth's atmosphere is largely or partially transparent. Its low value zones (2.5–3.0 μm, 4.0–4.5 μm) indicate absorption bands. Figure 3 shows engine exhaust spectra at different flight altitudes.

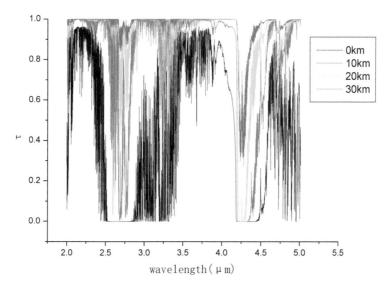

Figure 2. Atmospheric transmittance spectra from different altitudes to the outer atmosphere.

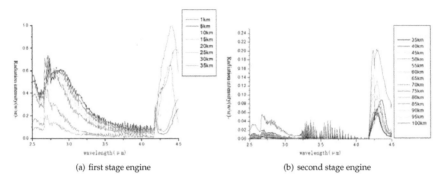

(a) first stage engine (b) second stage engine

Figure 3. Normalization of engine flame radiation.

From the comparison of Figures 2 and 3, it can be seen that there are two radiation peaks at 2.5–3 μm and 4.0–4.5 μm. The two bands in the atmosphere are the absorption bands, which can effectively shield the earth background radiation to the sensor. Therefore, the simulation band is determined to be 2.5–3 μm and 4.0–4.5 μm. Four typical backgrounds are selected to simulate the cities, deserts, mountains and waters. The simulation time is daytime (solar zenith angle 15 degrees, relative azimuth 180 degrees), night and cloudless sky. Atmospheric model: Mid-latitude summer.

Figure 4 shows the infrared radiation image of the jet and the earth's atmosphere at the altitude of 30 km. For the sake of intuitive description, 3D surfaces of the local area (red rectangular box) including the flame are drawn, corresponding to the 2D graylevel image of the upper parts, respectively.

3.2. Detectability Analysis of Flame Radiation

Space sensors usually output infrared images. Generally speaking, there are several commonly used evaluation indicators to evaluate the imaging quality of an infrared image, such as peak signal-to-noise ratio (PSNR), signal-to-clutter ratio (SCR), information entropy (En), contrast (Contrast), structural similarity index measurement (SSIM), homogeneity (Hom), smoothness (Smo), variance (Var), skewness (Skew), kurtosis (Kur) and so forth. These indicators are defined as follows.

(1) Signal Effectiveness

    I. Peak signal-to-noise ratio

    The calculation of the peak signal-to-noise ratio (PSNR) is based on the mean square error (MSE) and is defined by:

$$PSNR = 10\log_{10}\left(\frac{MAX_t^2}{MSE}\right) \tag{2}$$

    where MSE is defined as:

$$MSE = \frac{1}{mn}\sum_{i=0}^{m-1}\sum_{j=0}^{n-1}\|t(i,j) - b(i,j)\|^2 \tag{3}$$

    where t and b represent the target area and background area, respectively.

    II. Signal-to-clutter ratio

    The signal-to-clutter ratio is defined as:

$$SCR = \frac{|\mu_t - \mu_b|}{\sigma_b} \tag{4}$$

where μ_t is the average gray value of target pixels, μ_b is the average gray value of the background pixels, σ_b is the standard deviation of the gray value of the background area.

III.   Contrast

The contrast describes the gradual change of image brightness [18]. A larger contrast value represents a richer gray level change of an image. The contrast is defined as:

$$C = \sum_{\delta} \delta(i,j)^2 P_\delta(i,j) \tag{5}$$

where $\delta(i,j) = \|i - j\|$ represents the gray difference between adjacent pixels. $P_\delta(i,j)$ represents the distribution probability of pixels whose gray difference is equal to δ.

(2)   Statistical Characteristics

I.   Variance

Image variance is a measure of gray contrast and a measure of uniformity of sample distribution [19].

$$Var = \sum_{i=0}^{L-1} (z_i - m)^2 p(z_i) \tag{6}$$

where z is a random variable representing gray level, $p(z_i)$ is the corresponding histogram distribution, m is the mean of z.

II.   Skewness

The skewness of an image is defined by the third-order statistical moments [18]:

$$Skew = \sum_{i=0}^{L-1} (z_i - m)^3 p(z_i) \tag{7}$$

III.   Kurtosis

The kurtosis of an image is defined by the fouth-order statistical moments [18]:

$$Kur = \sum_{i=0}^{L-1} (z_i - m)^4 p(z_i) \tag{8}$$

(3)   Texture

I.   Homogeneity

The homogeneity describes the variance of pixels within a region. It is defined as follows [18]:

$$Hom = \sum_{i=0}^{L-1} p^2(z_i) \tag{9}$$

where z is a random variable representing gray level, $p(z_i)$ is histogram distribution, L is the number of different gray levels.

II.   Smoothness

The smoothness of an image is defined by the second-order statistical moments [18]:

$$Smo = \sum_{i=0}^{L-1} (z_i - m)^2 p(z_i) \tag{10}$$

where m is the average gray value of an image.

(4)   Information

The image entropy is the average number of bits per pixel in the gray level set of the image. The greater the image entropy, the more uniform the gray distribution of the image. The definition of image information entropy is as follows [20–24]:

$$En\,(z) = -\sum_{i=0}^{L-1} p\,(z_i)\,\log_2 p\,(z_i) \tag{11}$$

where z is a random variable representing gray level, $p\,(z_i)$ is histogram distribution, L is the number of different gray levels.

(5)   Structural similarity

The structural similarity index measurement(SSIM)is designed to improve on traditional methods such as PSNR and mean squared error(MSE) and is based on the image light, contrast and structure and is defined as [25]:

$$
\begin{aligned}
l\,(T,B) &= \frac{2\mu_t\mu_b + C_1}{\mu_t^2 + \mu_b^2 + C_1} \\
c\,(T,B) &= \frac{2\sigma_t\sigma_b + C_2}{\sigma_t^2 + \sigma_b^2 + C_2} \\
s\,(T,YB) &= \frac{\sigma_{tb} + C_3}{\sigma_t + \sigma_b + C_3} \\
SSIM\,(T,B) &= l\,(T,B) \times c\,(T,B) \times s\,(T,B)
\end{aligned}
\tag{12}
$$

where μ_t is the average gray value of target pixels, μ_b is the average gray value of the background pixels, σ_t is the standard deviation of the target, σ_b is the standard deviation of the background. C_1, C_2 and C_3 are constants.

According to the above evaluation metrics, we calculated the simulated data of frozen lakes scene at different altitudes (10–100 km). The data are divided into two parts, the area including the flame by overlapping (the upper parts of Table 1) and without flame (the bottom of Table 1). The results are shown in Table 1.

Table 1. Detectability results of different indicators on the lake scene.

Altitude \ Indicator		PSNR	Contrast	SSIM	En	Hom	Smo	Var	Skew	SCR	Kur
Including Flame	10km	43.32	36.08	0.97	4.70	7.2×10^{-2}	0.40	210.20	5.9×10^3	0.55	2.7×10^5
	20 km	40.04	38.00	0.95	4.71	7.0×10^{-2}	0.41	214.60	5.9×10^3	1.23	2.7×10^5
	30 km	39.99	64.09	0.93	4.96	6.1×10^{-2}	0.63	330.90	1.1×10^4	1.91	6.1×10^5
	40 km	38.39	68.41	0.91	5.06	5.4×10^{-2}	0.65	345.10	1.1×10^4	2.14	6.2×10^5
	50 km	37.91	94.92	0.88	5.08	5.5×10^{-2}	0.71	399.80	1.4×10^4	3.59	8.6×10^5
	60 km	36.19	150.77	0.84	5.28	4.7×10^{-2}	0.85	602.10	3.2×10^4	5.54	2.8×10^6
	70 km	35.63	167.29	0.83	5.33	4.4×10^{-2}	0.88	704.30	4.3×10^4	6.38	4.2×10^6
	80 km	34.86	239.91	0.81	5.53	3.6×10^{-2}	0.93	962.40	8.1×10^4	8.11	1.0×10^7
	90 km	34.96	224.68	0.81	5.52	3.8×10^{-2}	0.93	900.70	7.0×10^4	7.61	8.4×10^6
	100 km	33.01	213.19	0.77	5.66	3.1×10^{-2}	0.96	1.2×10^3	1.1×10^5	10.21	1.4×10^7
Without Flame	10 km	38.11	43.95	0.93	5.56	2.6×10^{-2}	0.29	161.90	-1.9×10^3	1.74	1.2×10^5
	20 km	37.54	46.07	0.92	5.61	2.6×10^{-2}	0.31	171.20	-1.7×10^3	2.46	1.3×10^5
	30 km	37.80	56.53	0.90	5.63	2.6×10^{-2}	0.35	188.30	-841.50	3.74	1.8×10^5
	40 km	36.99	60.16	0.89	5.66	2.5×10^{-2}	0.41	211.70	290.90	4.14	2.5×10^5
	50 km	36.54	87.18	0.87	5.70	2.4×10^{-2}	0.56	288.60	6.7×10^3	6.47	8.4×10^5
	60 km	35.36	119.96	0.84	5.74	2.4×10^{-2}	0.78	476.30	2.4×10^4	8.15	2.7×10^6
	70 km	34.91	121.29	0.83	5.79	2.3×10^{-2}	0.81	533.50	2.8×10^4	8.07	3.0×10^6
	80 km	34.03	140.35	0.82	5.83	2.2×10^{-2}	0.85	600.80	3.4×10^4	7.99	3.7×10^6
	90 km	34.18	139.90	0.82	5.83	2.2×10^{-2}	0.85	597.80	3.3×10^4	8.15	3.6×10^6
	100 km	32.53	153.77	0.79	5.75	2.4×10^{-2}	0.92	856.80	6.4×10^4	11.19	7.1×10^6

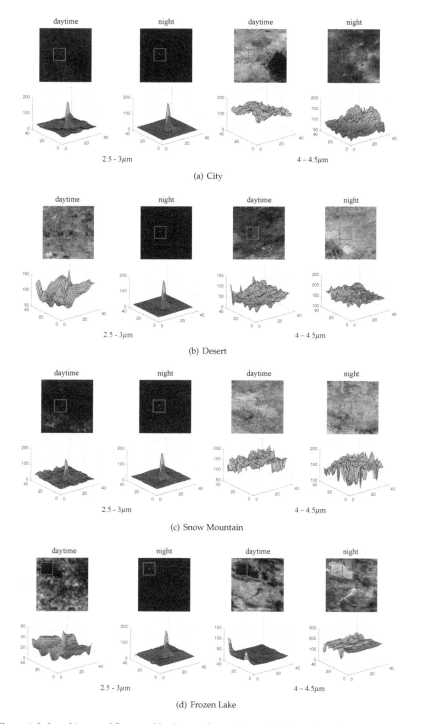

Figure 4. Infrared image of flame and background superimposed. The surface plots show the gray level of the target areas, which are labeled by the red bounding boxes in images.

4. Simulation Results

According to the analysis results of simulation experiment, three indicators: En, Var and SCR can reflect the infrared radiation response best. By using these three indicators, the differences (intensity in imagery) between background and target are well represented, that they indicate Information richness, texture characteristics and radiation intensity, respectively. Thus, these indicators are chosen to evaluate the detectability of the plume. The detectability of the infrared radiation of the plume at different flight altitudes is investigated below. The calculation area is a rectangular area of 61×21 size centered on the jet target. The size of the calculation window depends on the actual size of the simulation target.

From Figure 5, we can see that the three curves in 2.5–3 μm band (a,b,c) have obvious variation regularity, which indicates that the infrared radiation of the flame and background can be distinguished obviously. However, the curves in 4–4.5 μm band (d,e,f) is basically flat and excessive. The difference and discrimination of radiation indices are very small.

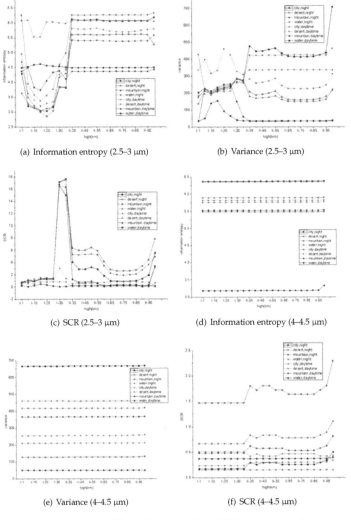

(a) Information entropy (2.5–3 μm)

(b) Variance (2.5–3 μm)

(c) SCR (2.5–3 μm)

(d) Information entropy (4–4.5 μm)

(e) Variance (4–4.5 μm)

(f) SCR (4–4.5 μm)

Figure 5. Detectability evaluation results at different wave bands.

5. Conclusions and Discussion

The calculation models for infrared radiation of the earth background and missile plume are constructed in this paper. The image quality of the simulation data, that are formed by overlapping the plume on the background, are evaluated to study the detectability of the plume radiation. Based on the above analysis and investigation, we summarize as follows:

(1) In the detection period, the observation condition at night is better than that at daytime. The difference is mainly reflected in the observation of level II targets (second-stage engine exhaust with the same propellant), which can be clearly observed at night at all heights but it is difficult to observe the level II targets at all heights during the day. On the one hand, the infrared radiation of level II engines is less than that of level I engines, on the other hand, the background radiation in the day is greater than that in the night.

(2) We can see from the simulation results, three indicators response, such as entropy, SNR and variance, is relatively sensitive in 2.5–3 μm detector band.

(3) In terms of altitude, the visibility of the target at a higher altitude is higher. The infrared radiation of the engine can be clearly observed at or above 10 km in both day and night conditions. The reason is that the attenuation effect of the atmosphere above 10 km on the plume radiation is reduced.

The above conclusions are drawn from the selected typical surface scenes and cloudless conditions. In the actual flight process, the engine is located in a complex background and meteorological conditions, which require specific analysis and calculation according to the specific detection conditions.

Author Contributions: X.L. (Xia Li) proposed the original idea, performed the experiments, and wrote the manuscript. J.W. and X.L. (Xingrun Liu) reviewed and edited the manuscript. Z.P. contributed to the direction, content, and revised the manuscript. M.L. revised the manuscript.

Funding: This work is supported by National Natural Science Foundation of China (61571096, 61775030), Open Research Fund of Key Laboratory of Optical Engineering (2017LBC003) and Sichuan Science and Technology Program (2019YJ0167).

Acknowledgments: We are very grateful to the two anonymous reviewers for their constructive comments on the revision and improvement of this article. Thanks also to Yuelu Wei, who is pursuing a master's degree in IDIP lab, for her helping with the drawing figures.

Conflicts of Interest: The authors declare no conflict of interest.

References

1. Huang, K.; Mao, X. Detectability of infrared small targets. *Infrared Phys. Technol.* **2010**, *53*, 208–217. [CrossRef]
2. Florez-Ospina, J.F.; Benitez, H.D. From local to global analysis of defect detectability in infrared non-destructive testing. *Infrared Phys. Technol.* **2014**, *63*, 211–221. [CrossRef]
3. Hiasa, S.; Birgul, R.; Catbas, F.N. Effect of defect size on subsurface defect detectability and defect depth estimation for concrete structures by infrared thermography. *J. Nondestruct. Eval.* **2017**, *36*, 57. [CrossRef]
4. Liu, Y.; Peng, L.; Huang, S.; Wang, X.; Wang, Y.; Peng, Z. River detection in high-resolution SAR data using the Frangi filter and shearlet features. *Remote Sens. Lett.* **2019**, *10*, 949–958. [CrossRef]
5. Peng, L.; Zhang, T.; Liu, Y.; Li, M.; Peng, Z. Infrared Dim Target Detection using Shearlet's Kurtosis Maximization Under Non-Uniform Background. *Symmetry* **2019**, *11*, 723. [CrossRef]
6. Fan, X.; Xu, Z.; Zhang, J.; Huang, Y.; Peng, Z.; Wei, Z.; Guo, H. Dim small target detection based on high-order cumulant of motion estimation. *Infrared Phys. Technol.* **2019**, *99*, 86–101. [CrossRef]
7. Fan, X.; Xu, Z.; Zhang, J.; Huang, Y.; Peng, Z. Infrared Dim and Small Targets Detection Method Based on Local Energy Center of Sequential Image. *Math. Probl. Eng.* **2017**, *4572147*, 1–16. [CrossRef]
8. Zhang, T.; Wu, H.; Liu, Y.; Peng, L.; Yang, C.; Peng, Z. Infrared Small Target Detection Based on Non-Convex Optimization with Lp-Norm Constraint. *Remote Sens.* **2019**, *11*, 559. [CrossRef]
9. Zhang, L.; Peng, Z. Infrared small target detection based on partial sum of tensor nuclear norm. *Remote Sens.* **2019**, *11*, 382. [CrossRef]
10. Zhang, L.; Peng, L.; Zhang, T.; Cao, S.; Peng, Z. Infrared small target detection via non-convex rank approximation minimization joint l2,1 norm. *Remote Sens.* **2018**, *10*, 1821. [CrossRef]

11. Wang, X.; Peng, Z.; Kong, D.; He, Y. Infrared dim and small target detection based on stable multi-subspace learning in heterogeneou sscene. *IEEE Trans. Geosci. Remote Sens.* **2017**, *55*, 5481–5493. [CrossRef]

12. Wang, X.; Peng, Z.; Zhang, P.; He, Y. Infrared small target detection via nonnegativity-constrained variational mode decomposition. *IEEE Geosci. Remote Sens. Lett.* **2017**, *14*, 1700–1704. [CrossRef]

13. Wang, X.; Peng, Z.; Kong, D.; Zhang, P.; He, Y. Infrared Dim Target Detection Based on Total Variation Regularization and Principal Component Pursuit. *Image Vis. Comput.* **2017**, *63*, 1–9. [CrossRef]

14. Li, X.; Liu, J.; Dong, Y. Simulation of global infrared background based on remote sensing data. *Infrared Laser Eng.* **2018**, *47*, 1104004-1–1104004-7.

15. Wout, V. Simulation of hyper spectral and direction radiance images using coupled biophysical and atmospheric radiative transfer models. *Remote Sens. Environ.* **2003**, *87*, 23–41.

16. Liu, Z.; Shao, L.; Wang, Y.; Sun, X. Influence on afterburning on infrared radiation of solid rocket exhaust plume. *Acta Opt. Sin.* **2013**, *33*, 1–8.

17. Ruan, L.; Qi, H.; Wang, S.; Yang, C. Numerical simulation of the infrared characteristic of missile exhaust plume. *Infrared Laser Eng.* **2008**, *37*, 59–962.

18. Gonzalez, R.; Wintz, P. *Digital Image Processing*; Addison Wesley Publishing Company: Boston, MA, USA; 1977; p. 451.

19. Huber, S.; Hadar, O.; Rotman, S.; Huber, L.; Evstigneev, S. Improving variance estimation ratio score calculation for slow moving point targets detection in infrared imagery sequences. In Proceedings of the Signal and Data Processing of Small Targets, International Society for Optics and Photonics, San Diego, CA, USA, 28–29 August 2013; p. 885707.

20. Zhang, X.; Chi, J.; Hu, J.; Liu, L. Xing, Y. Infrared small target detection using modified order morphology and weighted local entropy. In Proceedings of the International Conference on Computer Engineering, Information Science & Application Technology (ICCIA), Beijing, China, 8–11 September 2017; pp. 368–377.

21. Mello Román, J.C.; Vázquez Noguera, J.L.; Legal-Ayala, H.; Pinto-Roa, D.P.; Gomez-Guerrero, S.; García Torres, M. Entropy and contrast enhancement of infrared thermal images using the multiscale top-hat transform. *Entropy* **2019**, *21*, 244. [CrossRef]

22. Chi, J.; Fu, P.; Wang, D.; Xu, X. A detection method of infrared image small target based on order morphology transformation and image entropy difference. In Proceedings of the International Conference on Machine Learning and Cybernetics (ICMLC), Providence, Guangzhou, China, 18–21 August 2005; pp. 5111–5116.

23. Deng, H.; Liu, J.; Chen, Z. Infrared small target detection based on modified local entropy and EMD. *Chin. Opt. Lett.* **2010**, *8*, 24–28. [CrossRef]

24. Zhang, H.; Zhang, L.; Yuan, D.; Chen, H. Infrared small target detection based on local intensity and gradient properties. *Infrared Phys. Technol.* **2018**, *89*, 88–96. [CrossRef]

25. Wang, Z.; Bovik, A.C.; Sheikh, H.R. Image quality assessment: From error visibility to structural similarity. *IEEE Trans. Image Process.* **2004**, *13*, 600–612. [CrossRef] [PubMed]

 © 2019 by the authors. Licensee MDPI, Basel, Switzerland. This article is an open access article distributed under the terms and conditions of the Creative Commons Attribution (CC BY) license (http://creativecommons.org/licenses/by/4.0/).

Article

Detection of Salient Crowd Motion Based on Repulsive Force Network and Direction Entropy

Xuguang Zhang [1,*], Dujun Lin [1], Juan Zheng [2], Xianghong Tang [1], Yinfeng Fang [1] and Hui Yu [3,*]

1   School of Communication Engineering, Hangzhou Dianzi University, Hangzhou 310018, China; 172080035@hdu.edu.cn (D.L.); tangxh@hdu.edu.cn (X.T.); yinfeng.fang@hdu.edu.cn (Y.F.)
2   School of Electrical Engineering, Shandong Huayu University of Technology, Dezhou 253034, China; zjj@sdhyxy.com
3   School of Creative Technologies, University of Portsmouth, Portsmouth PO1 2DJ, UK
*   Correspondence: zhangxg@hdu.edu.cn (X.Z.); hui.yu@port.ac.uk (H.Y.); Tel.: +44(0)23-9284-5470 (H.Y.)

Received: 8 May 2019; Accepted: 18 June 2019; Published: 20 June 2019

Abstract: This paper proposes a method for salient crowd motion detection based on direction entropy and a repulsive force network. This work focuses on how to effectively detect salient regions in crowd movement through calculating the crowd vector field and constructing the weighted network using the repulsive force. The interaction force between two particles calculated by the repulsive force formula is used to determine the relationship between these two particles. The network node strength is used as a feature parameter to construct a two-dimensional feature matrix. Furthermore, the entropy of the velocity vector direction is calculated to describe the instability of the crowd movement. Finally, the feature matrix of the repulsive force network and direction entropy are integrated together to detect the salient crowd motion. Experimental results and comparison show that the proposed method can efficiently detect the salient crowd motion.

Keywords: crowd behavior analysis; salient crowd motion detection; repulsive force; direction entropy; node strength

1. Introduction

Video surveillance plays an important role in monitoring crowd safety, which is one of the key concerns in our daily life. Since the traditional human-computer interaction between video surveillance and crowd safety is time-consuming and labor-intensive, intelligent video surveillance issues such as target tracking, target detection and crowd analysis have become popular research topics. Crowd motion detection and analysis are essential for crowd behavior understanding [1,2]. It is thus very important to detect the salient motion in the crowd to monitor any potential threats or even damage to social safety. Salient motion has been defined as motion that is likely to result from a typical surveillance target as opposed to other distracting motions [3]. According to this definition, salient crowd motion usually indicates areas that are inconsistent with the mainstream pedestrians' movement. For video surveillance, these areas deserve more attention.

In recent years, due to the rapid development of computer vision technologies, progress has been made in detection of crowd saliency. For example, Lim et al. [4,5] proposed a method for automatically detecting a salient region using time variation of a crowd scene flow field by detecting the fluid activity in a given scene and detecting saliency with a minimum amount of observation region. Some methods for detecting globally salient motion regions for spectral singularity analysis of motion regions in video [6,7] have been also presented. Zhou et al. [8] studied the invariance of coherent neighbors as coherent motion priors, and proposed an effective clustering technique to detect crowd saliency. Solmaz et al. [9] overlaid the scene from the particle grid of the dynamic system defined by the optical flow, and proposed a method to identify the behavior of five people in the visual scene through time integration. Zhang et al. [10] surveyed physics-based methods for crowd video analysis and sorted out the existing public database of crowd video analysis. Although many methods have shown good performance in crowd salient motion detection, the internal mechanism of crowd movement still needs to be explored. The pattern of crowd movement depends on both individual movement and interaction between individuals. It is of great value to explore a method to describe individual interaction and apply it to crowd salient motion detection.

In this paper, we propose a salient crowd motion detection method based on a direction entropy and a repulsive force network. The optical flow is first obtained using the pyramid-based Lucas-Kanade optical flow algorithm. Then, the weighted network is constructed by the repulsive force and the node strength matrix is obtained by using the node degree as the characteristic parameter. Finally, the particle motion direction entropy is used to optimize the node strength matrix and to detect salient movements of the crowds. The framework of the proposed method is shown in Figure 1. A motion vector field is established by giving each pixel a velocity vector in each image through the Pyramid Lucas-Kanade optical flow algorithm. Each vector in the crowd vector field is treated as a moving micro-particle. In order to build a complex network model, we regard each particle and the relationship between two particles as node and edge in the network, respectively. In order to show whether there is a connection between two particle nodes, we use the interaction force to construct the network. After calculating by optical flow method, the position and velocity parameters of each particle can be determined. Whether there is an edge between the nodes depends on the value of repulsive force between these nodes. The repulsive force can be described by the inertial centrifugal force. The value of the inertial centrifugal force is the weight of the edge and a velocity vector node can be selected accordingly. In the neighborhood of the node, the relevancy between the two velocity vectors is taken as a condition to determine the relevancy between the corresponding nodes.

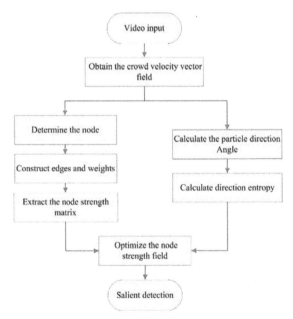

Figure 1. Framework of salient crowd motion detection based on repulsive force network and direction entropy.

A weighted crowd network model is constructed to obtain the adjacency matrix representing the crowd motion information. In order to obtain a complete boundary of salient motion region, the velocity field is reversed and the repulsive force between particles is calculated repeatedly to construct the repulsive force network model. Then, the edge and weight are constructed by the repulsive force model, and the results of the superposition are taken as a construction step. Once all nodes are traversed, the strength of each crowd-weighted network node is extracted as a characteristic parameter to construct the strength matrix of the nodes. By calculating the direction of the velocity entropy of each node in the neighborhood, we can obtain the direction entropy matrix of the node. Then, the normalized direction entropy matrix and the strength matrix of the node are used to further optimize the strength matrix of the node. Once the node strength matrix is obtained, the salient region in crowd movement can be detected.

2. Calculation of Crowd Velocity Vector Field

To calculate the velocity vector field, the crowd video is decomposed into image sequences. Then, each pixel of the image is given a velocity vector calculated using an optical flow algorithm. A motion vector field is thus established. In this paper, considering the spatio-temporal information in motion detection [11], we adopt an improved algorithm based on Lucas-Kanade optical flow algorithm [12] for this task, namely pyramid optical flow algorithm [13].

Lucas-Kanade optical flow, in the process of moving the picture, assumes that a pixel (x, y) on the image has a brightness of $I(x, y, t)$ at time t. After a small time interval of Δt, the brightness of the point becomes $I(x + \Delta x, y + \Delta y, t + \Delta t)$. The Taylor formula is used to expand and when Δt is small enough to approach zero:

$$I(x + \Delta x, y + \Delta y, t + \Delta t) = I(x, y, t) + \frac{\partial I}{\partial x}\Delta x + \frac{\partial I}{\partial y}\Delta y + \frac{\partial I}{\partial t}\Delta t \tag{1}$$

The optical flow constraint equation can be obtained from the brightness constant:

$$\frac{\partial I}{\partial x}\frac{dx}{dt} + \frac{\partial I}{\partial y}\frac{dy}{dt} + \frac{\partial I}{\partial t} = \frac{\partial I}{\partial x}u + \frac{\partial I}{\partial y}v + \frac{\partial I}{\partial t} = Ixu + Iyv + It = 0 \tag{2}$$

According to the uniformity of optical flow, we can establish the optical flow equations:

$$\begin{aligned} Ix1u + Iy1v + It1 &= 0 \\ Ix2u + Iy2v + It2 &= 0 \\ &\vdots \\ Ixnu + Iynv + Itn &= 0 \end{aligned} \tag{3}$$

Then use the least square method to gain the Lucas-Kanade optical flow, where u is the horizontal velocity and v is the vertical velocity:

$$\begin{bmatrix} u \\ v \end{bmatrix} = \begin{bmatrix} \sum\limits_{i=1}^{n} Iix^2 & \sum\limits_{i=1}^{n} IixIiy \\ \sum\limits_{i=1}^{n} IixIiy & \sum\limits_{i=1}^{n} Iiy^2 \end{bmatrix}^{-1} \begin{bmatrix} -\sum\limits_{i=1}^{n} IixIt \\ -\sum\limits_{i=1}^{n} IiyIt \end{bmatrix} \tag{4}$$

The basic ideas of Lucas Kanade optical flow algorithm are mainly based on three assumptions: (1) constant brightness; (2) time continuous or movement is "small movement"; (3) spatial consistency. If an object is moving fast, the second assumption is not fully satisfied. The value calculated by traditional Lucas-Kanade optical flow will have a larger deviation. Pyramid optical flow algorithm reduces the offset of the target motion by reducing the image layer by layer, which satisfies the hypothesis of optical flow calculation better and weakens the influence of fast target motion. In this paper, the crowd velocity field Q is obtained by the pyramid optical flow algorithm. All the velocity values in the horizontal direction and vertical direction are rounded up. The velocity vector field calculated using pyramid Lucas-Kanade optical flow algorithm for a crowd scene is shown in Figure 2.

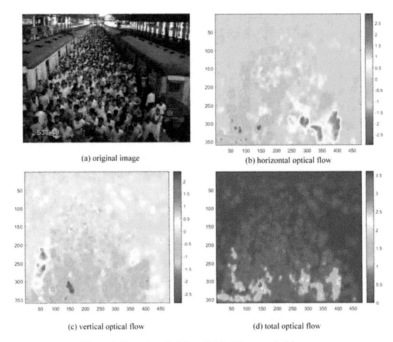

(a) original image

(b) horizontal optical flow

(c) vertical optical flow

(d) total optical flow

Figure 2. Crowd optical flow field of the sampled frame.

3. Construction of Repulsive Force Network

3.1. Establishment of a Network Node

Complex network is a useful tool for describing a complex system. Each element in the system unit is regarded as a node, and the relationship between elements is regarded as a connection. A complex system can be represented as a network [14]. The crowd velocity vector field can be described as a complex network, in which each velocity vector is a node, and the relationship between the velocity vectors is connected. If the properties of the velocity vectors are measured separately, information stored in the velocity vector cross-correlation cannot be obtained, because the correlation of velocity vectors carries more information than the nature of each velocity [15].

We use the interaction force between particles to construct the network [16,17]. After applying the optical flow method, each vector in the obtained crowd vector field is regarded as a moving microscopic particle. The position and velocity parameters of each particle can be then determined. In our crowd complex network, each particle is treated as a node, and the interaction force between two particles is treated as an edge in the network. Whether there is an edge between the nodes depends on the repulsive force between the nodes, the repulsive force can be described by the inertial centrifugal force, and the value of the inertial centrifugal force is the weight of the edge. A weighted undirected network G^w node set $Q = \{q_1, q_2, \cdots, q_n\}$ can be generated, where n is the total number of nodes. The number of network nodes is equal to the number of particles in the crowd velocity field.

3.2. Establishing the Network Edges Using Repulsive Force Model

In the whole particle field, the size and direction of particle velocity are instantaneous, and the motion of the next times is random. If the moving particle is assumed as an agent, there is a possibility of interaction and collision between particles in motion. Imagine that each particle is an agent. In order to avoid collision between agents, an agent adds a repulsive force element to prevent them from colliding with each other. This repulsive force can be described by inertial centrifugal force [18]. For a given crowds particle field Q (M, N) in the column N and row M, selecting a particle $q_{x_0 y_0}$ as the node, constructing a two-dimensional neighborhood δ, the size is $(x_0 \pm \varepsilon, y_0 \pm \varepsilon)$. In this region, the connection between $q_{x_0 y_0}$ and other nodes $q_{xy}(x \neq x_0, y \neq y_0)$ can be described as $e(q_{x_0 y_0}, q_{xy})$. Whether this connection exists is determined by the following formula:

$$e(q_{x_0 y_0}, q_{xy}) \begin{cases} \exists, & \vec{F}_{ij} \neq 0, q_{xy} \in \delta \\ \nexists, & \text{otherwise} \end{cases} \tag{5}$$

The formula for calculating the inertial centrifugal force is as follow:

$$\vec{F}_{ij} = -m_i k_{ij} \frac{v_{ij}^2}{dist_{ij}} \vec{e}_{ij} \tag{6}$$

where $\vec{e}_{ij}$ is the direction vector and m_i is the mass of particle q_i. In this paper, the mass of all particles is set as unit 1. v_{ij} is the relative velocity of two particles, $dist_{ij}$ is the distance between two particles, k_{ij} is a coefficient, the calculation of v_{ij} and k_{ij} is determined by the following formula:

$$v_{ij} = \begin{cases} (\vec{v}_i - \vec{v}_j) \cdot \vec{e}_{ij}, & (\vec{v}_i - \vec{v}_j) \cdot \vec{e}_{ij} > 0 \\ 0, & \text{others} \end{cases} \tag{7}$$

$$k_{ij} == \begin{cases} (\vec{v}_i \cdot \vec{e}_{ij})/v_i, & \vec{v}_i \cdot \vec{e}_{ij} > 0, v_i \neq 0 \\ 0, & \text{others} \end{cases} \tag{8}$$

Then, we can obtain the joint weight, which can be expressed by the magnitude of the repulsive force:

$$We = \left| \vec{F}_{ij} \right| \tag{9}$$

According to the repulsive force formula, if the particle moves away from the affected particle, the repulsive force will be very low. As shown in Figure 3, the arrow represents the moving optical flow, and the blue line represents the repulsive force generated. Figure 3a is the schematic diagram of the repulsive force in the original direction, and Figure 3b is the schematic diagram of the repulsive force in the opposite direction. Thus, for some application of salient region detection, only half of the boundary can be detected.

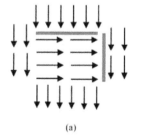
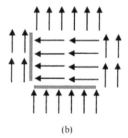

(a) (b)

Figure 3. Schematic diagram of repulsive force: (**a**) original motion flow; (**b**) motion flow reversed.

In order to get a complete boundary, the velocity field is reversed and the repulsive force between particles is calculated repeatedly. Thus, the repulsive force model can be used to construct the edge and weight of the repulsive force network. The results of the superposition of the two are taken as a construction step. Figure 4 shows an example. If we construct the repulsive force network for the optical flow field of the original video sequence, only half of the boundary can be obtained. If we construct the repulsive force network again after reversing the optical flow field, the other half of the boundary can be obtained.

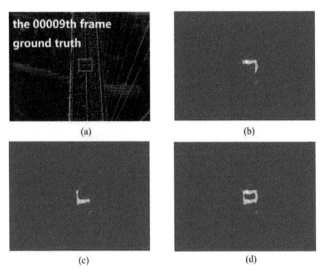

Figure 4. Expressing the effect of optical flow reversal: (**a**) original sample frame; (**b**) detection result using original optical flow; (**c**) detection result using reversed optical flow; (**d**) detection result after the integration of optical flow.

The two-dimensional crowd velocity field is transformed into weighted undirected network model $G^w(Q, E, We)$ by repeating the above steps for each node. The corresponding weighted undirected network node is set as $Q = \{q_1, q_2, \cdots, q_n\}$ and the network edge is set as $E = \{e_1, e_2, \cdots, e_m\}$. In the crowds weighted network model, the connection between nodes and the degree of connection between nodes can be expressed by the following adjacency matrix:

$$
A = \begin{bmatrix} \left|\vec{F}_{11}\right| & \left|\vec{F}_{12}\right| & \cdots & \left|\vec{F}_{1n}\right| \\ \vdots & \vdots & \ddots & \vdots \\ \left|\vec{F}_{n1}\right| & \left|\vec{F}_{n2}\right| & \cdots & \left|\vec{F}_{nn}\right| \end{bmatrix},
\tag{10}
$$

3.3. Calculation of Node Strength

Statistical characteristic parameters of a network can be used to represent the characteristics of a network, such as node degree, average path length, clustering coefficient. In this paper, node strength is chosen to describe the characteristics of the crowd complex network. In the complex network model, node strength is the generalization of node degree, which integrates the strength between edges and nodes [19,20]. From the adjacency matrix, the node strength $s(q_i)$ of node q_i can be expressed as follows:

$$
s(q_i) = \sum_{n=1}^{j} \left|\vec{F}_{ij}\right|,
\tag{11}
$$

After calculating each point in the crowd velocity field, we can get the node strength of all nodes. The node strength field $S(M, N)$ is also a two-dimensional matrix containing M rows and N columns. There is also a one-to-one correspondence between the node strength field and the crowd speed field:

$$
S = \begin{bmatrix} S_{11} & S_{12} & \cdots & S_{1N} \\ \vdots & \vdots & \ddots & \vdots \\ S_{M1} & S_{M2} & \cdots & S_{MN} \end{bmatrix},
\tag{12}
$$

In order to facilitate the node strength field optimization operation in later stage, the node strength field is normalized as follows:

$$
S\prime = \frac{S - S_{min}}{S_{max} - S_{min}} \begin{bmatrix} S_{11} & S_{12} & \cdots & S_{1N} \\ \vdots & \vdots & \ddots & \vdots \\ S_{M1} & S_{M2} & \cdots & S_{MN} \end{bmatrix},
\tag{13}
$$

where S_{max} and S_{min} are the maximum and minimum values of the nodes in all node strengths.

4. Optimizing Node Strength Field Using Direction Entropy

4.1. Establishment of Vector Direction Entropy Matrix

For a crowd motion field $Q\ (M, N)$ of the M row and N column, one particle $q_{xo\,yo}$ is selected, and thus, the direction angle of particle motion is divided into eight directions at 45 degrees interval. The calculation of velocity direction angle and direction grade is determined by the following formula:

$$
\theta = \arctan \frac{q_{yo}}{q_{xo}},
\tag{14}
$$

$$
d = \begin{cases} 1 & 0 \leq \theta < \frac{\pi}{4} \\ \vdots & \vdots \\ 8 & \frac{7\pi}{4} \leq \theta < 2\pi \end{cases},
\tag{15}
$$

Choose a two-dimensional neighborhood δ with the same edge and weight as the repulsive force model with the size of $(x_0 \pm \varepsilon, y_0 \pm \varepsilon)$. For a sub-image region, because of the different motion forms of particles, the direction of particle motion is uncertain at eight angles. Shannon entropy is a classical method to measure the uncertainty of information, and is the basis of communication science [21–23]. In this paper, Shannon entropy is used to measure the uncertainty of particle motion direction. In this paper, we employ Shannon entropy to describe the chaotic degree of crowd motion. In a neighborhood δ, each particle can be calculated by direction rank formula to get a direction rank d. Each direction rank occupies a certain probability p_i in all direction ranks. According to the definition of Shannon entropy [21] and [23], we can assign the velocity direction entropy between the central particle q_{xoyo} and other particles $q_{xy}(x \neq x_0, y \neq y_0)$ neighboring the central particle. The calculation is determined by the following formula:

$$
H_{xoyo} = -\sum_{i=1}^{n} p_i \log p_i, \qquad n = \varepsilon^2,
\tag{16}
$$

For each position, in the crowd particle field $Q\ (M, N)$, the entropy can be calculated by repeating the steps mentioned above. Therefore, the direction entropy of each particle in the crowd particle field can be obtained. The two-dimensional crowd velocity vector field can be transformed into a particle direction entropy matrix:

$$H = \begin{bmatrix} H_{11} & H_{12} & \cdots & H_{1N} \\ \vdots & \vdots & \ddots & \vdots \\ H_{M1} & H_{M2} & \cdots & H_{MN} \end{bmatrix},$$ (17)

where, $H_{11}, H_{12} \ldots \ldots H_{MN}$ is the entropy at the corresponding position of the crowd particle field. In order to facilitate the node strength field optimization operation in later stage, the direction entropy matrix is normalized as follows:

$$H\prime = \frac{H - H_{min}}{H_{max} - H_{min}} \begin{bmatrix} H_{11} & H_{12} & \cdots & H_{1N} \\ \vdots & \vdots & \ddots & \vdots \\ H_{M1} & H_{M2} & \cdots & H_{MN} \end{bmatrix},$$ (18)

H_{max} and H_{min} are the maximum and minimum values in the entropy matrix for all directions.

4.2. Optimizing the Node Strength Field

The direction entropy matrix of crowd movement can describe the degree of changes in the direction of movement of the nodes. Furthermore, the strength field of the repulsive force node describes the degree of repulsion of each node and the surrounding nodes. In order to reduce the noise caused by other interference motion, this paper combines these two kinds of model to optimize the node strength field. It is very important to choose an effective way to integrate these two features, e.g., node strength and entropy. There are many ways to integrate features, such as multiplication and addition. For the application of salient crowd motion detection, the way of feature fusion requires significant expression of specific crowd motion regions and adaptation to the changes of scene. We analyzed the feature of node strength and entropy. The saliency region can be detected by combining the two features by multiplying or add. However, the saliency region obtained by addition is more effective. Because the range of the two features is quite different and there are great changes in different scenarios, it is difficult to determine the combined weights. Therefore, this paper applies a normalized processing of the two features before adding the two features together. Although there are differences in dimension between them, as a normalized feature, it works well when integrating them at the application level.

The direction entropy matrix of crowd motion is in one-to-one correspondence with the strength field of nodes; thus, we have made a comparison according to the following formulas:

$$P_{ij} = \begin{cases} S_{ij}\prime + H_{ij}\prime & S_{ij}\prime \neq 0, H_{ij}\prime \neq 0 \\ 0 & \text{others} \end{cases},$$ (19)

The optimized node strength field is:

$$P = \begin{bmatrix} P_{11} & P_{12} & \cdots & P_{1N} \\ \vdots & \vdots & \ddots & \vdots \\ P_{M1} & P_{M2} & \cdots & P_{MN} \end{bmatrix},$$ (20)

Then, for nomalizing the optimized node strength field, the specific calculation formula is as follows:

$$P\prime = \frac{P - P_{min}}{P_{max} - P_{min}} \begin{bmatrix} P_{11} & P_{12} & \cdots & P_{1N} \\ \vdots & \vdots & \ddots & \vdots \\ P_{M1} & P_{M2} & \cdots & P_{MN} \end{bmatrix},$$ (21)

After normalizing the strength field of the nodes, we smoothed the node strength field with a 3×3 mean filter template. It can eliminate the negative effects of the node strength caused by too high or too low values on the experimental results. In order to intuitively describe and observe the value of node strength, we use a pseudo-color image display method to visualize node strength. Pseudo-color image shows the pixel value corresponding to the node strength value. In a crowd scene, it is obvious that the node pixel values in salient regions are higher than those in other regions.

5. Experimental Results and Analysis

In our experiments, we tested three crowded scene video sequences from Crowd Saliency dataset [5] and a video sequence in [24] to show the performance of the proposed method. Retrograde and instability regions of a crowd were detected in the experiment. For different crowded scenes, the scale of the velocity field $Q(M,N)$ and the parameters ε (the size of neighborhood) in the experiment are shown in Table 1. The proposed method is effective for images used in this experiment, which do not have a high resolution. If it is used to deal with high resolution images, there are two ways to processing the data. One is to reduce the high-resolution image using interval sampling and local mean, and the other is to process optical flow data by interval sampling.

Table 1. Different scenes and parameter values.

Crowded Scenes	Symbol of Parameter	The Value
Train station scene in Figure 5	ε	13
	$M \times N$	480×360
Single retrograde scene in Figure 6	ε	15
	$M \times N$	480×360
Marathon scene in Figure 7	ε	11
	$M \times N$	640×480
Pilgrimage scene in Figure 8	ε	15
	$M \times N$	640×480

5.1. Crowd Retrograde Behavior Detection

In this experiment, we used the train station scene and the single retrograde scene to show the salient detection for retrograde behavior. As shown in Figures 5 and 6, some pedestrians do not conform to the flow of the mainstream crowd, hence, a retrograde motion was formed instead. The particles will thus have a larger repulsion force and direction entropy in this region. This proposed method can effectively detect human retrograde movement. Figures 5a and 6a shows the original video frame. The node strength field calculated from the repulsive force network is shown in Figures 5b and 6b. It can be clearly seen that the regions with high node strength represents the retrograde motion. However, there are still some disturbances. As shown in Figures 5c and 6c, though the entropy value of the retrograde region is large, there are still some noise regions. Fortunately, the disturbance regions detected by node strength and direction entropy are different. Therefore, we can optimize the saliency detection results by integrating node strength and direction entropy, as is shown in Figures 5d and 6d. In order to illustrate the detection performance, we overlap the saliency detection results with the original video frames in Figures 5e and 6e. Experiments show that our method can detect pedestrians who even move oppositely to the flow of mainstream crowd.

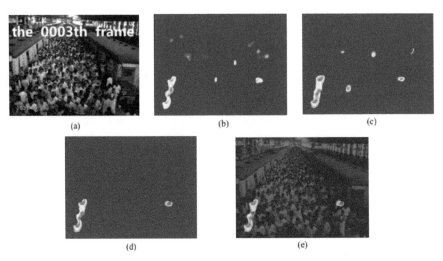

Figure 5. Retrograde motion detection in train station scene: (**a**) input frame; (**b**) node strength field of repulsive force network; (**c**) detection result using direction entropy; (**d**) salient region detection after optimized; (**e**) overlap the salient region with input frame.

Figure 6. Retrograde motion detection in single retrograde scene: (**a**) input frame; (**b**) node strength field of repulsive force network; (**c**) detection result using direction entropy; (**d**) salient region detection after optimized; (**e**) overlap the salient region with input frame.

5.2. Crowd Motion Instability Region Detection

In the crowd surveillance system, the instability area of crowd movement often deserves attention. In this experiment, we used two scenes, including the marathon scene (Figure 7) and the pilgrimage scene (Figure 8) to show the performance of the proposed method for detecting the instability crowd motion.

The sample frames for the two scenarios are shown in Figures 7a and 8a. There are instability motion regions (some pedestrians are different from the mainstream crowd) in these two crowds. The results of node strength fields of two scenes are shown in Figures 7b and 8b, respectively. We can see that the node strength of the repulsive force model is larger in instability motion regions.

The direction entropy fields of two scenarios are shown in Figures 7c and 8c, respectively. The entropy values of the instability region are clearly large. However, there is some noise in the unstable region detected by any single method. After integrating these two methods of node strength and direction entropy, the saliency detection results are optimized and the interference areas are effectively removed, which can be seen in Figures 7d and 8d. Figures 7e and 8e show saliency detection results after overlapping with the original video frame. Experimental results show that the proposed method can detect the salient crowd instability motion in large-scale crowded scenes.

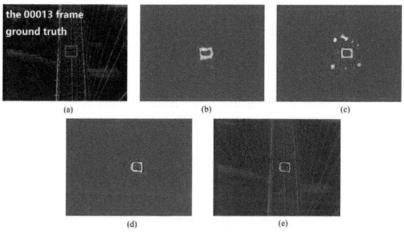

Figure 7. Salient crowd instability motion detection in marathon scene: (**a**) original video frame and ground true (red box); (**b**) node strength field of repulsive force network; (**c**) detection result using direction entropy; (**d**) salient region detection after optimized; (**e**) overlap the salient region with original video frame.

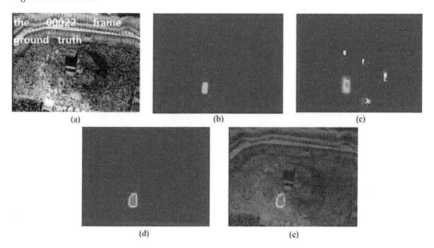

Figure 8. Salient crowd instability motion detection in pilgrimage scene: (**a**) original video frame and ground true (red box); (**b**) node strength field of repulsive force network; (**c**) detection result using direction entropy; (**d**) salient region detection after optimized; (**e**) overlap the salient region with original video frame.

5.3. Detection Results Using Different Neighborhood Size

It is very important to select a suitable neighborhood size ε to construct complex network. A neighborhood that is too small will not be bias to salient motion, while a neighborhood with too large a scale will introduce more noise. In this section, the salient crowd motion will be detected using different neighborhood sizes ε. For retrograde motion detection, the train station scene and single retrograde scene is used to show the performance of the proposed method. From Figure 9, we can see that the salient motion region detected by applying the size of 5×5 neighborhood is slightly scattering, while the area detected by the size of 13×13 is more complete. A larger the neighborhood 23×23, can cause more noise in the detection results. As for the Figure 10, the salient motion region size detected by using the 5×5 neighborhood is small, while the area detected by using the 15×15 neighborhood size is more complete. When choosing a larger neighborhood of 23×23, the result includes more noise. For instability motion detection, two scenes were used in this experiment. For the marathon scene, the detection result was usually not closed if the neighborhood size was too small (5×5 neighborhood). Applying a neighborhood size of 23×23, noise interference will be introduced, although closed salient motion regions can still be obtained. After the experiment, the closed salient motion region can be obtained using a neighborhood size of 11×11 (Figure 11). For another pilgrimage scene, although the saliency region can also be detected with a 5×5 size neighborhood, the result obtained with a 15×15 size neighborhood is closer to the ground truth (Figure 12).

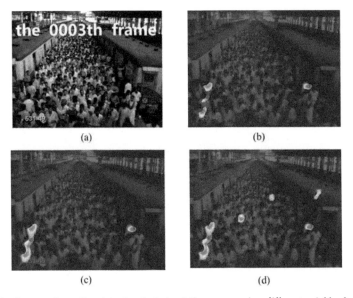

Figure 9. Retrograde motion detection in train station scene using different neighborhood size: (**a**) original video frame; (**b**) detection result using 5×5 neighborhood; (**c**) detection result using 13×13 neighborhood; (**d**) detection result using 23×23 neighborhood.

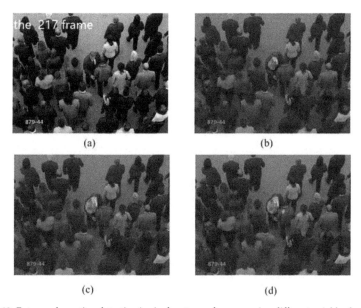

Figure 10. Retrograde motion detection in single retrograde scene using different neighborhood size: (**a**) original video frame; (**b**) detection result using 5 × 5 neighborhood; (**c**) detection result using 15 × 15 neighborhood; (**d**) detection result using 23 × 23 neighborhood.

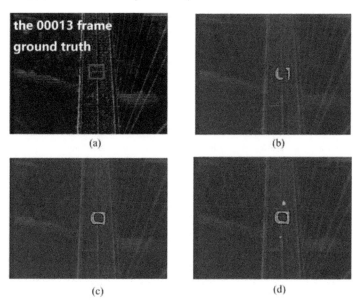

Figure 11. Instability motion detection in marathon scene using different neighborhood size: (**a**) original video frame and ground truth region; (**b**) detection result using 5 × 5 neighborhood; (**c**) detection result using 11 × 11 neighborhood; (**d**) detection result using 23 × 23 neighborhood.

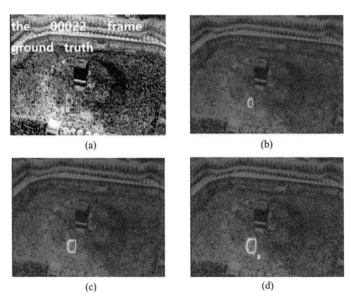

Figure 12. Instability motion detection in pilgrimage scene using different neighborhood size: (**a**) original video frame and ground truth region; (**b**) detection result using 5 × 5 neighborhood; (**c**) detection result using 15 × 15 neighborhood; (**d**) detection result using 25 × 25 neighborhood.

5.4. Performance Evaluation and Comparison

The ground truth of crowd salient detection for pilgrimage and marathon scene has been given in the Crowd Saliency dataset [5]. The ground truth is given using a rectangular area. In order to evaluate the performance of the proposed method, we calculate the minimum enclosing rectangle of the detected salient motion region. To quantitatively evaluate the performance of the method, two indicators (precision and recall) are calculated in our experiments. In this paper, precision is the ratio of the number of pixels in the detected region that belong to the ground truth to the number of pixels in the detected area, indicating whether the number of pixels in the detected local motion instability area is accurate, expressed by Pr. Recall is the ratio of the number of pixels in the detected results region that belonging to the instability motion region to the number of all pixels of the ground truth, represented by R [25]. The precision and recall can be calculated as:

$$Pr = \frac{TP}{TP + FP} \tag{22}$$

$$R = \frac{TP}{TP + FN} \tag{23}$$

where TP indicates that both the detection result and the ground truth are positive. FP indicates that the detection is positive and the actual is negative. TN indicates that both the prediction and the ground truth are negative. FN indicates that the prediction is negative but the actual is positive.

The precision and recall calculated from the pilgrimage and marathon scene using different parameters (the size of neighborhood) are given in Table 2. Obviously, according to the parameters selected in this paper, satisfactory detection accuracy can be obtained. If the neighborhood size is too large or too small, the detection accuracy will be seriously affected. Figure 13 shows the detection results of the pilgrimage and marathon scene using different methods. From Figure 13 we can see that both the proposed method and the methods mentioned in [26] can detect the salient region correctly. However, the rectangular region obtained by the proposed method is closer to the ground truth.

Table 2. The measurement of the accuracy of the detection results using different parameters.

Crowded Scenes	Statistics	Size of Neighborhood	Results
marathon	Pr	5 × 5	0.862
		11 × 11	0.910
		23 × 23	0.531
	R	5 × 5	0.841
		11 × 11	0.909
		23 × 23	0.877
pilgrimage	Pr	5 × 5	1
		15 × 15	1
		25 × 25	0.684
	R	5 × 5	0.244
		15 × 15	0.867
		25 × 25	0.656

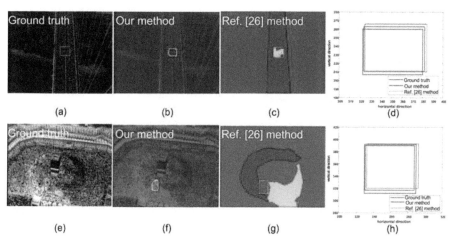

Figure 13. Comparison of the method in this paper with the article [26]: (a,e) are the ground truth of marathon and pilgrimage scene; (b,f) are the results gained by our method; (c,g) are courtesy of reference [26]; (d,h) are the local enlarged displays of the results.

6. Conclusions

In this paper, we proposed a method for crowd salient motion detection based on a direction entropy and a repulsive force network. This paper focused on how to detect saliency regions in crowd movement effectively. Firstly, the crowd video sequence frames are processed by the optical flow algorithm followed by the crowd velocity vector field calculation. Secondly, according to the repulsive force model, the interaction force between two particles is determined as a certain condition. The repulsive force network is obtained and the strength of the crowd weighted network node is extracted as the characteristic parameter to construct a two-dimensional feature matrix. Finally, the velocity vector direction entropy is combined with the repulsive force network characteristic matrix to detect the salient crowd motion structure. The experimental results of four crowd video sequences show that the proposed method can not only detect the region of retrograde behavior of crowd movement but also the region of unstable crowd movement in large-scale crowd scenes. For future work, we will focus on the development of a method for an adaptive threshold and neighborhood calculation.

Author Contributions: X.Z. proposed the idea in this paper; X.Z., D.L., J.Z. and X.T. conceived and designed the experiments; D.L., J.Z. and Y.F. performed the experiments; X.Z., D.L., and H.Y. analyzed the data; X.Z. and

D.L. wrote the paper; X.Z., Y.F., X.T. and H.Y. edited and reviewed the paper; All authors read and approved the final manuscript.

Funding: This research was supported by National Natural Science Foundation of China (no. 61771418).

Conflicts of Interest: The authors declare no conflict of interest.

References

1. Challenger, R.; Clegg, C.W.; Robinson, M.A. Understanding crowd behaviours: Supporting evidence. In *Understanding Crowd Behaviours*; Leigh, M., Ed.; Emergency Planning College: Easingwold, UK, 2009; pp. 1–326.
2. Zhang, X.; Shu, X.; He, Z. Crowd panic state detection using entropy of the distribution of enthalpy. *Phys. A Stat. Mech. Appl.* **2019**, *525*, 935–945. [CrossRef]
3. Wixson, L. Detecting salient motion by accumulating directionally-consistent flow. *IEEE Trans. Pattern Anal. Mach. Intell.* **2002**, *22*, 774–780. [CrossRef]
4. Lim, M.K.; Chan, C.S.; Monekosso, D.; Remagnino, P. Detection of salient regions in crowded scenes. *Electron. Lett.* **2014**, *50*, 363–365. [CrossRef]
5. Lim, M.K.; Kok, V.J.; Loy, C.C.; Chan, C.S. Crowd Saliency Detection via Global Similarity Structure. In Proceedings of the 22nd International Conference on Pattern Recognition, Stockholm, Sweden, 24–28 August 2014; pp. 3957–3962.
6. Loy, C.; Xiang, T.; Gong, S. Salient motion detection in crowded scenes. In Proceedings of the 5th International Symposium on Communications, Control and Signal Processing (ISCCSP), Rome, Italy, 2–4 May 2012; pp. 1–4.
7. Jian, M.; Zhag, W.; Yu, H.; Cui, C.; Nie, X.; Zhang, H.; Yin, Y. Saliency detection based on directional patches extraction and principal local color contrast. *J. Vis. Commun. Image Represent.* **2018**, *57*, 1–11. [CrossRef]
8. Zhou, B.; Tang, X.; Wang, X. Coherent Filtering: Detecting Coherent Motions from crowd clutters. In Proceedings of the 12th European Conference on Computer Vision (ECCV), Firenze, Italy, 7–13 October 2012; Volume 7573, pp. 857–871.
9. Solmaz, B.; Moore, B.E.; Shah, M. Identifying behaviors in crowd scenes using stability analysis for dynamical systems. *PAMI* **2012**, *34*, 2064–2070. [CrossRef] [PubMed]
10. Zhang, X.; Yu, Q.; Yu, H. Physics Inspired Methods for Crowd Video Surveillance and Analysis: A Survey. *IEEE Access* **2018**, *6*, 66816–66830. [CrossRef]
11. Hao, Y.; Xu, Z.J.; Liu, Y.; Wang, J.; Fan, J.-L. Effective Crowd Anomaly Detection Through Spatio-temporal Texture Analysis. *Int. J. Autom. Comput.* **2019**, *16*, 27–39. [CrossRef]
12. Kanade, L. An Iterative Image Registration Technique with an Application to Stereo Vision. In Proceedings of the 7th International Joint Conference on Artificial Intelligence, Vancouver, BC, Canada, 24–28 August 1981; pp. 674–679.
13. Bouguet, J.Y. Pyramidal implementation of the Lucas Kanade feature tracker description of the algorithm. *OpenCV Doc.* **1999**, *22*, 363–381.
14. Chen, G.; Wang, X.; Li, X. *Introduction to Complex Networks: Models. Structures and Dynamics*; Higher Education Press: Beijing, China, 2012.
15. Joldos, M.; Technical, C.C. A parallel evolutionary approach to community detection in complex networks. In Proceedings of the IEEE International Conference on Intelligent Computer Communication and Processing IEEE, Cluj-Napoca, Romania, 7–9 September 2017; pp. 247–254.
16. Boccaletti, S.; Latora, V.; Moreno, Y. Complex networks: Structure and dynamics. *Phys. Rep.* **2006**, *424*, 175–308. [CrossRef]
17. Dzwinel, W.; Yuen, D.A.; Boryczko, K.B. Diverse physical scales with the discrete-particle paradigm in modeling colloidal dynamics with mesoscopic features. *Chem. Eng. Sci.* **2006**, *61*, 2169–2185. [CrossRef]
18. Mohcine, C.; Seyfried, A.; Schadschneider, A. Generalized Centrifugal Force Model for Pedestrian Dynamics. *Phys. Rev.* **2010**, *82*, 046111. [CrossRef]
19. Zou, Q.; Sun, W.; Xing, L. Interest Point Detection in Images Based on Topology Structure Features of Directed Complex Network IEEE. In Proceedings of the 36th China Control Conference, Dalian, China, 26–28 July 2017; pp. 1507–1511.
20. Wu, Z.; Lu, X.; Deng, Y. Image edge detection based on local dimension: A complex networks approach. *Phys. A Stat. Mech. Appl.* **2015**, *440*, 9–18. [CrossRef]

21. Shannon, C.E. A mathematical theory of communication. *Bell Syst. Tech.* **1948**, *27*, 379–423. [CrossRef]
22. Shannon, C.E.; Wyner, S.A. (Eds.) *Collected Papers*; IEEE Press: New York, NY, USA, 1993.
23. Tribus, M.; McIrvine, E.C. Energy and information. *Sci. Am.* **1971**, *225*, 179–188. [CrossRef]
24. Mehran, R.; Oyama, A.; Shah, M. Abnormal crowd behavior detection using social force model. In Proceedings of the IEEE Conference on Computer Vision and Pattern Recognition, Miami, FL, USA, 20–25 June 2009; pp. 935–942.
25. Goutte, C.; Gaussier, E. A Probabilistic Interpretation of Precision, Recall and F-Score, with Implication for Evaluation. *Int. J. Radiat. Biol. Relat. Stud. Phys. Chem. Med.* **2005**, *51*, 952.
26. Ali, S.; Shah, M. A Lagrangian Particle Dynamics Approach for Crowd Flow Segmentation and Stability Analysis. In Proceedings of the IEEE Conference on Computer Vision and Pattern Recognition, Minneapolis, MN, USA, 17–22 June 2007; pp. 1–6.

 © 2019 by the authors. Licensee MDPI, Basel, Switzerland. This article is an open access article distributed under the terms and conditions of the Creative Commons Attribution (CC BY) license (http://creativecommons.org/licenses/by/4.0/).

MDPI
St. Alban-Anlage 66
4052 Basel
Switzerland
Tel. +41 61 683 77 34
Fax +41 61 302 89 18
www.mdpi.com

Entropy Editorial Office
E-mail: entropy@mdpi.com
www.mdpi.com/journal/entropy

CPSIA information can be obtained
at www.ICGtesting.com
Printed in the USA
LVHW070009211020
669275LV00018B/1010